Problems of Petroleum Migration

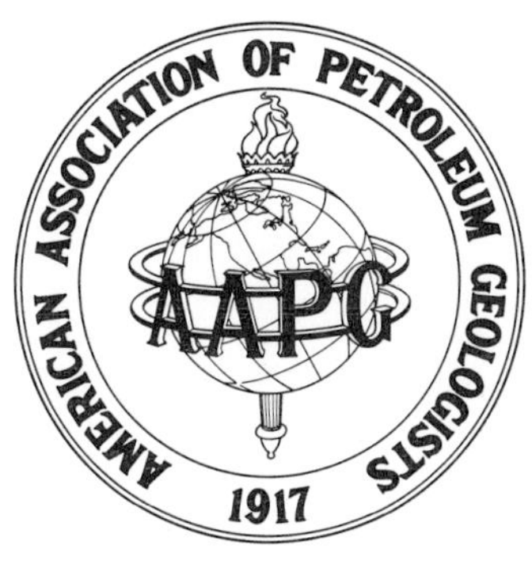

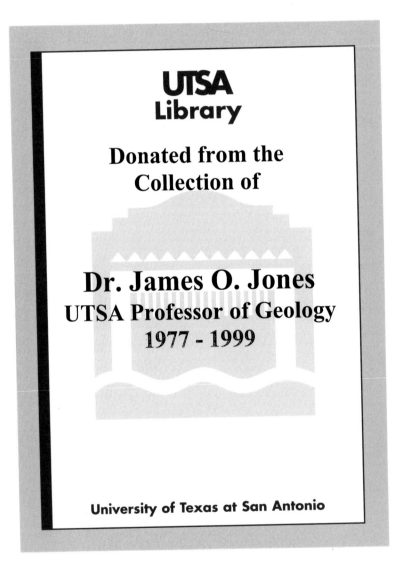

AAPG Studies in Geology No. 10

Problems of Petroleum Migration

Edited by
W. H. Roberts, III
and
Robert J. Cordell

**Published by
The American Association of Petroleum Geologists
Tulsa, Oklahoma, 74101, U.S.A.**

Published May 1980
Library of Congress Catalog Card No. 80-80879
ISBN: 0-89181-014-5

The AAPG staff responsible:
Ronald Hart, project editor
Laura Denson, production
Barbara Gariepy, production

Printed by
Edwards Brothers, Inc.
Ann Arbor, Michigan, U.S.A.

The American Association of Petroleum Geologists

AAPG Associate Editors

Table of Contents

Problems of Petroleum Migration: Introduction

by W. H. Roberts, III and R. J. Cordell

The title of this volume might be taken as a confession of ignorance. The problems referred to are not problems of nature. They are problems of people. They are not problems in the geological world of rocks and time and fluids and space. Nature has solved those problems. Accumulations of hydrocarbons are a reality. Migration has occurred and the results are in place. The problems exist in our minds where fallibility has marked our explanations of how these things happen.

At their 1976 annual meeting at New Orleans, the AAPG Committee on Research chose "Problems of Petroleum Migration" as the subject for the 1978 Research Symposium. That symposium and a companion short course entitled "Physical and Chemical Constraints on Petroleum Migration" were presented at Oklahoma City. This volume includes most of that material and makes it available to a greater number of petroleum explorationists where it can be considered and criticized for the benefit of all concerned. We regret that business pressures have prevented the inclusion of several other excellent papers which also were presented in Oklahoma City.

In introducing this collective effort which deals with *our problems* in understanding petroleum migration, we are obligated to express some honest agnosticism on behalf of the authors. No one of us presumes to have said the last word on anything. We know there is little we can prove, and we do not excuse ourselves for assumptions or conclusions which may later prove to be erroneous. Without embarrassment we admit that our ideas and our arguments are based on the *interpretation* of factual observations, sponsored and enhanced by imagination. In terms of time and space (both geologically dimensioned), our observations, and our samplings are quite limited. What little we do see is biased by what we look for with our minds. Thus, it can be said truly that the problems, as well as the oil-finding ideas, are in the minds of people. The sharing of these problems and these ideas is important to all of us. It may save any one of us from going too far down the wrong road with an erroneous interpretation of factual evidence.

The subject of petroleum migration is central to all thinking about origin and accumulation. No expert on any aspect of petroleum geology can afford to disregard it. Migration is the *sine qua non*—yet so elusive, so mysterious, so controversial, and of course, so frustrating. It is like an invisible moving target. None of us claims to have seen it happening. We are denied real-time observations of the actual movement of oil or gas under natural conditions at depth. The movement is either invisible or too slow to be observed. Therefore, we have devised concepts and models to represent our interpretation of certain evidence as we see it, fixed in place. We cannot actually prove when and whence a particular show of oil or gas came or how long it will remain where we see it. To be completely honest, we can only conjecture about oil and gas movement.

To most of us, "conjecture" is sort of a dirty word. It deflates the ego. It implies the compounding of concepts or unproven ideas, and we like to think at least some of our ideas are pretty well rooted in fact. But we have to live with concepts, and we have to live with conjecture. Where petroleum migration is concerned, let's face it—that's all we've got.

Probably the worst problem we can have is if our concepts and conjectures become inflexible. Then we have conceptual inertia, of the fixed, mind-made-up kind. Extra work is required to get around or through those fixed ideas.

There is general agreement that the collection of significant volumes of hydrocarbons in localized reservoir space requires some kind of transport—which we commonly refer to as *migration*. Wallace E. Pratt has said that "great oil pools contain such large volumes of hydrocarbons that the constituents must originally have been widely disseminated." It is when we try to be too specific about the kind of transport needed to collect those disseminated constituents that we run into problems. A nonspecific overview has some advantages.

The individual authors in this volume agree on the essentiality of migration. However, among them, there are some interesting and rather fundamental differences of viewpoint. Some think of migration comprehensively to include all stages of transfer between the primary source and the ultimate accumulation, all problems and all principles of transfer being (in degree) common to all stages. Others think of migration in the fine-grained, more compactible sediment as primary and migration in the course-grained, less compactible sediment as secondary, each proceeding in accordance with a special set of environmental conditions. Several argue that the pyrolysis of kerogen within certain temperature limits is required to initiate hydrocarbon movement, while others question the essentiality of the kerogen-source linkage as well as the temperature window for hydrocarbon generation. There are those who think that water has very little to do with the transfer of organic material from source to trap, and there are those who think water is almost totally responsible for the transfer. Some believe that oil and gas mixtures of hydrocarbons are formed as distinct fluids at the source end of the system (before migration) and others believe that the oil and gas mixtures become distinct fluids at the trap end of the system (after migration).

The existence of these different viewpoints on migration is not counterproductive. It is useful in several ways. It acknowledges that there is more work to do — that we still have problems. And it suggests the formation of oil and gas may take place in many places and in many ways between the beginning and the end of the migrational continuum. Moreover, it provides good reason for caution in adopting any logic which is too specific and leads to inflexible polarity of viewpoint.

In support of a broad, nonpolar overview and the accepted essentiality of migration, it is probably useful to define briefly the three basic modes in which the migration of organic materials between source and trap is variously conceived to occur by the present authors. The quantitative importance of each, of course, is assumed to vary with circumstances and may receive different emphasis by different authors. It appears unwise to assume that any one of the three is unique or exclusive. The three modes are: (1) distinct oil and gas fluids, with or without coordinated water movement; (2) molecular or colloidal dispersion of hydrocarbons in moving water; and (3) soluble organic compounds (e.g., organic acids) in moving water, transformable to hydrocarbons en route or at destination.

An overview of the migration scene should take some notice of the known distribution of the petroleum which is presumed to have "migrated." Although the materials and the conditions which probably attend the generation and migration of hydrocarbons appear to have existed for at least 600 m. y. and possibly for 2 b. y., and some hydrocarbons are found in rocks representing all of that time span, most of the known oil and gas is now trapped in rocks less than 200 m. y. old. From this we could conclude that the processes of migration, or remigration, tend to move the petroleum in general from older to younger and from deeper to shallower strata. A significant implication might be that in the geologic time frame petroleum accumulations are never permanent. It would reasonably follow that migration and the reworking of oil and gas deposits can be ongoing processes. Even if the organic endowment of the system is continuously or periodically recharged by biologic and geologic processes, it looks as if the whole thing (old and new) tends to go forward in time. Obviously this cannot be explained only by the mechanics of early migration keyed to compaction.

A brief topical rundown on the papers to follow may be useful. In the first paper, **Hinch** expresses a concern that the generation of hydrocarbons from organic-rich Gulf Coast Tertiary shales occurs too late for their migration to be involved with compaction water movement. He suggests that the hydrocarbons may diffuse through the shale pore system in a molecular interaction between rock grains and pore fluids. Next, **McAuliffe** declares the primary migration of crude oil in aqueous solution improbable, based on a pronounced mismatch between the relative aqueous solubility of various hydrocarbon groups and the proportionality of those groups in typical crude oils. Because residual oil content in source rock appears insufficient to support continuous phase movement, he proposes that hydrocarbon fluid may flow as a molecular film, following the continuity of kerogen surfaces.

With the general, compacting, sedimentary condition as a backdrop, **Barker** emphasizes the many variables and complexities of hydrocarbon generation, transformation, and movement. Acknowledging that migration mechanisms are nonunique, he offers the probability that a separate, pore-center oil filament can be extruded from a source rock by thermal expansion of the structured water surrounding the filament. **Magara** also addresses the conditions of "primary" hydrocarbon movement within source rock which lead to hydrocarbon expulsion therefrom. He favors four sources of propulsion energy: compaction, thermal expansion of water, clay mineral conversion, and osmosis. He cites the requirement, however, of oil phase continuity for migration to occur.

R. W. Jones stresses the energy and the diversity of migration processes, all of which have a general tendency to assist the hydrocarbons in escaping from subsurface confinement. His assessment of important petroliferous basins shows that to account for the known oil-in-place requires separate, continuous-phase oil movement in most cases, and may depend on water-borne hydrocarbon relocation in others. **Bonham** cites the importance of water movement for hydrocarbon migration whether in separate phase or in water solution. Using the Gulf Coast area as a model, he calls attention to the relative vertical component of compaction water movement in relation to depth, to the depositional surface, and to buried sedimentary levels. Also described are the significant effects of temperature on the solution and exsolution of hydrocarbons in relation to water depth.

HITCHON gives a broad view of the natural role of water as a universal vehicle and solvent in the forming and reforming of organic (e.g., petroleum) as well as inorganic mineral deposits. He points out that man's imitations of nature in his use of water for the artificial migration of minerals (e.g., flooding, leaching, steaming) may involve environmental risks which require special management. The role of deep-circulating meteoric waters in moving or reworking hydrocarbons into positions of minimum potential energy in mature (pre-compacted) basins is described by **TÓTH**. The energy is shown to be generated by meteoric recharge at high elevations, whereas water-borne hydrocarbons are carried toward and into upward-trending traps for petroleum by ascending waters in discharge areas.

A solution to the problem of carrying significant quantities of organic raw materials in water from source to trap is offered by **HODGSON**. Laboratory work shows that common organic acids (humic, fulvic) dissolved in waters are readily transformed into a suitable variety of hydrocarbons at moderate temperatures. Thus dispersed in water, the hydrocarbons are believed able to move through porous media to points of accumulation without requiring an exclusively thermogenic source-rock history.

HEDBERG offers a rather extensive analysis of the presence and the function of methane in mobilizing oil in petroliferous systems. Included are some interesting ideas on the dynamics of pressures, fracturing, and diapirism in relation to ubiquitous methane. One is reminded of an earlier observation by Wallace E. Pratt that we have discovered greater reserves of energy in the form of gas than in the form of oil — by volume and by weight.

P. H. JONES explains the distribution of hydrocarbon and mineral-moving hydraulic forces in deep sedimentary basins by describing three pressure-depth zones: (1) hydropressure zone, (2) geopressure zone, and (3) transitional zone. The variation of temperature, sediment stability, water salinity, and the mass transfer of water-solubles including minerals and hydrocarbons is keyed to these zones.

The producing end of the migration scene is treated by **ROBERTS**, who that typical oil and gas traps have the ability to separate hydrocarbons and other water-borne organic compounds from waters passing through them. If continuous-phase oil and gas mixtures can readily be formed in traps, the need for mechanically difficult, separate-phase migration of preformed oil and gas is greatly reduced.

HORVITZ reports on his experience with migrated hydrocarbons at the end-of-the-line in near-surface sediment samples taken from onshore and offshore. Recognizable hydrocarbon distribution patterns were found in unproven areas, a number of which have since been found productive. In some cases the shallow hydrocarbons check out to be isotopically similar to trapped hydrocarbons at depth.

At this point, we would like to repeat in essence the introduction to AAPG Course Note Series 8 covering the 1978 Oklahoma City meeting short course on Physical and Chemical Constraints on Petroleum Migration:

In the business of petroleum exploration, it's hard to think of anything compared to *migration* which has enjoyed a greater bounty of ideas and opinions and at the same time suffered such a famine of facts. True, we have some good, solid information about the organic endowment of fine-grained rocks, young and old. And we've documented in many ways in many places the mode of occurrence of oil and gas in course-grained or broken rocks. Thus we seem to feel rather sure of the beginning and the end of the *migration* scene. Why, then, have we not been able to make factual observations of petroleum in the act of moving from one place to another? Certainly we have tried. We may have to give up the idea of witnessing the movement. But in lieu of direct observation, why can we not interpret the "signs" of movement unequivocally?

Anyone interested in this volume probably feels as we do that if we knew more surely how oil and gas moved, we could more surely track them to their resting places. But the *migration* enigma has baffled a lot of honest, hard-working experts. The authors of these papers are recognized for their work in this field. We cannot expect them to agree with each other on all points of interpretation, but we value highly the factual data they offer to enable us to think or rethink the *migration* problem objectively. We assume that as we tie more and more factual reference points into our framework of thinking, our inferences, deductions, and interpretations will become increasingly valid and dependable. The probability of a useful breakthrough in understanding *migration* (with useful applications to both *origin* and *accumulation*) is definitely increased by the attention we give it.

As we express our gratitude to the authors represented herein, may we also express the hope that all readers will share the benefits of improved understanding of the *migration* problem. In the event that any material in this volume raises a question, or unexpectedly reveals the importance of some dormant information, the reader is urged to submit documented comment to the AAPG Editor or to communicate directly with one of the authors.

The Nature of Shales and the Dynamics of Hydrocarbon Expulsion in the Gulf Coast Tertiary Section [1]

By Henry H. Hinch [2]

Abstract It has been recognized for a number of years that shales are the most probable source beds of hydrocarbons. And it has been natural to attribute shale compaction as the means of expulsion of those hydrocarbons into adjacent carrier and reservoir beds. The carrier medium for the hydrocarbons was assumed to be pore water expelled under compaction. A few geologists now recognize that it is difficult to explain expulsion of hydrocarbons from shales in this manner.

In the last 10 years, many reasonable doubts have arisen as to the mechanism of hydrocarbon expulsion. By the time hydrocarbons are generated in significant amounts, most of the pore water has been expelled and it is questionable whether the amounts of pore water remaining are sufficient to flush hydrocarbons, either in solution or as a separate phase, from the shale source beds.

Recent studies of Gulf Coast Tertiary shales have cast further doubts on the generally accepted mechanisms of hydrocarbon migration within the shale pore system. Hydrocarbon expulsion from Gulf Coast Tertiary shales may be due to diffusion of hydrocarbon molecules through the shale pore system rather than flushing of the hydrocarbons by water expelled during compaction. This diffusion process is the result of mechanisms related to the physical properties of the shales and their pore fluids and the molecular interaction between rock grains and pore fluids.

INTRODUCTION

Early in this century, many geologists (Sorby, 1908; Blackwelder, 1920; Monnett, 1922; Hedberg, 1926; and Athy, 1930) observed that shales compact and lose porosity with depth of burial. The curves in Figure 1 include some of the first documentation of porosity loss in a quantitative manner.

The early recognition that shales are a primary source of hydrocarbons led to the belief that expulsion of petroleum and natural gas depended on shale pore water as a carrier medium and that shale compaction was necessary for expulsion (Beckstrom and Van Tuyl, 1928). Many petroleum geologists visualized that the driving force of the fluids, gravitational loading, would be directed outward from the compacting fine-grained shales into the much less compactable coarser-grained carrier beds and reservoirs that have a grain-supported framework. However, some geologists soon realized that it might be difficult to explain the expulsion of petroleum and natural gas from shales by means of the shale compaction model because of the large quantities of expelled shale pore water required for expulsion.

In the last 10 years, additional serious doubts have arisen regarding compaction as the mechanism of oil and gas expulsion. The generally accepted thermochemical concept of petroleum generation requires that expulsion occur on a major scale at depths where shale compaction has appreciably diminished. Application of this theory indicates that it is unrealistic to associate oil expulsion with the shallow first flush of water from the newly deposited sediment. In the later expulsion required by the thermochemical concept, the fine pore size of the shales and the low solubility of hydrocarbons in water are major obstacles to an effective expulsion mechanism. Amoco's recent work has cast even further doubt, not only on some of the generally accepted processes and effects of shale compaction, but also on some of the suggested mechanisms of internal migration of fluids within shales, and expulsion of those fluids, especially oil and natural gas, from Gulf Coast Tertiary shales.

[1]Manuscript received, July 26, 1978; accepted, August 20, 1979.

[2]Amoco Production Company Research Center, Tulsa, Oklahoma, 74135. The writer wishes to thank Amoco Production Company for permission to publish and would also like to thank his many colleagues who contributed ideas and information to this study.

Many of the observations included in this report were made by scientists within Amoco Production Company other than the writer. In some cases, these observations are documented in internal Amoco reports only, and thus are not specifically referenced. Therefore, all such acknowledgements in this report are simply referenced as oral communications from the person who made the observations and the year in which the observation was made (in the cases when this information is known). Any omission of acknowledgement is certainly not intentional. The writer also wishes to thank J. A. Momper for his comments and assistance in preparing this manuscript.

Article Identification Number:
0149-1377/79/SG10-0001/$03.00/0.

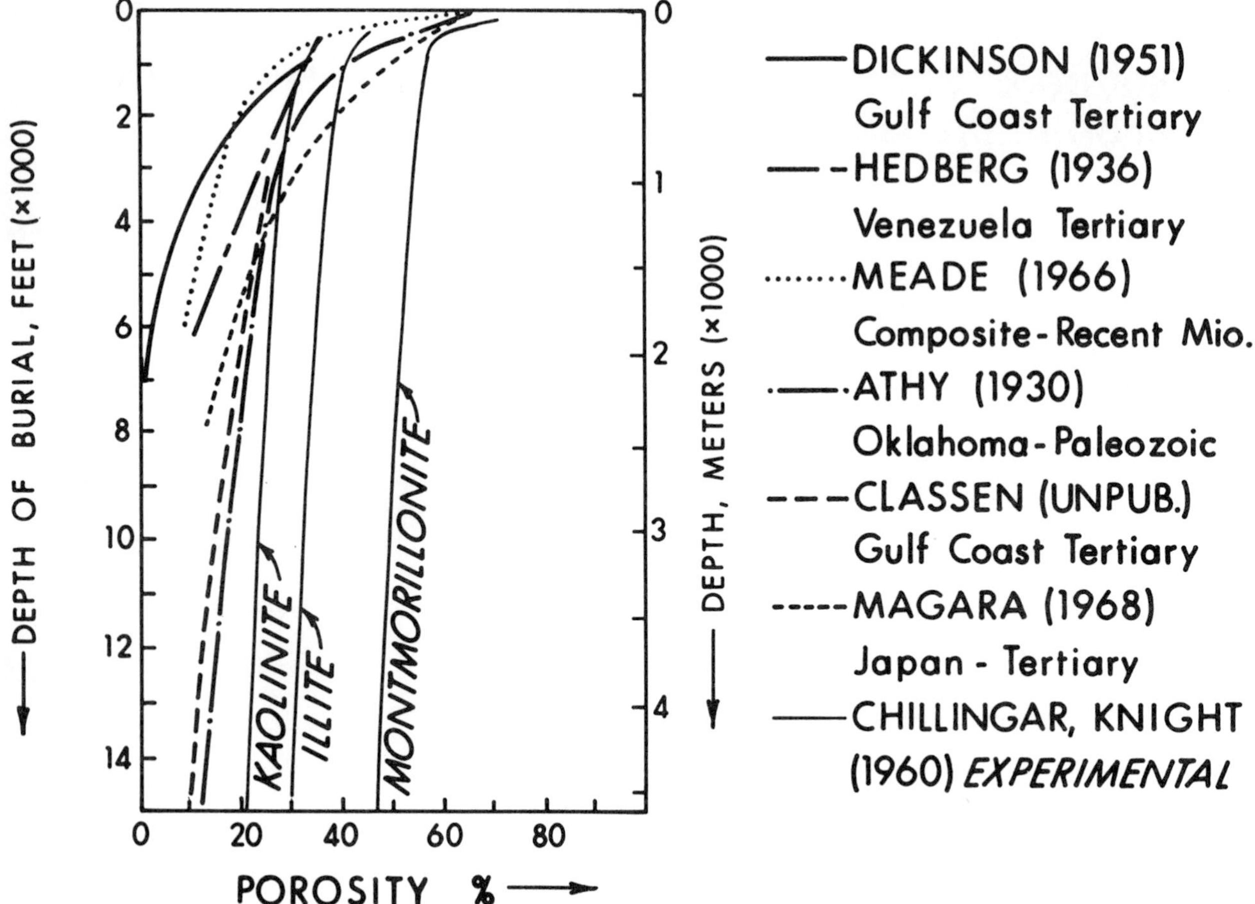

FIG. 1—Selected curves showing changes in shale porosity with depth.

Purpose of Study

1. To evaluate the relative importance of compaction, cementation, pressure solution, and recrystallization in reducing shale porosity and contributing to the expulsion of pore water, oil, and natural gas in the Gulf Coast Tertiary section.

2. To present a summary of thoughts on the internal geometry and physical properties of the "shale-water system."

3. To present the principles involved in the behavior of the "shale-water system" and to speculate on the relevance of this behavior to porosity loss in the Gulf Coast Tertiary section and the expulsion of pore fluids, including pore waters containing dissolved solids, petroleum, and natural gas.

PROCESSES CAUSING POROSITY REDUCTION IN SHALES

Shale Compaction

Geologists accept the fact that shales compact but they do not always agree on the definition of the term *compaction*. Compaction is defined for the purposes of this paper as the mechanical rearrangement of shale mineral grains to form a more consolidated, denser, more ordered fabric, thereby causing expulsion of water with a consequent reduction in porosity and rock-unit thickness. Geologists do not always agree on the definition of the term *shale* either. Therefore, to avoid confusion, shale is defined for the purposes of this paper as simply any lithologic unit in which (1) the average grain size is less than that of fine silt, (2) porosity is less than ~50%, and (3) the mineralogy is predominantly quartz and clay.

Compaction Stages

Early in the history of compaction studies, geologists noted that shale porosity did not decrease uniformly with increasing depth, but decreased in stages. Hedberg (1936) recognized four stages, each determined by the rate at which porosity decreased with depth. In accordance with what he believed to be the dominant process controlling porosity loss in each stage, Hedberg called them: (1) the mechanical rearrangement stage, (2) the dewatering stage, (3) the mechanical deformation stage, and (4) the recrystallization stage.

Porosity-depth profiles in shale can be considered as consisting of as many as five distinct straight-line seg-

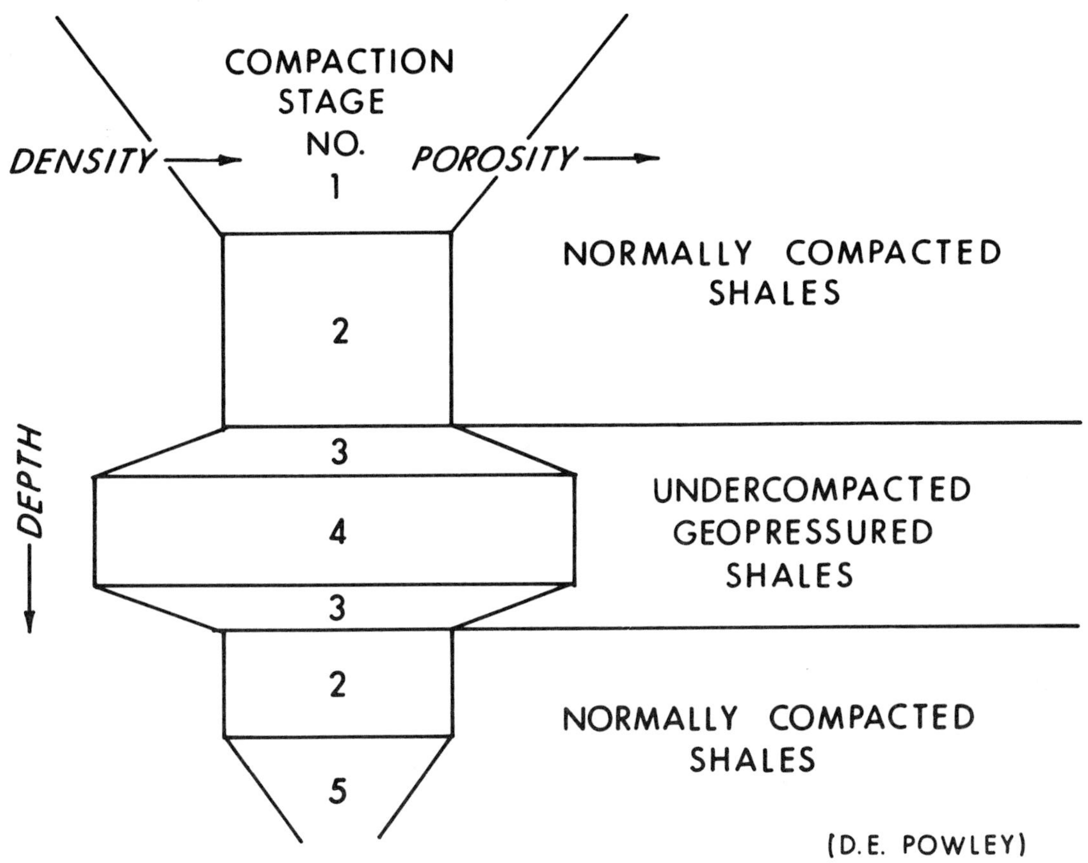

FIG. 2—Generalized shale compaction stages showing characteristic changes in porosity and bulk density with depth (after D. E. Powley).

ments, or compaction stages (Figs. 2 and 3), according to D. E. Powley (oral commun.). Each of these compaction stages has been related, as shown in Table 1, to porosity changes with depth, and to reservoir pressure, reservoir water salinity, and to petroleum occurrences in the Gulf Coast Tertiary section by B. G. Newton (oral commun.).

Above Compaction Stage 1, a shallow zone extending to the sediment surface corresponds to Hedberg's (1936) mechanical rearrangement stage in which there is a rapid decrease of porosity from that of the newly deposited sediment, which may contain 80 or 90% water. This rapid porosity decrease slows in the range of 25 to 30% porosity at approximately 2,500 to 3,000 ft (762 to 914 m) in depth in the Gulf Coast Tertiary. With the exception of biogenic gas, therefore, this zone is of relatively little importance in the consideration of oil and gas expulsion.

Other Porosity Reducing Processes in Shales

The preceding discussion of compaction stages points out that porosity loss in shales with depth is not always as straightforward as might be expected. If it is assumed that the loss of porosity with depth is exclusively the result of compaction, one may tend to overlook other factors possibly involved in porosity loss, namely cementation, recrystallization, and pressure solution.

Figure 4 illustrates that where porosity is reduced as the result of cementation, much of the original fabric and thickness may be maintained and a part of the porosity is filled by secondary minerals.

Recrystallization, or recrystallization combined with pressure solution, reduces porosity by "chemically" rearranging the grain framework which results in a net increase in the size of individual grains, either through solution and redeposition of a single mineral constituent or the combination of more than one mineral into a single crystal.

Effectiveness of Porosity-Reducing Processes

What evidence is there that compaction, cementation, recrystallization, or pressure solution are responsible for the observed loss of shale porosity with depth? Laboratory compaction experiments (Fig. 1), in which wet clays are squeezed under very high pressures, superficially depict naturally occurring loss of porosity through the shallow zone of rapid porosity loss and Compaction Stage 1 (Chillingar and Knight, 1960). However, porosities of the laboratory-compacted clays were higher than those of naturally compacted shales at an equivalent depth of burial. In

Table 1. Generalized Changes Noted in Compaction Stages with Increasing Depth of Burial in the Gulf Coast Tertiary Section (See Figs. 2 and 3)

Compaction Stage	Shale Porosity	Reservoir Fluid Pressure	Reservoir Water Salinity	Petroleum and Natural Gas Occurrences[5]
1	Decreases linearly to 10%	Hydrostatic	Increases	Minor; 25%
2	Unchanged at 10%[1]	Hydrostatic or slightly greater than hydrostatic[2]	High; decreases or remains constant	Major; 60%
3[3-4]	Increases from 10%	Increases	Decreases	Minor; 10%
4[3-4]	Unchanged; higher than in Stage 2	Geopressured	Low; unchanged	Minor; 5%
5[6]	Decreases below 10%	Hydrostatic or geopressured	Low or high	Very minor

[1] A porosity of ~10% in Compaction Stage 2 is typical of Miocene shales. As a general rule, the average porosity in Stage 2 decreases with increasing geologic age.
[2] Geopressures have been observed in a few wells in Compaction Stage 2, though this is not common.
[3] Stages 3 and 4, which are characteristic of geopressure shales, are not always present.
[4] When Stage 4 is developed, there is often a *reversed* Stage 3 underlying Stage 4 (Fig. 2) in which porosity is greater than that in Stage 2 and decreases with depth. Reservoir fluids in *reversed* Stage 3 may remain geopressured or may decrease with depth, and reservoir water salinity may remain low or increase with depth. *Reversed* Stage 3 may be underlain by either a repeated Stage 2 or Stage 5.
[5] Percentages are only approximate. There may be significant differences in occurrences in local areas.
[6] Stage 5 is not always present. Note that Stage 5 is missing in Fig. 3.

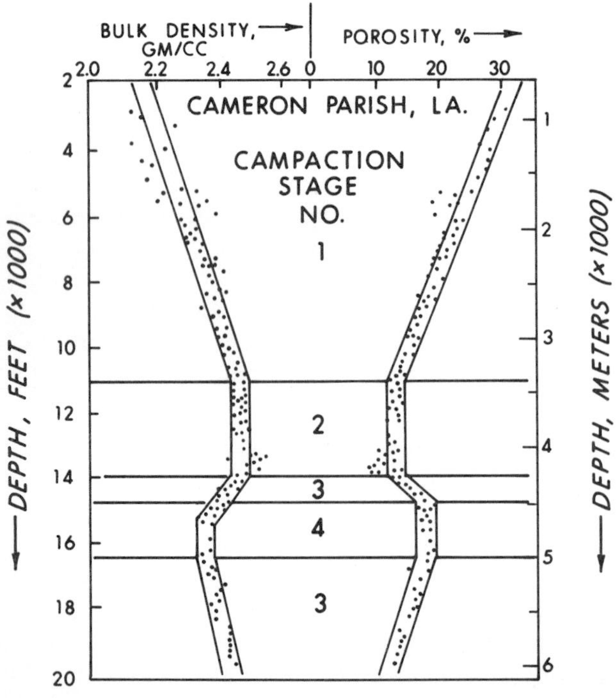

FIG. 3—Gulf Coast Miocene example of shale compaction stages, composited density and porosity data from six wells in one field, Cameron Parish, Louisiana.

the laboratory, Chillingar and Knight found that a montmorillonitic clay retained 27% porosity at a pressure of 200,000 psi, which is roughly equivalent to 400,000 ft (121,920 m) of burial, and kaolinite retained a porosity of 10% at similar pressures. In contrast, natural shales are found with porosities as low as 8 to 15% at much shallower depths. It is unlikely that appreciably lower pressures in nature would reduce porosity much more than those in the laboratory experiments, merely by maintaining the pressures for long intervals of geologic time. Thus, it is difficult to explain the loss of porosity observed in nature by gravitational compaction alone. Gravitational compaction probably is the dominant factor in the rapid loss of porosity at depths shallower than 3,000 ft (914 m), but some mechanism other than simple mechanical compaction is needed to account for the observed rates of porosity loss and fluid expulsion in Stages 1 through 5 (Bradley, 1975).

Cementation is suggested as a significant factor in porosity reduction of shales (Bradley, 1975). If cementation is a significant factor, it is not evident from X-ray diffraction data for Miocene shales from the Gulf Coast (Fig. 5). X-ray data show no evidence of the introduction of significant amounts of cementing minerals with increasing depth. Calcite is the dominant carbonate mineral present but calcite is almost invariably

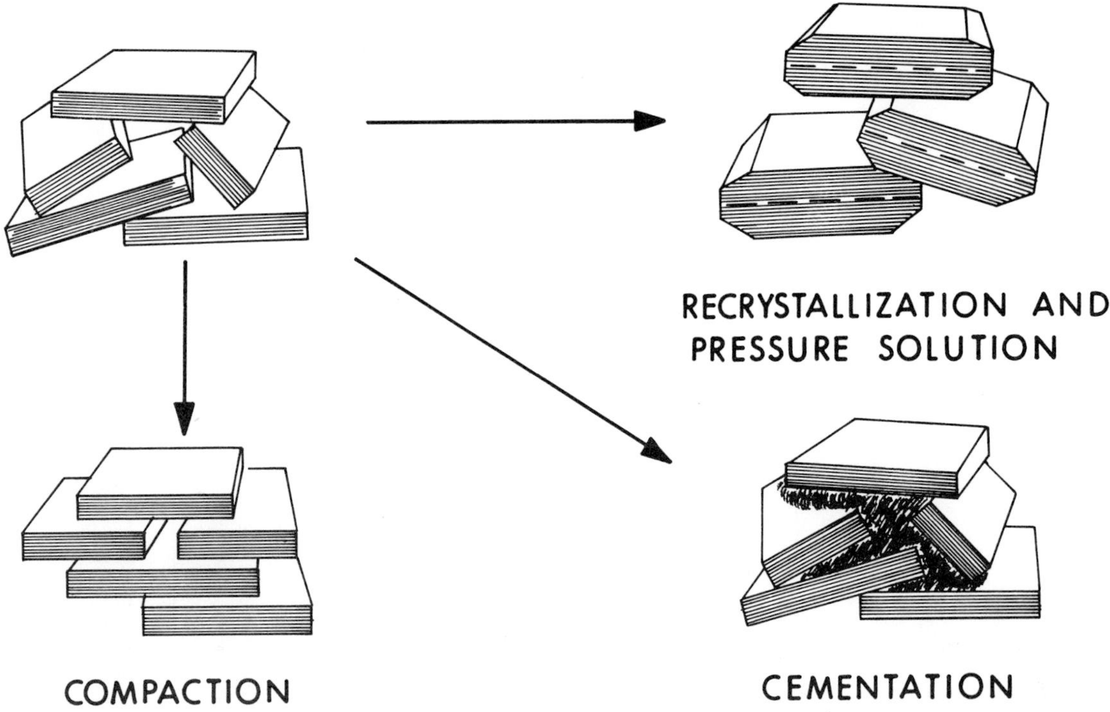

FIG. 4—Potential processes for porosity reduction in shales.

present as primary detritus (i.e., molluscan fragments or foram tests). Dolomite is authigenic but occurs only at shallow depths. Siderite is also authigenic but commonly is present only in localized lenses or siderite nodules. Quartz content increases somewhat with depth but the increase does not correlate with porosity reduction.

From studies of sandstone, it is known that pressure solution and recrystallization are factors in porosity loss. Both of these processes tend to cause a net increase in mineral grain size which can be detected by a decrease in the ratio of total surface area to total grain volume. The surface area to grain volume ratios of Gulf Coast upper Tertiary shales (Fig. 6A) do decrease with depth, down to the top of the geopressured zone, suggesting enlargement of the shale grains. However, comparison of this plot with that of the expandable clay, expressed as weight percent of total shale (Fig. 6B), indicates a close parallelism in that both surface to grain volume ratios and the amounts of expandable clay decrease to about one-fifth of their original values between 3,000 and 13,000 ft (914 and 3,962 m) in the example shown. Studies conducted by the writer prior to 1973 suggest that the net increase in grain size, evidenced by a decrease in surface area to grain volume ratio with depth, is a function only of the diagenetic decrease in expandable clays (Fig. 6C). However, the net increase in grain size associated with the recrystallization of expandable clay does not correlate with porosity loss. Note in Figure 7 that the percentage of montmorillonite in the mixed-layered illite/montmorillonite clay fraction decreases within Compaction Stage 2, where porosity is not changing with depth, and continues to decrease in Stages 3 and 4 where porosity is higher than at shallower depths. This indicates that, although diagenetic recrystallization and grain enlargement are occurring, they are not controlling factors in porosity loss and fluid expulsion from shales.

Data discussed up to this point lead to the conclusion that compaction, cementation, pressure solution, and recrystallization are not significant processes in the reduction of shale porosity and fluid expulsion in the Gulf Coast. What, then, are the factors controlling porosity reduction and fluid expulsion from shales? In seeking alternatives, it is necessary to first reconsider the internal geometry of the shale-water system.

INTERNAL GEOMETRY OF THE SHALE-WATER SYSTEM

The processes of porosity loss and fluid expulsion considered thus far (compaction, cementation, pressure solution, and recrystallization) are modeled on the concept that shales are simply very, very fine-grained sandstones and that they behave in essentially the same way as sandstones. This concept is valid only in a very superficial way. There are subtle, yet significant, consequences of the fine-grained texture of shales which must be considered in explaining the behavior of shales, both in terms of compaction and the

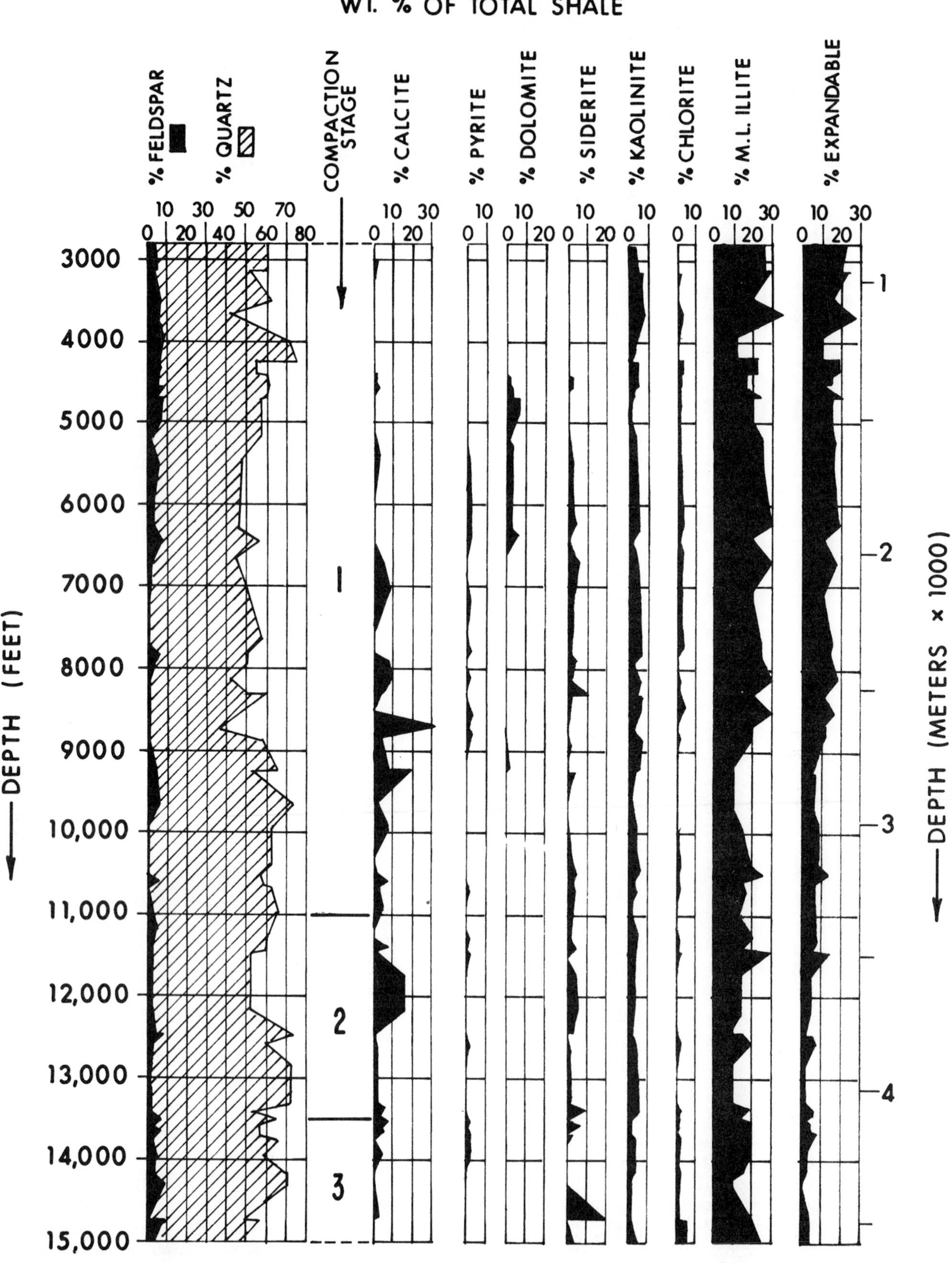

FIG. 5—Typical example of Miocene shale mineralogy determined from X-ray diffraction analyses of sidewall cores from Well-A in the Cameron Parish, Louisiana field.

migration and expulsion of pore water, oil, and gas. As will be seen, the internal geometry of shales is reduced to a molecular level because of the small size of shale grains; at this level, intermolecular forces between mineral grain surfaces and pore fluids play a significant role in controlling the physical properties and behavior of shales and their pore fluids.

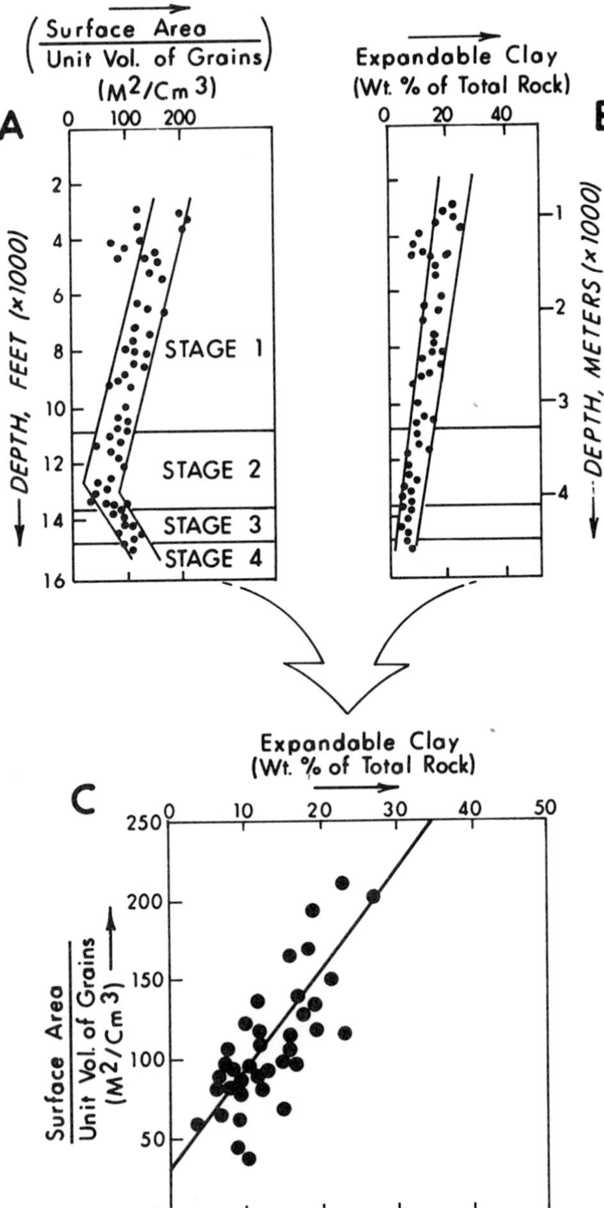

FIG. 6—Relationships among surface to grain volume ratio, expandable clay content, and depth. Miocene shales from Well-A in the Cameron Parish, Louisiana field. Surface to grain volume ratios, in sq m of surface area per cu cm of dry shale grains were determined from nitrogen adsorption. The expandable clay is mixed-layered illite/montmorillonite determined from X-ray diffraction analyses of shale sidewall cores (after H. H. Hinch, J. T. Robison).

The writer examined the relevance of the fine-grained sandstone concept for shales by comparing it to a model for the internal geometry of shales, based on laboratory measurements of porosity and surface area; this model more adequately explains the compaction behavior of Gulf Coast shales and the manner in which shale pore fluids are expelled.

Shale Grain Size and Surface Area

A primary difference between a sandstone composed of large, rounded mineral grains, and a shale composed of exceedingly small plate-shaped grains is the surface area of the grains (Fig. 8). Surface area increases as grain size decreases, with the result that the surface area of what the writer considers to be a "typical" model for a Gulf Coast Tertiary shale is approximately 80,000 times that of the same volume of sandstone. For example, the surface area of approximately 50 cu cm of sand grains (Fig. 8) is approximately that of a chair seat (0.3 sq m), whereas the surface area of the same volume of shale grains is about equal to 5⅓ football fields or 23,500 sq m.

Shale Pore Size

Shale pores are extremely small (Fig. 9). Median shale pore diameters (calculated from porosity and surface area data) decrease from approximately 100Å (1,000 nannometers) at 3,000 ft (914 m) to, typically, about 25Å (250 nannometers) as porosity decreases with depth to the top of Compaction Stage 2. In Compaction Stage 2, the median shale pore diameter of approximately 25Å is only slightly larger than the individual molecules of fluids, especially the bitumen compounds, in the pores (molecular diameters of $H_2O \cong 2.7$Å; $CH_4 \cong 4.2$Å;, n-paraffins range, $\cong 4.2$Å across and 4.2Å to 40.0Å long; complex ring structures range from 15Å to 20Å across; asphaltenes, >50Å).

Effects of Adsorption on the Nature of Shale Pore Fluids

The most significant consequences of the fine grain size of shale are the result of molecular interactions between shale grains and pore fluids. Very strong, short-range forces along these grain surfaces cause the pore fluids to become adsorbed onto the grain surfaces. These intermolecular forces are electrostatic in nature and molecules with high dielectric constants (such as water) tend to be preferentially adsorbed over molecules with lower dielectric constants, such as nonpolar compounds comprising the bulk of oil and gas. Therefore, any rock sequence containing both water and mobile hydrocarbons in solution would tend to be water-wet due to preferential adsorption of water on grain surfaces.

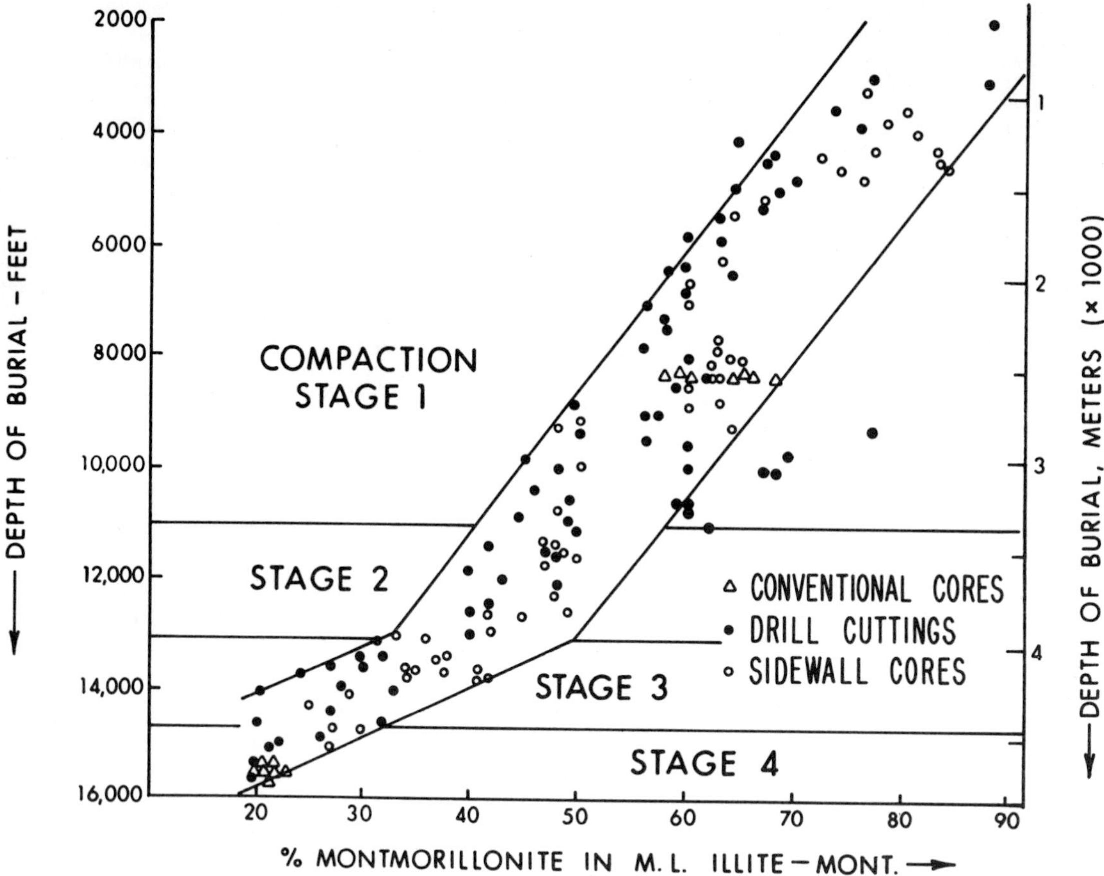

FIG. 7—Typical example of the diagenetic change with depth in the mixed-layered illite/montmorillonite clay fraction of Miocene shales, Well-B in the Cameron Parish, Louisiana field. Weight% montmorillonite (i.e., expandable layers) in the mixed-layered illite/montmorillonite clay fraction was determined from the position of the 002/003 X-ray diffraction peak (after J. T. Robison, A. J. Nash, H. H. Hinch).

The effect of adsorption is to cause "dynamic" structuring[3] of the water close to the mineral grain surfaces, as shown conceptually in Figure 10. In coarse-grained rocks characterized by low surface areas, the amount of structured pore water is small compared to the total volume of pore water. As a result, the behavior of pore water in coarse-grained rocks is not significantly affected by adsorption. However, in shales, all of the pore water is either adsorbed or close enough to grain surfaces to be structured to some degree. Therefore, the behavior of shale pore water is essentially controlled by molecular interactions.

Because of the short range of the intermolecular forces along grain surfaces, the pore water closest to the grain surfaces is more strongly adsorbed and more highly structured than is the water farther from the grain surfaces. As a result, water that is confined in smaller shale pores is more highly structured than water in the larger pores. These differences in the degree of pore water structure within the shale pore system are important because hydrocarbon molecules tend to be excluded from the more highly structured pore water in much the same way as dissolved salts tend to be excluded from the structure of ice when a brine is frozen. The net effect of this exclusion process in a source rock would be to cause hydrocarbon molecules in solution to diffuse in a directional manner. Initially, the diffusing molecule would move away from grain surfaces, then into larger pores, and eventually into silt laminae, bedding surfaces, or fractures before being expelled into the much larger pores of adjacent carrier and reservoir beds, following paths of least resistance. The rates of diffusion of individual molecules are related almost directly to molecular size (Witherspoon and Saraf, 1964). Therefore, diffusion would be more efficient for gases than for petroleum liquids, but the effect would be the same in terms of "gathering" oil and gas from fine pores and concentrating them in larger pores from which they would eventually be expelled, either in solution or as separate phases.

In the foregoing source rock model, therefore, pore water and shale grains should not be considered as a

[3]The term "dynamic" structuring is used here to dispell the implication that structured pore water is completely immobile. Adsorption does cause the structured water close to the grain surfaces to be immobile in a hydrodynamic sense, but not in a thermal sense. Structured water close to grain surfaces is highly mobile on a molecular scale as indicated by the arrows in Figure 10, though not as mobile as water farther from grain surfaces.

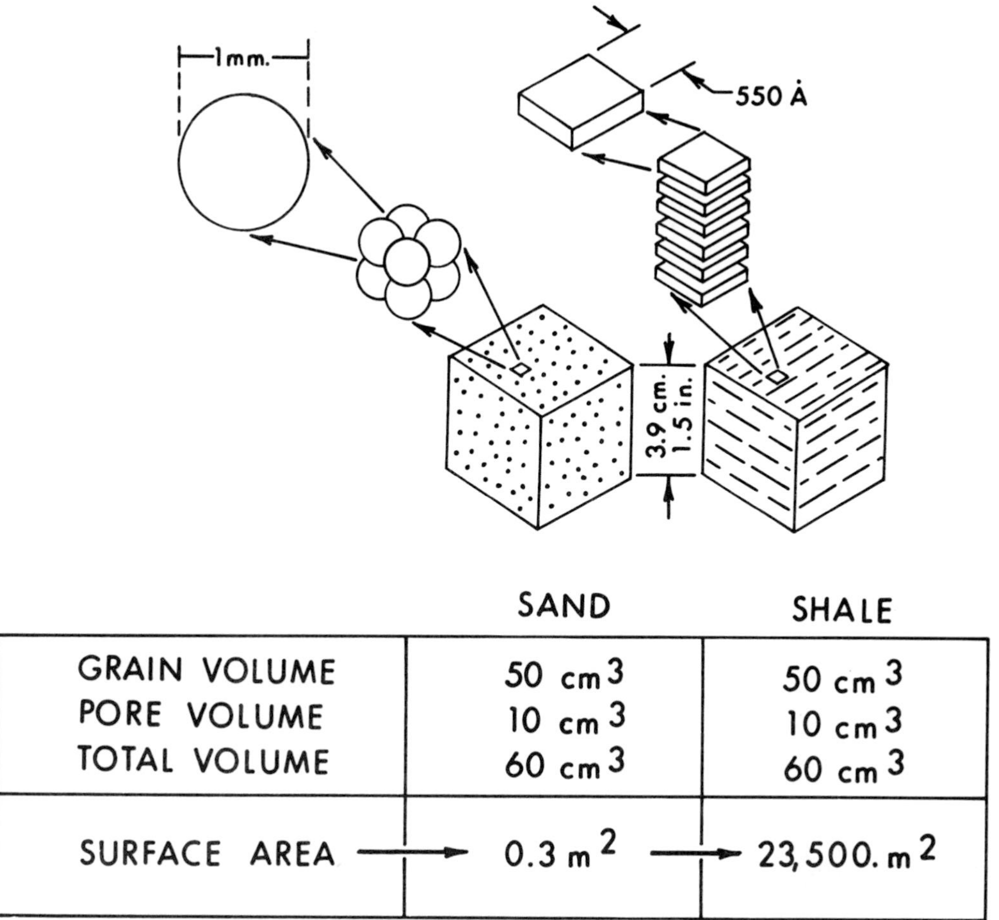

	SAND	SHALE
GRAIN VOLUME	50 cm3	50 cm 3
PORE VOLUME	10 cm 3	10 cm3
TOTAL VOLUME	60 cm3	60 cm 3
SURFACE AREA	0.3 m 2	23,500. m 2

FIG. 8—Comparison between surface areas of equal volumes of sandstone and shale. The surface area of the sandstone was calculated from the surface to grain volume ratio for 50 cu cm of 1 mm spheres. The surface area of the shale was calculated for 50 cu cm of grains from the average of approximately 150 surface to grain volume ratio measurements made on shale samples from the Cameron Parish, Louisiana field. Surface to grain volume ratios were determined from ethylene glycol monoethyl ether adsorption.

two-phase system, as in the conventional view of water in a sandstone. They must be considered as interactive through adsorption, that is, as a single shale-water system in which the behavior of pore water, oil, and gas are essentially controlled by intermolecular forces.

THERMAL STABILITY OF ADSORBED PORE WATER IN SHALES—IMPLICATIONS WITH RESPECT TO SHALE COMPACTION

Compaction due to mechanical loading appears to be an inadequate explanation for porosity loss and fluid expulsion from shales in Compaction Stages 1 through 5 (Figs. 2 and 3). However, all the loading experiments of which the writer is aware have been conducted at laboratory temperatures, whereas during natural compaction, shales are at elevated temperatures and in communication with coarser-grained aquifers that contain brines.

Studies by the writer suggest that the process controlling porosity reduction in shales may be thermophysical as well as mechanical. This process involves the reversible thermal desorption of the structured water in shales as temperature increases with depth of burial through Compaction Stage 1. An increase in thermal energy would overcome the tendency of the shale to retain its pore water via adsorption and would make pore water available for expulsion by overburden pressure.

At a depth of 3,000 ft (914 m) and with a porosity of 30%, the median number of molecular layers of water separating the clay-size grains in a shale is approximately 40. At the top of Compaction Stage 2, where the porosity is typically 9 to 12%, an average of only ten layers of water molecules separate the grain surfaces. Apparently, water molecules at distances greater than five molecular diameters away from grain surfaces are thermally desorbed in Compaction Stage 1.

However, the last molecular layers remaining in Compaction Stage 2 apparently are thermally stable under most subsurface temperatures encountered in an un-

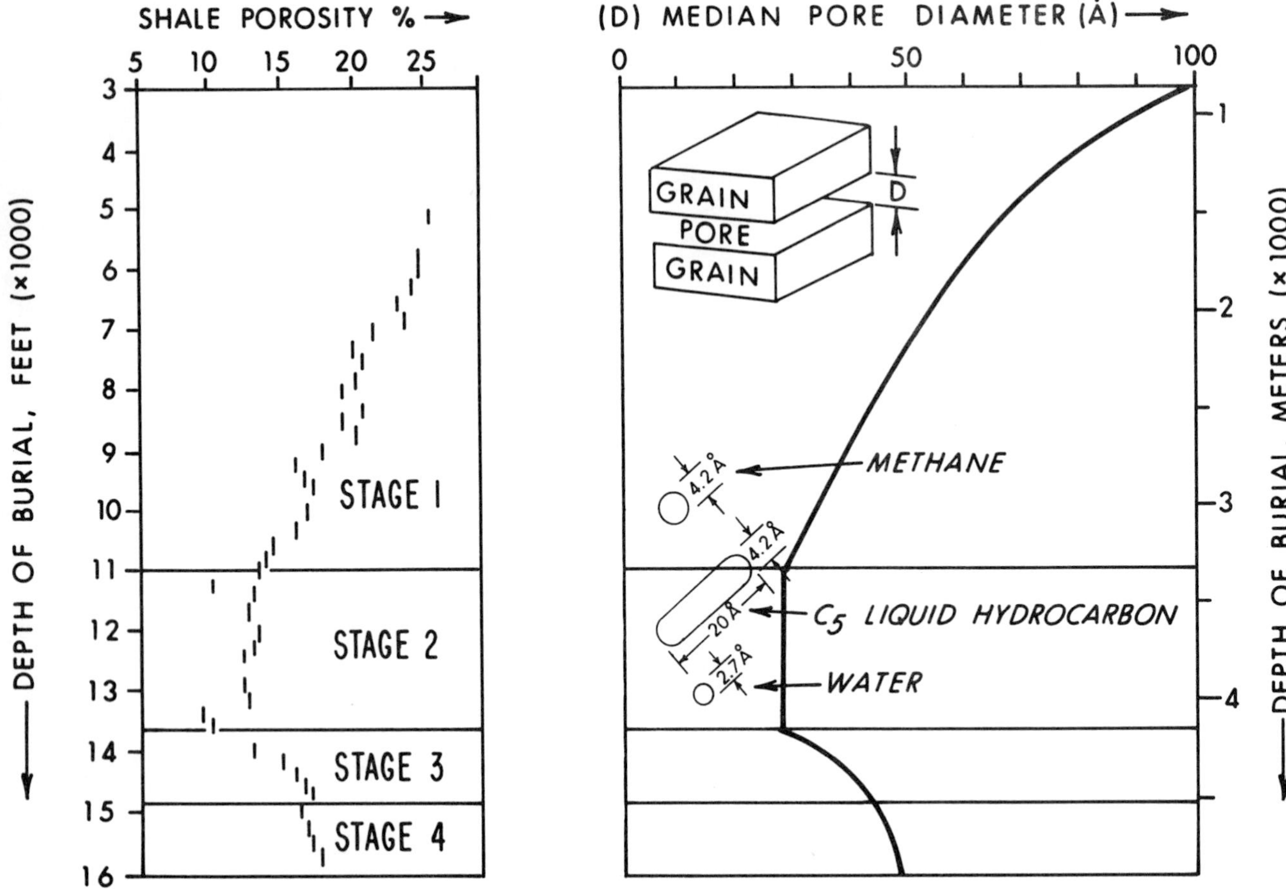

FIG. 9—Changes with depth in median shale pore diameter calculated from porosity and surface area measurements on shales from the Cameron Parish, Louisiana field.

metamorphosed sequence such as is present in the Gulf Coast province. This may explain why the porosity in Compaction Stage 2 is essentially constant.

In the Gulf Coast Tertiary section, the top of Compaction Stage 2 (where porosity is typically 9 to 12%), can occur at a wide range of depths between 6,000 and 14,000 ft (1,829 and 4,267 m) indicating that porosity reduction is not related simply to mechanical overburden stress. However, plots of many data points as shown in Figure 11 (D.E. Powley, oral commun.) indicate that a time-temperature relationship localizing the boundary between Stages 1 and 2 exists which suggests the thermophysical aspect of porosity reduction in shales.

STRATIGRAPHIC TIMING OF SHALE COMPACTION AND HYDROCARBON GENERATION IN THE GULF COAST TERTIARY SECTION

The temperature at the top of Compaction Stage 2 in Gulf Coast upper Tertiary deposits is frequently close to the temperature needed for initiating the main phase of petroleum generation (LaPlante, 1974; Fig. 12), and peak petroleum generation commonly occurs within Compaction Stage 4 (Momper, oral commun.). This near-coincidence of the main oil-generation phase with Stages 2 through 4 leads to an apparent dilemma in understanding expulsion of petroleum from shale source beds in the Gulf Coast because, in Stages 2 through 4, the release of water apparently has ceased, for practical purposes, and porosity is stabilized.

At this point, four major constraints seem evident with respect to feasibility of internal migration of petroleum within shales and petroleum expulsion from shales. First, the water remaining in the shales when hydrocarbons are thermochemically generated is insufficient to flush the generated hydrocarbons into the carrier system or reservoir beds. Second, at the stage when significant amounts of hydrocarbons are being generated within the shales, fluid expulsion is greatly reduced or ceases altogether, judging from the lack of porosity loss through Compaction Stages 2 through 4. Third, the water remaining in the shales is highly structured in adsorbed layers on the grain surfaces and cannot readily participate hydrodynamically in petroleum expulsion. Finally, the median size of shale pores is only slightly larger than migrating hydrocarbon molecules, thus adding to the difficulty of petroleum movement through the pores.

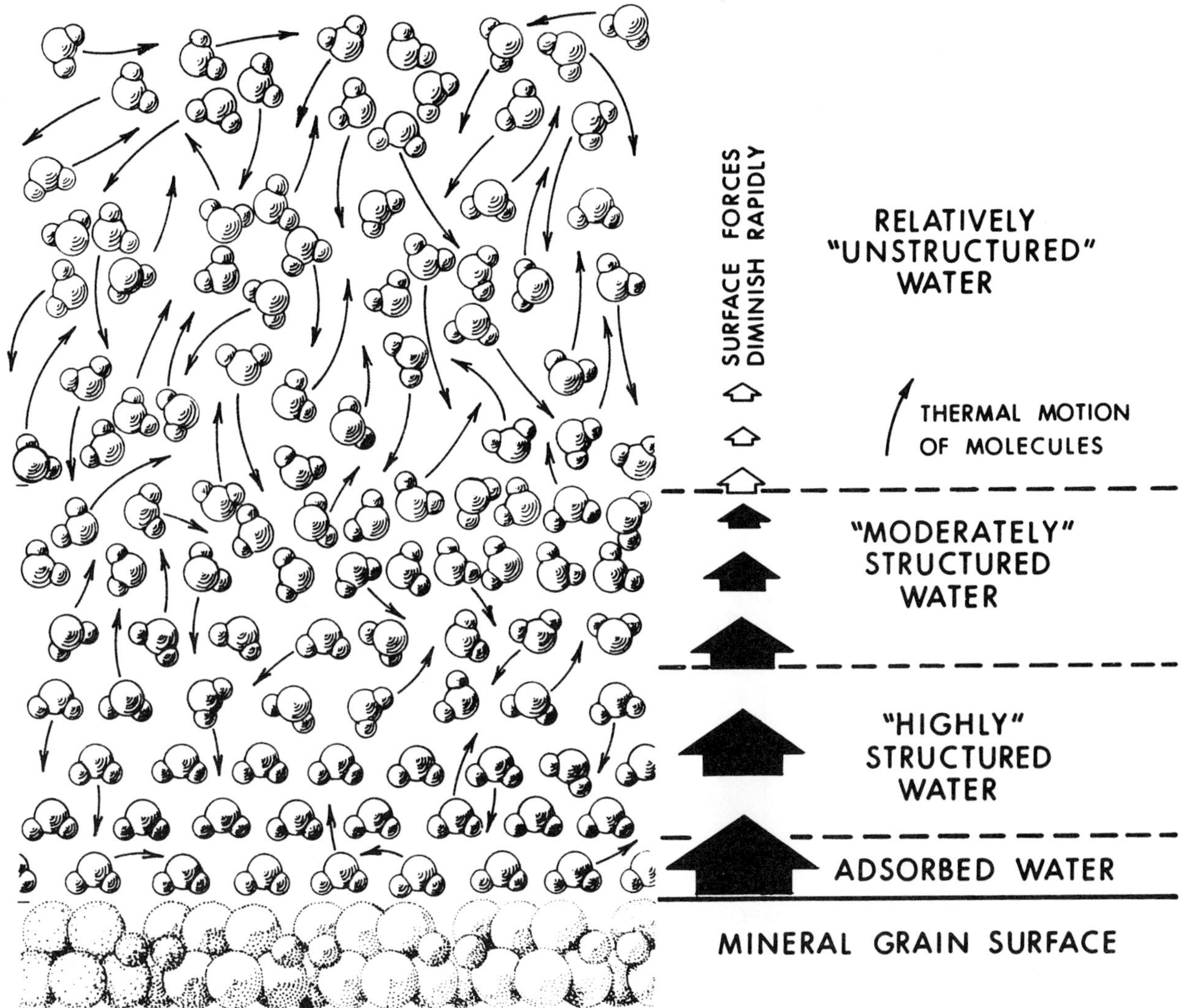

FIG. 10—Conceptual model for the "dynamic" structuring of water due to adsorption close to a mineral grain surface.

MECHANISMS FOR OIL AND NATURAL GAS EXPULSION NOT RELATED TO POROSITY LOSS

In light of the previous four constraints, it is not surprising that the expulsion efficiency of source rocks generally is low. Obviously, these limitations are overcome to some degree in nature. This part of the text considers mechanisms for internal migration and expulsion of oil and gas that may operate in nature despite the four problems enumerated above. However, adequate consideration of these mechanisms requires emphasis of three key points:

(1.) In Compaction Stages 2 through 4, fluids are being expelled from the shales even though porosity is not being reduced with increasing depth. However, the fluid expulsion is not due to compaction, cementation, or pressure solution/recrystallization. In a sense, shale porosity is simply a measure of the volume of the shale "container." If pore fluids expand with depth, as they do, then fluids can "overflow" the container and fluids can be exchanged between shales and adjacent formations without there having to be a reduction in porosity (i.e., container volume).

(2.) The net effect of adsorption in a water-wet shale is a tendency of the shale to retain its pore water under stress, but not to retain the relatively nonpolar hydrocarbons which are generated within the shale pores. Nonpolar hydrocarbons are relatively free to diffuse through the shale pore system.

(3.) The structuring of shale pore water may be an essential factor in concentrating oil and gas in large pores, bedding planes, and fractures. From these larger openings, the oil and gas is more readily expelled into adjacent carrier beds in response to an increase in pore fluid pressure within the shale.

Three phenomena are believed capable of increasing the shale pore fluid pressure, preliminary to causing expulsion of oil and gas into adjacent carrier and reservoir beds, without any associated reduction in shale porosity. These are: (1) thermochemical generation of fluids, including oil and natural gas, resulting in a volume increase of fluids and kerogen within the shale, relative to the volume of the original organic matter;

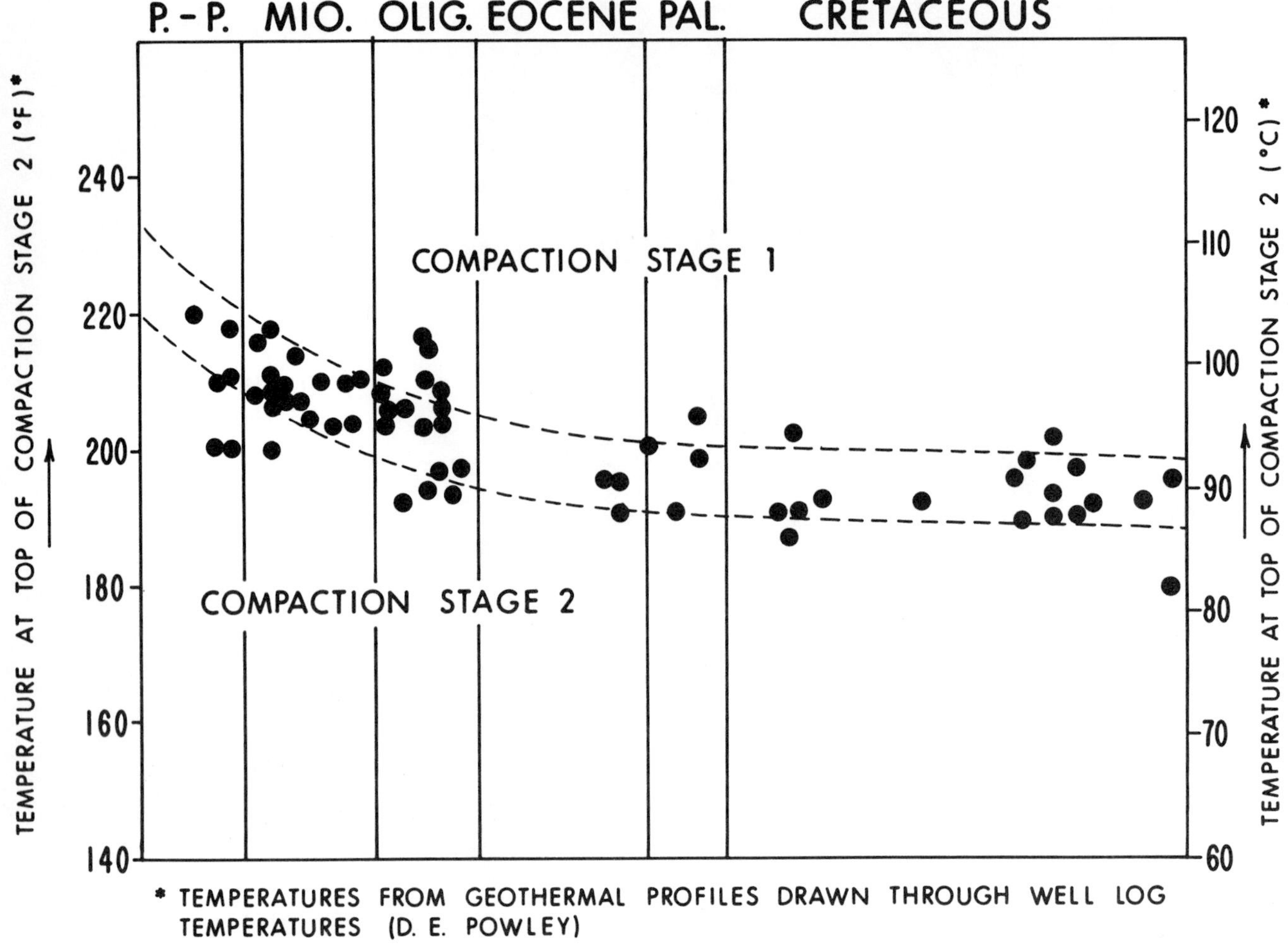

FIG. 11—Temperature at the top of Compaction Stage 2 in the Gulf Coast province showing the relationship between temperature and geologic age of shales in Compaction Stage 2 (D. E. Powley).

(2) thermal expansion of pore water and other fluids, due to temperature increases with burial, greater heat flux with time, or reduced heat loss resulting from diminished fluid loss; (3) exchange of fluids between shales and sandstones due to the development of an osmotic imbalance between shale pore water and low salinity waters in associated reservoir and carrier beds.

Pressure Increase Due to Hydrocarbon Generation

Volumetric changes associated with the diagenesis of organic matter are of particular interest because these changes are directly associated with the thermochemical generation of fluids in a source bed, including bitumen, water, and gases. Furthermore, the volumetric increase of generation products relative to original organic matter could be a major factor in fluid expulsion. These volume changes can be calculated but the computation is complicated by consideration of the phases. Basing the computation on results from pyrolysis experiments on organic matter performed in 1974 (Harwood, 1977), it appears that there is a net increse in volume of liquids due to bitumen and water generation of approximately 34 to 39 bbl per acre-foot for a typical oil source bed averaging 1.0 wt% organic carbon (Yarborough, oral commun.). This volume increase is equivalent to 4.5 to 5% of the total pore volume of a shale with 10% porosity. Therefore, a significant increase in shale pore fluid pressure would be caused by fluid generation. This would, in turn, expedite expulsion of fluids from the source bed. The tendency of shale to retain its pore water (due to adsorption) but not its nonpolar hydrocarbon molecules would eventually result in diffusion of hydrocarbon molecules out of the shale pore system into adjacent reservoir and carrier beds. Thus, the process of hydrocarbon generation may provide the mechanism for expulsion as well.

Pressure Increase Due to Thermal Expansion of Shale Pore Fluids

Once hydrocarbons are generated within the shale source bed, they would tend to be expelled in response to increases in shale pore fluid pressure resulting from thermal expansion. If we assume that the thermal expansion of the adsorbed pore water and bitumens in shales is comparable to liquid water, the volume in-

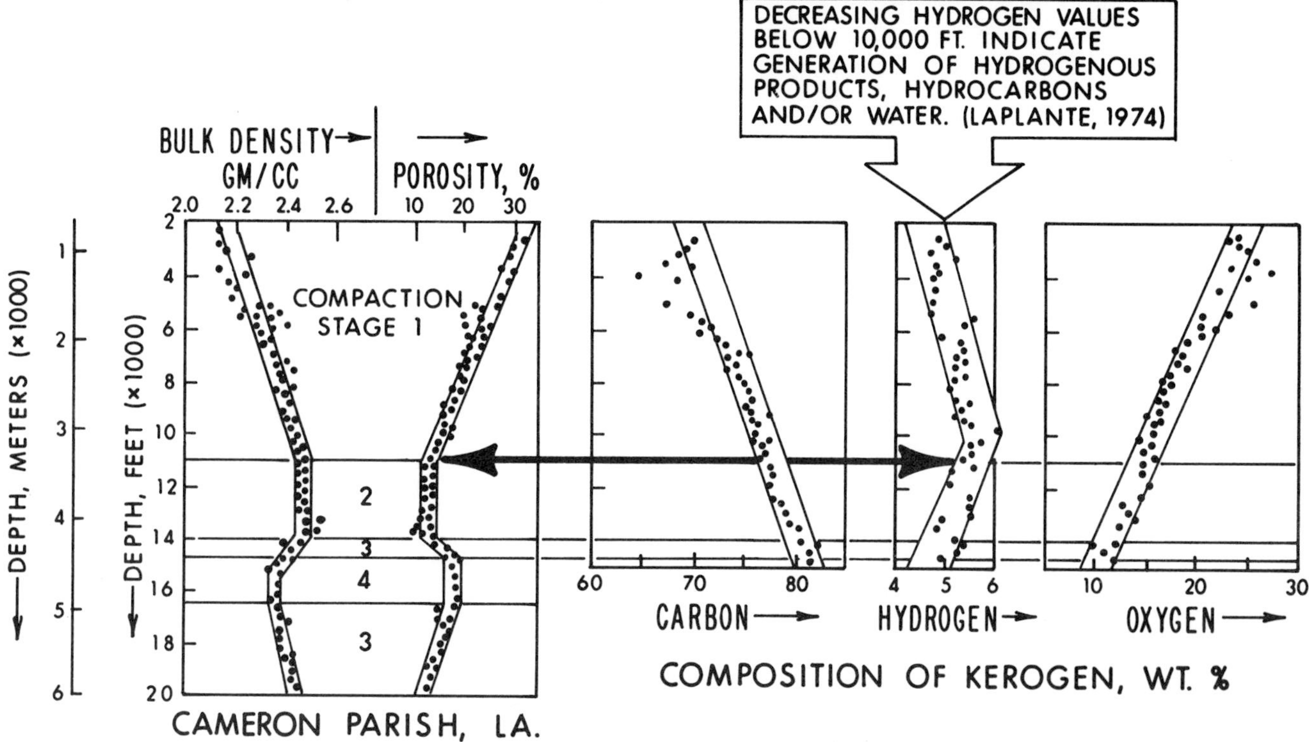

FIG. 12—Typical relationships among shale bulk density, shale porosity, and the composition of kerogen, showing the correspondence between the top of Compaction Stage 2 and the level of organic diagenesis corresponding to the threshold of intense oil generation. Data from Miocene shale, Well-C in the Cameron Parish, Louisiana field.

crease due to thermal expansion can be computed. For a shale with a porosity of 10%, the volume increase is approximately 4 bbl per acre foot for each 1,000 ft (305 m) of increased burial within Stage 2 or approximately 0.5% of the shale pore volume. This increase in volume would tend to cause both water and hydrocarbons to be expelled; but again, the tendency of the shale to retain its pore water, but not hydrocarbons, would result in preferential diffusion of hydrocarbon molecules out of the shale.

Pressure Increase Due to Osmotic Imbalances

The primary migration and expulsion of hydrocarbons may be allied to the expulsion of dissolved ions from the shales. It has been determined that shale waters are less saline than sandstone waters (Schmidt, oral commun., 1971; Schmidt, 1973). The higher salinity of sandstone waters has been attributed to ionic filtration as water moves from the sandstones through the intervening shales during the overall expulsion of water from the section (White, 1965). However, with most of the water absorbed within the shale, hydrodynamic flow across shale beds is most unlikely. The escape of the water through the geologic section as a whole presumably is dominated by flow through interconnected permeable beds, fractures, and faults, but movement of water out of the shale beds is much more impeded, and is best accomplished by molecular diffusion. Studies by the writer in 1972 suggest that the relatively low salinity of shale water may be due to the preferential expulsion of ions from the shales during compaction in response to osmotic imbalances between shale pore water and that in associated sandstones.

The most familiar demonstration of osmotic diffusion is that of the flow of water through a semipermeable membrane, separating two solutions of different salinities (Fig. 13A). This flow is in the direction which tends to equalize the salinity of the two aqueous systems, and it occurs because of an imbalance in vapor pressure across the membrane. The dissolved ions are bonded with water molecules, thereby lowering the vapor pressure of the solutions. The more concentrated the solution, the more water molecules are bonded, and the lower is the vapor pressure.

The vapor pressure of an aqueous system (in this case a solution) relative to that of distilled water at the same temperature is called the aqueous activity of the system. The higher the salinity of the solution, the lower the vapor pressure, and the lower the aqueous activity. Osmotic flow is from high to low aqueous activity and tends to balance aqueous activities (Fig. 13B). Osmotic flow ceases when activity equilibrium is established.

The osmotic relationship between shale and sandstone pore waters in nature is best understood by observing that a water molecule adsorbed on a shale mineral grain surface is bound in much the same way

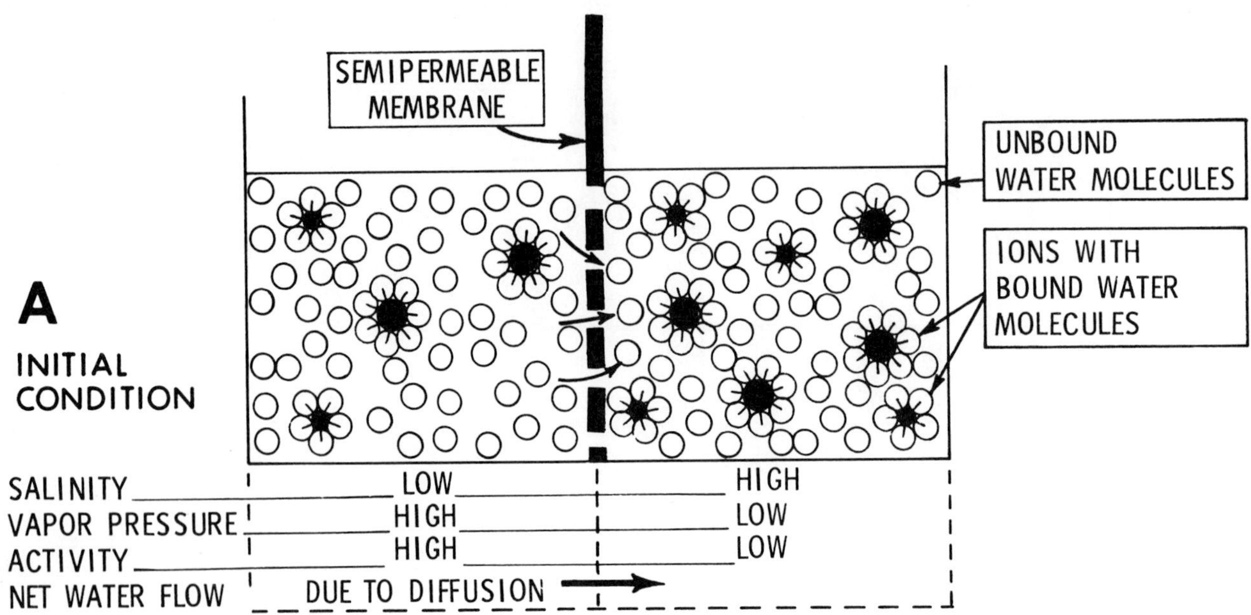

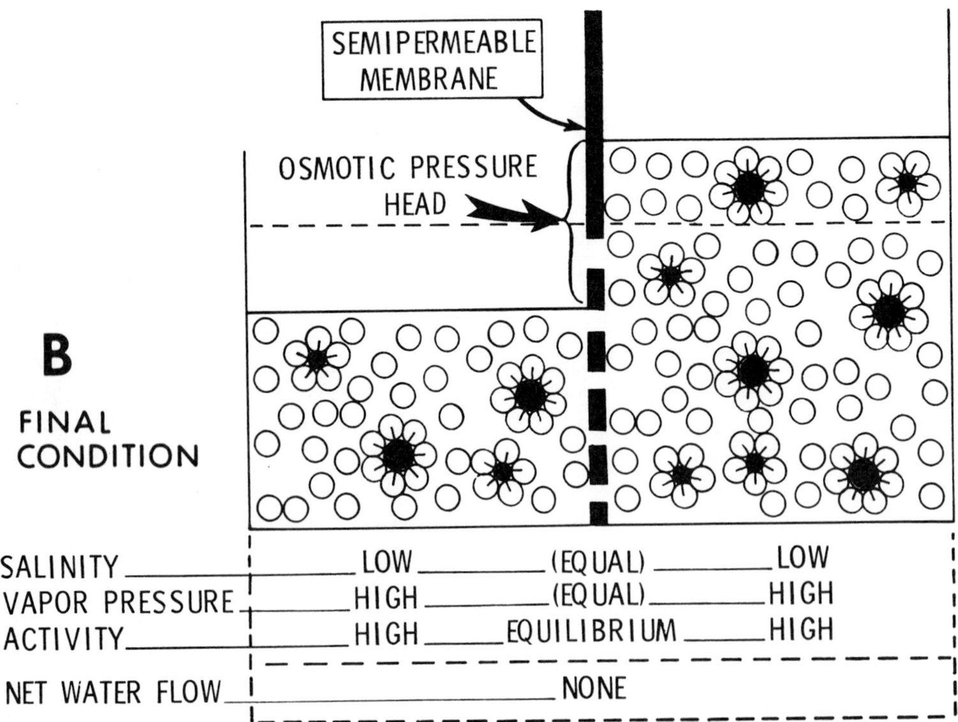

FIG. 13—Principles of osmotic diffusion through a semi-permeable membrane.

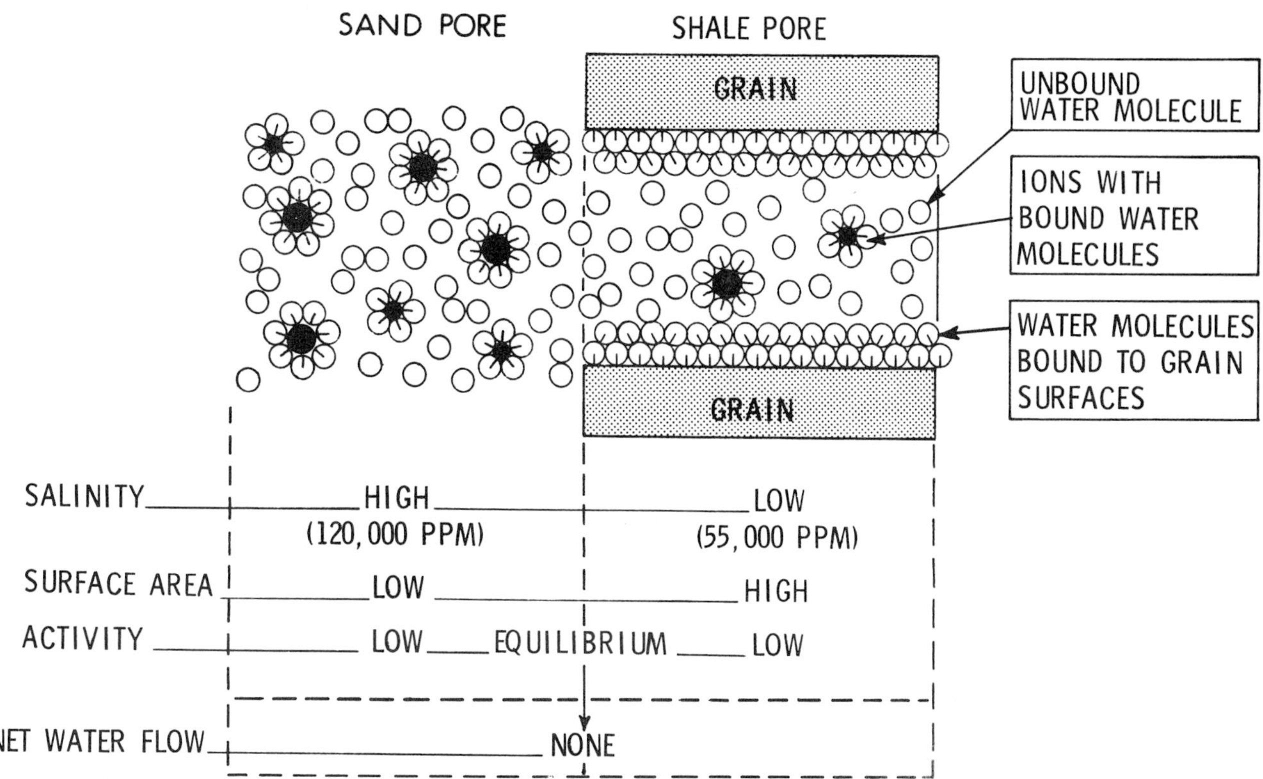

FIG. 14—Effect of adsorption on the aqueous activity of shale pore water showing that relatively fresh shale pore water can be in activity equilibrium with high salinity sandstone pore water.

as a water molecule associated with a dissolved ion (Fig. 14). The activity of the shale pore water is low due to this bonding on the grain surfaces. Computations by the writer indicate that shale waters within Compaction Stage 2 in the Miocene section of the Gulf Coast, which have an approximate salinity of 55,000 ppm, are in activity equilibrium with sandstone waters, which have an average salinity of 120,000 ppm. The bonding of water molecules on the large surface of the shale grains compensates for the bonding of water molecules on the additional dissolved ions in the sandstone waters.

At an early stage of burial, salinities of the shale and sandstone waters are approximately equal (Fig. 15A). However, due to adsorption of water molecules on the shale grain surfaces, the activity of the shale-water system is lower than the sandstone-water system. Sandstone pore water would therefore tend to diffuse into the shale, but diffusion of sandstone water into the shale is overbalanced by increasing overburden pressure as the shale compacts. Both water and dissolved ions can move from the compacting shale into the sandstone (Fig. 15B). However, the movement of some water molecules is inhibited by their adsorption on the shale grain surfaces, while the dissolved ions are relatively free to move into the sandstones. This movement of the ions into the associated sandstones lowers the activity of the sandstone water by increasing its salinity and brings the two systems toward activity equilibrium.

When compaction has continued until an average of only ten layers of water molecules separate the shale grains (a typical situation in Compaction Stage 2 and the top of the geopressured shale), the average size of the pores (25Å to 27Å) is so reduced that the dissolved ions, with their shell of adsorbed water molecules, are no longer completely free to move out of the shale. This movement is inhibited because the sizes of the hydrated ions (i.e., $Na^+ \cong 7$Å; $Cl^- \cong 9$Å) are close to the average pore size (Fig. 15C). The sandstone waters, deprived of the continuing supply of ions, tend to freshen with increasing depths of burial and continuing expulsion. Both these factors, the high proportion of adsorbed water molecules in the shale and the increased activity of the sandstone water relative to the shale water, tend to prevent further fluid loss from the shales. This condition may be one of the reasons for the stability of Compaction Stage 2.

B. G. Newton (oral commun.) observed that some of the hydrocarbon accumulations in the Gulf Coast are associated with sandstone waters of slightly subnormal salinity. Subnormal sandstone-water salinity may be a characteristic of Compaction Stage 2 in the Gulf Coast, and there is evidence to suggest that this is due to the introduction of fresher waters from the underlying geopressured shales as well as Compaction

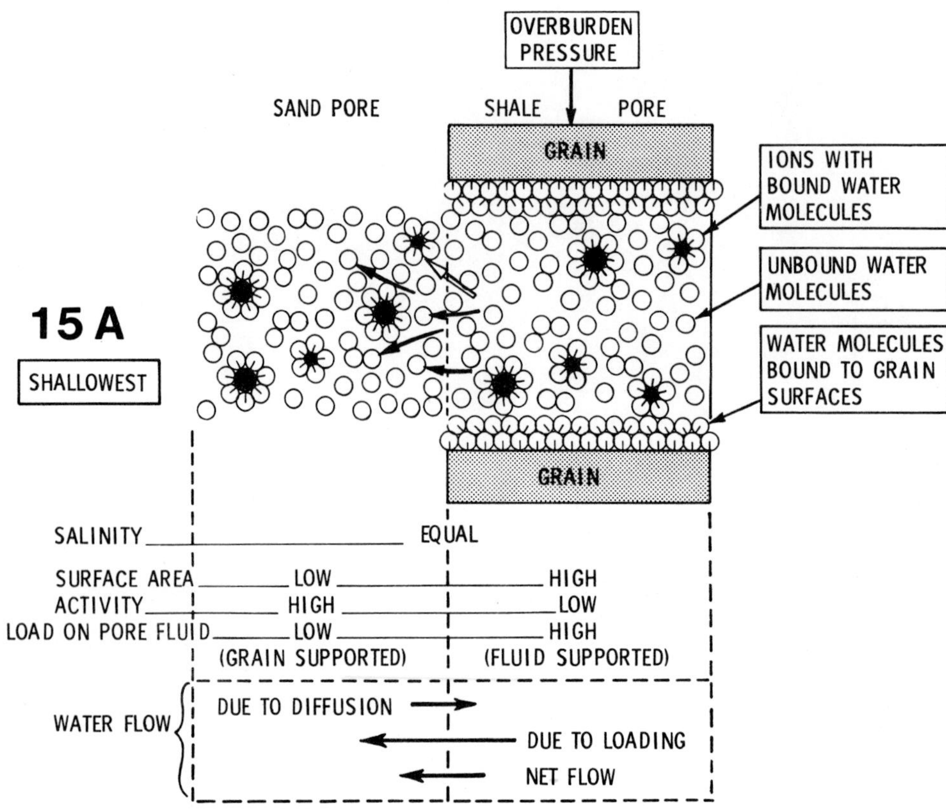
OVERBURDEN PRESSURE
SAND PORE
SHALE PORE
GRAIN
IONS WITH BOUND WATER MOLECULES
UNBOUND WATER MOLECULES
WATER MOLECULES BOUND TO GRAIN SURFACES
15 A
SHALLOWEST
GRAIN
SALINITY EQUAL
SURFACE AREA LOW HIGH
ACTIVITY HIGH LOW
LOAD ON PORE FLUID LOW HIGH
(GRAIN SUPPORTED) (FLUID SUPPORTED)
WATER FLOW
DUE TO DIFFUSION
DUE TO LOADING
NET FLOW

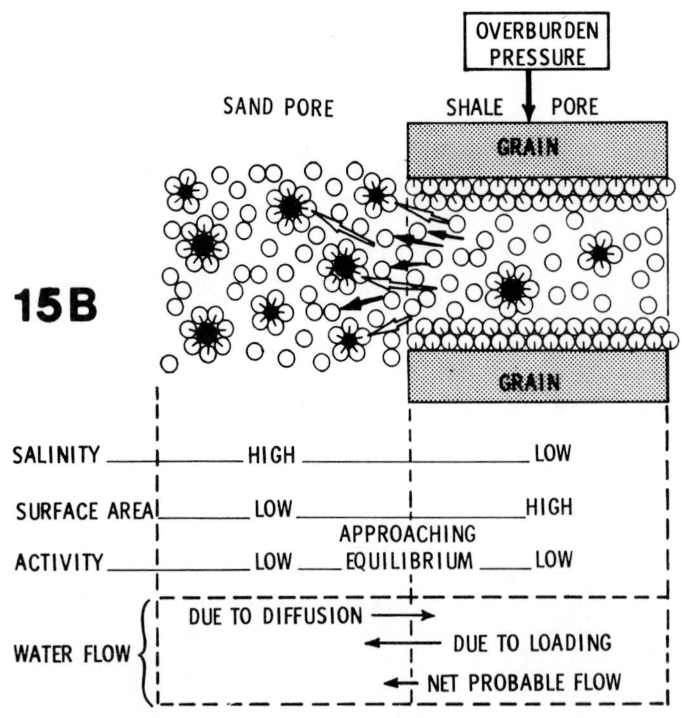
OVERBURDEN PRESSURE
SAND PORE
SHALE PORE
GRAIN
15B
GRAIN
SALINITY HIGH LOW
SURFACE AREA LOW HIGH
APPROACHING
ACTIVITY LOW EQUILIBRIUM LOW
WATER FLOW
DUE TO DIFFUSION
DUE TO LOADING
NET PROBABLE FLOW

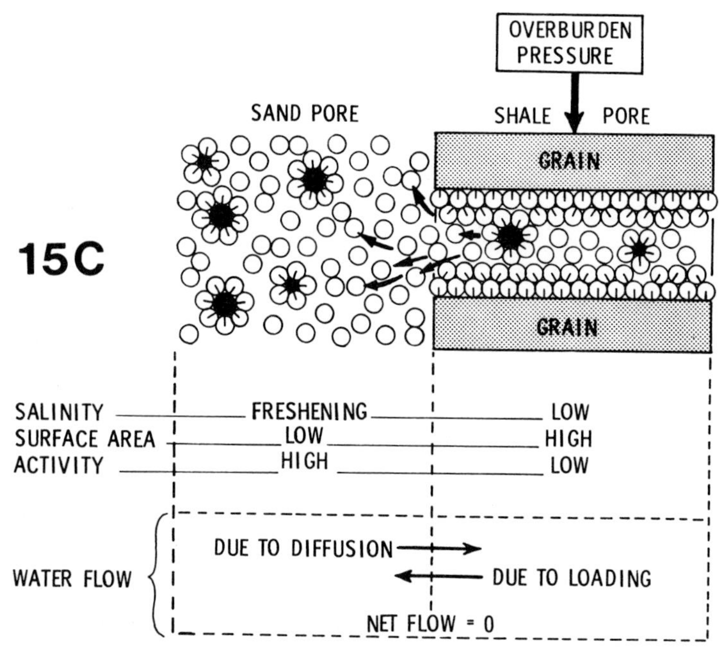

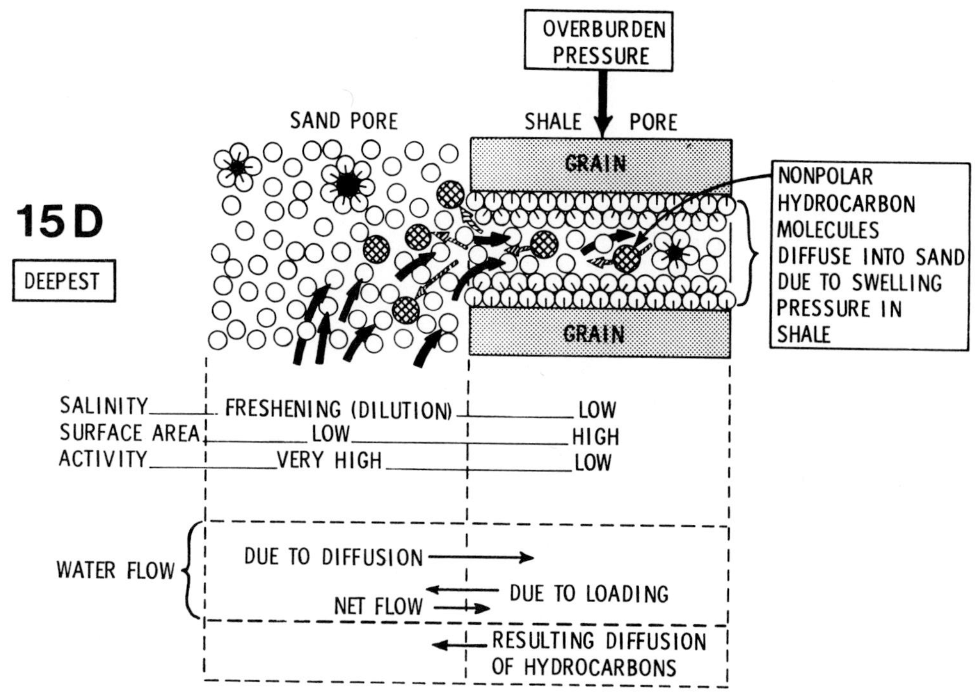

FIG. 15—Conceptual view of the probable diffusion of water and dissolved solids in response to activity disequilibria which develop between the water in compacting Gulf Coast shales and that in adjacent sandstones during a typical burial sequence. Also shown are the changes in shale and sandstone water salinity that would result, as well as the probable effect of activity disequilibrium on hydrocarbon expulsion from shales.

Stage 2 shales.

Any further dilution of the sandstone waters (e.g., flushing by relatively fresh waters expelled from a geopressured shale section) will further increase the activity of the sandstone waters (Fig. 15D), further encouraging a flow of water from the sandstones to the shales. If the aqueous activity of the sandstone water becomes sufficiently overbalanced, water will diffuse from it into the shale as shown in Figure 15D, and hydrocarbons will tend to diffuse out of the shales to compensate for the increase in osmotic pressure within the shale ("swelling pressure") produced by the diffusion of water into the shale from associated sandstones.

It is thus possible that hydrocarbon expulsion from a source bed into associated sandstones might be triggered or enhanced by the flushing of associated sandstones with relatively fresh water. The effect would be a molecular replacement of hydrocarbon molecules in the shale by sandstone waters.

CONCLUSIONS

(1) Expulsion and migration of petroleum still is the least understood facet of petroleum geology. The ideas presented require extensive testing and criticism.

(2) Hydrocarbon expulsion from shale source beds does not result from the flushing of very small quantities of oil or gas by large volumes of water; the movement and concentration of hydrocarbons within shales is essentially independent of water movement.

(3) Fluid expulsion from a shale pore system is conceptualized as principally diffusion of pore water, dissolved ions, and hydrocarbon molecules, directionally influenced by pressure, temperature, and activity gradients.

(4) Phenomena contributing to fluid expulsion from Gulf Coast upper Tertiary shales include: volume increase within the shale pore system due primarily to hydrocarbon generation; thermal expansion of shale pore fluids with increased depth of burial; and increased pore pressure in shales due to the effect of an imbalance of activities between sandstone and shale water systems.

REFERENCES CITED

Athy, L. F., 1930, Density, porosity, and compaction of sedimentary rocks: AAPG Bull., v. 14, p. 1-24.

Beckstrom, R. C., and F. M. Van Tuyl, 1928, Compaction as a cause of the migration of petroleum: AAPG Bull., v. 12, p. 1049-1055.

Blackwelder, E., 1920, The origin of the central Kansas oil domes: AAPG Bull., v. 4, p. 89-94.

Bradley, J. S., 1975, Abnormal formation pressure: AAPG Bull., v. 59, p. 957-973.

Chillingar, G. V., and L. Knight, 1960, Relationship between pressure and moisture content of kaolinite, illite, and montmorillonite clays: AAPG Bull., v. 44, p. 101-106.

Dickinson, G., 1951, Geological aspects of abnormal reservoir pressures in the Gulf Coast region of Louisiana, U.S. A.: AAPG Bull., v. 37, p. 410-432.

Harwood, R. J., 1977, Oil and gas generation by laboratory pyrolysis of kerogen: AAPG Bull., v. 61, p. 2082-2102.

Hedberg, H. D., 1926, The effect of gravitational compaction on the structure of sedimentary rocks: AAPG Bull., p. 1035-1072.

——— 1936, Gravitational compaction of clays and shales: Am. Jour. Science, v. 31, no. 184, p. 241-287.

LaPlante, R. E., 1974, Hydrocarbon generation in Gulf Coast Tertiary sediments: AAPG Bull., v. 58, p. 1281-1289.

Magara, K., 1968, Compaction and migration of fluids in Miocene mudstone, Nagaoka Plain, Japan: AAPG Bull., v. 52, p. 2466-2501.

Meade, R. H., 1966, Factors influencing the early stages of the compaction of clays and sands—review: Jour. Sedimentary Petrology, v. 36, no. 4, p. 1085-1101.

Monnett, V. E., 1922, Possible origin of some of the structures of the Mid-Continent oil fields: Econ. Geology, v. 17, no. 3, p. 194-200.

Schmidt, G. W., 1973, Interstitial water composition and geochemistry of deep Gulf Coast shales and sandstones: AAPG Bull., v. 57, p. 321-337.

Sorby, N. C., 1908, On the application of quantitative methods to the study of the structure and history of rocks: Geol. Soc. London Quart. Jour., v. 64, p. 171-232.

White, D. E., 1965, Saline waters of sedimentary rocks, *in* Fluids in subsurface environments: AAPG Memoir 4, p. 342-366.

Witherspoon, P. A., and D. N. Saraf, 1964, Diffusion of methane, ethane, propane, and N-butane in water: Am. Chem. Soc. Div. Petroleum Chem. Preprints, v. 9, no. 3 (August).

Primary Migration: The Importance of Water-Mineral-Organic Matter Interactions in the Source Rock[1]

By Colin Barker[2]

Abstract It now is generally accepted that some of the petroleum generated in organic-rich source rocks can move to reservoir rocks and accumulate. Movement of hydrocarbons (or their percursors) through an aqueous medium involves complex interactions among the mobile organic materials, kerogen, rock matrix and water, and all of them change in composition and/or amount with increasing depth of burial. Water is present in large quantities during the early stages of compaction but its availability diminishes with depth unless supplied by the smectite-illite conversion. The average pore size in shales also decreases with depth. The organic material available for migration initially contains high concentrations of compounds with oxygen, nitrogen, and sulfur (which impart enchanced solubility in water). However, as depth increases these become quantitatively less important and increased amounts of hydrocarbons are generated from the kerogen. Shallow oils are relatively rich in nitrogen, sulfur, and oxygen compounds, but with increasing depth these give way to more mature oils, then condensates, and finally to dry gas. Thus, the nature of the material migrating changes with depth as do the rock and water system through which it moves, and it seems very likely that there is not one mechanism of migration, but many, and that for different conditions different mechanisms will operate. Mechanisms involving transport of hydrocarbons, precursors, or micelles in water require some process to cause exsolution. Because the biggest discontinuities in chemical and physical properties are at the sand-shale contact this seems to be the most likely place for exsolution. Where sufficient hydrocarbons are generated in the source rock a separate oil phase separates, and if the water is structured close to clay surfaces the hydrocarbons will accumulate in the centers of the pores, ultimately forming a pore center network. This will be squeezed as the water expands thermally and droplets will be expressed from the ends of the network. This mechanism operates even where the network is discontinuous. Comparison of crude oil composition with the source rock extract shows that in general saturate materials move in preference to the aromatics, with the NSOs being least mobile. This is the reverse of the solubility trend but is the sequence of adsorption. It appears that adsorption of bitumens in the source rock has an important influence on the composition of the material that migrates. The various migration mechanisms probably all operate to some extent, with one or more being dominant under any set of geologic conditions.

INTRODUCTION

Petroleum is found reservoired in rocks which have high values for porosity and permeability. The processes which generate these characteristics also act to destroy or remove much of the organic matter originally present. In sand, wave action winnows away the finer-grained clay particles and with them the organic matter. At the same time oxygen is introduced which also acts to destroy organic matter. These processes leave the potential reservoir rocks with too little organic matter to generate commercial quantities of crude oil. Fine-grained material, which eventually becomes shale, settles where there is little wave action and where conditions are conducive to the preservation of organic material. On the average, shales contain 1% organic matter (Hunt, 1972), which is enough to generate commercial quantities of crude oil (Dow, 1977), but this crude oil cannot be produced unless it has accumulated in a reservoir. A considerable body of geologic and geochemical data now supports the "source-rock concept" which is the theory that petroleum forms in organic matter-rich source rocks (such as shales), migrates, and subsequently accumulates in permeable reservoir rock (such as sandstones). Many of the features of petroleum generation in the source rock are understood, at least in principle, but the time and mechanism of migration are imperfectly known.

Because migration links the source rock to the reservoir an understanding of the migration mechanism is important in many aspects of exploration. Source rocks, for example, contain the residue left after migration has occurred, and this residual material can be interpreted only if the nature of the migrated material is known. The correlation of reservoired crude oils to their source rocks is based on the assumption that the

[1]Manuscript received, September 6, 1978; accepted, December 11, 1978.

[2]The University of Tulsa, Tulsa, Oklahoma 74104.

Article Identification Number:
0149-1377/79/SG10-0002/$03.00/0.

composition of the source rock extract should closely resemble that of the reservoired oil. This is true if the migration mechanisms cause minor or no chemical fractionation, but if the composition of the material which migrates and accumulates in the reservoir is very different from that left behind in the source rock then correlation will not be possible. Also, some knowledge of the time of migration is important, especially its relationship to the time of generation and the development of traps.

In spite of its obvious importance in the process of petroleum accumulation, the migration of petroleum from source rock to reservoir is not understood. Many migration mechanisms have been proposed (Cordell, 1972) and, although all explain some of the observed features, none seems generally applicable. Problems arise because the petroleum must move through an aqueous medium yet hydrocarbons have very low solubilities in water, and because movement through the very small pores of a fine-grained source rock puts severe limitations on the form in which hydrocarbons travel. The situation is complicated further by the lack of adequate terminology—*primary migration*, for example, probably involves several different processes acting in sequence—and because different mechanisms may dominate under different geologic conditions. In the following discussion the main emphasis is on the movement of organic compounds through and out of fine-grained source rocks and less with the subsequent movement through more permeable beds and accumulation in the reservoir.

MIGRATION MECHANISMS

Many mechanisms have been suggested for moving petroleum out of a source rock and they can be divided into those which involve movement of an aqueous medium and those which operate independently of water movement. We start by considering mechanisms involving movement in solution.

True Solution

Hydrocarbons have a very low solubility in water (Table 1), but even a low solubility may be enough to move large amounts of petroleum if sufficient quantities of water are available. In general, aromatics are most soluble, paraffins least soluble, and naphthenes intermediate, with solubilities decreasing as the size of the molecule increases. Most of the published hydrocarbon solubility data were obtained in fresh water at 25°C and one atmosphere pressure, but solubilities decrease with increasing salinity and increase with increasing temperature, particularly at high temperatures (Fig. 1). Unfortunately, little data for solubilities under realistic subsurface conditions of salinity, temperature, and pressure are available (Price, 1976).

Exsolution of hydrocarbons from the migrating aqueous solution poses a major problem—why are the dissolved hydrocarbons not swept through the trap with the water? Among the explanations suggested are those involving changes in salinity, temperature, pressure, and carrier-bed pore size. It seems that the major discontinuity along the migration path occurs at the sandstone-shale contact. Here there are abrupt changes in water salinity and composition (Baharlou, 1973; Schmidt, 1973) so that pore water saturated with dissolved hydrocarbons will be supersaturated in the more saline waters of the adjacent sandstones (Fig. 2). The exsolved droplets of hydrocarbons could then be carried by moving waters through the sandstones until they accumulate by buoyancy in a structural trap or are collected at a permeability barrier in a stratigraphic trap.

Table 1. Solubility of Selected Hydrocarbons in Water[1] (McAuliffe, 1963, 1966)

Hydrocarbon	Solubility (ppm)
Benzene	1,780.0
Toluene	538.0
Cyclopentane	156.0
Cyclohexane	55.0
Methylcyclopentane	42.6
Methylcyclohexane	14.0
1,2-Dimethylcyclopentane	6.1
2,3-Dimethylbutane	18.4
3-Methylpentane	12.8
2-Methylpentane	18.8
n-Pentane	38.5
n-Hexane	9.5
n-Heptane	2.4

[1] These values were obtained for fresh water at 25°C.

Enhanced Solubility

Several mechanisms have been suggested which involve the movement of species more soluble than simple hydrocarbons. In these models moving water still is required but smaller amounts can transport enough material to form a commercial accumulation. The problems of exsolution remain but major effects can be expected at the sandstone-shale interfaces.

Accommodation—Accommodation is a term used for the suspension of minute colloidal particles in water and Peake and Hodgson (1967) showed that n-alkanes can be accommodated in distilled water in amounts much greater than their true solubility. The n-alkanes are accommodated in direct proportion to their abundance, and any odd predominance present in the source will be preserved in the water medium, and hence in any subsequently formed oil accumulation. However, the violent aggitation needed to produce accommodation is unlikely to occur in source rocks.

Petroleum precursors—It has been suggested that it is not the petroleum hydrocarbons themselves which migrate but some more water-soluble precursors. Compounds containing NSO-containing [3] functional groups, such as acids and alcohols, have high solubilities in water (Table 2) but the timing of the formation of hydrocarbons by loss of the function groups would be critical. If the compounds retain their functional groups too long they will remain in solution and be swept through the trap with the moving water.

Micelles—Polar organic molecules may form small colloidal aggregates called "micelles" (Fig. 3) with the water-compatible polar ends oriented outward into the water. The hydrocarbon-like volume in the center can incorporate other organic molecules enhancing their apparent solubility. For this reason micelle-forming compounds are often referred to as "solubilizers."

Separate Oil Phase

If hydrocarbons are generated in quantities sufficient to saturate the water as well as the adsorptive capacities of the shales and organic matter, discrete droplets of crude oil may form in the pore spaces. As water is expelled from the shale, oil droplets will be carried out into the coarser-grained carrier beds. This mechanism can operate only where enough moving water is available and where the diameter of the oil drop remains less than the diameter of the pore. Where the droplet encounters a constriction narrower than the drop diameter, deformation must occur for it to pass (Fig. 4). Deformation involves an increase in surface area and hence an increase in surface energy. The only forces operating on the droplet are buoyancy, which acts vertically upwards, and the hydrodynamic gradient. Hill (quoted by Levorsen, 1954) has shown that the hydrodynamic gradient is many orders of magnitude too small to force the droplet through the constriction, while the buoyancy forces have only a small component in the direction of movement through near-horizontal pores. It commonly has been concluded that droplets cannot negotiate the pore system of a shale. However, this simple model may have to be modified if the oil droplet traps and isolates a volume of water (as illustrated in Fig. 5) because the temperature rise which accompanies increasing depth of burial (and causes generation) results in high overpressure in the isolated volume. The high pressure differential produced might force the droplet through the constriction. A pressure differential of about 1 atm is needed and this will be generated aquathermally by a depth increase of about 10 ft (3 m; Barker, 1972). For this mechanism to be important the droplet must effectively block the constriction, the volume of isolated water must be large compared to the droplet, and a separate repressuring is needed at each constriction. These conditions are probably too restrictive for the mechanism to be generally important. However, the

[3]NSOs is a convenient shorthand way of indicating compounds containing Nitrogen, Sulfur, and/or Oxygen atoms.

Table 2. Solubilities of Paraffins, Acids, and Alcohols in Water (ppm)

	$R=C_5H_{11}$	$R=C_6H_{13}$	$R=C_7H_{15}$
R – H	39	9.5	3
R – COOH	10,700	2,200	––
R – OH	26,000	5,600	1,800

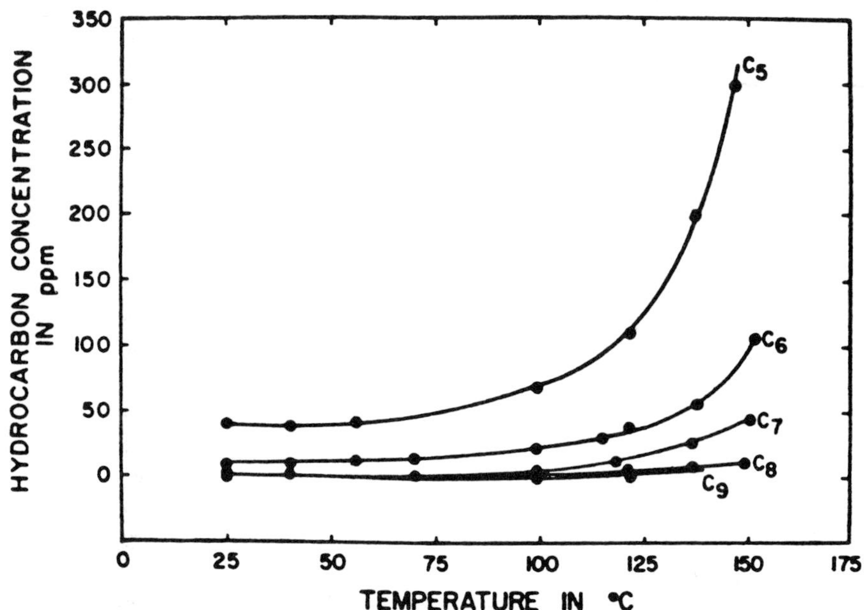

FIG. 1—Solubilities of pentane (C_5), hexane (C_6), heptane (C_7), octane (C_8), and nonane (C_9) in water as function of temperature at systems' pressure (Price, 1976).

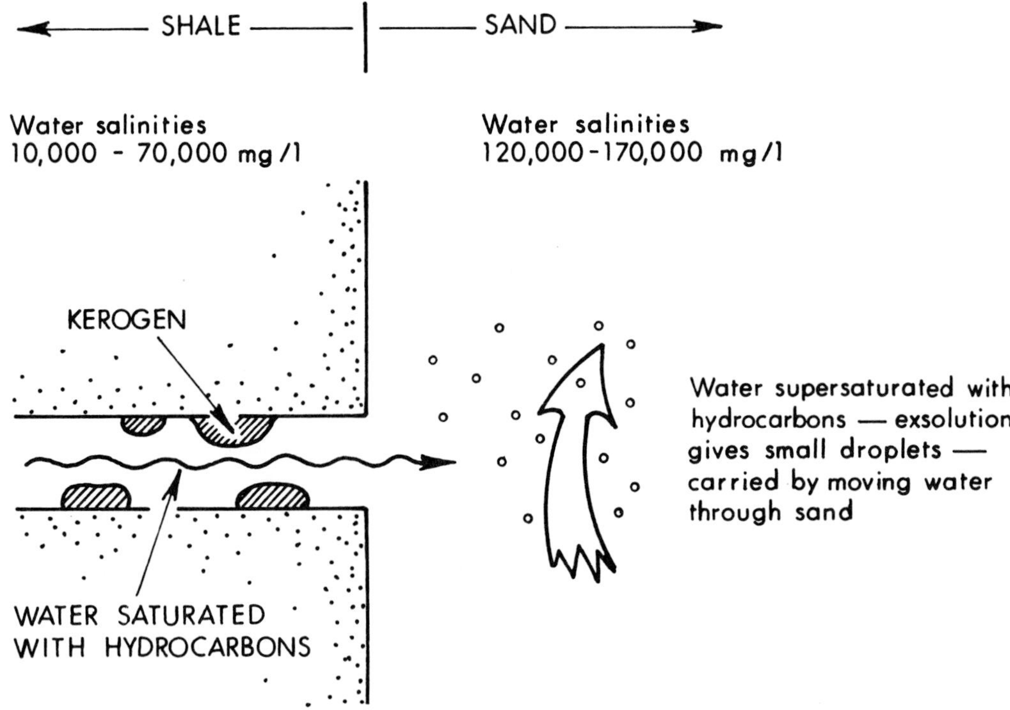

FIG. 2—Exsolution of dissolved hydrocarbons at a shale-sandstone contact. Salinity data from Schmidt (1973).

theoretical treatment is independent of scale and this mechanism may become important if the edges of a thick shale unit dewater preferentially so that average pore size is reduced near the edges. The center can then overpressure and force the release of both water and hydrocarbon droplets. Overpressuring effects caused by hydrocarbon generation are added to those produced by aquathermal pressuring and the combined effects may be enough to cause fractures in the source rock and provide avenues for escaping petroleum.

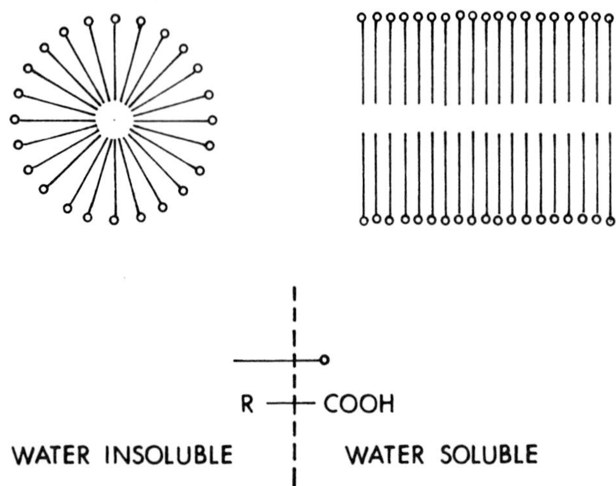

FIG. 3—Two configurations of micelle as proposed by Baker (1960).

Pore-center Network

The development of a separate oil phase depends on the relative amounts of hydrocarbons and water. Estimation of the effective water volume is not straightforward because in the pores of the source rock the clay surfaces interact strongly with the adjacent water through hydrogen bonds. These bonds between the oxygen of the clay SiO_4 tetrahedra and the hydrogen of the water molecules orient the water close to the surfaces and "structure" it (Drost-Hansen, 1969; Roberts and Zundel, 1979; Fig. 6). The degree of structuring decreases with distance from the clay surface. An isolated hydrocarbon molecule in a pore will assume a configuration of least energy, and this will be in the center of the pore where the water is least structured (Fig. 7A). If a second hydrocarbon molecule enters the system it may lie alongside the first (Fig. 7B) or end-to-end with it (Fig. 7C). Additional hydrocarbon molecules should eventually lead to the situation rep-

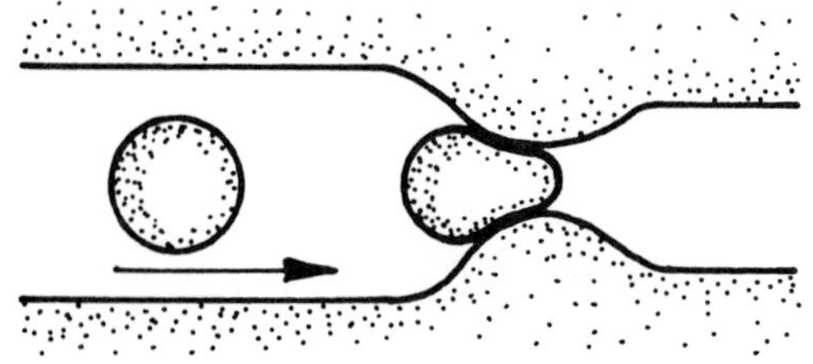

FIG. 4—Deformation of an oil drop as it enters a constriction.

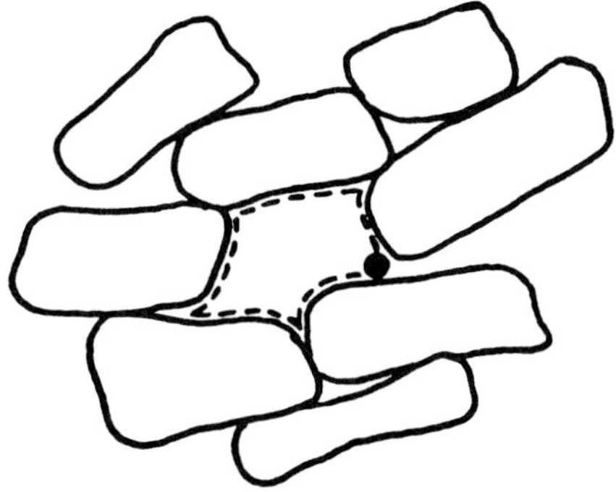

FIG. 5—Schematic diagram showing how a droplet of oil may isolate a volume of water.

resented in Figure 7D. Because hydrocarbon molecules do not suddenly appear in the pores, but are generated from the kerogen, a possible sequence of events may be similar to that shown in Figure 8. The generated hydrocarbons are initially adsorbed onto the kerogen but as generation continues they become relatively more abundant and build out into the water. Preferential replacement of the less structured water causes lateral development until finally they merge with hydrocarbons generated by other pieces of kerogen to form a pore-center network (Barker, 1973).

The networks developed in this way will probably be discontinuous because the irregular packing of clay platelets causes large variations in pore diameter throughout the shale, and in the large cavities there will be insufficient hydrocarbons to continue the network. This situation is shown schematically in Figure 9A where hydrocarbon networks have developed in the pores but do not extend across the larger voids.

Even after the formation of pore center networks some pressure differential or other force is needed to produce a net movement of hydrocarbons out of the rock. Again we must recognize that the system is not isothermal, and that petroleum generation generally is caused by rising temperature. The water and other fluids in the pores will expand as the temperature rises and create an internal presure in the shales and this may be augmented by the process of hydrocarbon generation.

Because the water is structured it will exhibit non-Newtonian behavior and will resist movement, while the hydrocarbons will be squeezed out. This process operates even when the network is not continuous. Consider for example, Figure 9B—if the temperature rises from 85 to 115°C (185 to 239°F); (equivalent to increasing the depth of burial from 10,000 to 14,000 ft [3,048 to 4,267m] in the Gulf Coast), the water density changes from 1.0165 to 1.317 cu cm/g corresponding to a 1.5% expansion at constant pressure. If the water in volume A expands by 1.5% it will force an equivalent volume of oil through the narrower pore structure into the next large opening B, so that this receives an influx of oil equivalent to 1.5% of the volume of A. However, the water in B is also subjected to increasing temperature and its water expands by 1.5% of the volume of B. This increase in volume is added to that produced in volume A so that the volume of oil expelled to the sand is approximately 3% of the volume of B if cavities A and B are of equal volume.

It appears that the volume of oil squeezed from the rock by the thermal expansion of water is equal to 1.5% of the total pore space when the temperature rises from 85 to 115°C (185 to 239°F) and the pressure from 5,000 (1,000 ft; 3,048 m) to 7,000 psi (14,000 ft; 4,267 m) even where the pore network is discontinuous. If this mechanism operates with 100% efficiency, it leads to unrealistically high values for the percent of the bitumens that migrate. For an average shale with 1% kerogen this will occupy about 2.5% of the rock volume and generate bitumens that occupy roughly

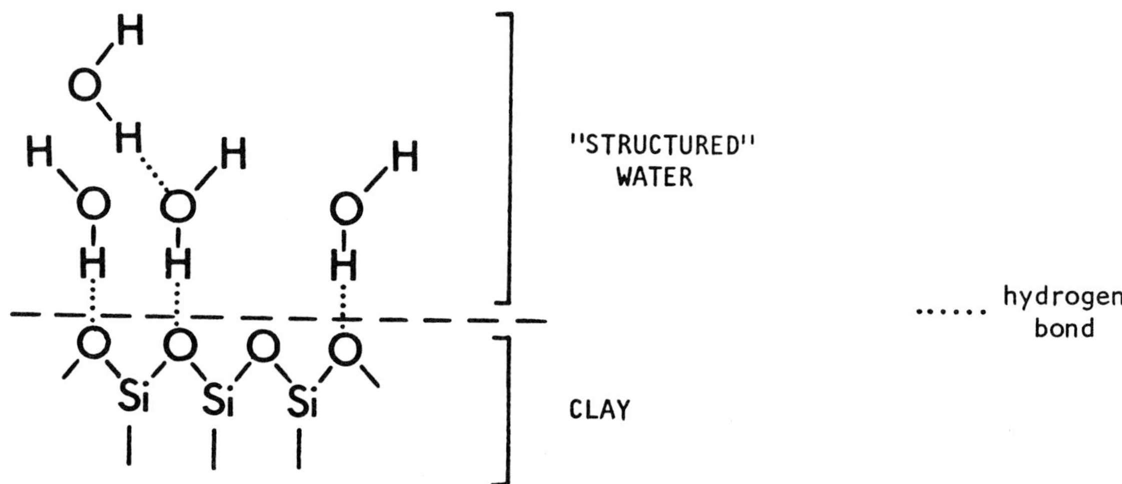

FIG. 6—Water molecules near clay surfaces are held by hydrogen bonds and no longer behave like bulk liquid water.

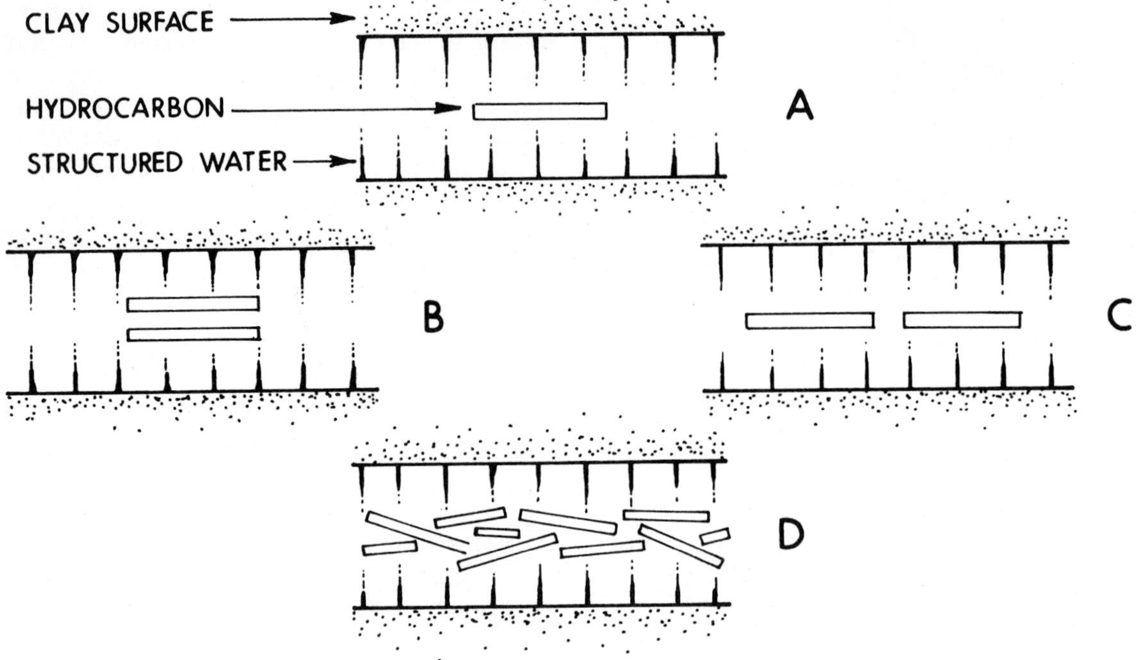

FIG. 7—Organic molecules in water-filled pores.

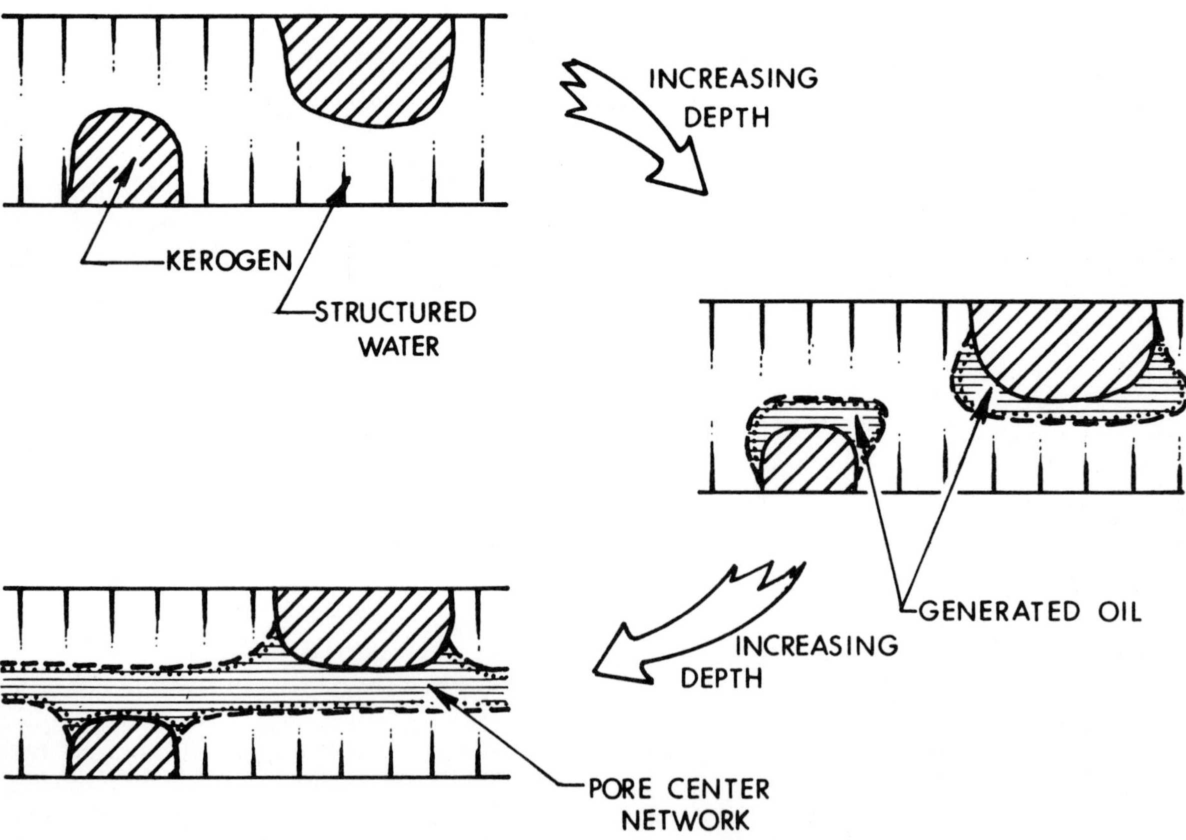

FIG. 8—Development of a pore center network of hydrocarbons as depth increases.

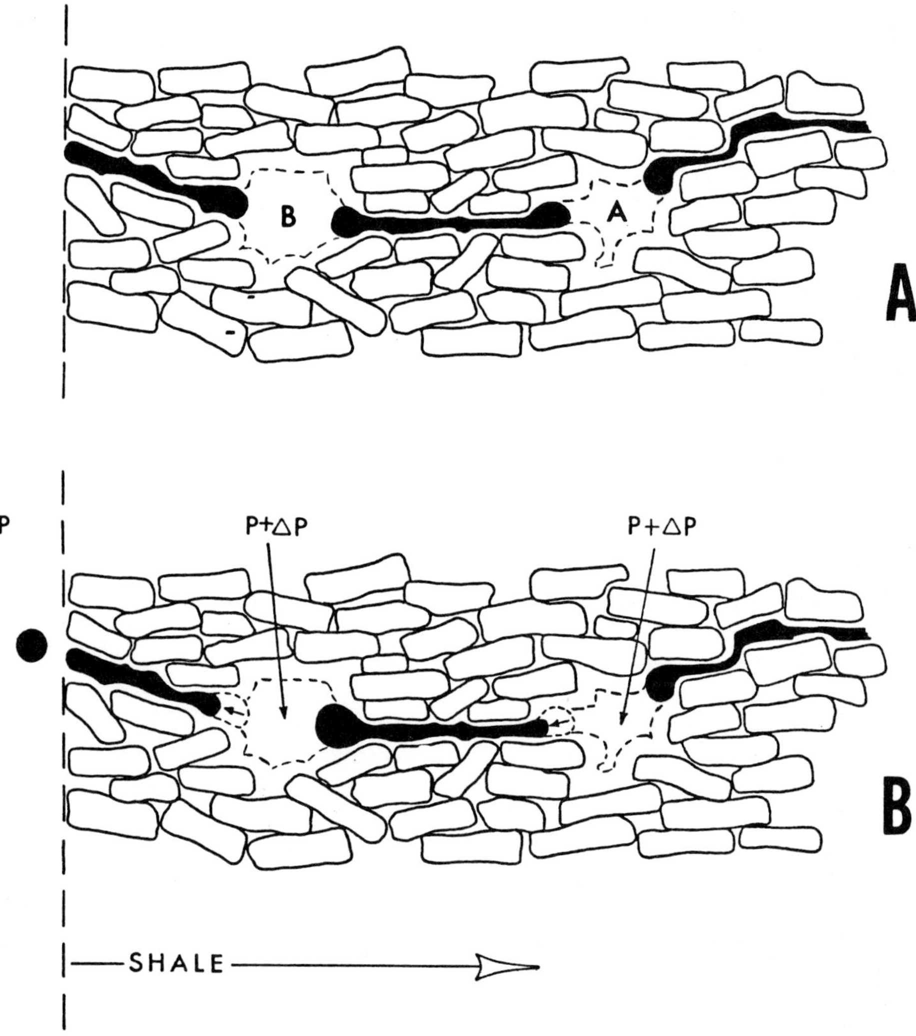

FIG. 9—Possible distribution of oil as a pore center network in a shale adjacent to a sand. **A.** Distribution of oil and water at 10,000 ft (3,048 m) and 85°C (185°F)
B. Burial of the shale to 14,000 ft (4,267 m) and 115°C (239°F) causes thermal expansion of the water in volumes A and B, and leads to the expulsion of part of the hydrocarbons.

0.25% of the rock. For a shale with 10% porosity the increase in volume of the water on heating (again using 85 to 115°C [185 to 239°F] and 10,000 to 14,000 ft [3,048 to 4,267 m]) will be 1.5%, or 0.15% of the volume of the rock. If only bitumen moves, 0.15/0.25 of it will be expelled, which is 60% of the available bitumen. Previous studies of migration efficiency (Hunt, 1977, for example) showed that only about 10% or less of the bitumens migrated. This leads to the conclusion that the pressuring mechanism only needs to be about 10% efficient so that if 90% of the water bypasses oil-filled pores enough bitumen can be expelled to produce a commercial accumulation in an associated reservoir rock.

It should be noted that the theory for aquathermal pressuring was developed for liquid-filled systems (Barker, 1972). If the systems (such as volume A, Fig. 9B) contain free gas they may not overpressure (Barker, 1979b), and no net pressure differential will be developed to drive migration. Although simple thermal expansion of a gas-containing isolated volume does not overpressure, the continued generation of gas from the thermal maturation of kerogen may develop sufficient pressure (Momper, 1978).

Miscellaneous Mechanisms

A wide variety of other mechanisms have been discussed in the literature. Some of these may be important under certain situations. For example, the role of high pressure gas in transporting other hydrocarbons presumably will be most important in deeply buried rocks where gas generation is most pronounced (Sokolov and Mironov, 1962), but the role of gas at shallower depths where it could be generated from woody-type kerogens is unknown. Diffusion of hydrocarbons, which permits migration without a moving water phase, is generally considered to be incapable of moving hydrocarbons over large enough distances. However, it could have a role in moving materials short distances to fractures or sandy stringers where some other mechanism may take over. Continuous networks of kerogen distributed through the source rock have been postulated as one way of permitting transfer of hydrocarbons to the adjacent reservoir rocks (McAuliffe, 1979). It is difficult to see how a network would be established initially especially in rocks containing only half a percent of organic matter (which is widely considered to be the lower limit for a source rock). Survival of the network during the rearrangement of grains caused by compaction and diagenesis poses a problem since a single break in the network would stop the whole migration process.

SIGNIFICANCE OF CORRELATION STUDIES

In many basins crude oils can be related to the specific rock units which generated them because there is a close similarity between the composition of the bitumens in the source rock and the composition of the crude oil (Barker, 1975). This implies that, at least in these cases, migration causes only minor changes in chemical composition. This, in turn, puts severe constraints on the mechanism of migration.

In a detailed study of crude oils and possible source rocks in the Williston basin, Williams (1974) identified three types of crude oils which could be correlated to Winnipeg, Bakken, and Tyler source rocks respectively. Correlation was based on C_4-C_7 hydrocarbons, C_{15+} hydrocarbons, carbon isotope values, optical rotation, and Correlation Index plots. Figure 10 shows how closely the compositions of the source rock extracts resemble those of the crude oils when plotted on a straight chain-branched chain-saturated ring compound ternary diagram. The solubilities of hydrocarbons in these three compound types vary considerably (Table 3), and any migration mechanism involving a solubility mechanism should lead to marked differences in the chemical composition of the oil relative to the source rock extract. Because this is not observed it strongly suggests that a mechanism involving solution in water is not appropriate in the Williston basin. Data for the C_{15+} range of straight chain hydrocarbons also supports this conclusion because the source rock extract and the oils have very similar abundance patterns (Williams, 1974).

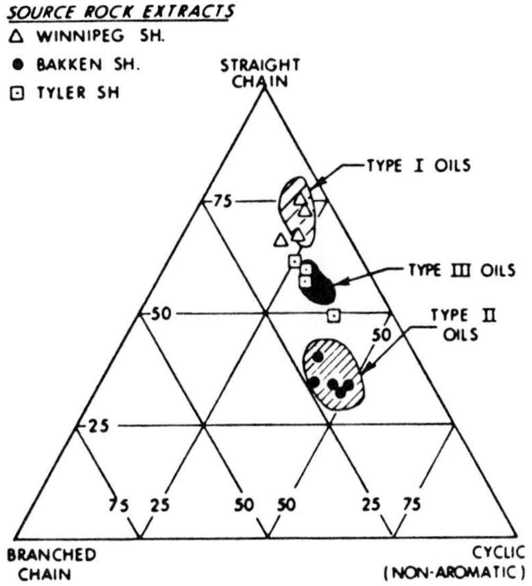

FIG. 10—Distribution of straight chain, branched, and cyclic paraffins in the C_4-C_7 fraction of Williston basin crude oils and source rock extracts (Williams, 1974).

It is unwise to generalize from a single example and, because the Williston basin data are for saturate com-

Table 3. Solubilities in Water (ppm) of Some Saturated Hydrocarbons with Five, Six or Seven Carbon Atoms (Data from McAuliffe, 1963, 1966)

	Straight Chains	Branched Chains	Cyclic Compounds
C_5	38.5 (Pentane)	47.8 (iPentane) 33.2 (DMPropane)	156 (c Pentane)
C_6	9.5 (Hexane)	13.8 (2MPentane) 12.8 (3MPentane) 18.4 (2,2DMButane)	55 (cHexane) 42.6 (McPentane)
C_7	3 (Heptane)	4.1 (2,4DMPentane)	29.6 (cHexane) 14.0 (McHexane)

i=iso-; c=cyclo-;M=methyl; DM=dimethyl.

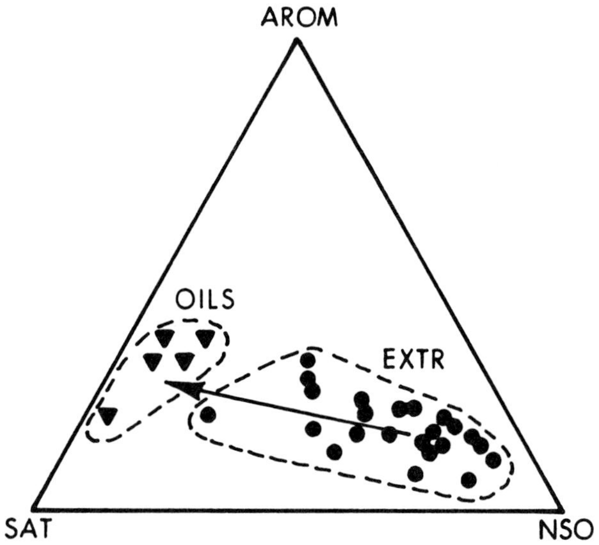

FIG. 11—Distribution of saturate, aromatic, and NSO compounds in Jurassic crude oils and source rock extracts from the Parentis basin, France (Deroo, 1976).

pounds only, it may not serve as a reliable guide for other areas. Many source-rock extracts show very different percentages of the various compound types compared with the oils they sourced. In general the crude oils are enriched in saturate hydrocarbons (relative to the aromatics) and commonly have less NSO compounds than the source-rock extracts. Data for the Parentis basin are presented in Figure 11 and show that, relative to the source-rock extract, the oils are enriched in compounds in the sequence: saturates > aromatics > NSOs. A similar relation is reported for the Mahakam delta (Combaz and deMatheral, 1978).

In these studies, like all basin studies, samples of the source rock were selected from larger units and the reservoired crude oil was probably an average product of the whole shale, so that local variations in composition may be masked. To eliminate some of the uncertainties associated with long distances and large rock volumes, the writer has analyzed the hydrocarbons in a narrow (2 mm) sandstone stringer encased in shale, and compared the composition with those for the extracts from the directly adjacent shales. The extract from the sandstone had a much higher ratio for (saturates)/(total organic carbon) than the shales. The aromatics were somewhat enriched in the sandstone and the NSOs were also slightly higher (Fig. 12A). The extract data normalized to one of the shales (Fig. 12B) show that the normal paraffins are enriched over 50 fold in the sandstones compared with adjacent shales. The sequence of enrichment then decreases from the normal paraffins to the branched-plus-cyclics, then the aromatics, and then the NSOs. Again, this is the same sequence observed in many basins including the Parentis basin discussed previously. Interestingly this sequence is exactly the opposite of the aqueous solubilities—the oils are enriched in the least soluble materials.

Although it might be argued that the more soluble compounds are less likely to accumulate and continue to move with the transporting aqueous solution, it seems much more likely that adsorption in the source rock is playing a major role. Normal paraffins are adsorbed less strongly than aromatics or NSOs by both organic matter and mineral surfaces and they will move out of the rock preferentially. An example of the interaction of an oxygen-containing compound (an alcohol) with a clay surface is shown in Figure 13. Moving hydrocarbons through a rock is, in some ways, similar to moving a mixture through a chromatographic column and separations can be expected due to the different relative rates of movement.

OVERVIEW OF PRIMARY MIGRATION

It follows from the previous discussion of migration mechanisms that the movement of hydrocarbons, or their precursors, out of a fine-grained source rock involves complex interactions between petroleum, rock matrix and water. In Figure 14 an attempt has been made to summarize the important variations in these parameters as depth and temperature increase.

Water from compaction of fine-grained sediments is abundant at shallow depths but decreases with depth and is probably minor below 10,000 ft (3,048 m). Water can also be produced from smectite (montmorillonite), first by release of interlayer water and, at greater depth, by conversion to illite through reaction with potassium ions. Only a generalized water-loss curve is shown in Figure 14. Not all source rocks contain abundant smectite and this deep source of water may not be available. The question of whether the extra water helps, hinders, or has no role in migration is unanswered, although it is certainly not essential for migration because many areas with little or no smectite have abundant petroleum.

The rock is the matrix through which migrating materials move. The pore structure of the rock, particularly the average pore size, puts constraints on the form of the migrating material. Much less is known about the pore structure of shales than sandstones but average pore size is reduced as compaction proceeds and diagenetic effects can have a role. The chemical interaction of mineral surfaces (and kerogen) with migrating organic matter is not well documented for subsurface conditions even though the movement of organics through an interacting rock is comparable to moving them through a chromatographic column and separations may occur. The limited data suggest a preferential movement of paraffins.

Depth-related changes in the composition of the organic materials in rocks are well documented (Tissót

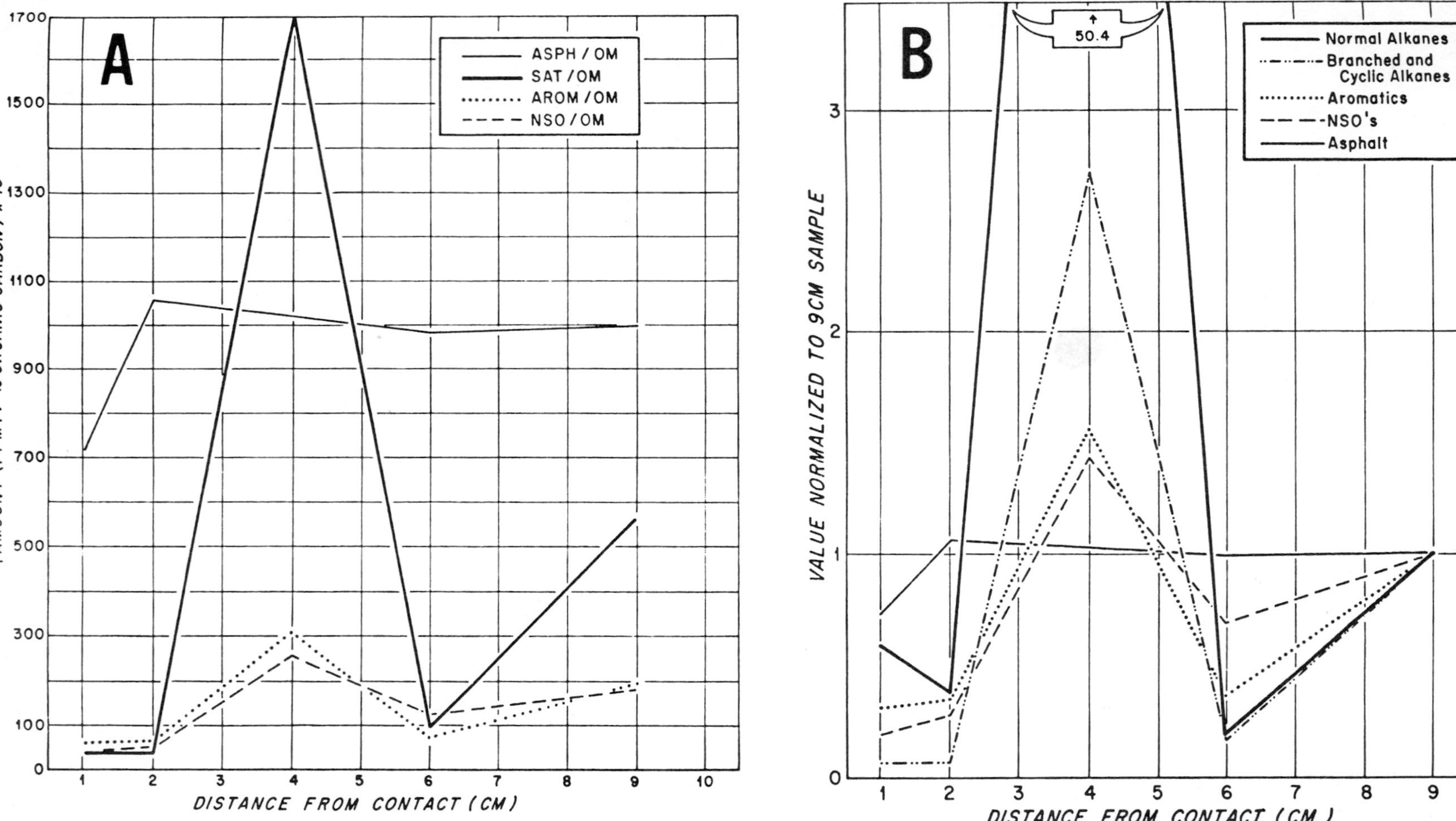

FIG. 12—Distribution of various classes of compounds in a shale core containing a sandy stringer 4 cm from a contact with a major sandstone unit. The core is from 6,529 ft in the South Glenrock field, Converse County, Wyoming. **A.** Amounts of various organic matter classes relative to the amount of organic carbon.

B. Amounts of various organic matter classes normalized to the values for the shale at 9 cm.

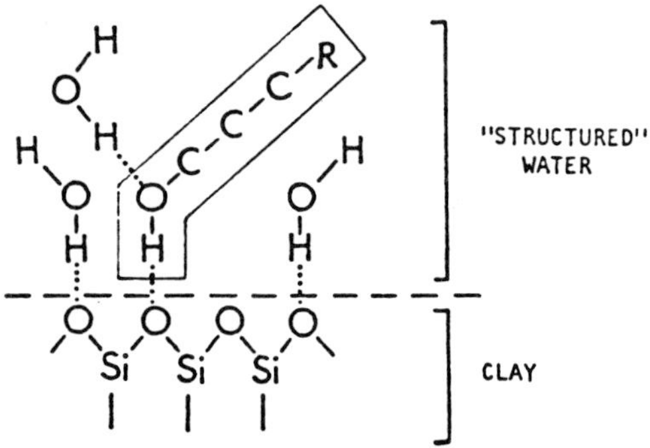

FIG. 13—Interaction of the -OH group of the R-C-C-C-OH alcohol with a clay surface. Hydrogen bonding is involved in the same way as it is for the -OH groups in water molecules.

and Welte, 1978; Barker, 1979a). At shallow depths there may be some bacterial methane along with the biologically produced lipids (bitumens) but hydrocarbons generated thermally from the kerogen do not become important until depth of burial exceeds about 6,000 ft (1,829 m). The bitumens are initially NSO-rich but become lighter with depth and give way to condensate and finally gas. Thus, the material available for migration changes with depth. Compounds which can form micelles or move as petroleum percursors generally have functional groups, like -COOH or -OH, but these are lost as temperature rises and the concentration of such compounds, and any possible role in petroleum migration (such as micelle formation), decreases with depth.

All these changes commonly occur while temperature, pressure, and time (age) are increasing and the physical properties of organic materials such as solubilities, viscosities, and densities are being altered. We can discuss the proposed migration mechanisms outlined against this background information for the rock-organic matter-water system.

In the first few thousand feet, water from compaction is abundant and pore size is relatively large, but tempertures are too low for significant generation of petroleum and the only hydrocarbons present are those preserved from once-living systems. These contain no C_2-C_{10} hydrocarbons but many of the larger molecules have functional groups which enhance their solubilities and they can move with the water. Trapping efficiency seems to limit accumulation. Kidwell and Hunt (1958) found that early accumulations in

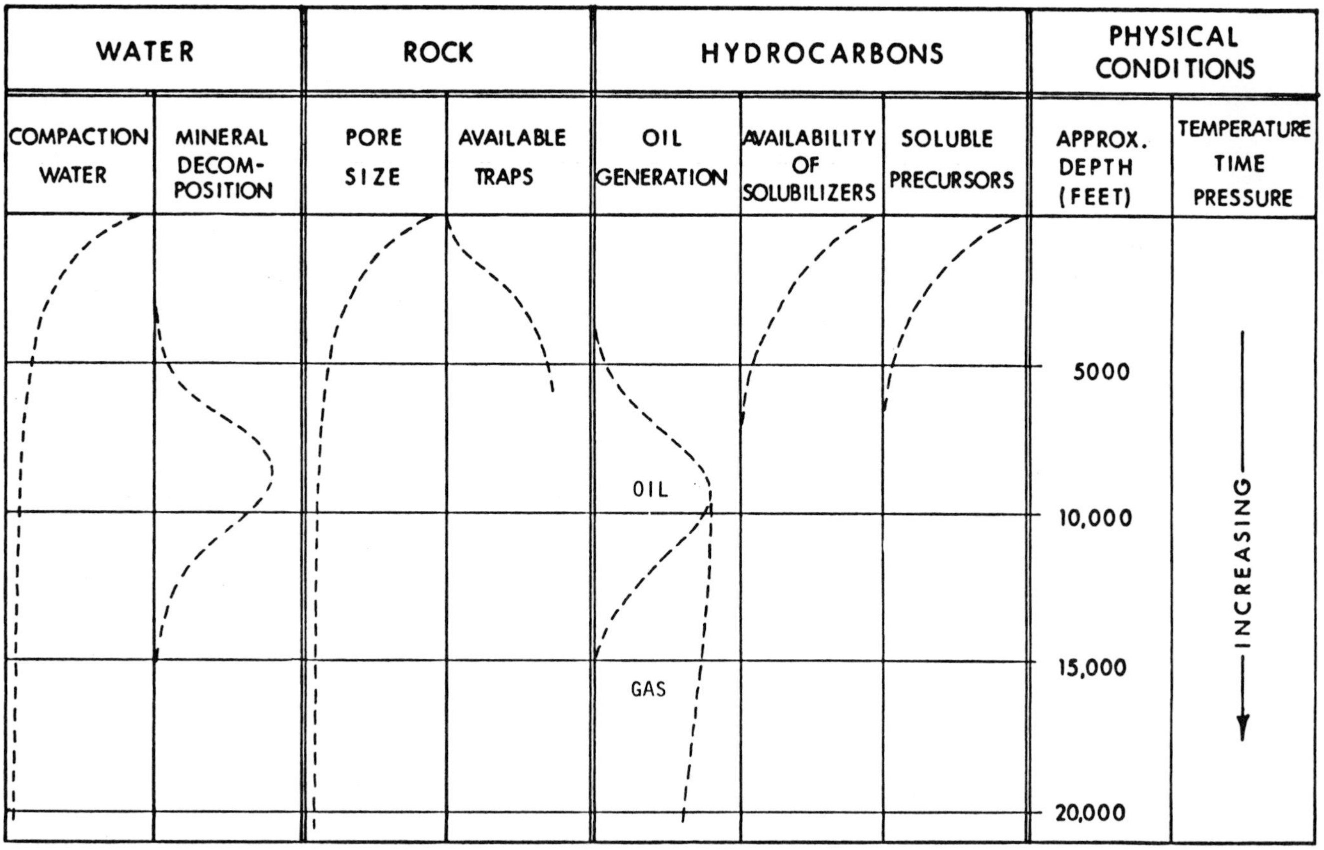

WATER | ROCK | HYDROCARBONS | PHYSICAL CONDITIONS

COMPACTION WATER | MINERAL DECOMPOSITION | PORE SIZE | AVAILABLE TRAPS | OIL GENERATION | AVAILABILITY OF SOLUBILIZERS | SOLUBLE PRECURSORS | APPROX. DEPTH (FEET) | TEMPERATURE TIME PRESSURE

OIL
GAS
5000
10,000
15,000
20,000
INCREASING

FIG. 14—Variation with depth for the available water, rock character, nature of the organic matter, and physical conditions, for a given rock.

the Pedernales area of the Orinoco delta were much enriched in aromatics relative to paraffins. This is the expected relation for water-transported oganics because the aromatics in general have higher solubilities. Bacterially-generated methane also may be present in shallow sediments and some commercial accumulations have been formed from this. Presumably the high solubility of the methane enhances migration but movement of a separate gas phase by buoyancy is also possible.

As depth of burial exceeds five or six thousand feet, thermal generation of hydrocarbons from kerogen becomes important and traps are common, but the available water is limited, average pore size is diminished, and the concentration of micelle formers and soluble hydrocarbon precursors has decreased. The commercial accumulations which form at these depths tend to contain heavy oils with high nitrogen, sulfur, and oxygen contents because thermal diagenesis of kerogen has not yet produced appreciable quantities of the low molecular weight hydrocarbons. These are formed in abundance as depth and temperature continue to increase and at a depth of 10,000 ft (3,048 m), higher °API gravity crudes are common. The water available from compaction at these depths is minor although mineralogic changes can provide large quantities. This water will be available only in certain locations but, petroleum occurs in areas where this source of water has not been available. At these depths simple molecules with functional groups (such as acids and alcohols) will no longer be present to form micelles or move as precursors because reactions such as decarboxylation remove the functional groups which confer water solubility. Micelles are unlikely to be important anyway because the average pore size is now less than the diameter of a micelle which would restrict their movement to the larger pores and fractures. This will leave large amounts of potentially productive source rock undrained. Lack of water and progressing generation may lead to the separation of a separate crude oil phase in the pore. Pore center network formation may also develop at this stage allowing transfer of hydrocarbons out of the shale.

In the deepest parts of basins crude oil gives way to condensate and ultimately gas so that the material to be moved changes greatly in physical properties. It is here that mechanisms involving gas solution seem most plausible. Increasing temperatures also enhance the solubilities of hydrocarbons.

It should be obvious from this brief discussion that the nature of the material to be moved varies considerably and the physical nature of the medium through which the migration occurs also shows marked changes with depth. There is probably no single mechanism of migration for petroleum hydrocarbons and several different mechanisms may operate at different times and at various stages of burial and generation.

CONCLUSIONS

The idea that there is a single mechanism for the migration of petroleum out of a source rock is almost certainly wrong. There is probably not one mechanism but many, so that the only realistic approach to the problem of petroleum migration becomes (1) to establish all possible mechanisms by which petroleum may migrate, and (2) to establish the conditions under which each of these mechanisms operates. The various mechanisms proposed for petroleum migration should not be regarded as alternative suggestions, but rather as complementary ones, and the question should be asked as to what conditions are required for each mechanism to become quantitatively significant.

It is particularly important to distinguish between mechanisms which move organic components through fine-grained source rocks and those which act to transport the materials through the carrier system to the reservoir. These two steps commonly are discussed together as "primary migration" yet they probably involve different processes. Only where all the major processes of migration are understood, and the chemical, physical, and geological limitations on their operation are determined, will it be possible to examine a specific rock unit and make geologically important deductions. These will include the ability to recognize that migration has occurred (i.e., it is a source rock), and to predict whether the reservoired oil will chemically resemble the source rock extract (i.e., oil-source correlation is possible).

REFERENCES CITED

Baharlou, A., 1973, A comparison of the chemical composition of interstitial waters of shales and associated brines: Ph.D. thesis, Univ. of Tulsa.

Baker, E. C., 1960, A hypothesis concerning the accumulation of sediment hydrocarbons to form crude oil: Geochim. et Cosmochim. Acta, v. 19, p. 309-317.

Barker, C., 1972, Aquathermal pressuring—role of temperature in development of abnormal-pressure zones: AAPG Bull., v. 56, p. 2068-2071.

——— 1973, Some thoughts on the primary migration of hydrocarbons: Paper presented at the AAPG Ann. Mtg. (San Antonio).

——— 1975, Oil source-rock correlation aids drilling site selection: World Oil (Oct. 1975), p. 121-126, 213.

——— 1979a, Organic geochemistry in petroleum exploration: AAPG Course Note Series 10, 159 p.

——— 1979b, Effect of temperature changes on pressure in reservoirs containing free gas: Log Anal., March-April, p. 34-35.

Combaz, A., and M. deMatharel, 1978, Organic sedimentation and genesis of petroleum in Mahakam Delta, Borneo: AAPG Bull., v. 62, p. 1684-1695.

Cordell, R., 1972, Depths of oil origin and primary migra-

tion: a review and critique: AAPG Bull., v. 56, p. 2029-2067.
Deroo, G., 1976, Correlations of crude oils and source rocks in some sedimentary basins (in French): Bull. Centre Rech. Pau-SNPA, v. 10, p. 317-335.
Dow, W. G., 1977, Kerogen studies and geological interpretation: Jour. Geochem. Expl., v. 7, p. 79-99.
Drost-Hansen, W., 1969, Structure of water near solid interfaces: Ind. Eng. Chem., v. 61, p. 10-47.
Hunt, J. M., 1972, Distribution of carbon in crust of earth: AAPG Bull., v. 56, p. 2273-2277.
——— 1977, Ratio of petroleum to water during primary migration in Western Canada basin: AAPG Bull., v. 61, p. 434-435.
Kidwell, A. L., and J. M. Hunt, 1958, Migration of oil in recent sediments of Pedernales, Venezuela, *in* L. G. Weeks, ed., Habitat of Oil: Tulsa, AAPG, p. 790-817.
Levorsen, A. I., 1954, Geology of petroleum: San Francisco, W. H. Freeman, Co., p. 703.
McAuliffe, C. D., 1963, Solubility in water of C_1-C_9 hydrocarbons: Nature, v. 200, p. 1092-1093.
——— 1966, Solubility in water of paraffin, cycloparaffin, olefin, acetylene, cyclo-olefin, and aromatic hydrocarbons: Jour. Phys. Chemistry, v. 70, p. 1267-1275.
——— 1979, Oil and gas migration—chemical and physical constraints: AAPG Bull., v. 63, p. 761-781.
Momper, J. A., 1978, Oil migration limitations suggested by geological and geochemical considerations, *in* Physical and chemical constraints on petroleum migration: AAPG Course Note Series 8, p. B-1 to B-60.
Peake, E., and G. W. Hodgson, 1967, Alkanes in aqueous systems: The accommodation of C_{12}-C_{36} n-alkanes in distilled water: Jour. Am. Oil Chemists, v. 44, p. 696-702.
Price, L. C., 1976, Aqueous solubility of petroleum as applied to its origin and primary migration: AAPG Bull., v. 60, p. 213-244.
Roberts, N. K., and G. Zundel, 1979, Infrared studies of long-range surface effects—excess proton mobility in water in quartz pores: Nature, v. 278, p. 726-728.
Schmidt, G. W., 1973, Interstitial water composition and geochemistry of deep Gulf Coast shales and sandstones: AAPG Bull., v. 57, p. 231-337.
Sokolov, V. A., and S. I. Mironov, 1962, On the primary migration of hydrocarbons and other oil components under the action of compressed gases, *in* The chemistry of oil and oil deposits: Acad. Sci. USSR, Inst. Geol. and Exploit. Min. Fuels, p. 38-91 (in Russian); 1964 Engl. trans. by Israel Progr. Sci. Trans., Jerusalem.
Tissót, B. P., and D. H. Welte, 1978, Petroleum formation and occurrence: Springer-Verlag, 538 p.
Williams, J. A., 1974, Application of oil-correlation and source rock data to exploration in the Williston basin: AAPG Bull., v. 58, p. 1243-1252.

Agents for Primary Hydrocarbon Migration: A Review[1]

By Kinji Magara[2]

Abstract Time of oil migration is a controversial matter in petroleum geology; modern geochemical data suggest that a relatively long geologic time is necessary to generate oil and therefore to migrate it, whereas many petroleum geologists still believe in relatively early migration. A new concept of rate of oil generation is proposed on the basis of geochemical source rock analysis. This new concept suggests that the time of generation and migration of oil must be earlier than the time which the geochemical data of accumulated oil indicate.

Most oil must migrate in its own phase because of its low solubility in water; however, most gas may migrate in solution in water. To ensure the migration of oil in the oil phase, the importance of semi-solid or structured water was proposed. In compacted shales, a larger proportion of water may be in semi-solid state. Concentration of oil in the remaining liquid water may be relatively high. If the shales compact further, the oil may migrate along with water, provided the oil saturation is more than that for critical oil migration in the oil-phase.

Four possible agents of primary migration are considered—compaction, aquathermal, clay mineral conversion, and osmosis. These agents govern the movement and/or generation of liquid water. It must be noted that there is a large quantity of water and a relatively small amount of hydrocarbons at the time of primary migration. In other words, movement of some kind of water would be very important in understanding the movement of hydrocarbons at the primary migration stage.

INTRODUCTION

Movement of hydrocarbons from nonreservoir rocks to reservoir rocks is called primary migration, and is distinguished from concentration and accumulation within the reservoir rocks (known as secondary migration; Levorsen, 1967). Primary migration of oil and gas is one of the most controversial subjects in petroleum geology. There are at least four different problems as follows: (1) time of primary migration; (2) form of hydrocarbons at the time of primary migration, such as aqueous solution or separate hydrocarbon phase; (3) agents of hydrocarbon movement from source rock to reservoir rock; (4) source of water, cause of its movement, and direction and volume of its movement.

In the discussion of the first three examples, a clear distinction between gaseous and liquid hydrocarbons must be made. Because physical and chemical properties of these two hydrocarbons are quite different, it would be much easier to discuss these problems if they were treated separately, rather than grouped together.

Modern petroleum geologists are familiar with the fact that oil can be generated within a relatively narrow temperature-geologic time range, called an oil window, but gas (especially methane) can be generated at almost any geologic time in most drillable depth sections of sedimentary basins. Therefore, the discussion of time of generation and migration for oil may be completely different from that for gas.

Solubility of gaseous and liquid hydrocarbons in water is quite different. In discussing the form of hydrocarbons at the primary migration stage, the understanding of the aqueous solubility of hydrocarbons is very important.

The form of hydrocarbons at the time of primary migration may restrict the means of migration. This is because if hydrocarbons migrate in water solution, the behavior of the water is an important factor; if hydrocarbons migrate in their own phase, the behavior of water in source rocks may have a different type of influence on primary hydrocarbon migration.

In any case, the four different problem-examples listed above are interrelated. To discuss one problem, the other factors or problems must be understood as well.

TIME OF PRIMARY MIGRATION

Regarding the first problem of time of primary migration, Wilson (1975) stated that "many geochemists advocate that oil is expelled from source rocks only

[1]Manuscript received, August 9, 1978; accepted, August 20, 1979.

[2]Reservoir Studies Institute, Department of Geosciences, Texas Tech University. Presently with the Bureau of Economic Geology, The University of Texas at Austin, Austin, Texas 78712.

Article Identification Number:
0149-1377/79/SG10-0003/$03.00/0.

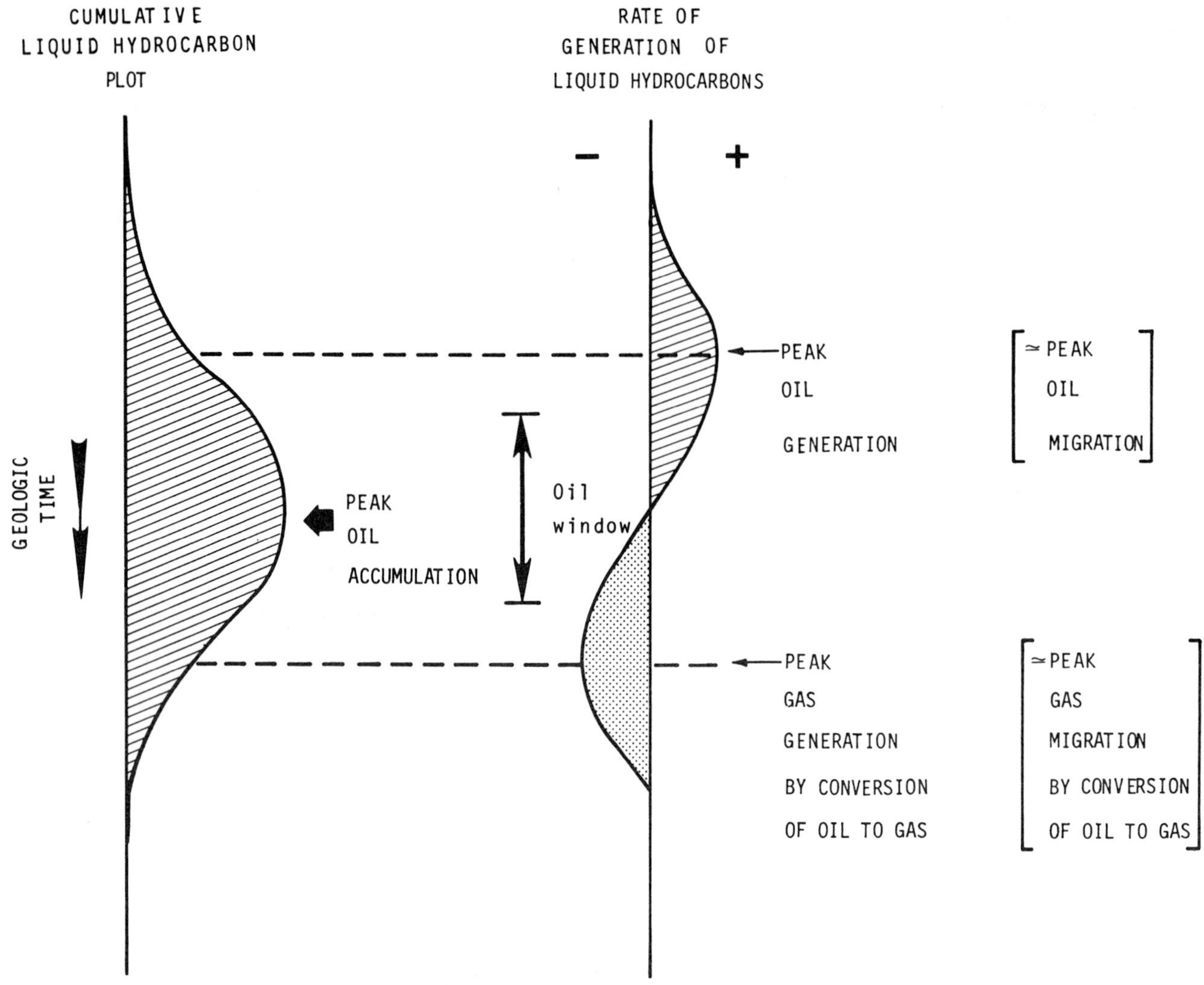

FIG. 1—Comparison of peak oil accumulation and peak rate of oil and gas generation.

when crude oil-like hydrocarbons are formed in source rocks through the heat and pressure imposed by deep burial. Conversely, many experienced petroleum geologists point out that empirical data on habitat of oil accumulations call for early, short-lived expulsion of hydrocarbons shortly after the burial of source and reservoir sediments."

The problem of the time of oil/gas migration is important in prospect evaluation, because early or late migration will cause a significant difference in evaluating an exploration target.

The left side of Figure 1 shows a schematic plot of the geochemical source-rock analysis of a well. It indicates the changing amount of liquid hydrocarbons remaining in the source rocks. Commonly, the amount increases with burial depth until the peak is reached, then decreases due to the conversion of liquid hydrocarbons to gas. The zone of the highest amount of liquid hydrocarbons is called the "oil window." The geologic time and temperature commonly increase with depth.

Although the geologic time and temperature necessary in accumulating oil in source rocks or maybe in reservoir rocks can be estimated from an oil window, the time of maximum rate of oil generation must be expressed differently.

The right side of Figure 1 indicates the rate of generation of liquid hydrocarbons, which corresponds to the plot on the left. The rate increases rapidly up to the inflection point of the curve on the left, then decreases to zero at the "peak oil accumulation" point on the left. Below this point, the rate has negative values, suggesting that the liquid hydrocarbons generated were converted to gas. The peak generation of gas due to the conversion will take place after the peak oil accumulation.

It must be noted that biogenic gas can be generated at much earlier stages. In addition, natural gas which can be generated directly from organic matters, especially of a woody type, is not included in this plot.

The plot on the right side of Figure 1 suggests that the stage of peak oil generation would be earlier than the stages indicated by the oil window, which is based on accumulated liquid hydrocarbons. It is reasonable to assume that the peak oil migration took place when peak oil generation was reached.

Examples of plots of source-rock analysis in Japan are shown in Figure 2 (Asakawa and Fujita, 1979).

In summary, the time for the peak rate of oil generation must be earlier than that given by the peak oil accumulation. This conceptual difference may be important in reducing the conceptual gap between geochemists and petroleum geologists on time required for active oil generation and migration.

FORM OF HYDROCARBONS AT PRIMARY MIGRATION

The form of hydrocarbons at the time of primary migration, such as molecular solution, micellar solution, and separate hydrocarbon phase, is another controversial matter in petroleum geology.

The molecular solubility of liquid hydrocarbons in water at relatively high temperatures was extensively studied by Price (1976). He showed a tendency that the solubility increases with increasing temperature (Fig. 3). The solubility of the Farmer's whole oil at 160°C is approximately 150 ppm, and the trend shows that it increases with further increase in temperature. However, these higher temperature values are much higher than the temperature range of 60°C to 150°C for most oil accumulations.

Micellar solution was proposed by Baker (1962) as another solution mechanism. He suggested that solubility of hydrocarbons will be substantially higher if the water contains micelles formed by soaps of organic acids. However, there are several problems associated with this proposed mechanism. First of all, that such micelles exist in large quantity in shales has not been proved. Even if they do exist, they would not move easily in shales because their size is relatively large. Even then, the micelles would increase the solubility of the heavier hydrocarbons in water only to a few parts per million (Dickey, 1975). In other words, it may not increase solubility substantially.

Another problem in believing micellar solution to be important in primary oil migration is that there is no satisfactory explanation for unloading the hydrocarbons carried by the fluid containing micelles at the final trapping position in the reservior.

Regarding the concentration of liquid hydrocarbons in the compaction stream, Dickey suggested values more than 10,000 ppm, on the basis of his own estimate of the compaction water and the migrating oil at the time of primary migration. These figures are at least two orders of magnitude higher than those given by Price in a realistic temperature range of petroleum generation.

Another approach to estimating the required concentration of oil in the compaction stream is, as follows:

Tissót and Pelet (1971) analyzed the amounts of hydrocarbons, resins, and asphaltenes in shales adjacent to a reservoir in Algeria. The results of their analyses in mg/g organic carbon are shown in Figure 4. It is seen that the amounts of resins and asphaltenes in shales remain relatively constant; however, the amount of hydrocarbons decreases toward the reservoir, indicating possible primary migration of hydrocarbons. The difference in hydrocarbon contents at the 14 m point and at the near-reservoir point is about 40 mg/g organic carbon. Within this short interval of 14 m, the level of organic maturation would be relatively uniform. In other words, this 40 mg represents the lowest possible amount of hydrocarbons which were expelled per gram of organic carbon from the shale closest to the reservoir.

If the density of this shale is 2 g/cc and the concentration of organic carbon is 1 wt.%, 1 cc of this shale lost 0.8 mg of hydrocarbons. The figures used for this estimate would be the lowest possible values to produce the lowest possible hydrocarbon yield. If the porosity difference between these two points is 10%, which seems to be the largest possible figure under these conditions, the amount of hydrocarbons in the compaction stream is estimated to be about 8,000 ppm.

The above estimate is based on the lowest possible estimate of hydrocarbons and the highest possible estimate of compaction water. Thus, the value given here is the lowest possible concentration of hydrocarbons in compaction water; the real value could be higher. In any situation, this figure is surprisingly close to the values greater than 10,000 ppm given by Dickey (1975), and is at least one order of magnitude higher than the highest molecular-solubility figure in the temperature range for oil generation. It must be noted that the porosity data for the shales in Algeria are not readily available, so they had to be assumed.

There is further evidence that compaction fluid contains more hydrocarbons than can account for by aqueous solubility. Vyshemirsky et al (1973) made experiments of squeezing the mixture of clay, liquid hydrocarbons, and water up to 300 atm of pressure. They found that the amount of hydrocarbons squeezed with water was more than could be accounted for by the solution mechanism alone. Therefore, it may be difficult to believe that most oil migrates as molecular solution in water.

Migration of gas may be completely different. Figure 5 (Dodson and Standing, 1944) shows that the solubility of natural gas in water ranges from 4 cu ft (0.112 cu m) per barrel of water at 500 psi to about 30 cu ft (0.84 cu m) at 10,000 psi. In other words, compaction water could contain a relatively large quantity of natural gas in solution in water. A pressure differ-

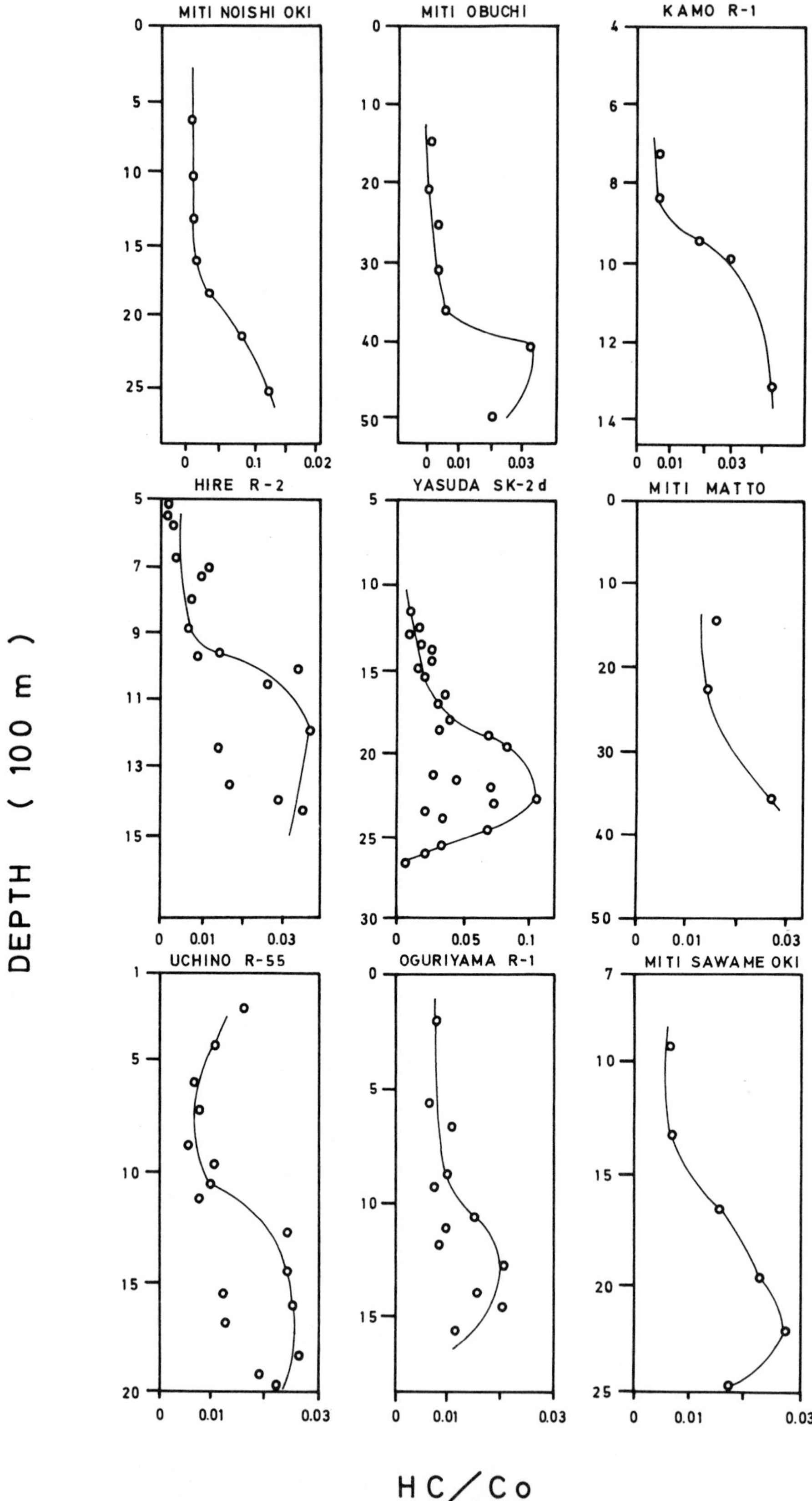

FIG. 2—Examples of source-rock analysis in Japanese sedimentary basins (from Asakawa and Fujita, 1979). HC: hydrocarbons; Co: organic carbon.

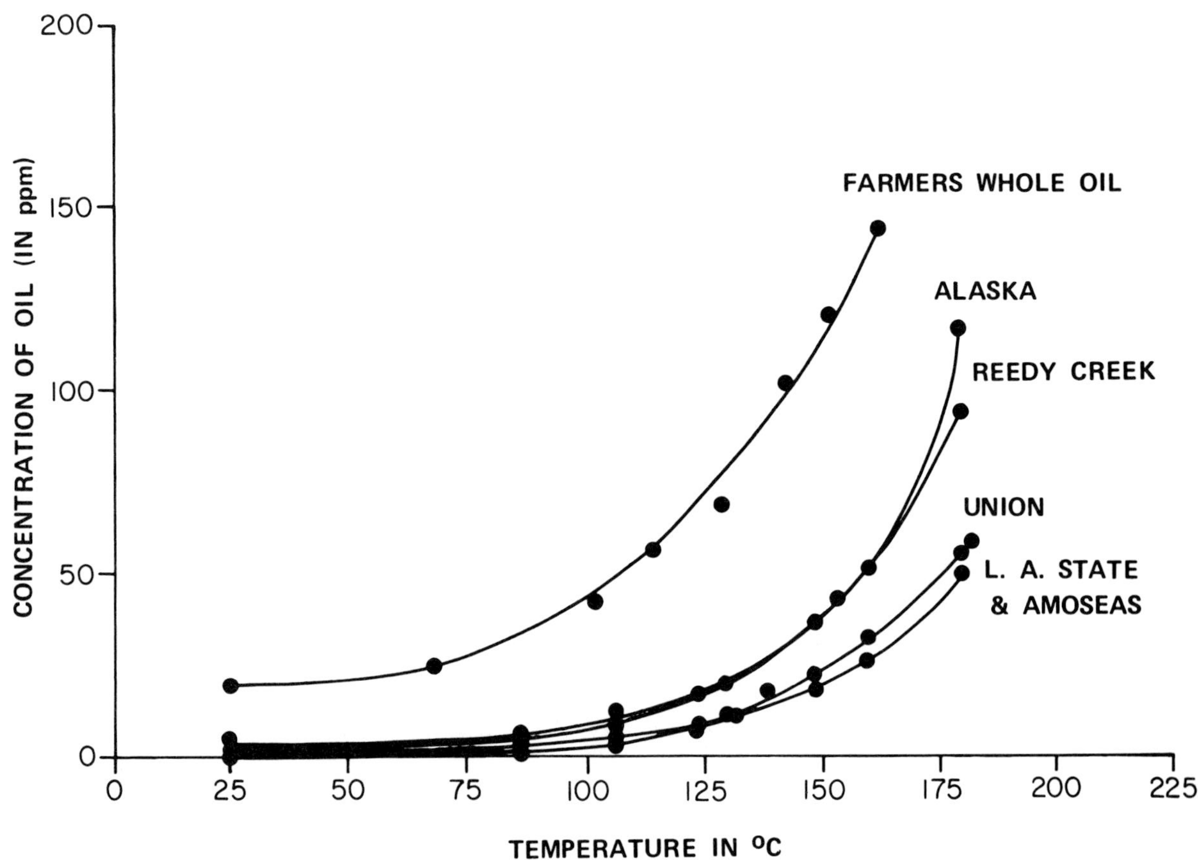

FIG. 3—Solubilities of two whole oils (Wyoming Farmers and Louisiana State) and four topped oils (Amoseas Lake, Reedy Creek, Alaska, and Union Moonie) as functions of temperature in water. Topping temperature is 200° C (392° F), from Price (1976).

ential between synclinal and anticlinal areas may be sufficient in releasing a significant volume of free gas in the latter area and causing significant gas accumulations in many cases.

On the basis of the above observations, it may be concluded that the greater proportion of liquid hydrocarbons must move in their own phase. However, most gas could migrate in aqueous solution because of its higher solubility.

The most common criticism against the movement of liquid hydrocarbons in their own phase is that it would be quite difficult to move separate hydrocarbon droplets in fine-grained rocks because of their capillary restrictions. The liquid hydrocarbons in their own phase are probably in the form of very tiny globules in the source rocks. To solve this problem, Dickey (1975) proposed the importance of the double water layer concept.

In compacted shales, the larger proportion of water is electrically charged at the clay surfaces, and has a relatively high viscosity. Figure 6 shows the result of estimates (Low, 1976) of water viscosity in montmorillonite. The viscosity changes from about 1 to 8 centipoises. In other words, the water in montmorillonite can be 8 times as viscous as the normal drinking water in the laboratory conditions. Dickey used the term *"semi-solid"* for this viscous water.

The amount of liquid (or free water) in the compacted shales is probably not great. Under these circumstances, if the shales compact further, the oil as well as the liquid water will migrate provided the oil saturation in the liquid phase is higher than the critical value for oil migration.

If, for example, the oil saturation in the total pore water is 100 ppm or 0.01 wt.% and if only 1% of the water is in liquid phase and 99% is semi-solid, the oil saturation in the liquid phase will be 10,000 ppm or 1 wt.%. If some of the liquid water is expelled by the compaction effect, then the oil saturation in the liquid phase will increase further, approaching and even exceeding the critical saturation for oil migration, which may be assumed to be a value between 1 and 10% (Dickey, 1975). If the oil saturation reaches such a level, the oil will move in the oil phase along with the water.

Although most oil may migrate in its own phase, the movement of water is important in causing the oil migration in the oil phase, because the effective water expulsion from shales would cause a relatively dry condition in shales. This dry condition means the con-

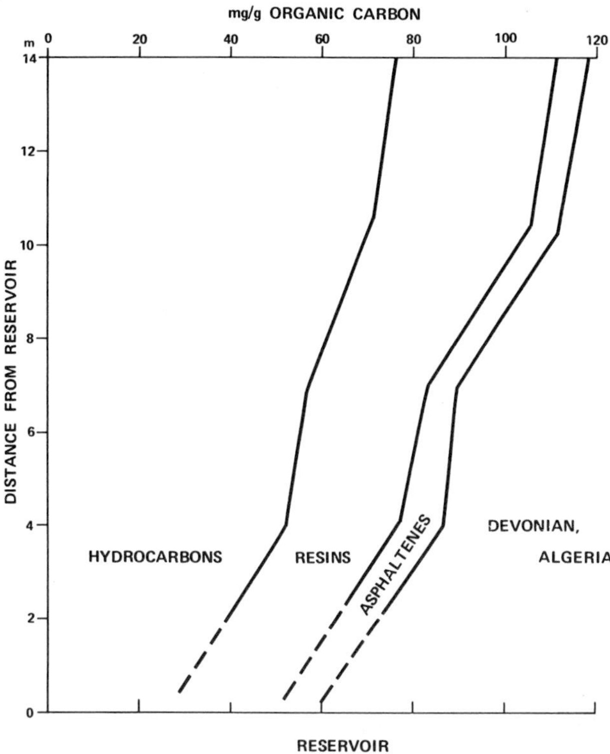

FIG. 4—Plot of amounts of hydrocarbons, resins, and asphaltenes over organic carbon (g) of Devonian shales underlying a reservoir in Algeria. Original data derived from Tissót and Pelet (1971).

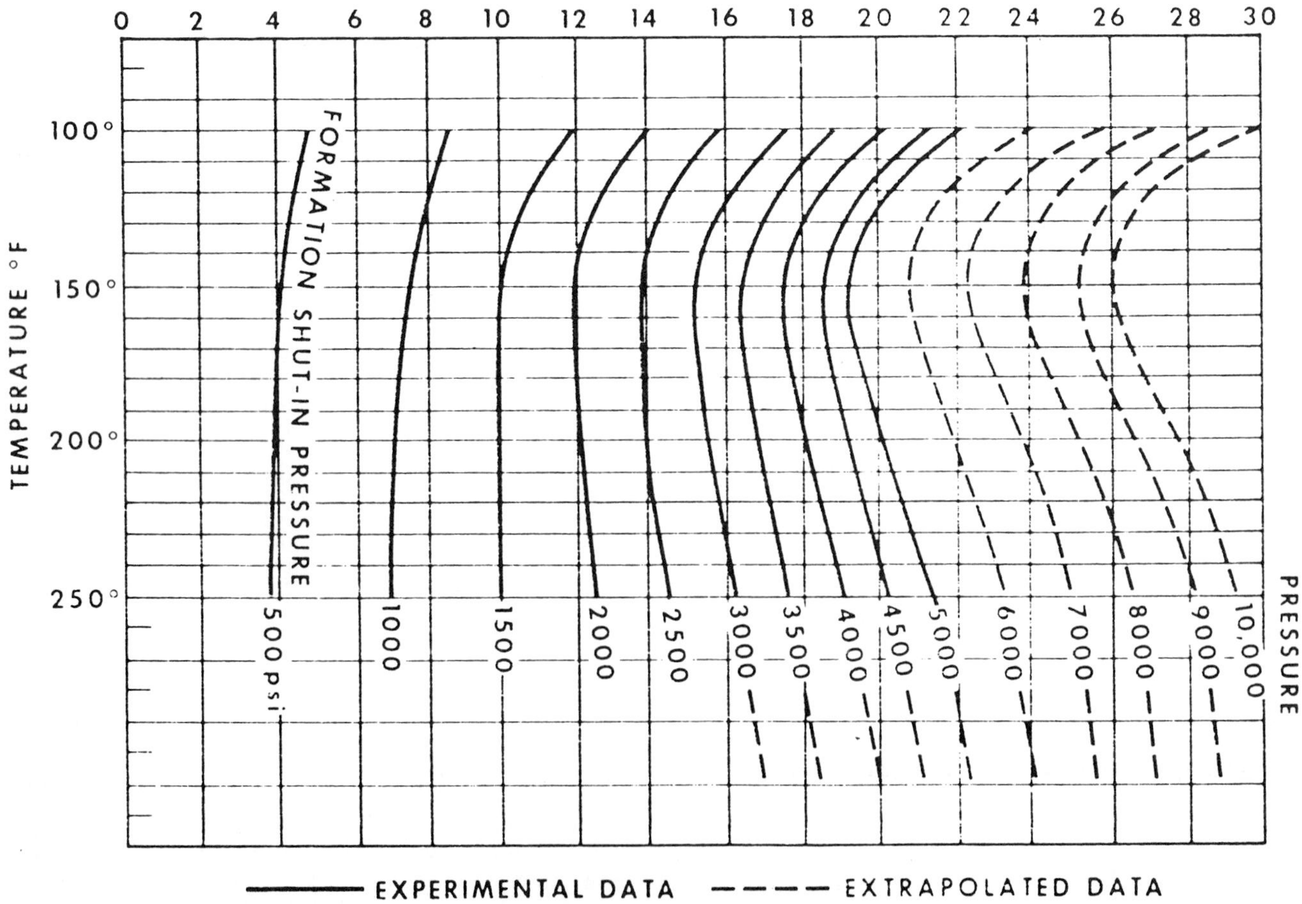

FIG. 5—Solubility of natural gas in formation water (from Dodson and Standing, 1944).

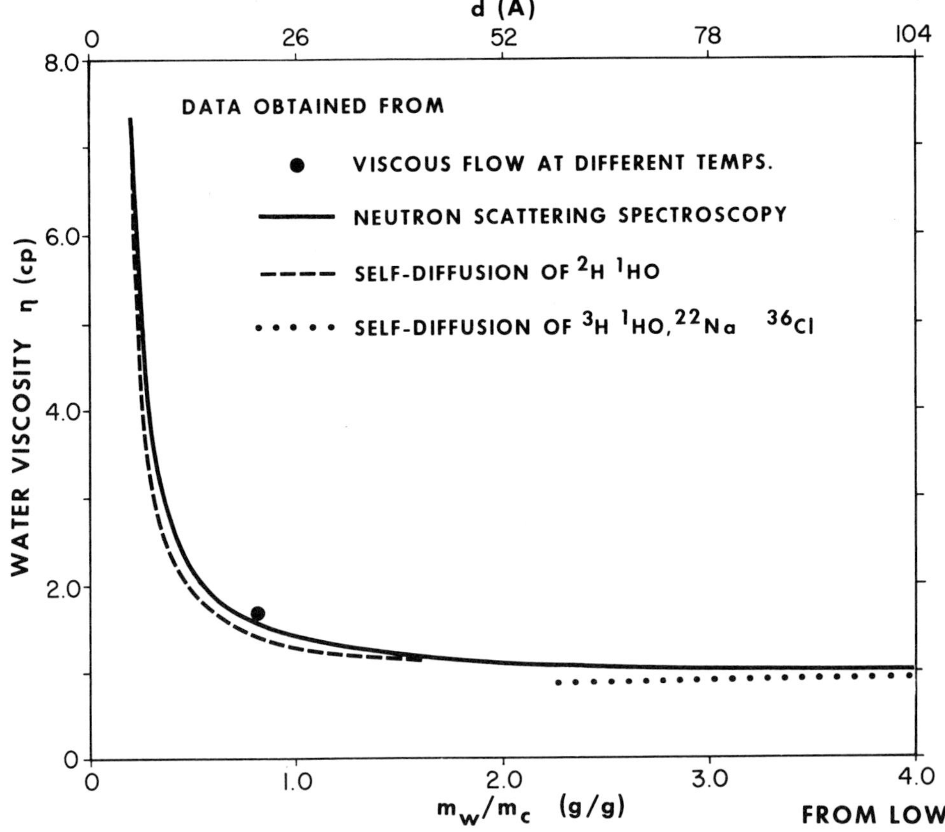

FIG. 6—Relationship between viscosity of water in clay and ratio of amount of water (mw) over amount of clay (mc) for distance (d) from clay surfaces (from Low, 1976).

dition at which the shales contain mostly semi-solid or structured water. If such shales are compressed further, oil globules will be squeezed out easily. This mechanism may be something like squeezing a tube of tooth paste.

If the water was not effectively expelled from the fine-grained rocks, undercompaction and abnormal pressure will result. On the other hand, if the water was effectively expelled as overburden load increased, the normal hydrostatic pressure will be maintained.

Timko and Fertl (1971) showed statistically that most oil accumulations in the Gulf Coast area occur in relatively low-pressured sections. In these sections, there must have been effective drainage of water that helped to move hydrocarbons from the source rocks to the reservoirs. In the abnormal pressure zones, where the drainage was not effective, there probably was no effective primary migration of oil in the oil phase. However, a very small quantity of oil could move in solution in water, even under these circumstances.

AGENTS OF PRIMARY HYDROCARBON MIGRATION

Compaction

From the previous discussions, it may be obvious that one of the most important agents of primary migration is the sediment compaction and the movement of compaction water. Figure 7 shows examples of shale porosity-depth curves in several sedimentary basins. As this figure shows, the rate of shale porosity reduction or compaction commonly decreases as the shales become more deeply buried. In other words, the rate of pore-fluid expulsion decreases with burial. The most common conclusion from these observations is that by the time oil has been generated at depth under a relatively high temperature, there may not be enough water available in the shales for primary migration. However, this conclusion may not be valid because we do not know how much water is necessary for effective primary migration. If all the oils are assumed to have moved in solution water, it may be possible to evaluate the amount of water needed to move the known oil reserve in a given field; however, if all or a part of the oil has migrated in its own phase, it is not possible to calculate the amount of water necessary to move a certain amount of oil.

In other words, even if most of the water in a shale has been lost before the hydrocarbons were generated, who knows that what remains is insufficient for effective primary migration? This type of discussion or conclusion must be avoided unless one knows exactly how much water is necessary for primary migration, or unless one can prove that the shales concerned did

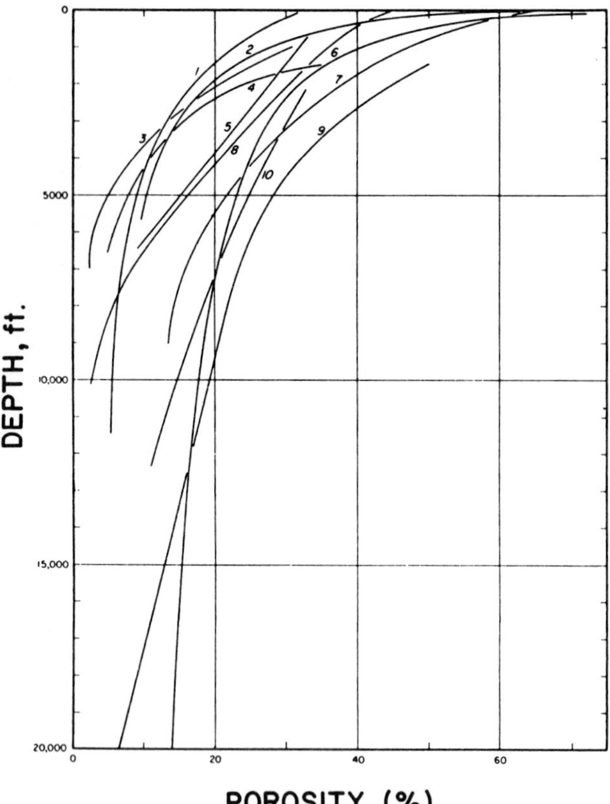

FIG. 7—Relation between porosity and depth of burial for shales and argillaceous sediments (from Rieke and Chilingarian, 1974). Data plots: **1,** Proshlyakov (1960); **2,** Meade (1966); **3,** Athy (1930); **4,** Hosoi (1963); **5,** Hedberg (1936); **6,** Dickinson (1953); **7,** Magara (1968a); **8,** Weller (1959); **9,** Ham (1966); and **10,** Foster and Whalen (1966).

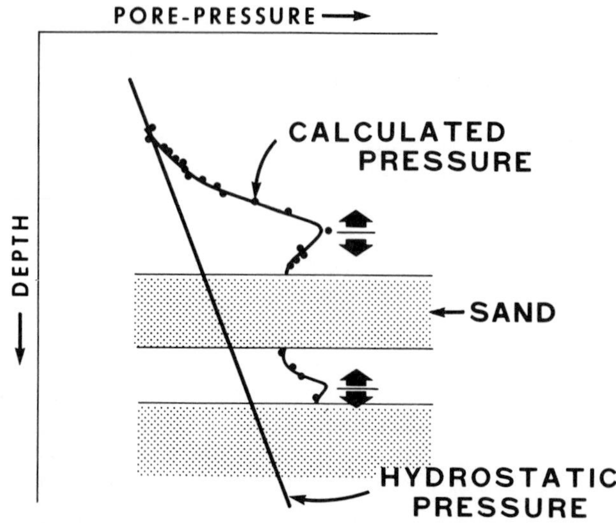

FIG. 8—Example of calculated shale pressure-depth plot of a well in Mackenzie Delta, Canada.

not expel any fluids after the oil generation. Shales commonly have some permeability even if the value is quite low and there are vertical and horizontal fluid pressure gradients in most shales. In these circumstances, the fluids must be moving all the time; the rate is commonly quite slow, however. Figure 8 shows an example of the shale fluid pressure-depth plot for a well drilled in the Mackenzie delta. It shows strong vertical excess fluid pressure[3] gradients in shales.

A good evidence of fluid movement in shales is the tendency of shale porosity reduction with burial depth, such as shown by the porosity-depth plots of many sedimentary basins (Fig. 7). Dickey, for example, estimated that about 11% of porosity loss occurred during burial between 5,000 and 15,000 ft (4,572 m) corresponding to the temperature range of oil generation (Dickey, 1975). To cause this level of porosity reduction, there must have been significant fluid expulsion. If hydrocarbons move in their own phase along with the compaction water as discussed above, the amount may be significantly large.

For example, if a block of source rock, whose area is 100 sq mi (160 sq km) and whose thickness is 100 ft (30 m), lost only 10% porosity after the oil generation, approximately 4 trillion barrels of compaction water would have been expelled. If a 1% (or 10,000 ppm) hydrocarbon figure in total stream is used, about 40 billion barrels of oil can be flushed in separate phase from the shales with water.

Although the compaction water itself is believed to be an important cause of primary migration, there is additional help. This is the aquathermal effect proposed by Barker (1972).

Aquathermal

Fluid pressure and temperature commonly increase with burial. The pressure tends to reduce the volume of a given weight of water and the temperature tends to increase the volume. However, the temperature effect is more pronounced, resulting in the water expansion with burial.

Figure 9 shows such increases of specific volume of water during burial, for the three geothermal gradients, 18°C/km (1°F/100 ft), 25°C/km (1.37°F/100 ft), and 36°C/km (2°F/100 ft) respectively. This figure shows that, if the geothermal gradient is higher, the rate of expansion is more.

For a given geothermal gradient, the rate of expansion increases with burial. This is important because the rate of compaction water expulsion decreases with depth. However, the aquathermal effect becomes more significant at deep burial. Hence, such an aquathermal effect will help to push more hydrocarbons from the source rocks to reservoirs in relatively deep section.

[3]Fluid pressure in excess of hydrostatic pressure.

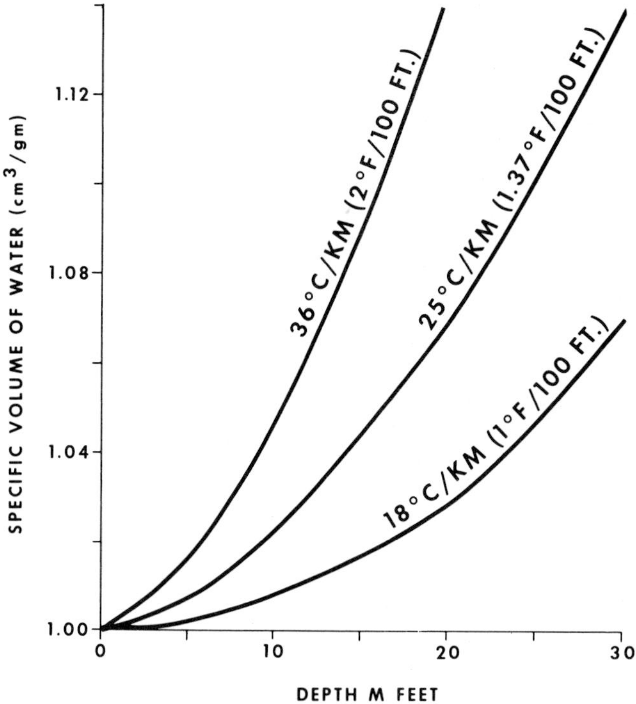

FIG. 9—Specific volume (of water)-depth relations in normally pressured zones for three geothermal gradients of 25°, 18°, and 36°C/km (from Magara, 1974b).

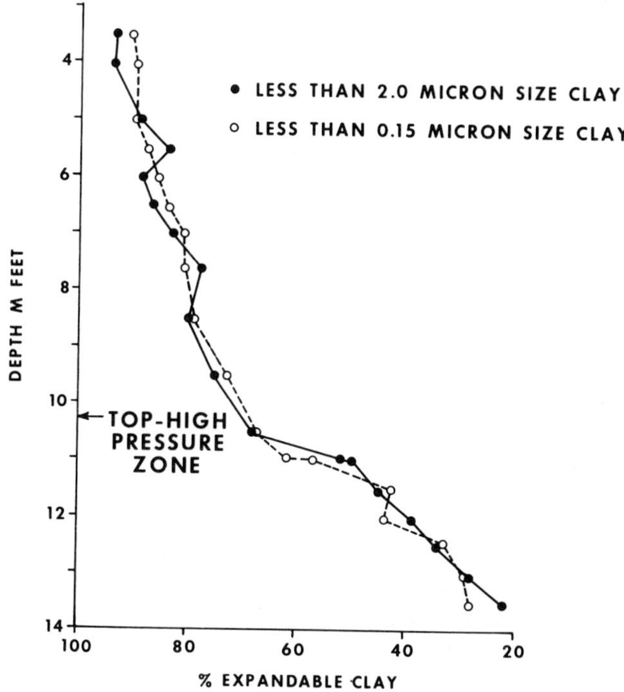

FIG. 10—Plot of per cent expandable clay versus depth showing accelerated increase in diagenesis of smectite to mixed-layer clay (from Schmidt, 1973).

Clay Mineral Conversion

Smectite-illite conversion discussed by Powers (1967) and Burst (1969) is another source of water at depth. According to Burst, the second stage dehydration which seems to be most significant in primary migration occurs at an average temperature of 221°F (105°C). This statement may give us an impression that such a conversion may be quite drastic and limited in a relatively narrow depth range.

However, the plot made by Schmidt (1973) indicated that the change of the expandable clay content is continuous and gradual from the surface to depth in the Gulf Coast (Fig. 10). At about 10,000 ft (3,048 m), at which the temperature is approximately 200°F (93°C), the rate of change of the expandable clay content is increased, but geothermal gradient is also increased at that depth (Fig. 11).

The above data lead to the conclusion that the smectite-illite conversion is not a drastic, but probably a very continuous process. The water generated by the clay mineral conversion is an addition to the total water, but this does not always mean effective expulsion of water. If the generated water cannot move out of the source rocks, and stays there, it will reduce the concentration of liquid hydrocarbons in the liquid phase. Under these circumstances, there would be little chance of primary migration of liquid hydrocarbons in their own phase. Even hydrocarbons in solution in water cannot migrate effectively because the movement of water is restricted.

In summary, the smectite-illite conversion will create additional water at depth, but the generated water must be expelled effectively to cause significant hydrocarbon migration. Thus, a good drainage condition is more important than the mineral conversion itself. The drainage condition commonly is excellent if the shales are interbedded with permeable beds such as sandstones of relatively large areal extent. The effectiveness of the drainage may be studied from the fluid pressure profiles made within the shaly sections (see Fig. 8; and Magara, 1978).

Osmosis

In many sedimentary basins, the salinity of formation water increases with burial depth or compaction. The subsurface salinity values commonly are higher than that of sea water (35,000 ppm). The principal cause of this salinity increase is considered to be ion-filtration by shales.

Ion-filtration by clays or shales was also documented by laboratory methods (McKelvey and Milne, 1962; Englehardt and Gaida, 1963), that showed that the clays and the shales filter salt ions from a solution. The water moving through the shales is fresher than the original solution.

Figure 12 shows the chloride concentration (ppm) versus shale porosity plots for the Burgen field in Ku-

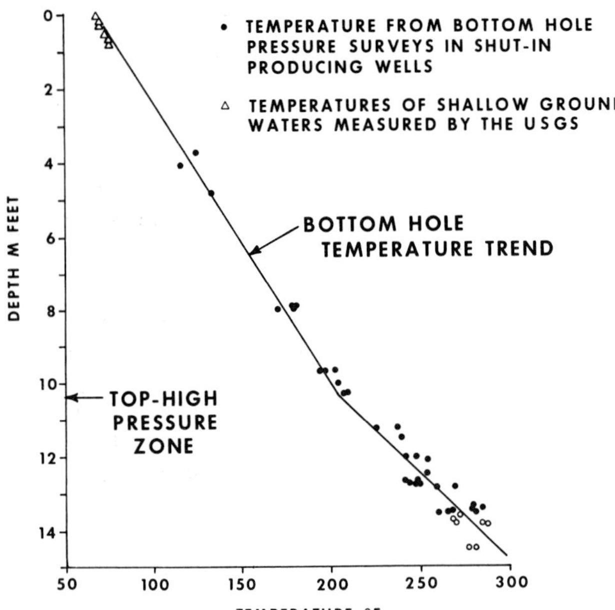

FIG. 11—Plot of measured temperatures versus depth, which shows increased temperature gradient at top of high-pressure zone (from Schmidt, 1973).

wait and several oil fields in Texas determined by Hedberg (1967). Salinity may be estimated from the chloride concentration by multiplying by a factor of 1.65. There may be a hyperbolic relationship between the chlorinity and porosity—the chlorinity increases as the porosity decreases. Such chlorinity or salinity variation due to changing shale porosity may cause osmotic fluid movement in the subsurface.

Agents in Combination

Figure 13 combines the concepts of the compaction, the aquathermal, and the osmotic fluid movements in an interbedded section of sandstones and shales. Figure 13A is a plot of shale porosity vs depth; Figure 13B shows the corresponding fluid pressure-depth plot. The compaction fluid flow would occur from the center to the edges of each shale bed, because there is the maximum excess fluid pressure at the center of the shale bed and the pressure drops toward the upper and lower edges.

The expansion of water due to the aquathermal effect must be highest if the shales contain the largest amount of water or porosity. The fluid flow due to the aquathermal effect would take place in the same direction as that of compaction fluid flow. In other words, the aquathermal effect may simply increase the effectiveness of compaction fluid flows.

As discussed previously, the formation water salinity tends to increase as the shales compact. Hence, the salinity would increase from the center to the edges of each shale bed due to ion-filtration. Because osmosis would move water from a fresher to a more saline side, the fluid flow direction due to osmosis may be inferred as shown by the arrows in Figure 13C.

In summary, the directions of fluid movements due to the compaction, the aquathermal, and the osmotic effects are the same—from the center to the edges of each shale bed or from shale to sandstone.

There are several other possible causes of primary migration, such as capillary pressure, buoyancy, diffusion, and generation of hydrocarbons (especially gas). Most of these causes are not closely related to the movement of water. Because we are dealing with a relatively large amount of water and a relatively small amount of hydrocarbons in sediments, understanding water movement of a certain kind must be important in understanding primary hydrocarbon migration.

DIRECTION OF WATER MOVEMENT

From the preceeding discussions, it may be clear that the water movement is important in moving hydrocarbons, whether hydrocarbons are in solution in water or separate.

Water movement in a vertical direction was discussed in the previous section. The movement in a horizontal direction due to the compaction effect is mainly controlled by the sediment loading pattern. This problem was recently discussed by Magara (1977), who showed that fluid moves from a point of more loading (or thicker bed) to a point of lesser loading. Therefore, if a sediment loading pattern (which can be made of an isopach and a bulk density of a bed) is combined with a vertical fluid pressure plot discussed above, we are able to study the three dimensional fluid flow.

The large-scale fluid flow direction due to the aquathermal effect would be from the basin center to its edges and/or from deep to shallow, because the fluid would move from a point of higher temperature to a point of lower temperature due to this effect.

MODEL FOR PRIMARY MIGRATION

Based on the concept of the oil migration in the oil phase, Magara (1978) proposed a model for primary migration. The top diagram of Figure 14 shows a schematic of relative permeability versus degree of shale compaction. As the shale compacts, the relative permeability to water decreases and the relative permeability to oil increases, because the proportion of semi-solid or structured water will increase with compaction.

Although the relative permeability to oil increases with compaction, the absolute permeability of the shale will continually decrease as the shale loses more liquid water (middle diagram, Fig. 14). By looking at the top diagram of Figure 14 only, one might conclude that the oil migration in the oil phase increases with compaction. However, this is not true because the absolute permeability of the shale decreases with compaction. The actual volume of migrating oil is

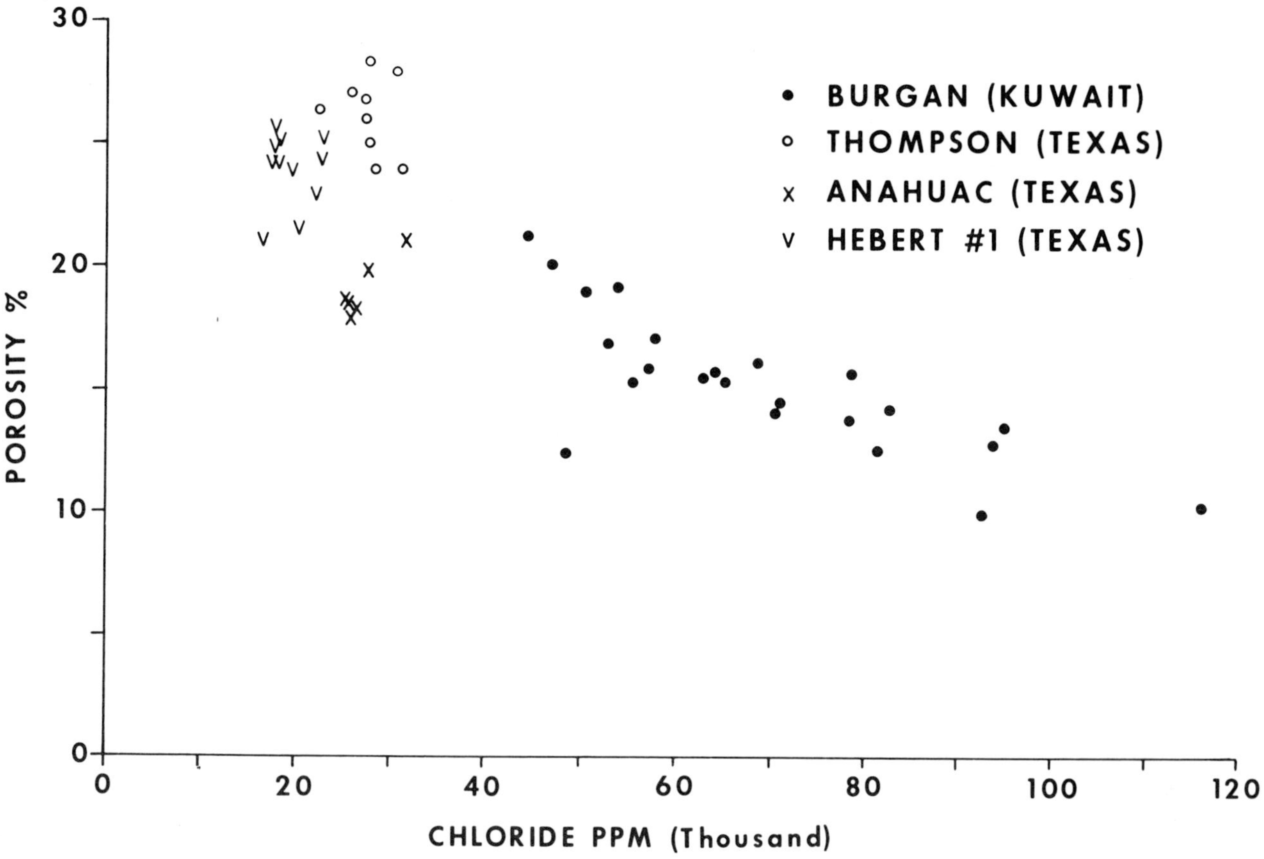

FIG. 12—Shale porosity-chlorinity relationship, Kuwait and Texas (data from Hedberg, 1967).

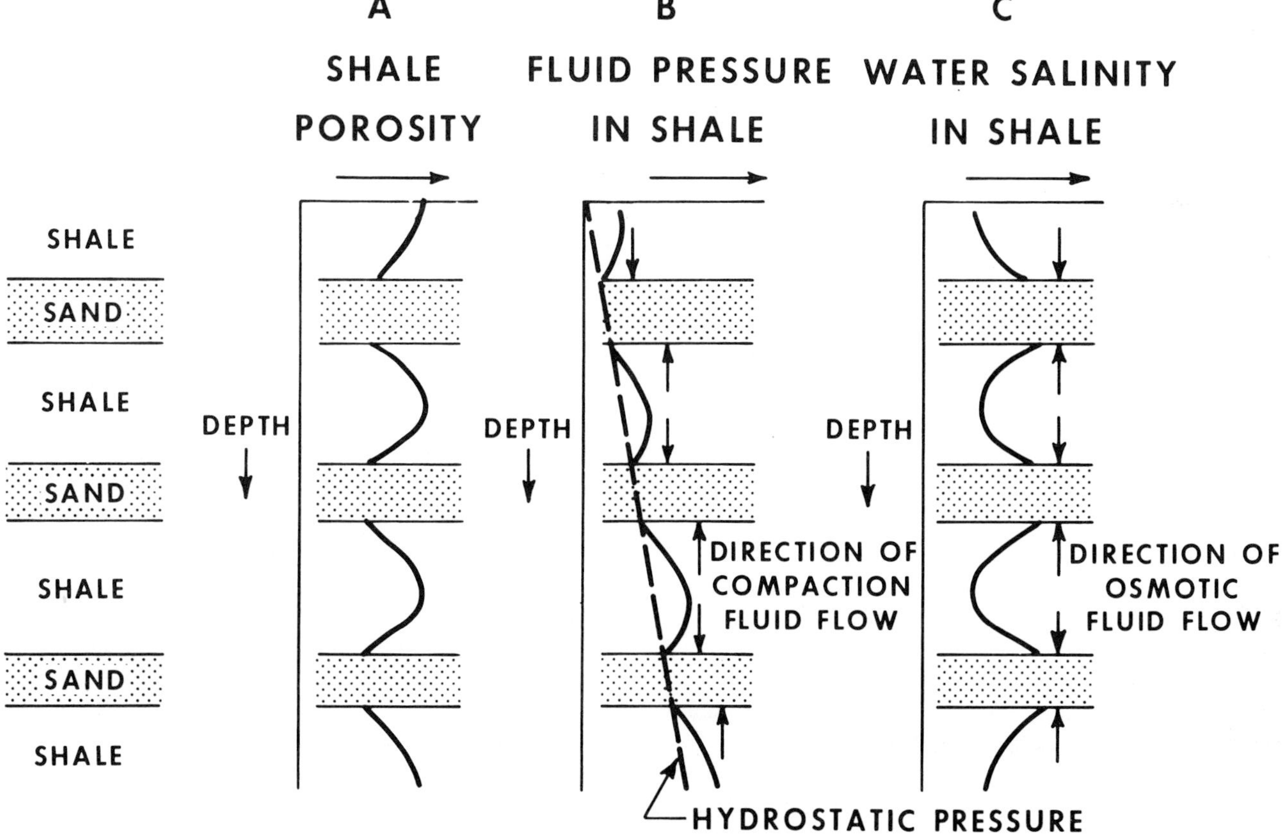

FIG. 13—Schematic diagram showing shale porosity, fluid pressure and pore-water salinity distributions in interbedded sandstone-shale sequence (from Magara, 1974a).

controlled by the product of the relative permeability and the absolute permeability, provided that the other factors such as pressure gradient and fluid viscosity stay constant. In other words, the maximum oil migration may take place at an intermediate compaction stage (bottom diagram, Fig. 14). If this maximum migration stage, or timing, is not far from the effective oil generation stage as discussed earlier, we may be able to have a significant oil accumulation.

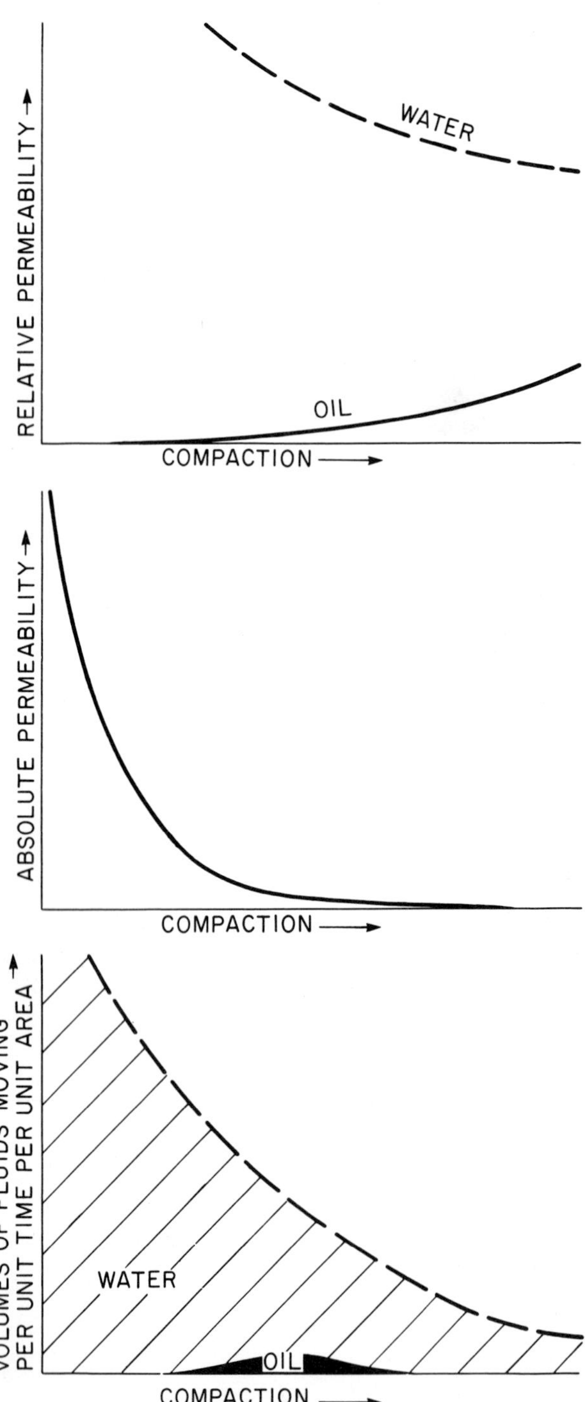

FIG. 14—Hypothetical relationships of relative permeability, absolute permeability, and fluid movement-degree of compaction of shale (from Magara, 1978).

CONCLUSIONS

1. Liquid hydrocarbons must have begun migration from the source rock earlier than the geologic time that corresponds to the maximum accumulation of oils in the reservoir. A new concept of the rate of generation of liquid hydrocarbons on the basis of the geochemical source rock analysis is introduced and discussed in this paper, in order to improve our understanding about the time of generation and migration of liquid hydrocarbons.

2. Because of relatively low solubility of liquid hydrocarbons in water, most oil may have had to migrate in its own phase. On the other hand, most gas may migrate in aqueous solution because of its higher solubility.

3. In compacted shales, a larger proportion of water may be in the form of semi-solid. Very tiny globules of oil in the shales may stay close to the remaining liquid water. In these circumstances if the shales compact further, the oil as well as the liquid may migrate, provided that the oil saturation in the liquid phase is higher than the critical level for oil migration.

4. Although a larger proportion of oil may move in its own phase, the movement of water is very important. This is because a good drainage condition for water will cause a relatively dry condition in the shales. This dry condition means that the shales contain mostly semi-solid water; there will be a very little amount of liquid water there. The concentration of oil in the liquid will be relatively high, because the amount of the water is so small.

5. Four possible causes of primary migration—compaction, aquathermal, clay mineral conversion, and osmotic effects—are considered in this paper. All of these causes are related to water movement. Several other causes not associated with movement of water have been proposed by other workers as well. The writer believes that the movement of water is important in primary migration because we are dealing with a large amount of water and a relatively small amount of oil at the primary migration stage.

6. The previous reasoning may lead the reader to conclude that the effective removal of liquid water from shales is essential in primary migration of oil in the oil phase. A good drainage condition will increase the relative permeability for oil, because the concentration of oil in the liquid phase increases as the liquid water is effectively expelled. If a good drainage condition covers a relatively long geologic period, there would be a good chance of migrating the generated oil in its own phase.

7. Although the relative permeability for oil in a good drainage section may increase as shales are compacted, the absolute permeability continuously de-

creases. This suggests that there may be a peak oil-phase migration stage at which the relative permeability is fairly high and yet the absolute permeability is not too low. If this peak oil-migration stage is not very far from the peak oil-generation stage, significant oil accumulation may be expected.

REFERENCES CITED

Asakawa, T., and y. Fujuta, 1979, Organic metamorphism and hydrocarbon generation in sedimentary basins of Japan, *in* Proceedings of the seminar on generation and maturation of hydrocarbons in sedimentary basins; C.C.O.P., p. 142-162.

Athy, L. F., 1930, Density, porosity and compaction of sedimentary rocks: AAPG Bull., v. 14, p. 1-24.

Baker, E. G., 1962, Distribution of hydrocarbons in petroleum: AAPG Bull., v. 46, p. 76-84.

Barker, C., 1972, Aquathermal pressuring: role of temperature in development of abnormal-pressure zones: AAPG Bull., v. 56, p. 2068-2071.

Burst, J. F., 1969, Diagenesis of Gulf Coast clayey sediments and its possible relation to petroleum migration: AAPG Bull., v. 53, p. 73-93.

Dickey, P. A., 1975, Possible primary migration of oil from source rock in oil phase: AAPG Bull., v. 59, p. 337-345.

Dickinson, G., 1953, Geological aspects of abnormal reservoir pressures in Gulf Coast, Louisiana: AAPG Bull., v. 37, p. 410-432.

Dodson, C. R., and M. B. Standing, 1944, Pressure-volume-temperature and solubility relations for natural gas-water mixtures: Am. Petroleum Inst. Drilling and Production Practice, p. 173-178.

Engelhardt, W. V., and K. H. Gaida, 1963, Concentration changes of pore solutions of clay sediments: Jour. Sed. Petrology, v. 33, p. 919-930.

Foster, J. B., and H. Whalen, 1966, Estimation of formation pressures from electrical surveys—offshore Louisiana: Jour. Petroleum Technology, v. 18, p. 165-171.

Ham, H. H., 1966, New charts help estimate formation pressures: Oil and Gas Jour., v. 64, p. 58-63.

Hedberg, H. D., 1936, Gravitational compaction of clays and shales, Am. Jour. Sci., v. 31, p. 241-287.

Hedberg, W. H., 1967, Pore-water chlorinities of subsurface shales: Ph.D. thesis, Univ. Wisconsin (Univ. Microfilms, Ann Arbor, Michigan)..

Hosoi, H., 1963, First migration of petroleum in Akita and Yamagata Prefectures: Japanese Assoc. Mineralogists, Petrologists and Econ. Geologists Jour., v. 49, p. 43-55, p. 101-114.

Levorsen, A. I., 1967, Geology of Petroleum: San Francisco, W. H. Freeman Co., (2nd ed.), 724 p.

Low, P. F., 1976, Viscosity of interlayer water in montmorillonite, Soil. Sci. Soc. America Proc., v. 40, p. 500-505.

Magara, K., 1968a, Compaction and migration of fluids in Miocene mudstone, Nagaoka plain, Japan: AAPG Bull., v. 52, p. 2466-2501.

——— 1968b, Subsurface fluid pressure profile, Nagaoka plain, Japan: Japan Petroleum Institute Bull., v. 10, p. 1-7.

——— 1974a, Compaction, ion-filtration and osmosis in shales and their significance in primary migration: AAPG Bull., v. 58, p. 283-290.

——— 1974b, Aquathermal fluid migration: AAPG Bull., v. 58, p. 2513-2516.

——— 1975a, Reevaluation of montmorillonite dehydration as cause of abnormal pressure and hydrocarbon migration: AAPG Bull., v. 59, p. 292-302.

——— 1975b, Importance of aquathermal pressuring effect in Gulf Coast: AAPG Bull., v. 59, p. 2037-2045.

——— 1976, Water expulsion from clastic sediments during compaction— directions and volumes: AAPG Bull., v. 60, p. 543-553.

——— 1977, A theory relating isopachs to paleocompaction-water movement in a sedimentary basin: Bull. Canadian Petroleum Geology, v. 25, p. 195-207.

——— 1978, The significance of the expulsion of water in oil-phase primary migration: Bull. Canadian Petroleum Geology, v. 26, p. 123-131.

McKelvey, J. G., and I. H. Milne, 1962, The flow of salt solutions through compacted clay, *in* Clays and clay minerals: New York, Pergamon Press, Earth Sci. Ser. Mon. 11, p. 248-259.

Meade, R. H., 1966, Factors influencing the early stages of compaction of clays and sands—review: Jour. Sed. Petrology, v. 36, p. 1085-1101.

Powers, M. C., 1967, Fluid-release mechanisms in compacting marine mudrocks and their importance in oil exploration: AAPG Bull., v. 51, p. 1240-1254.

Price, L. C., 1976, Aqueous solubility of petroleum as applied to its origin and primary migration: AAPG Bull., v. 60, p. 213-244.

Proshlyakov, B. K., 1960, Reservoir properties of rocks as a function of their depth and lithology: Geol. Neft. Gaza, v. 12, p. 24-29.

Rieke, H. H., III, and G. V. Chilingarian, 1974, Compaction of argillaceous sediments: Amsterdam, Elsevier, 424 p.

Schmidt, G. W., 1973, Interstitial water composition and geochemistry of deep Gulf Coast shales and sandstones: AAPG Bull., v. 57, p. 321-337.

Timko, D. J., and W. H. Fertl, 1971, Relationship between hydrocarbon accumulation and geopressure and its economic significance: Jour. Petroleum Technology, v. 22, p. 923-930.

Tissót, B., and R. Pelet, 1971, Nouvelles donnees sur les mecanismes de genese et de migration du petrole simulation mathematique et application a la prospection: 8th World Petroleum Cong. Proc., p. 35-46.

Vyshemirsky, V. S., et al, 1973, Bitumoids fractionation in the process of migration, *in* B. Tissót and F. Rienner, eds., Advances in organic geochemistry: Paris, Technip, p. 359-365.

Weller, J. M., 1959, Compaction of sediments: AAPG Bull., v. 43, p. 273-310.

Wilson, H. H., 1975, Time of hydrocarbon expulsion, paradox for geologists and geochemists: AAPG Bull., v. 59, p. 69-84.

Some Mass Balance and Geological Constraints on Migration Mechanisms [1]

By R. W. Jones[2]

Abstract Oil and gas are not at rest in the sedimentary mantle of the earth. They are not in equilibrium, whether they are finely dispersed in a potential source rock or whether they are concentrated in a trap in a reservoir rock. A wide variety of possible escape mechanisms exists; these include diffusion, continuous single phase flow, solution of oil in gas or gas in oil, and solution in water derived from compaction, clay diagenesis, or meteoric sources. The problem is to quantify the possible mechanisms and to rank their relative importance under a given set of physical, chemical, and geologic conditions. The quantitative importance of the various proposed mechanisms can vary by orders of magnitude, depending on the physical, chemical, and geologic conditions.

During the past decade, oil-to-source correlations have become reliable and the timing of peak generation and concomitant migration has been sufficiently quantified to allow the geologist/geochemist to make estimates of when and how much petroleum moved from one location to another. Combined with a knowledge of the physical, chemical, and geologic conditions at the time of migration, such quantitative descriptions of subsurface petroleum transfer permit an empirical test of the applicability of the various proposed migration mechanisms. The application of this technique to selected areas suggests that most of the major commercial oil accumulations of the world left their source rock in a continuous oil phase. When bitumen concentrations in the rock are too low for continuous phase flow to exist, other migration mechanisms, which always are operative, will increase in both absolute and relative intensity. Solution of oil in gas may become significant in thick Tertiary delta systems, and meteoric water may be a surprising asset in some very specific geologic settings. However, it is unlikely that solution of oil in water derived from compaction or from dehydration of clay has much to do with the origin of many of the major oil accumulations of the world.

INTRODUCTION

Many mechanisms have been proposed as the primary cause of migration out of the source into the reservoir. Conceptually, most of them fall into two end member types. One depends on water to move the petroleum or petroleum precursors out of the source rock; the other moves the petroleum out of the source rock in a separate phase independently of any associated water movement. In fact, removing or immobilizing most of the water probably is a prerequisite of separate phase movement. Some proposed migration mechanisms and carriers for water-controlled primary migration include: oil in water solution, gas in water solution, other organics in water solution, micellar solution, emulsion, diffusion, convection, meteoric water, compaction water, and clay dehydration water. Some proposed migration mechanisms and carriers for separate phase transport within the pore system include: oil, gas, solution of gas in oil, and solution of oil in gas. Proposed migration mechanisms acting within the kerogen network include diffusion and pressure gradients.

Many of the specific migration mechanisms mentioned were explicitly advocated in the papers on migration presented at the AAPG National Convention in 1978 and included in this volume. It would simplify the problem if we could believe that only one of the authors is correct, and that only one of the proposed mechanisms is "the correct one." However, it is very unlikely that such is the case. All of these mechanisms exist in nature on some scale. Their relative quantitative importance at a given point in space and time is a function of a very large number of factors. Listed below are some of the rock, fluid, and kerogen properties of the source which could be expected to have a significant influence on the migration characteristics of a source rock.

Rock Properties Include: porosity, permeability, pore-size distribution, tortuosity, capillary pressure, wettability, absorption, adsorption, hydrated minerals,

[1]Manuscript received, August 29, 1978; accepted, August 1, 1979. This manuscript was derived from a paper originally presented in the AAPG Migration Short Course given at the Oklahoma City Annual Meeting, April 4, 1978. Published with permission of Chevron Oil Field Research Company.

[2]Chevron Oil Field Research Company, La Habra, California 90631.

Article Identification Number:
0149-1377/79/SG10-0004/$03.00/0.

particle-size distribution, fracture transmission, and heterogeneity.

Fluid Properties Include: pressure and pressure gradient, fluid potential, temperature and temperature gradient, water salinity, interfacial tension, water bound to clay surfaces, viscosity, saturation, compressibility, density, and buoyancy.

Kerogen Properties Include: quantity, type (generation products), type (absorption characteristics), maturation, and distribution within the source rock.

Even if we could measure all of the rock, fluid, and kerogen properties of source rocks, the possible permutations and combinations are obviously sufficient to challenge a thousand physicists and chemists for a thousand years. In the interim, empirical solutions must be sought. Constraints on possible solutions are provided by what nature has done regarding the subsurface transfer of petroleum. Although, it is not yet possible to quantitatively evaluate the large number of combinations of physical and chemical factors that control migration, nature is and has. It might be expected that the amounts and distribution of oil and gas in traps relative to those in the source rock will reflect the migration mechanism that ejected them from the source. Indeed, that is a major thesis of this paper.

To apply mass balance considerations to the problems of evaluating possible migration mechanisms, it is necessary to study basins in which there is in-depth knowledge about: (1) the amount, distribution, and characteristics of the oil and gas, bitumen, and kerogen; (2) the amount, distribution, characteristics, and movement, past and present, of the water; and (3) the structural and stratigraphic history. Our knowledge of the listed factors in several basins is combined with recognized geochemical principles in the next two sections of the paper to argue that (1) continuous phase flow is the mechanism by which petroleum left the source rocks of many of the major oil accumulations of the world; and (2) little of the commercial petroleum of this world left its source dissolved in water derived from compaction or clay diagenesis.

In this paper the concepts of continuous phase flow and aqueous solution in water derived from compaction and clay dehydration are treated as the main adversaries. Although somewhat artificial, the adversary position is convenient due to the writer's belief in continuous phase flow as the mechanism of primary migration accounting for the majority of the world's large oil fields, and to the strong advocacy in the American literature of aqueous solution in water derived from compaction and clay dehydration as the dominant mechanism of primary migration. Possible contributions from other mechanisms generally are not evaluated or discussed in this paper, except in the last part where two problem areas for any theory of primary migration, the Gulf Coast Tertiary and the Athabasca oil sands, are discussed.

CONTINUOUS PHASE PRIMARY MIGRATION

Introduction

Continuous phase primary migration refers to the movement of oil within and out of the source rock in a continuous oil phase. The movement is believed to be caused by high differential pressures within the rock. The pressure differentials are caused by local supernormal pressure (SNP) generated by normal compaction of a fine-grained rock relative to its more permeable neighbors, and by the continuous increase in volume of the totality of the organic matter (OM) in the rock by generation of organically derived material of lower density. Thus generation is not only the *sine qua non* of migration, it is the cause.

Historically, the combined effects of small pore and capillary size, a water wet pore system, and surface tension effects between water and oil have been thought to preclude the possibility of continuous phase oil flow in and out of a fine-grained source rock (Baker 1960; McAuliffe 1979). The physical principles involved are well known and irrefutable; they probably are the primary reason why oil generated in most shales never leaves the shale as oil. However, in organic-rich rocks there are several factors operating which can override the effects of surface tension and capillary pressure. We are not dealing with buoyance effects alone to move the oil, but high differential pressures caused by compaction and hydrocarbon generation. In addition, the pore system need not be water wet due to the large amount of organic matter in the rock, the tendency for the organic matter to be squeezed into the pore system (Bradley, *in* Hoots et al, 1935), and the fact that the organic matter is oil wet. Perhaps of even greater importance is the heterogeneity of most excellent source rocks. Source rocks are typically varved due to their low energy depositional environment and the lack of bioturbation required of good source rocks (Demaison and Moore, 1979). Recently Momper (1978) discussed in detail the source rock characteristics such as partings, laminae, various irregular distribution of grains, crystals, kerogen, and microfractures which can contribute to the development of relatively large pores and micropathways which are exploited by the pressure buildup due to hydrocarbon generation. The writer agrees with his evaluation. Because of the thoroughness of Momper's discussion, and other excellent discussions on the possible mechanisms of continuous phase migration published by Snarskii (1970), Dickey (1975), Magara (1978a) and Tissót and Welte (1978), this will not be pursued further here, but instead the writer will turn to some mass balance considerations which are believed to support the continuous oil phase concept independently of our knowledge of the details of the mechanisms.

Numbers and discussions are given for some formations and basins in which primary migration in continuous phase flow probably occurred. No claim of high precision or accuracy is made for the numbers. However, the examples chosen can stand an error by a factor of 2-5 and not invalidate the point presented. In addition, some arguments supporting the concept of continuous phase primary migration are presented based on the distribution of kerogen and hydrocarbons in source rocks and on the timing of primary migration.

Specific Areas/Formations

Bakken Shale, Williston basin—The Bakken Shale is a thin (0 to 100 ft; 0 to 32 m), organic rich, transgressive marine shale of latest Devonian/earliest Mississippian age which is the source of at least 3×10^9 bbl of in-place oil in the Williston basin (Dow, 1974). In a perceptive study of the Sanish Pool, Antelope Field, North Dakota, Murray (1968) pointed out that the visual and log characteristics of the Bakken shales indicated that their pore space was hydrocarbon saturated. Subsequently, Meissner (1976; 1978) described the physical characteristics of the Bakken which support the continuous oil phase concept. These include: (1) no formation water is ever recovered during drillstem tests or initial well completions; (2) where thermally mature, the Bakken is overpressured with respect to both the regional gradient and reservoir beds separated from it by tight strata; and, (3) the electrical resistivity increases abruptly to nearly infinite values where the Bakken enters peak oil generation. Meissner inferred this last relationship from temperature versus resistivity plots, but the proposed relationship has been confirmed by Chevron geochemical data. Meissner used the three facts listed above in combination with Williams' (1974) demonstration that oils in the Bakken, the stratigraphically lower Devonian Nisku Formation, and stratigraphically higher Mississippian Mission Canyon Formation are identical, to argue for continuous phase primary migration in and out of the Bakken, both upward and downward. Meissner further believed that the outward migration of oil takes place through a spontaneously generated fracture system created by high overpressuring caused by hydrocarbon generation. The data in Table I support Meissner's conclusions. At low resistivities the hydrogen-to-carbon (H/C) ratios of the organic matter (OM) are high, and the ratio of generated to remaining-to-be-generated hydrocarbons is low. At the high resistivities, the amount of generated hydrocarbons is much higher, and the remaining generative potential much lower. In addition, the total hydrocarbon potential per gram of organic carbon in the mature rock is approximately half that of the immature one. Because the type of organic matter is the same in both samples, this suggests that $\cong$167 mg of hydrocarbons per gram of organic carbon have migrated out of the mature rock. For the mature rock described in Table I, this $\cong$50,000 ppm (vol.) or $\cong$5% of the rock volume, a figure only compatible with continuous phase migration.

Given below are mass balance calculations that demonstrate the impossibility of the Bakken Shale sourced oil in the Williston basin having moved out of the Bakken in aqueous solution. Dow (1974) provided the data for the Bakken Shale in the Williston basin necessary to calculate the minimum solubility required if the Bakken-sourced oils were carried out of the Bakken by compaction water during peak generation. The Bakken is the source of 3×10^9 bbl of in-place oil. The average thickness of mature, organic-rich Bakken $\cong$40 ft (12.2 m) and the volume of mature Bakken $\cong 100 \text{ mi} \times 200 \text{ mi} \times (40/5{,}280) \text{ mi} \cong 150$ cu mi $\cong$626 cu km. Thus, minimum oil expelled per cu mi (cu km) of Bakken equals 3×10^9 bbl/150 cu mi $\cong$20

Table 1. Some Contrasts Between Mature and Immature Bakken Shale

	Immature	Mature
Sample Depth (ft)	7,570	11,260
Total Organic Carbon = TOC (wt. %)	8.8	12.5
Hydrogen/Carbon Ratio (H/C)	1.23	0.83
Resistivity (ohm-meters)	10	> 100
*Hydrocarbons (mg) in bitumen/gm of rock	0.9	6.6
*Hydrocarbons (mg) generated/gm of rock	31.8	18.6
*Hydrocarbons (mg) in bitumen/gm of TOC	10	53
*Hydrocarbons (mg) generated /gm of TOC	360	150
*Hydrocarbon potential = Hydrocarbons (mg)/gm of TOC	370	203

*Measured by programmed temperature pyrolysis with a Rock-Eval pyrolysis device (Espitalié et al, 1977).

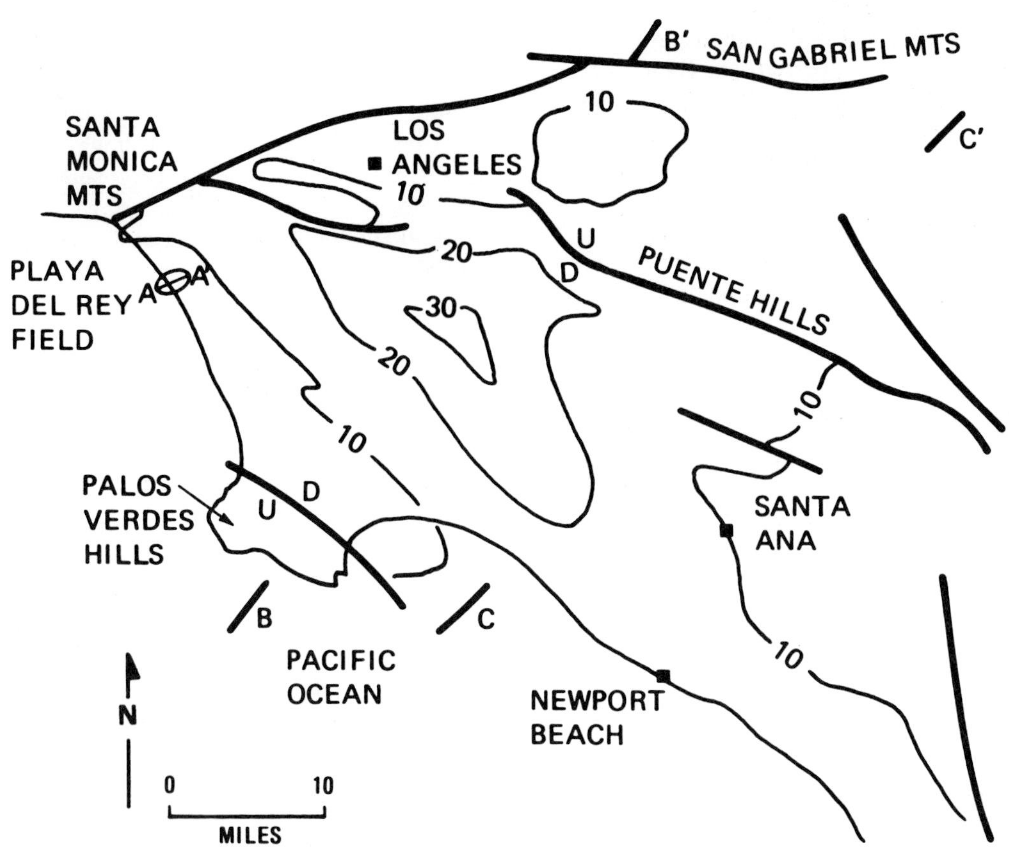

FIG. 1—Generalized basement contours (kilofeet), Los Angeles basin, California. A-A′, B-B′, C-C′, show locations of cross sections depicted in Figures 2 and 3.

$\times 10^6$ bbl/cu mi $\cong 4.8 \times 10^6$ bbl/cu km $\cong$770 ppm. If expulsion of oil occurred while Bakken porosity dropped from 10% to 5%, necessary water solubility would be $\cong$15,000 ppm. If 3.3 times as much oil was expelled from Bakken as was pooled (Dow, 1974), the necessary water solubility would be $\cong$50,000 ppm. Price (1976) presented experimental data which indicated a maximum oil solubility of 50 to 200 ppm at the temperatures at which the Bakken probably reach peak generation (70°C to 120°C). Thus the minimum numbers of required solubilities are at least two orders of magnitude too high for the oil to have migrated in aqueous solution in Bakken compaction water at the maximum temperatures reached in the Bakken.

Los Angeles basin, California—A case for the continuous phase migration from a source rock in the Los Angeles basin can be made by using geochemical data and astute microscopic observations provided in a perceptive 43-year-old paper by Hoots et al (1935), which dealt with the origin of the oil in the Playa del Rey field (Figs. 1, 2). The source rock is the "nodular shale," a laminated organic and phosphate-rich rock that directly overlies an oil-filled conglomeratic sandstone resting on a schist basement. Several shale samples yield an average of 1.8 wt% or $\geq$4.0 vol% chloroform extractables. Playa del Rey field is an excellent example of the upward and downward movement of commercial amounts of oil from an organic-rich, bitumen-saturated, oil source rock.

The oil in the upper zone is very similar to the oil below the "nodular shale" but contains more saturates as would be expected from chromatographic effects of migration. The upper zone oil was undoubtedly emplaced from the "nodular shale" along faults that, when cored, are associated with free oil. Hoots et al (1935), argued very convincingly that the types of compositional differences between the extract from the "nodular shale" and the underlying oil indicate that the oil migrated from the overlying shale and that the high chloroform extractable content of the shale is not the result of oil migrating from the reservoir into the shale. In fact, the geochemical data showing a higher percentage of paraffins and naphthenes in the oil than the source reported by Hoots et al, probably are the earliest published data which show the differences between oil and source which have frequently been described in recent years (Tissót and Welte, 1978). Perhaps the most fascinating aspect of this study are the microscopic observations of the organic matter by W. H. Bradley, which were included in the Hoots et al (1935) paper. Bradley recognized four types of organic matter microscopically, and described them in detail.

> "The first. . . is. . . spore exines. The second. . . which is the most abundant . . . is deep reddish

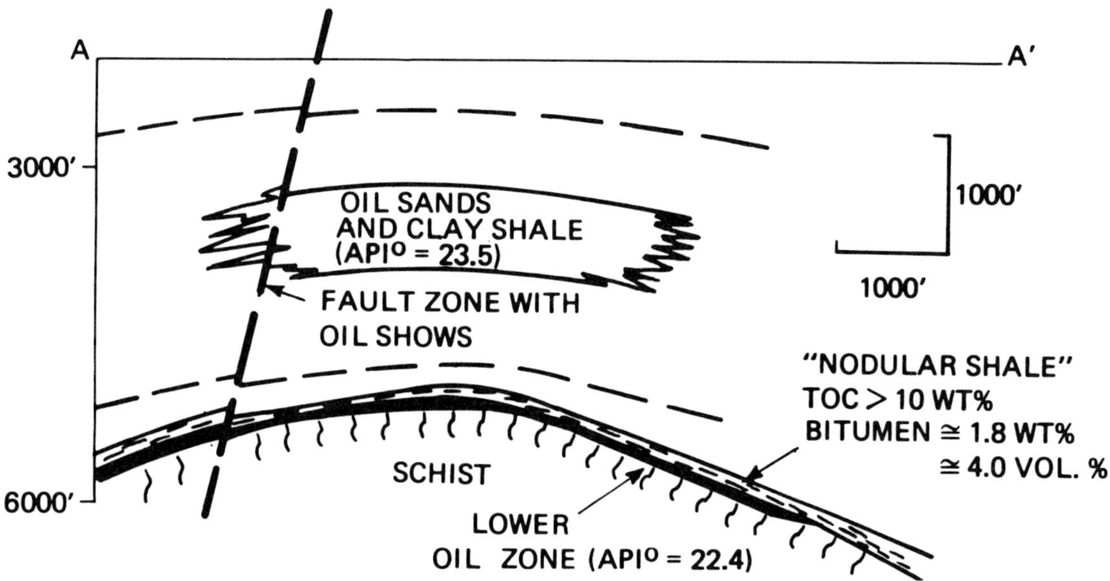

FIG. 2—Generalized structure section (after Hoots et al, 1935), Playa del Rey field, Los Angeles basin, California (line A-A' from Fig. 1).

brown. . . is essentially structureless. . . occurs in thin stringers and small irregular flat flakes or sheets and in small irregular flocculent masses. . . range greatly in size. . . is less altered than most of the organic matter. . . has undoubtedly undergone some change from its original composition. The third kind of organic matter is light amber in color and is more or less evenly diffused through the rock. It is. . . perfectly translucent. . . has been at some time during its history either a liquid or a very fluid gel. . . was *forced into all available pores during compaction of the sediment* (italicizing is the writer's). The material of the second sort mentioned appears almost to grade into this pale yellowish material. As the size of the reddish brown flocculent material decreases, it approaches both in color and clarity the perfectly clear yellow stuff of the third class. The fourth kind of organic matter is a dark reddish brown opaque homogeneous stuff. . . which occurs rather sparingly, filling cavities . . . appears to have been either liquid or a quite fluid gel. . . is quite closely related to the pale yellowish organic matter just described and like the yellowish material was probably derived by decomposition from the reddish brown fragmental organisms."

Probably the first two types of organic matter described by Bradley are kerogen with the exines being a very subordinate component, the third type is the hydrocarbon rich part of the bitumen, and the fourth type is the more asphaltic part of the bitumen which physically was separated in the source due to its less mobile properties. A more modern study of the "nodular shale" would clearly be of interest.

Bradley's work indicates that microscopy never has achieved the importance in source-migration studies which it deserves. Unfortunately, the development of "objective" geochemistry with its sophisticated technology pushed "subjective" microscopy with its 50-year-old technology into the background. Only recently with the development of fluorescence microscopy (see Teichmüller, 1975; Teichmüller and Ottenjann, 1977) and of the scanning electron microscrope (SEM) has microscopy started to fulfill its potential in migration studies that was implicit in Bradley's astute observations 40 years ago.

Mass balance calculations confirm the inferences from the geochemical and microscopic observation of Hoots et al (1935), that primary migration of oil in aqueous solution is not important in the Los Angeles basin. Figure 3 contains two northeast-southwest cross sections through the Los Angeles basin which show the basin configuration, location of major oil fields, and the approximate top of peak generation as determined by Philippi (1965) on the west side of the basin and extended across the basin on the basis of other data. The pertinent mass balance data are listed below in a sequential development.

1. Oil in place ≅25 × 10⁹ bbl ≅1.0 cu mi (4.2 cu km).

2. Rock volume related to the oil fields ≅1,600 cu mi (6,670 cu km; Barbat, 1958).

3. 1/1,600 or 0.06% or 600 ppm of pertinent basin volume is oil in traps.

4. Mature and post-mature source rock ≅200 cu mi (835 cu km).

5. 1/200 or 0.5% or 5,000 ppm of source rock volume or ≅2,000 ppm of source rock weight is amount of oil in traps.

6. Assume ≅25% of original kerogen converted to oil.

7. Original organic content of source rock ≅4.0 wt%.

8. Total oil generated ≅1.0% wt% = 2.5 vol. % of source rock.

9. Total oil generated ≅200 cu mi (835 cu km) × 0.025 ≅5 cu mi (21 cu km).

10. Oil in reservoir/oil generated ≅1/5 ≅20%.

11. Assume migration occurred during porosity decrease from 10% to 5% in source rock.

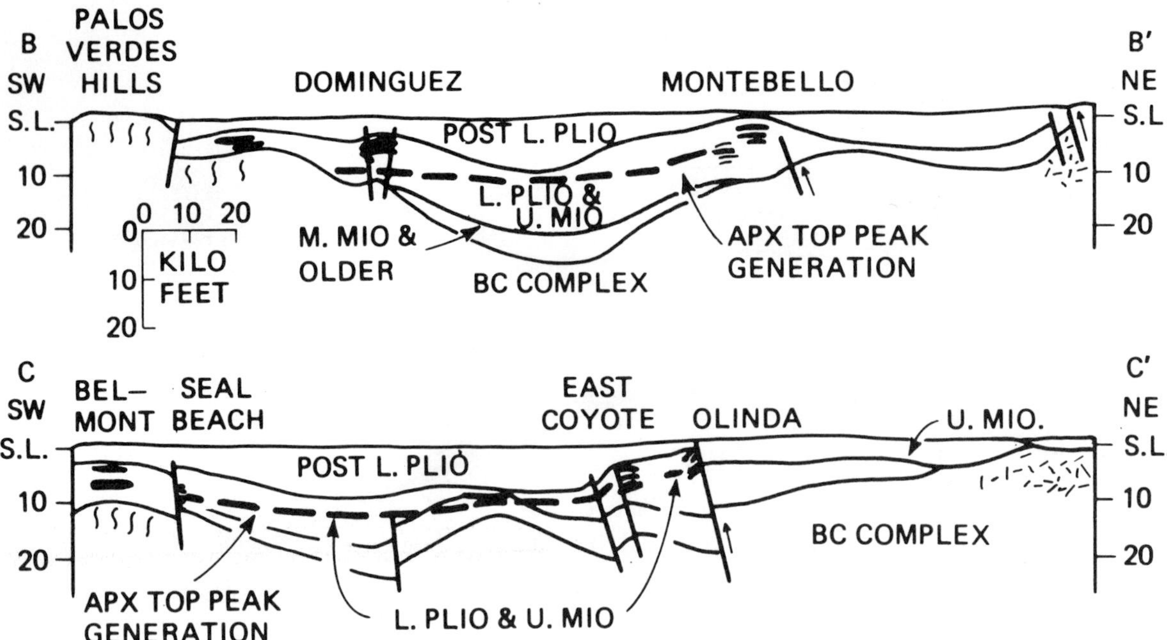

FIG. 3—Diagrammatic northeast-southwest cross sections (from Gardett, 1971), Los Angeles basin, showing oil fields and approximate top peak oil generation (lines B-B′ and C-C′ from Fig. 1).

12. Fluid loss during oil migration ≅10 cu mi (42 cu km).
13. Assume all migrated oil is in known traps.
14. Then oil/water ratio during migration from source ≅1/9.
15. Necessary solubility ≥100,000 ppm.

All of the numbers are impressive. Attention is directed particularly to item 8, which suggests that ≅2.5% of the source rock volume was converted to oil; item 10, which demonstrates a high migration and trapping efficiency; and item 15, which demonstrates that migration in aqueous solution did not have much to do with migration from the source rocks in the Los Angeles basin.

Those who support migration of oil in aqueous solution might argue that we are not dependent solely on indigenous water emitted from the source rocks, but that large amounts of water could pass through the source rocks from below and pick up hydrocarbons in the process. This is unlikely. When subsiding and compacting, even a normal nonsource shale with a permeable substratum will be overpressured in its interior and emit fluids both upward and downward (Smith et al, 1971; Evans et al, 1975). When additional overpressuring from the generation of substantial amounts of hydrocarbons is added to the normal overpressuring from compaction, the chances of significant amounts of water moving upward through an organic-rich source rock become remote. The fact that many of the excellent source rocks of the world act as both source and cap for oil sourced within them confirms this interpretation. Interbedded source and reservoir is a perfect combination for draining temporarily overpressured source rocks by continuous phase flow rather than upward moving water and this, of course, is the situation in the Los Angeles basin where turbidite reservoirs are interbedded with pelagic source shales.

There are some additional inferences regarding migration in the Los Angeles basin which can be made from Figure 3 and some additional data on the distribution of oils and their potential sources. The upper Pliocene contains both abundant reservoirs and potential sources by conventional standards (Philippi, 1965), and was structurally deformed during the Pleistocene orogeny which formed the anticlines that contain the large quantities of oil in the lower Pliocene–upper Miocene sequence. However, the upper Pliocene contains less than 0.1% of the basin's oil, and this only in association with major faulting in structures containing abundant oil in the deeper section. In addition, the abrupt cessation of reservoired oil near the top of the lower Pliocene section correlates well with the top of peak generation in the basin center for the same stratigraphic units. These facts suggest the following: (1) any compaction and clay diagenesis water moving out of the depocenter of the upper Pliocene carried negligible amounts of liquid hydrocarbons despite the presence of abundant oil-prone source rocks in the early generation stage of thermal diagenesis (Philippi, 1965;

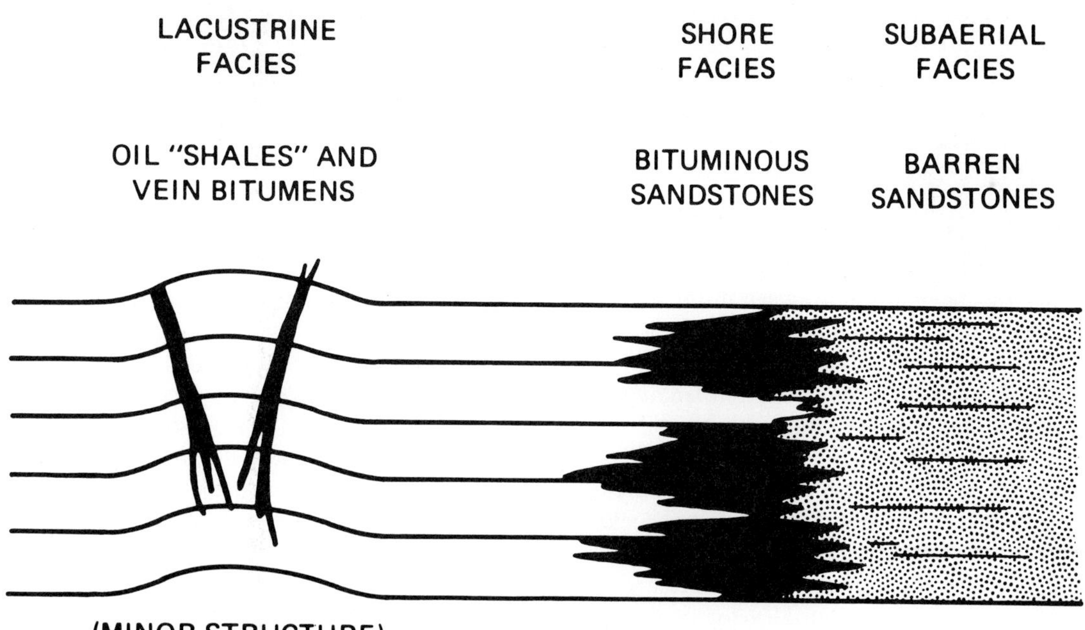

FIG. 4—Association of types of deposits of bitumens with sedimentary facies in the Uinta basin, Utah (from Hunt, 1963).

Bostick et al, 1978); (2) due to insufficient generation, continuous phase oil migration did not occur in the post–lower Pliocene section either; and (3) the faults are dominantly sealing rather than leaking. Thus, most of the oil in the lower Pliocene reservoirs probably originated in interlayered, thermally mature, lower Pliocene source rocks deeper in the basin, contrary to Philippi's (1965) hypothesis that the oil in the lower Pliocene reservoirs was emplaced along faults from underlying Miocene source rocks.

Green River Formation, Uinta basin, Utah—Almost everyone who has seen the bitumen dikes in the Green River and Uinta Formations of the Uinta basin is impressed by what nature can do with regard to continuous phase flow of bitumens when the conditions are right. Listed below is a summary of the conditions that probably existed in the Green River Formation and caused the emplacement of the bitumen dikes. The conditions are uncommon, but neither unique nor extreme.

1. 25+ vol. % hydrogen rich, amorphous organic matter (OM).
2. Varved, heterogeneous rock.
3. Fine-grained, brittle, carbonate matrix.
4. ≥6,000 ft of overburden—The overburden provided the driving force for fluid movement by placing the organic matter under lithostatic pressure, and decreased viscosity of the organic matter, but overburden was not sufficiently thick to raise the temperatures high enough to create conventional thermal maturity.
5. Tectonic tensile fracturing.
6. Release of pressure on the organic matter by intrusion of fractionated organic matter into fractures.

Figure 4 is a diagrammatic section in the Green River basin from Hunt (1963) which shows the gross stratigraphic relationships between the bitumen dikes and oil saturated sandstones at the updip edge of the basin. Subsequent to the dike formation and the end of tectonic tensile fracturing, additional maturation of the organic matter in the source rocks probably led to continuous phase movement of less viscous bitumens through heterogeneities in the fine-grained source rocks into the laterally equivalent sandstone reservoirs.

One reviewer of this manuscript argued that the demonstrable continuous phase movement from oil shales like the "nodular shale" and the Green River shale has little to do with primary migration from more normal source rocks. The writer rejects this argument on several grounds: (1) Total organic content of "oil shales" and conventional source rocks overlap. Any division is arbitrary. The Kimmeridgian oil shale of Scotland was famous long before its offshore equivalents became recognized as the source of the oil in the northern North Sea. (2) Both the "nodular shale" and the Green River "oil shale" have demonstrably sourced conventional oil accumulations, although pertinent chemical analyses of both oil and extract have shown them to be relatively immature (Hoots et al, 1935; Tissót et al, 1978). Clearly, further maturation of these rocks would cause them to generate and migrate large amounts of oil. (3) Oil shales are recognized as such because of their immaturity. I am not aware of any published study of former oil shales which are post mature. However, it is reasonable to believe that a total organic content of 4.38 wt.% for the Woodford Shale ≅15,000+ ft (4,570+ m) below the oil deadline at 27,725 ft (8,450 m) in the Anadarko basin in Oklahoma (Price, 1977) is the remnant of a former oil shale that made a significant contribution to commercial accumulations of hydrocarbons in that basin.

Fractured Shale Accumulations—There are many commercial accumulations in fractured organic-rich shale in the western United States. The reservoirs are brittle, relatively competent zones which commonly are encased in relatively plastic clay shales that seal the fluids within the fractured zones (Mallory, 1977). Water solubility can have no significance in the origin of these accumulations. The accumulations have low to no water recovery and many depend on gravity drainage. Most of the water either has migrated away or is tightly bound, and it is predominantly the oil that has access to the fractures.

The brittle rocks in which these accumulations are found are not oil shales, but simply good source rocks with total organic content typically ranging from 1 to 4 wt.%. The productive fracturing is due to external forces, not hydrocarbon generation. Development of tension fractures in a brittle rock must have created large pressure gradients between the fractures and nearby pores and literally sucked the oil into the fractures and thereby created a continuous oil phase in the fracture system. It is easy to imagine that where the fractured rock was not enclosed in plastic impermeable shales, the oil moved out in search of more conventional reservoirs and traps. This sequence of events may be important in continuous phase primary migration in certain geologic settings.

Organic Content of Source Rocks

Most of the major oil accumulations of the world originated in source rocks with total organic carbon contents (TOCs) $\geq$2.5 wt.% and occasionally with TOCs ranging past 10% by weight. Many examples come to mind: the Kimmeridgian of the northern North Sea, Silurian of Algeria, upper Miocene of the Los Angeles and San Joaquin basins of California, the Cretaceous black shale overlying the major unconformity of the Prudhoe Bay area in Alaska, the Bakken and Woodford Shales of the Paleozoic of the Mid-Continent, the Duvernay of the Devonian of Alberta, Canada, the La Luna limestones of the Cretaceous of Venezuela, and the source rocks of the Middle East.

The most notable exceptions to the previous generalization are the Tertiary delta systems of which the Niger and Mississippi are prime examples. In them, the source rocks underlie the reservoirs and consist of a 10,000+ ft (3,048+ m) section containing mature-post mature organic matter with TOCs ranging from 0.3 to 1.0 wt.%. The dominant migration process here is essentially unique to these geologic systems and is discussed later.

Before examining why high total organic content of most source rocks requires continuous phase migration, it is appropriate to examine the question of the minimum TOC necessary for a rock to become an effective source rock (a source of commercial accumulations). An answer to this question is important in understanding migration, but it has been treated in a rather cavalier fashion in most of the literature. Momper's (1978) observations are a clear-cut exception, and are an excellent discussion of how the prevailing number of 0.5 wt% organic matter became imbedded in the literature without either proof or in-depth discussion. The writer is a bit nonplussed why, after his illuminating discussion, Momper appears to fall back on the 0.5 wt% value as a generally valid criterion. Commercially significant migration probably is unlikely with organic matter contents $\cong$0.5 wt.% unless one is dealing with thick Tertiary delta systems. The reason is simple and straightforward. There are not enough liquid hydrocarbons generated relative to the available pore space to permit a continuous oil phase to form, and nature is forced to turn to less efficient migration mechanisms to move any liquid hydrocarbons out of the source rock.

The minimum amount of organic carbon necessary for a potential source rock to become an effective source is not simply an elusive constant. It is a wide-ranging variable dependent on many other variables. Not only does it depend on both the type of organic matter and its distribution in the rock, but also on the position of the potential source bed with respect to potential carrier beds, the maturation of the organic matter, the physical and chemical characteristics of the inorganic part of the source rock at peak generation, and, of course, the size of the fields which are economic. Thus, most coals with $>$90% organic content are not potential oil sources, despite the fact that they generate more heavy hydrocarbons per unit weight than most oil source rocks. However, vitrinite and inertinite, the dominant organic macerals in most coals, do not become oversaturated and release hydrocarbons until the liquid hydrocarbons are cracked to gas and condensate within the organic matter (Jones, in prep.). In addition, it is easy to visualize that the ease of migration, and hence, the amount of organic matter necessary to create an effective source, is partly controlled by whether the organic matter is homogeneously distributed in a potential source bed or distributed in varves of varying organic richness. Most excellent source beds are highly varved. In them the organic matter is likely to be distributed in rich layers which alternate with layers nearly barren of organic matter. This is particularly true of rocks containing an appreciable carbonate or siliceous content like the Green River or Monterey Formations. Such a layered distribution of both organic matter and rock properties facilitate the formation of a continuous oil phase in the rock at peak generation. The organic matter will compress and drive out the pore water and, at peak generation when the kerogen is exuding hydrocarbons, the increased pressure and rock heterogeneities should facilitate the development of microfractures and a continuous oil phase in the existing porosity. On the other hand, an organically rich zone surrounded by lean shales would undoubtedly have difficulty in emitting its liquid hydrocarbons to effective carriers, although if adequately fractured, it might become an effective reservoir and trap. The Sabym field in western Siberia with its 10,000+ sq km (3,8 61+ sq mi) of potentially productive Upper Jurassic bituminous shales (Auldridge, 1977) is the most notable example of such an accumulation. Thus, the minimum organic matter content necessary to make an effective source depends on a large number of variables which control the ease of primary migration.

A parallel question to the total organic content requirements of a source rock concerns the minimum hydrocarbon content required to eject hydrocarbons from a source rock. This is a legitimate question, but is also as unanswerable a question as what is the minimum TOC necessary for a rock to become an effective source rock. There are too many ifs, ands, and buts. For example: are hydrocarbons in the pore system or in the kerogen; is the rock a lime mud, an illitic shale, or a silty mixed-layer clay; are the hydrocarbons heavy or light; what percentage of the bitumen is hydrocarbons; are the hydrocarbons in thin, organic-rich layers or evenly dispersed through the rock? Momper (1978) handled this question with the statement that "the amount of extractable bitumen needed in a source bed before expulsion begins is about. . . 825 to 850 ppm. . . *or more*." (The italics are the writer's.) Momper did not discuss the evidence and geologic and geochemical setting of the 825 ppm minimum figure, although the numbers seem about right for peak generation in the Gulf Coast Tertiary and other major Tertiary delta systems. However, it seems clear from the literature that substantially larger figures exist, they are the norm in organic-rich rocks, and they often are not enough by themselves. Philippi's (1965) data indicate that the extractable C_{15}+ hydrocarbons now in the source rocks of the Los Angeles basin reach at least 2,500 ppm; Hunt (1961) gave a 3,000 ppm hydrocarbon content of samples of the Woodford and Duvernay (source) Shales; the Brooks and Thusu (1977) data indicate a hydrocarbon content of several thousand ppm for the Kimmeridgian Shale in the northern North Sea at peak generation; and Tissót et al, (1971) measured several thousand ppm C_{15}+ hydrocarbons in the Toarcian shales of the Paris basin, despite the fact that the Toarcian apparently was not the source of any of the oil fields in the Paris basin (Poulet and Roucaché, 1970). Hoots et al (1935) indicated that the "nodular shale" at Playa del Rey contains $\cong$18,000 ppm chlroform extractable organic matter of which $\cong$50% is probably hydrocarbons (petroleum ether extractable). Thus, minimum bitumen figures, like minimum TOC figures, probably are not very helpful unless they are closely tied to the detailed geology and geochemistry of an area. There are no magic numbers for calculations although there must be magic combinations of organic matter quantity, types, maturity, and distribution in different rock associations. Even with a hydrocarbon content of several thousand ppm the porosity of a mature source rock is several times larger than the hydrocarbon volume. Those who believe in continuous phase primary migration have developed a number of possible explanations of why oil will flow preferentially to water under such circumstances. The explanations involve the concepts of structured vs. liquid water (Miller and Low, 1963; Dickey, 1975) and of oil wetness. Dickey (1975) in a perceptive paper, and Magara (1978) in an extension of Dickey's arguments, suggested that in highly compacted source shales the ratio of oil to movable, liquid, water will be sufficiently high that oil will preferentially flow out of the shale. In addition, the existence of two or more layers of structured water can be expected to close off the smaller pores to any flow. Because the kerogen particles are, in general, preferentially located in layers and are larger than the associated clay particles (Momper, 1978), the larger connected pores probably

Table 2. Hydrocarbon Availability versus Pore Water Capacity

Oil Prone TOC (Wt. %)	Shale Porosity %	C_{15} + Hydro-carbons in in Rock (ppm, vol.)	Hydrocarbon Solubility (ppm)	Water Capacity HCs (ppm)	Excess HCs in Rock
0.5	40	30	10	4	8X
0.5	30	70	15	4.5	16X
0.5	20	200	25	5	40X
0.5(5)	10	1,000	100	10	100X (1,000X)
0.5	5	200	500	25	8X

are oil wet and contain a high percentage of movable oil. All of the above is somewhat conjectural and considerably more experimental and observational data are needed (Dickey, 1975).

Hydrocarbon Distribution in Source Rocks

At peak generation and slightly beyond, and during primary migration, most of the hydrocarbons are in the pore system. They are not in aqueous solution, not attached to clays, and not in the kerogen. That most of the hydrocarbons in the source rock are not in solution in the pore water at any time is clear from the data in Table II which show that even a very low total organic content rock always contains many times the hydrocarbons necessary to saturate the pore water. The numbers in Table II are hypothetical in the sense that each horizontal line of numbers does not indicate measured values of each parameter. They show my integration of depth versus porosity data summarized by Dickey (1975), generation data from Tissót and Welte (1978), solubility data from Price (1976), and an average geothermal gradient. No claim is made for accuracy in the numbers but, as in the discussion of solubilities in the case of the Bakken shale and Los Angeles basin, even an error by a factor of five will not invalidate the point the writer is trying to make.

Near peak generation and during primary migration a source rock with a TOC $\cong$2.5 wt% contains several hundred times the hydrocarbons required for saturation of the pore water. This in itself is a strong argument against aqueous solution being important in primary migration.

The question of the adsorption of hydrocarbons on clays at the temperatures of peak generation is difficult to answer quantitatively with available data and needs more research. However, the far greater polarities of water and NSO compounds than hydrocarbons, and the decreased adsorption at the temperature of peak generation strongly suggest that most of the hydrocarbons are elsewhere than adsorbed on the clays.

That most of the hydrocarbons in the rock-organic matter-pore system are not in the organic matter at peak generation is strongly indicated by the maturation tracks of unextracted oil-prone organic matter (Fig. 5). The sharp drop in hydrogen/carbon (H/C) ratio of the unextracted oil-prone organic matter with increasing maturation can only mean the ejection from the kerogen—by the kerogen—of the more hydrogen-rich generation products (i.e., the hydrocarbons). In this context the "coalification jumps" of the coal petrographers—which are basically times of rapid emission of volatiles from the oil-prone macerals—are worthy of study by those interested in petroleum migration (see Stach, 1953; Teichmüller, 1975). Figure 5 is more important in the search for an understanding of primary migration than the kerogen (extracted organic matter) H/C versus O/C diagrams which more

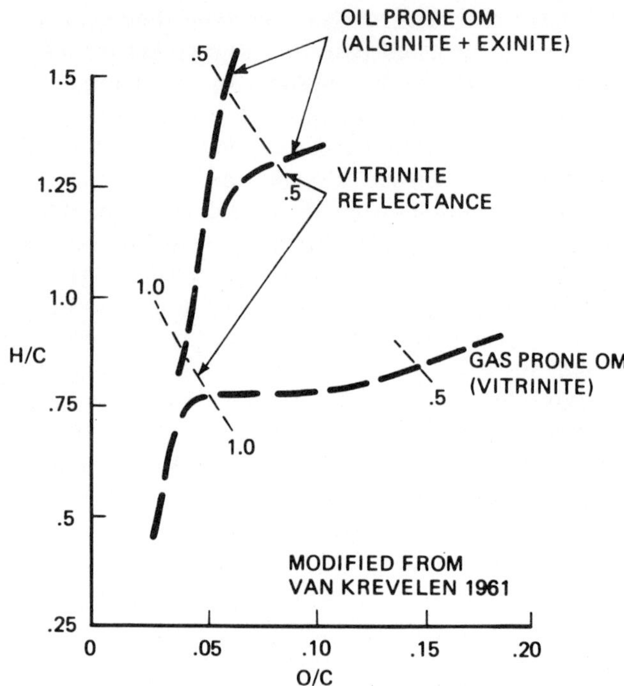

FIG. 5—Diagrammatic maturation tracks of unextracted oil and gas prone organic macerals (after Van Krevelen, 1961).

frequently appear in the petroleum literature. For example, the horizontal nature of the vitrinite curve within the oil generation window from 0.7 to 1.1 vitrinite reflectance clearly shows that vitrinite does not emit significant amounts of hydrocarbons until the condensate-wet gas stage of generation has been reached. During this same maturation interval the unextracted oil-prone organic matter is emitting liquid hydrocarbons into the pore system as shown by its near vertical pathway within the oil generation window.

The presumption from mass balance considerations that at peak generation most of the bitumen is in the pore space (or microfracture) receives confirmation from microscopic observations (Hoots et al, 1935) and resistivity measurements (Murray, 1968; Meissner, 1976, 1978).

Bitumen Redistribution in Source Rocks

The migration implications of the microscopic observations are supported by bitumen analyses of organic-rich, oil-prone shales which indicate that bitumen redistribution begins quite early in their thermal history. For example, in the WOSCO EX-1 core of the Green River Formation, Utah, USA, Robinson and Cook (1975) found greater than a tenfold variation in extraction ratio (bitumen/TOC) for samples with low maturation and similar organic type as indicated by H/C ratios in the 1.45 to 1.55 range. The movement of bitumen implied by the wide variation in extraction ratios is supported by the presence of four bitumen-impregnated tuffs in the same relatively immature section and by the generally inverse correlation between the extraction ratio and the TOC. In less organically rich rocks the redistribution of the bitumen in the source occurs at greater thermal alteration. Deroo et al, 1977, noted many "impregnations" in potential source rocks in their study of the Alberta basin and observed that their compositions, as would be expected from migration within the source, are intermediate between the normal bitumen and oils derived from the same source rock. The Russian literature is full of references to "parautochthonous bitumoids" which they interpret as indicators of redistribution of bitumen in the source rock. They have used variations of the composition and amounts of bitumens in source rocks to help quantify the amount of migration (Trofimuk et al, 1974).

Drainage Efficiency

Mature, organic-rich, oil-prone source rocks have a greater hydrocarbon drainage efficiency than less organically rich rocks. For example, in the Los Angeles basin the previous calculations indicated that $\cong$20% of the generated oil is now in the traps in the form of oil. In the Tertiary of the Gulf Coast probably <4.0% of the oil generated is now in traps[3] despite the abundance of traps and oil and the low TOCs. If water movement was more important than separate phase migration, the drainage efficiency of organically lean, oil-prone source rocks should be considerably higher than the organically rich ones. The increased drainage efficiency for the lean rocks would be expected because a given amount of hydrocarbon-saturated water would carry the same amount but a higher percentage of the originally available hydrocarbons. However, as illustrated in the contrast between the Gulf Coast and the Los Angeles basin, such a relationship is not observed.

Time of Primary Migration

At peak generation the bitumens in the source rock show their greatest compositional similarity to the genetically associated, reservoired oils as clearly shown by the various publications of Philippi and the Institut Francais du Pétrole. The compositions, of course, are not identical. However, the oils, where unaltered, commonly are composed of lighter, less viscous, less polar material than the bitumens (Tissót and Welte, 1978), a difference very compatible with the hypothesis of continuous phase flow and very much at odds with relative solubilities in aqueous solution (McAuliffe, 1979).

In some productive but rather lean basins the source rocks never reached full peak generation. Examples include the Paris basin, some parts of some California basins, parts of the Cretaceous in the Rocky Mountains, and the upper Paleozoic in parts of the Illinois and Michigan basins. The oil that was emitted prior to full peak generation in these areas came from relatively rich oil-prone rocks, not from rocks with low TOCs.

The first oils emitted from organically rich, oil-prone source rocks (and often most of the oils) commonly are undersaturated with gas, despite the great generating capacity the organic matter has for gas at high levels of thermal maturity. The northern North Sea and the Central Sumatra basin are good examples. This is a clear indication that neither water nor gas had much to do with the migration, and that the migration is triggered when the bitumen saturation in the pore space reaches a critical value.

The time of migration from the source rock of various types of compounds is consistent with the hypothesis of physical ejection from the source rock. The dikes of hydrocarbons and other bitumens in the Tertiary of the Green River basin are not only proof of the existence of separate phase migration from the source rock, but they also indicate such movement can occur early in the maturation history if certain other conditions are met. Additional thermal maturity created the immature, high pour point conventional crudes with a significant carbon preference index (CPI) which characterizes many Green River oils. In the Uinta basin the more mature oils are sourced from the deeper, less organic-rich part of the Green River shale because the richest part never reached full thermal maturity. However, separate phase migration is still indicated by the details of the stratigraphy and production history (Lucas and Drexler, 1975).

The Green River data illustrate several important points which are repeated in many areas of the world where organic-rich, oil-prone source rocks have yielded the major oil accumulations. Most of the oil is emitted during the time of increasing generation, it commonly shows some compositional signs of relative immaturity, can be easily tied to its source by chemical similarities, and it commonly is undersaturated with gas unless substantial post migration subsidence of source and/or reservoir can be demonstrated. Oils and formations in which these characteristics are common include most of the Tertiary oils of California, the Jurassic sourced oils in Cook Inlet, Alaska, Phosphoria oils in Wyoming, and most of the oils in the northern North Sea and in the Middle East. Because these oils moved out of the source rock during peak generation, a severe strain is placed on any proposed

[3]Assuming an average wt.% Total Organic Carbon of 0.64, a source rock density of 2.5, 50% of organic matter is oil prone, a 25% conversion of organic matter to oil, and that 50% of the basin is mature or post-mature source rock, then there was an average of 25 $\times 10^6$ bbl of oil generated per cu mi of sedimentary rock. Mason (1971) indicated a maximum recovery of 300,000 bbl/cu mi for the richest trends and average recovery for the Gulf Coast Tertiary of 80,000 bbl/cu mi. Assuming a 30 to 35% recovery, the in-place oil would be 1×10^6 and 0.25×10^6 bbl/cu mi respectively. Thus the ratio of oil in traps to oil generated would be 0.04 ($10^6/25 \times 10^6$) for the richest trends and 0.01 ($0.25 \times 10^6/25 \times 10^6$) for the entire basin.

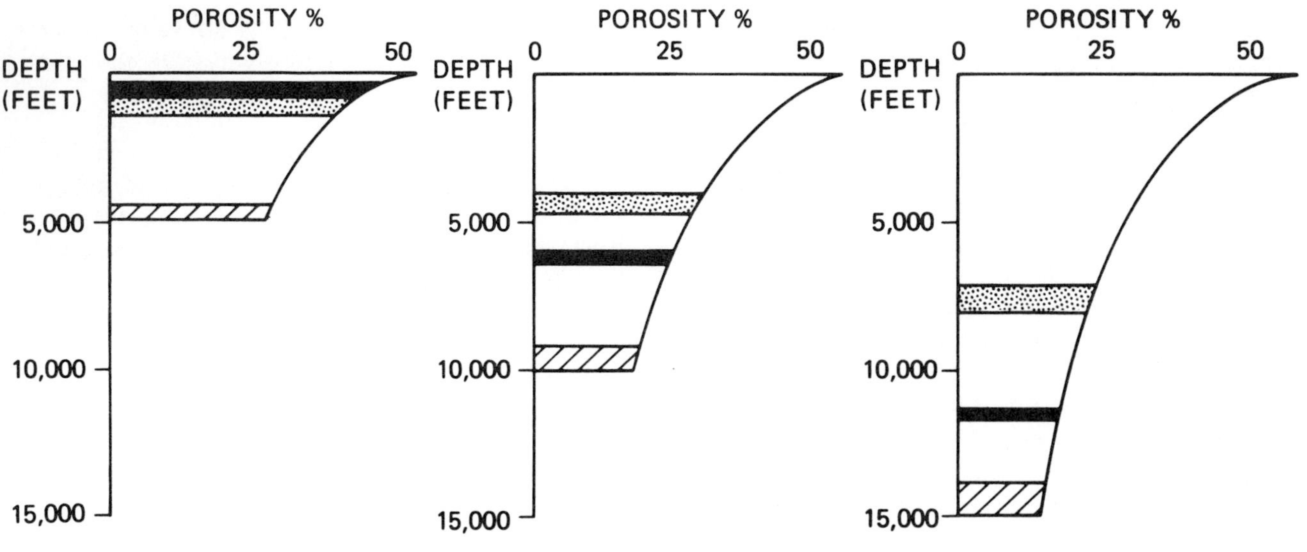

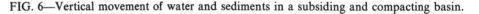

FIG. 6—Vertical movement of water and sediments in a subsiding and compacting basin.

mechanism of primary migration which lacks a strong genetic tie between generation and migration.

SOME PROBLEMS WITH MIGRATION OF OIL IN AQUEOUS SOLUTION

Some of the many and varied problems associated with the primary migration of commercial amounts of oil in aqueous solution in water derived from compaction or clay diagenesis are discussed below under four categories: (1) general—all depths; (2) shallow-immature organic matter; (3) peak generation; and (4) super hot (>300°C). Those aspects with minimum previous discussion in the literature are covered here.

General: All Depths

Water movement—The amount of water in a subsiding and compacting basin which is moving upward from a thermally mature source rock to a cooler reservoir section (the classic solution model for the Gulf Coast) is much less than often envisioned. Figure 6 shows the relative position of two water and one sediment layers as a basin with a relatively nonporous basement and a depositional interface near sea level subsides and compacts without the development of supernormal pressure (SNP). The overall water movement is into the basin and downward rather than out of the basin and upward. The water moves upward stratigraphically, but downward into higher temperatures. Assumptions inferred from Figure 6 are not as restrictive as one might think. For example, if SNP and undercompaction occur for whatever reason, the water will, on average, rise even less stratigraphically than indicated. Bypassing of SNP water along faults will only bring the average water distribution back to that shown on Figure 6. Lateral movement into the section on Figure 6 must be accompanied by lateral movement out someplace else. Bonham (1978, 1979) discussed the migration implications of such a model for the Gulf Coast Tertiary. He concluded that there are too many reservoired hydrocarbons and too little upward moving water for aqueous solution of hydrocarbons to be a significant factor in the origin of the major oil accumulations in the Gulf Coast. A variation of this model appears to have first been published by Hobson (1961) who explicitly used it as an argument against the aqueous solution theories of Baker (1960).

Coals as source rocks—It is well known that vitrinitic coals are not a source of commercial oil fields, although oily films sometimes are seen in locations of exinite enrichment. Although the generating capacity of vitrinite for liquid hydrocarbons is substantially less than for the more hydrogen-rich macerals, the much greater organic content of coal relative to a normal source rock means that for a given weight of rock, a coal often contains many times the ppm of hydrocarbons that a presumed source rock does. For example, in the Cherokee Formation of southeastern Kansas, Baker (1962) found a mean value of 129 ppm hydrocarbons for 37 samples of the presumed source rock, whereas two associated coals averaged 6,900 ppm. Coals and other vitrinitic organic matter commonly are associated with sandstone aquifers and reservoirs. If water solubility was important in forming most oil accumulations, it is difficult to understand why coal is not a prolific source, and why most coal-bearing sections are devoid of significant oil occurrence unless other types of potential source rocks are present.

Oil composition—The molecular composition of oil bears no relationship to variations in molecular solubility in water. (See McAuliffe [1979] for an in-depth discussion of this important fact.)

Micelles— The problems with micelles as an effective mechanism of primary migration have been thoroughly documented in the literature (most recently by McAuliffe, 1979). The two most condemning problems are the large amount of surfactant needed and the large size of the micelles relative to the pore system at the time of primary migration.

Total organic content of source rocks—With the exception of areas like the Tertiary of the Gulf Coast, which is underlain by 10,000+ ft (3,048+ m) of mature and post-mature source rock, rocks with low TOC are not the source of major oil accumulations. Because rocks with TOC <0.2 wt.% nearly always contain more than enough hydrocarbons to saturate the pore water (Table 1), one might expect, if transportation of hydrocarbons in water solution is an important factor regarding oil accumulations, for there to be virtually no such thing as a nonsource rock. At peak generation, where the oil-prone kerogen has emitted most of its hydrocarbons into the pore system, low TOC rocks should saturate their pore water as fully as high TOC rocks. Thus low TOC rocks should be as effective source rocks as high TOC rocks—if solution in water was the primary migration mechanism. But they are not.

The first step in primary migration occurs when generation has proceeded to the point where the absorption capacity of the kerogens is exceeded and bitumen is ejected into the pore system. At this point in the history of an effective source rock there are two to three orders of magnitude more bitumen in the pore system than can dissolve in the water even if we ignore "bound" water (Table 1). It is probably the inability of low TOC rocks to provide high concentrations that precludes significant primary migration from them.

Cap rocks as oil source rocks—Many of the best source rocks in the world are transgressive marine shales which are both cap rock and source rock for oils in reservoirs directly underlying them (see Fig. 2). It is unlikely that enough water could have penetrated these shales from above to form the oil accumulations directly below by a solution mechanism.

Secondary migration—Some writers (Hodgson, 1978; Roberts, 1978) have postulated that hydrocarbons, or other organics, are carried in solution in both the source and carrier rocks and are subsequently precipitated in the trap where the water carrier is presumed to enter the overlying shale. If the mechanism of precipitation in the trap was effective, the distribution of oil and water in a reservoir should not be so completely controlled by capillary effects caused by the pore geometry and the buoyancy of the oil column.

Shallow-Immature Organic Matter

The discovery of enough hydrocarbons in Holocene sediments (Smith, 1954; Kidwell and Hunt, 1958) to form commercial oil fields if the hydrocarbons could be sufficiently concentrated added substantial support to the theory of early and shallow hydrocarbon migration and accumulation. This is, of course, the time in a sediment's history when a maximum amount of water is moving about stratigraphically. The theory of early migration and accumulation that arose from the combination of these two facts is still with us (Wilson, 1975), in part because many notable petroleum geologists (Weeks, 1961; Hedberg, 1964) gave it written support prior to the full advent of modern organic geochemistry. Many explicit arguments have been developed in the last ten years against the theory of early migration and accumulation. Three are briefly outlined below.

Access to mature source rocks—There are no large oil fields whose reservoirs do not have access to mature source rocks. The Cretaceous of the Williston basin and the upper Pliocene of the Los Angeles basin are clear-cut examples of all the factors needed to make oil fields—except access to thermally mature source rocks.

Bitumen composition—The hydrocarbon composition of immature bitumens is very different from the composition of even the most immature oils. All oils contain a preponderance of thermally formed compounds and the CPIs of oil commonly are not reached in bitumens with vitrinite reflectances of ≤ 0.60. Although it has been argued (Cordell, 1972) that solution of hydrocarbons will not reflect a high CPI in the bitumen source, geochemical evidence does not support this concept. McAuliffe (1979) showed data from a variety of sources which demonstrated that, when water is in equilibrium with crude oils, the concentrations of hydrocarbons found in the water phase are controlled by both the water solubility of each hydrocarbon and its mole fraction concentration in the crude oil. Because n-paraffins which differ by only one carbon number have similar solubilities, the hydrocarbons in water in equilibrium with a bitumen with a high CPI should have essentially the same CPI as the bitumen. Observations of the subsurface support the experimental data. Often slight irregularities in the n-paraffin distribution in an oil will be duplicated in the source (Tissót et al, 1978). In addition, the pristane/phytane ratio commonly is a source indicator which it would not be if primary migration occurred in solution and solution effects smoothed out the original distribution of hydrocarbons with similar solubilities.

Hydrocarbon solubilities—Hydrocarbon solubilities

Table 3. Example of Gulf Coast Tertiary Solubility Requirements

Oil Volume (Barrels)	Water Volume (Barrels)	Water Volume (ϕ Units)	Average Solubility (ppm)
6 x 10^6	26 x 10^9	33.3	230
6 x 10^6	7.8 x 10^9	10.0	770
6 x 10^6	3.9 x 10^9	5.0	1,540

are very low and possible precipitation mechanisms are quantitatively inadequate at shallow depths and low temperatures to account for oil accumulations (McAuliffe, 1979; Price, 1976). In addition, the variations in solubilities with molecular type and weight are not reflected in crude oil composition (McAuliffe, 1979).

Peak Generation

As indicated earlier, many reasons exist for believing that primary oil migration and peak generation occur essentially simultaneously. Because the hydrocarbon compositions of the bitumens at peak generation are most similar to the reservoired oils, some of the arguments against aqueous solution do not apply at peak generation. However, the basic problems of low absolute solubilities and not enough water moving through the source rock remain.

Water availability—Little compaction or clay dehydration water is available at peak generation in many oil-rich areas, particularly those with carbonate source beds.

In several oil-rich late Tertiary basins where peak generation occurs $\geq$150°C, the shale porosity is low, SNP is minimal, and most of the compaction and clay dehydration water moved out of the source rock prior to peak generation (ex. Los Angeles basin).

If a section is undercompacted owing to the inability of the water to escape from a thick shale section, the stratigraphically upward movement of water will be even less than depicted in Figure 6, and the absolute downward movement of water in the basin will be greater.

Gulf Coast Tertiary—If solubility in compaction or clay-dehydration water is ever a viable mechanism of primary migration, the Gulf Coast Tertiary is a likely place because there are 10,000+ ft (3,048+ m) of mature post-mature source shales underlying the reservoir section. But in the Gulf Coast Tertiary there is not enough water moving upward from the source, even if we forget about the implications of Figure 6. Outlined below are some approximate mass balance requirements for a highly productive trend in the Gulf Coast Tertiary with an average oil in place estimate of one million bbl per cu mi (Mason, 1971).

Given: (1) Sedimentary prism, one mi sq (2.6 sq km), six mi (9.7 km) thick.

(2) 6×10^6 bbl of oil in place in traps in the upper part of the six cu mi (25 cu km) prism.

(3) Source rock of oil was the entire lower three cu mi (12.5 cu km) of sediment where the organic matter is mature or post mature.

(4) Oil left source in aqueous solution.

(5) *All* of the solubilized oil is in the traps.

Table 3 shows the oil solubilities necessary when varied amounts of water move out of the three cu mi (12.5 cu km) of source. The required solubility (ppm) $\cong$ oil volume/10^6 water volume.

Even with assumptions which drastically minimize the necessary solubilities, the required solubilities for a realistic water loss are too high. The loss of 5 ϕ units of porosity is roughly equal to the average water loss if a source section of SNP shale three cu mi (12.5 cu km) thick was collapsed to normal pressure and all of the released water driven upward through the overlying reservoirs (Bonham, 1980). Numbers like these and larger forced Price (1976, 1978) to call on temperatures of 300 to 350°C (572 to 662°F) to develop the necessary solubility.

Super Hot ($\geq$300°C)

Price (1976, 1977, 1978) has eloquently argued that most of the oil accumulations of the world originated from source rocks that were heated in the subsurface to temperatures in excess of 300°C. Some of the reasons for the general lack of acceptance of his theory are briefly discussed here.

Bypassing water—Essential to Price's hypothesis is the upward movement of considerable water into cooler and more saline regimes. Normal compaction in a subsiding basin does not accomplish this (Fig. 6). It is necessary to postulate water movement along deep-seated faults that extend into the reservoir section and bypass water around the normally compacting section. Such bypassing undoubtedly occurs and specific examples can be spectacular (Price, 1976). The question is whether it is quantitatively significant. I am unaware of any published numbers purporting to quantify the amount of bypassed water. The previously discussed model of Figure 6 suggests it is not very high and, in fact, there is no *a priori* reason for it to be significant.

Barker (1977) argued that: (1) the persistence of SNP limits the amount of bypassed water; (2) the existence of low grade metamorphic minerals in hydrous phases suggests the water staying put rather than moving upward at the temperatures required by Price; and (3) loss of only 6.6% of the pore water at 300°C and 30,000 ft would reduce a lithostatic pressure to hydrostatic. Although Price (1977) rebutted Barker's statements, both the increasing amounts of chlorite (wt.% water ≅12.0%) observed at depth in Gulf Coast well profiles, and the persistence of SNP at depth suggest that we need more definite evidence before the hypothesis of significant amounts of far ranging, upward moving, super hot water can be accepted.

Location of exsolution—The Gulf Coast Tertiary is a reasonable test area for Price's migration mechanism because continuous phase migration is difficult to accept owing to the low TOCs, and the fact that the required temperatures of 300°+C exist near the base of part of the sedimentary column. However, the postulated temperatures exist 10,000 to 20,000 ft (3,048 to 6,096 m) below the reservoirs that contain the oil accumulations. What happens in the intervening section? Because the increase in solubility is apparently exponential with temperature, one thing that happens is that any hot, saturated water moving upward through the shale in the 10,000 to 20,000 ft (3,048 to 6,096 m) below the reservoirs will start to precipitate hydrocarbons within the shale. Most of the hydrocarbons will probably be precipitated within the massive shale long before the reservoir carrier beds are reached (Bonham, 1979) and, in part, in shales that are approximately at the thermal maturity of conventional peak generation. Thus, the basic problem of moving the hydrocarbons from the pore system of mature source shales to the reservoir remains.

Bitumens at high temperatures—To demonstrate bitumen availability at high temperatures, Price (1977, 1978) examined many samples from deep wells and established to his satisfaction that uncommonly high amounts of bitumens occasionally exist at maximum temperatures in excess of 200°C. However, innumerable temperature (depth) versus wt.% bitumen/TOC plots exist in the literature which show severe, abrupt, and permanent drops in the wt.% bitumen/TOC ratio at maximum paleotemperatures well below 200°C. Many company geologists also have made their own unpublished plots of a similar nature which accompanied the demise of the hope for commercial oil accumulations in a given well. In most cases there simply are not enough bitumens in the rocks at >200°C to source commercial oil accumulations. In addition, no evidence has been presented to show that those bitumens which do exist at >200°C ever include the biological markers (ex., porphyrins, steranes, triterpanes) which are characteristic of most oil accumulations with API° ≤40°.

The writer has no quarrel with Price's observations that substantial amounts of indigenous bitumen occasionally exist in rocks with higher paleotemperatures than normally expected, although others (Barker, 1977) do. However, wherever the writer has observed this phenomenon it always has occurred in very tight rocks. Thus, the bitumen is there simply because it was not (and is not) available for migration. It is trapped in a minute, high-pressure pocket from which it can neither migrate nor, because of pressure-volume restrictions, further mature by cracking to gas and a carbon-rich residuum. Less dramatic retardation of thermal maturation is occasionally seen in very tight calcareous or siliceous rocks at shallower depths, as noted by Bostick and Alpern (1977).

Basin temperatures, past and present—The source rocks in many basins of the world never reached the 300 to 350°C temperatures required by Price's (1978) model. The Cretaceous basins of the Rocky Mountains, many of the Paleozoic basins of the Mid-Continent, and some of the California basins are obvious examples in the United States. Even in the Los Angeles basin where the stratigraphically deepest source rocks may approach 300°C near the center of the basin, the relative immaturity of the oils clearly indicates that migration of the reservoired oils occurred prior to such temperatures being reached. In some basins the paleogeothermal gradient was clearly higher than the present gradient (ex., Douala basin, Cameroon); however, in most basins the time-temperature-maturation relationships suggest that no significant change in gradient has occurred during the basins' history (Tissót and Welte, 1978).

Price stressed the role of faults and fractures in migration, developed needed solubility data at high pressures and temperatures, and pointed out that we have not adequately evaluated pressure effects in generation. The writer thanks Price, particularly for his provocative assaults on our orthodoxy; however, the writer does not believe that Price has solved the enigma of commercial migration, either in the Gulf Coast Tertiary or elsewhere.

TWO PROBLEM AREAS

Origin of Petroleum in the Gulf Coast Tertiary

Despite extensive literature on the subject, the origin of the oil in the Tertiary of the Gulf Coast remains an enigma. It is unlikely that separate phase migration of oil from the pores adjacent to the generating kerogen can be the dominant primary migration mechanism for the Tertiary of the Gulf Coast. There simply appears to be too little kerogen (TOCs ≅0.3 to 1 wt.%) and too much pore space (10 to 20%). However, as indicated earlier, if one migration mechanism is inoperative, others will automatically increase in both relative and absolute importance, as nature makes less dense and less viscous hydrocarbons available for migration. Several alternative migration mechanisms which might be applicable to the Gulf Coast Tertiary have been presented (during the migration course and symposium at the 1978 AAPG convention). These included: (1) solution in deep, hot >300°C) water (Price, 1978); (2) solution near peak generation (Roberts, 1978; Cordell, 1978); (3) diffusion (Hinch, 1978); and (4) transportation with methane (Hedberg, 1978). The reasons for not accepting either Price's thesis that the Gulf Coast petroleums originated at temperatures ≥300°C or the more conventional solution mechanisms, were previously discussed. Hinch's diffusion hypothesis is fascinating, in part because of the use he makes of the concept of bound water, an area where there is a need for more definitive data. However, the writer maintains reservations towards a diffusion hypothesis at this time for the following reasons: (1) diffusion is basically a mode of dispersion rather than of concentration; (2) diffusion coefficients of the heavier petroleum molecules are very low; (3) the multiple reservoirs of the structural traps in the Gulf Coast do not fill from the bottom up; (4) diffusion has low rates of mass transfer over substantial distances, particularly in shales. Thus, one can legitimately question whether diffusion was important in the emplacement of the large oil fields in the Pleistocene sandstones at Eugene Island offshore Louisiana. It is clear from such occurrences as the concentration of the oil in highly faulted intrusive salt domes (Spillers, 1965) and the apparently capricious alternation of oil and water in reservoirs interbedded with shales of similar source characteristics that many of the faults must have been conduits of petroleum at certain times in their history. Insufficient water moved up the faults to form the petroleum by exsolution in the reservoirs (Bonham, 1980). Therefore, the oil probably moved up the fault zones as a continuous oil phase or as the solute in a gas phase. The latter possibility is attractive because recent data on hot, deep, formation fluids in the Gulf Coast Tertiary indicate that they are saturated, and may be over-saturated with methane. In addition, Russian studies (Zhuze et al, 1963; Zhuze and Bourova, 1977) indicate that from both a compositional and quantitative viewpoint, solution of oil in gas is a much more viable migration mechanism than solution in water. However, we are still left with the problem of how the hydrocarbons moved from the source shales to the faults in the volume that they have. More data on the details of the faults and on the distribution of pressure, temperature, fluid, and rock properties near them might help resolve what seems to remain the enigma of primary migration in the Gulf Coast Tertiary. However, it is important to remember that the thick Mesozoic-Tertiary delta systems are unique among the oil basins of the world in many ways, including widespread SNP and

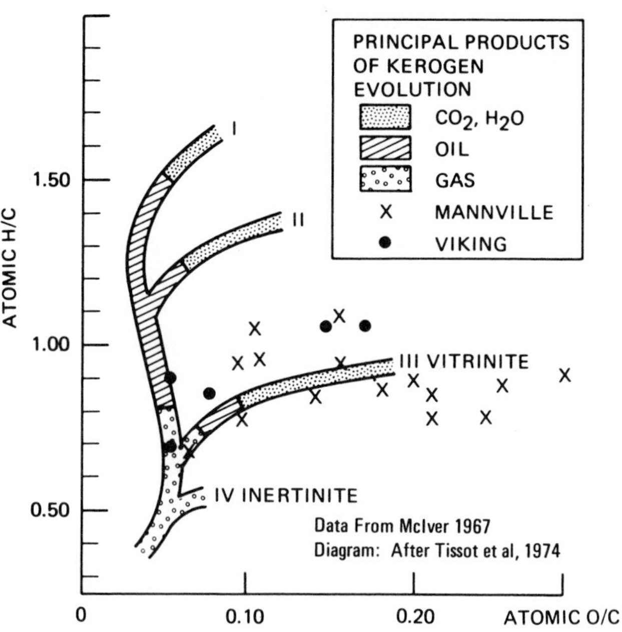

FIG. 7—H/C-O/C ratios of kerogens of lower Cretaceous shales, Alberta, Canada (after Tissót et al, 1974).

10,000+ ft (3,048+ m) of mature post-mature source rocks beneath the reservoirs, and that one of those unique ways is very apt to be the main mechanism of primary migration.

Origin of Oil Sands of the Lower Cretaceous of Alberta

Introduction—Anyone who theorizes the primary migration of petroleum must address the problem of the origin of the trillion bbls of heavy oil in the Lower Cretaceous of Alberta. After several decades of discussions, it has finally been agreed among most geochemists—and some geologists—that the oil in the Lower Cretaceous sandstones of Alberta originated within the Lower Cretaceous section (Vigrass, 1968; Deroo et al, 1977). Unfortunately, there are no chemical comparisons between the oil sandstones and the bitumens of the organic rich but volumetrically quite limited Jurassic Nordegg shale. Nevertheless, in recent years most of the published discussions have accepted the Lower Cretaceous origin, hotly debated whether the composition of the oil is more dependent on original immaturity or degradation, and completely ignored the compelling question of how this immense amount of oil got to its present location (Montgomery et al, 1972a; Montgomery et al, 1974b; Deroo et al, 1974; Deroo et al, 1977; George et al, 1977). Exceptions to the latter comment are the micellar solution theory of Vigrass (1968) and the molecular solubility theory of Hunt (1977). These will be briefly evaluated later in this paper. It is probable that the virtual ignoring of the question of the mechanism of primary migration of the Lower Cretaceous heavy oils simply reflects the intuitive recognition that all of the more popular migration theories are not applicable.

Continuous phase migration—Continuous phase flow of oil out of the source rock is probably not the answer. To the writer's knowledge, no one has identified an organically rich, highly oil-prone source rock in the Lower Cretaceous. Deroo et al (1977) were puzzled about how a section with TOCs ≅1 to 3%, but with organic matter of dominantly terrestrial origin, could source such large accumulations of oil. Data computed from McIver (1967) and plotted on an H/C–O/C kerogen maturation diagram from Tissót et al (1974) confirm the generally poor quality of the organic matter (Fig. 7). In addition, much of the Lower Cretaceous has not yet reached thermal maturity. If such a source section could generate and migrate a trillion bbls of petroleum by continuous phase flow, the world should be awash in oil.

Compaction water—Solution by compaction waters (Vigrass, 1968; Hunt, 1977) offers no better explanations. Where did all the micelles come from—and go to—(McAuliffe, 1979)? Movement by molecular solution means deriving 40 cu mi (167 cu km) of oil from ≅50,000 cu mi (209,000 cu km) of shale[4]. This is equivalent to 800 ppm (vol.) of oil emitted from the rock. Allowing a liberal 10% loss of porosity to be the compaction-water loss during oil migration, and assuming all the solubilized oil to be trapped with negligible loss to degradation, an average solubility ≅8,000 ppm is needed. Thus, solution in compaction—or clay dehydration water—is not an acceptable answer either.

Meteoric water—Currently, the only migration mechanism that appeals to the writer regarding the origin of the Alberta oil sandstones is the action of the meteoric water over a period of ≅50 million years. This proposed solution to the problem of the origin of the heavy oil sandstones of Alberta is implicit in several papers (Hitchon, 1969, 1974; Hodgson et al, 1964), but not explicitly applied with numbers and geology.

The argument for the dominance of meteoric water is indirectly based on the bankruptcy of other possible mechanisms, but is directly based on the present subsurface water flow patterns in Alberta, the reasons for them, and the probable persistence of very similar patterns since Laramide time.

Only a brief summary of the basic geology and arguments can be given here. The reader is referred to Vigrass (1968) for an excellent description of the oil deposits and their geologic setting. Figures 8 and 9 show the areal distribution of the Athabasca and Peace River heavy oil deposits with respect to the hogback of Paleozoic carbonates which strongly influenced the distribution of the thickness and facies of the basal Cretaceous deposits. It is almost certainly not a coincidence that Athabasca lies directly east of a broad sand-filled channel which breached the subdued ridge of Paleozoic carbonates. The possibilities of major long distance, focused migration within the basal Cretaceous are obvious. Undoubtedly much compaction water moved towards and through Athabasca during compaction of the Lower Cretaceous, but that compaction water was only a harbinger of the large volume of water which followed similar paths from early Tertiary time to the present. Figures 10 and 11 are hydraulic head cross sections through the Athabasca and Peace River deposits by Hitchon (1969, 1974) which show that the two oil accumulations are currently within the ground-water regime. In fact, the unconformity and the basal Cretaceous sandstones are potential sinks, and the break in the buried hogback of Paleozoic strata is essentially a straw which drains the fluids entering the sink into Athabasca. Although the hydraulic head contours and the flow lines generally indicate water flow towards the unconformity area from both above and below, the permeability of the Cretaceous clastic deposits is so much higher than that of the Paleozoic carbonates and evaporites that the eastward flowing water near the unconformity can be presumed to be virtually all meteoric water that descended from the present erosion surface. The salinity of <500 mg per liter for the waters associated with the Peace River deposits (Deroo et al, 1977) is consistent with this hypothesis.

It is very probable that similar water flow patterns have existed since early Tertiary time when the Rocky Mountains began to rise. Certainly, the unconformity and the distribution of the Lower Cretaceous sandstones were identical. The eastern slope of the erosion surface which is an important factor in the persistence of the necessary hydraulic head also has been present in varying degrees since the early Tertiary. Several thousand feet of Cretaceous-Tertiary rocks have been eroded from the plains area west of the oil sandstones since early Tertiary time. Much of this section consisted of coarse clastic rocks which would have easily imbibed rain water, limited surface runoff, and helped develop the hydraulic head necessary to move meteoric water through the finer grained, underlying Cretaceous clastic sediments. Everything considered, it is difficult not to believe that the gross patterns of subsurface water movement which exist today east of the mountain front have persisted for tens of millions of years. Table 4 shows how the concept of extracting the oil in the oil sandstones from their source by me-

[4]The volume of shale from which the oil deposits in the Lower Cretaceous were derived is, of course, not known. Hitchon (1968) indicated ≅55,000 cu mi (229,000 cu km) of Lower Cretaceous shale exist between the eastern edge of the disturbed belt and the erosional edge to the east and between 49° and 60° N lat. If we arbitrarily assume that oil-contributing shale in the disturbed belt is slightly less than the shale elsewhere which was not a contributor, a figure of 50,000 cu mi (209,000 cu km) of potentially contributing Lower Cretaceous shale seems a reasonable approximation.

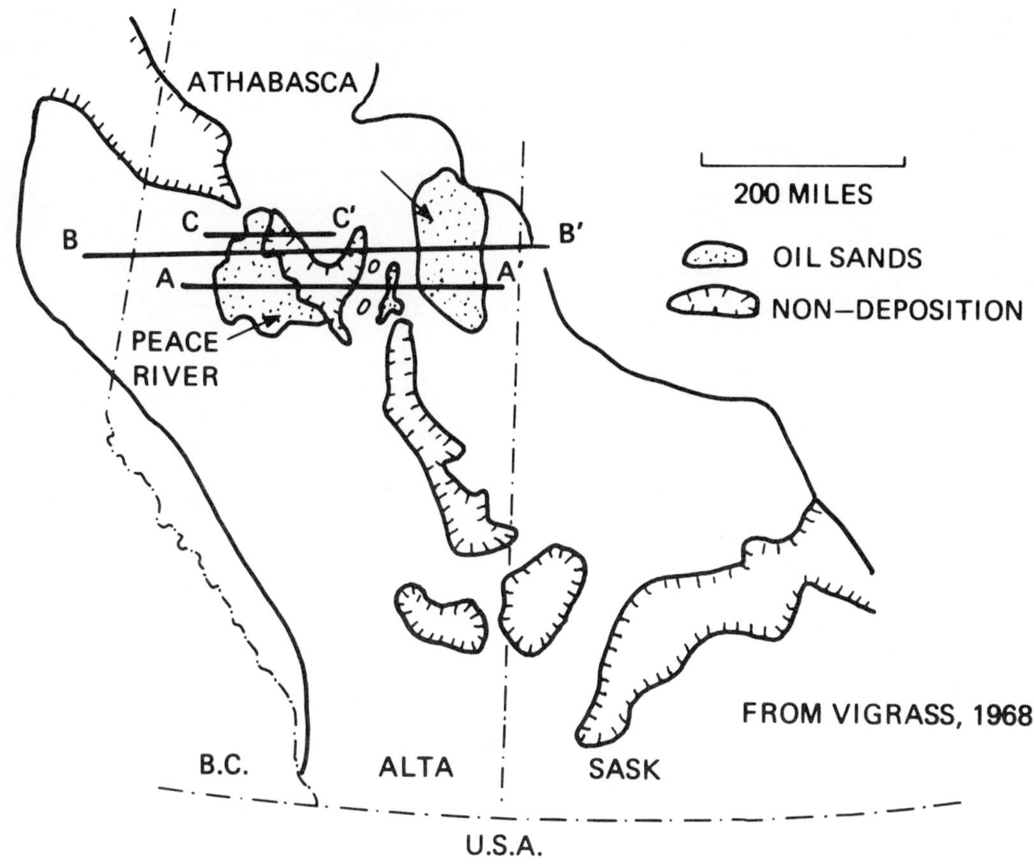

FIG. 8—Lower Mannville (basal Cretaceous) depositional framework, western Canada (from Vigrass, 1968).

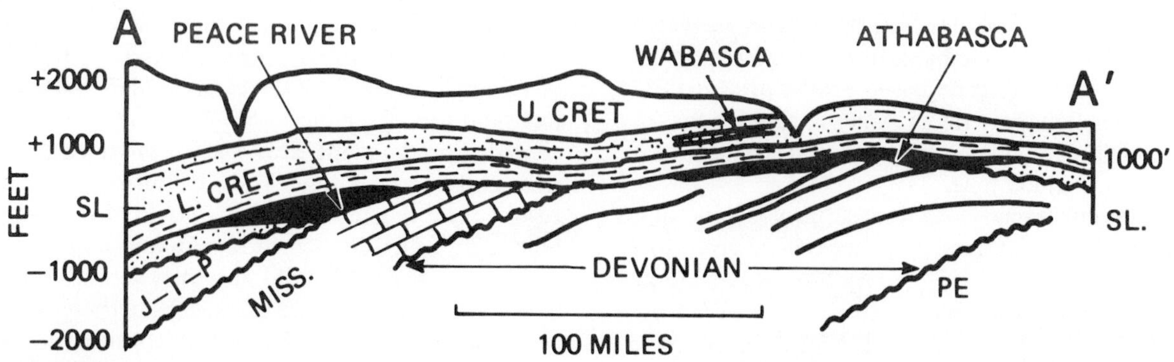

FIG. 9—Diagrammatic east-west cross section showing geologic setting of Lower Cretaceous oil sands, Athabasca-Peace River area (line A-A′ from Fig. 8).

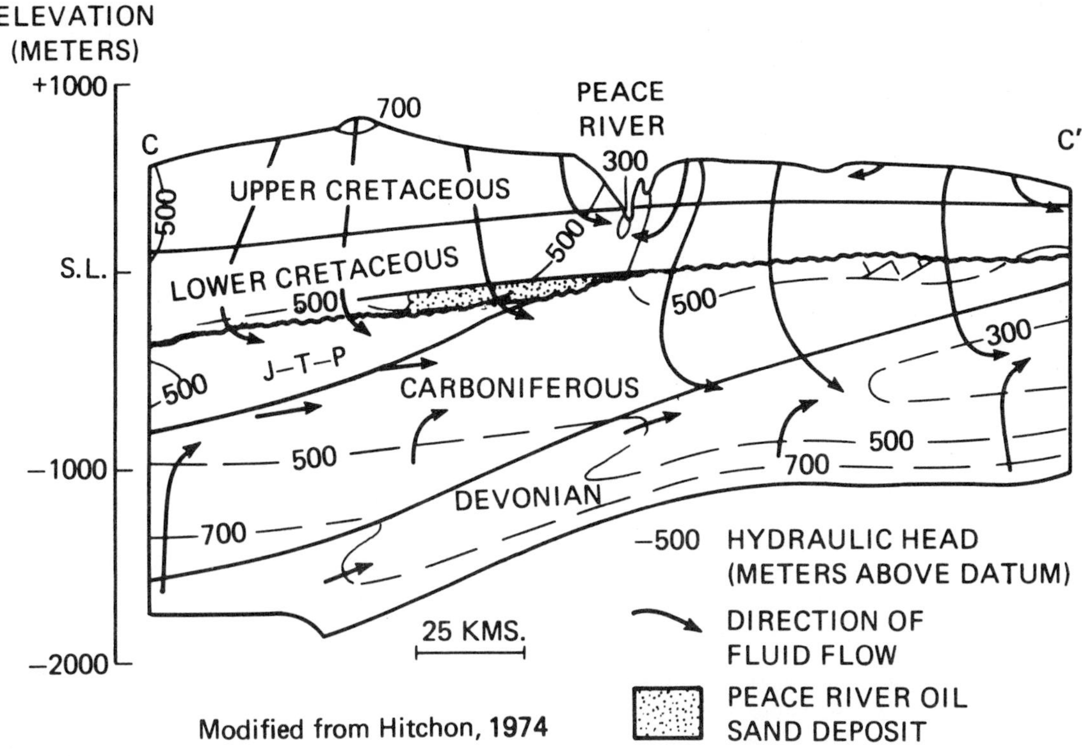

FIG. 10—Hydraulic head cross section, Peace River oil-bearing sandstone deposit (line C-C′ from Fig. 8; figure after Hitchon, 1974).

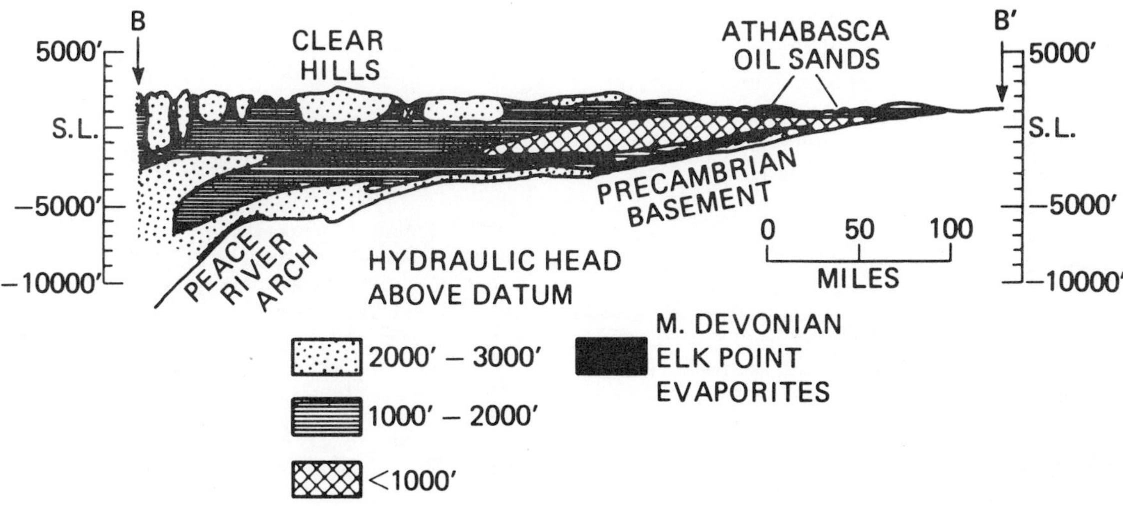

FIG. 11—Hydraulic head distribution, east-west cross section, northern Alberta (line B-B′ from Fig. 8; figure after Hitchon, 1969).

Table 4. Mass Balance Calculations Regarding Origin of Oil Sands of Lower Cretaceous of Alberta from Meteoric Water

1. Oil in place $\cong 10^{12}$ bbls $\cong$40 cu mi (167 cu km).
2. Surface drainage area $\cong 300 \times 10^3$ sq mi (777×10^3 sq km).
3. Assume 30 inches (76 cm) of rain for 50×10^6 years.
4. Total rain volume per sq mi (2.59 sq km) $\cong$30 inches/year $\times 16 \times 10^{-6}$ mi/in. $\times 50 \times 10^6$ years $\times$ 1 sq mi $\cong 24 \times 10^3$ cu mi (99×10^3 cu km).
5. Total rainfall on drainage area $\cong 7.2 \times 10^9$ cu mi (30×10^9 cu km).
6. Assuming 1% of rainfall passes through oil sandstones by subsurface flow, available water $\cong 72 \times 10^6$ cu mi (300×10^6 cu km).
7. $\therefore$ oil/water $\cong 40/72 \times 10^6 \cong 0.56 \times 10^{-6}$ or 0.56 ppm.
8. Could account for oil sands with 3 ppm solubility and 20% precipitation in reservoir.

teoric water flow over a period of $\cong 50 \times 10^6$ years simplifies the mass balance situation. As true of most of the numerical examples in this paper, the actual numbers can be varied substantially without changing their implication. Emplacement by meteoric water can handle the mass balance requirements of the origin of the oil sandstones whereas other migration mechanisms cannot.

It is possible that Cretaceous coals might have contributed to the oil in the oil sandstones by means of the meteoric water mechanism. The writer refers this interesting question (and the more general question of the compatibility—or lack thereof—of the chemical composition of the oil with an origin from meteoric water) to the geochemists.

SUMMARY AND CONCLUSIONS

Most major oil accumulations of the world left the source rock in a continuous oil phase. This conclusion is directly based on microscopic observations, log characteristics, and producing characteristics of fractured shale reservoirs; and is indirectly based on the 2 to 3 order of magnitude discrepancy between the amount of hydrocarbons that can be carried in solution by compaction and clay dehydration water and the amount which actually moved from source to traps. Continuous phase primary migration is also supported by bitumen redistribution in source rocks, the bitumen content of the pores at peak generation, the high TOCs of most source rocks (2 to 10 wt.%), the usual greater drainage efficiency of organic-rich source rocks, and the genetic tie between peak generation and primary migration.

When bitumen concentrations in the rock are too low for continuous phase flow to exist, other migration mechanisms, which are always operative, will increase in both absolute and relative intensity. Solution of oil in gas may become significant in thick Tertiary deltaic systems and meteoric water may be a surprising contributor in some very specific geologic settings. However, it is unlikely that solution of oil in water from compaction or from clay dehydration has much to do with many of the major oil accumulations of the world.

REFERENCES CITED

Albrecht, P., M. Vanderbroucke, and M. Mandengue, 1976, Geochemical studies on the organic matter from the Douala basin (Cameroon), I. Evaluation of the extractable organic matter and the formation of petroleum: Geochim. et Cosmochim. Acta, v. 40, p. 791-799.

Auldridge, L., 1977, Russia drives to sustain big gains in oil output: Oil and Gas Jour., v. 75, no. 53, p. 69-72.

Bajor, M., M. H. Roquehort, and B. M. Van Der Weide, 1969, Transformation de la matiere organique sedimentaire sous l'influence de la temperature: Bull. Centre Recherches Pau - SNPA, v. 3, no. 1, p. 113-124.

Baker, D. R., 1962, Organic geochemistry of Cherokee Group in southeastern Kansas and northeastern Oklahoma: AAPG Bull., v. 46, p. 1621-1642.

Baker, E. G., 1960, A hypothesis concerning the accumulation of sediment hydrocarbons to form crude oil: Geochim. et Cosmochim. Acta, v. 19, p. 309-317.

Barbat, W. F., 1958, The Los Angeles basin area, California, *in* L. G. Weeks, ed., Habitat of oil: AAPG, p. 62-77.

Barker, C., 1977, Aqueous solubility of petroleum as applied to its origin and primary migration; discussion: AAPG Bull., v. 61, p. 2146-2149.

Bonham, L. C., 1978, Migration of hydrocarbons in compacting basins (abs.): AAPG Bull., v. 62, p. 498-499.

——— 1980, Migration of hydrocarbons in compacting basins: (this volume).

Bostick, N. H., and B. Alpern, 1977, Principles of sampling preparation and constituent selection for microphotometry in measurement of maturation of sedimentary organic matter: Jour. Microscopy, v. 109, pt. 1, p. 41-47.

——— et al, 1978, Gradients of vitrinite reflectance and

present temperature in the Los Angeles and Ventura basins, California: Symposium on low temperature metamorphism of kerogen and clay minerals, Pacific Section SEPM, p. 65-96.

Brooks, J., and B. Thusu, 1977, Oil-source rock identification and characterization of the Jurassic sediments in the northern North Sea: Chem. Geology, v. 20, p. 283-294.

Cartmill, J. C., and P. A. Dickey, 1970, Flow of a disperse emulsion of crude oil in water through porous media: AAPG Bull., v. 54, p. 2438-2447.

Cordell, R. J., 1972, Depths of oil origin and primary migration: A review and critique: AAPG Bull., v. 56, p. 2029-2067.

——— 1978, Migration pathways in compacting clastic sediments: paper, AAPG Res. Symposium on Problems of Petroleum Migration, Ann. Mtg. (Oklahoma City).

Demaison, G. J., and G. T. Moore, 1979, Anoxic environments and oil source bed genesis: Organic Geochemistry (in press).

Deroo, G., et al, 1974, Geochemistry of the heavy oils of Alberta, *in* Oil sands, fuel of the future: Canadian Soc. Petroleum Geologists, Memoir 3, p. 148-167.

——— et al, 1977, The origin and migration of petroleum in the western Canadian sedimentary basin, Alberta: Canada Geol. Survey Bull. 262, p. 1436.

Dickey, P. A., 1975, Possible primary migration of oil from source rocks in oil phase: AAPG Bull., v. 59, p. 337-345.

Dow, W. G., 1974, Application of oil-correlation and source rock data to exploration in Williston basin: AAPG Bull., v. 58, p. 1253-1262.

Durand, B., and J. Espitalié, 1976, Geochemical studies on the organic matter from the Douala basin (Cameroon), II. Evolution of kerogen: Geochim. et Cosmochim. Acta, v. 40, p. 801-808.

Espitalié, J., et al, 1977, Methode rapide de caracterization des roches meres de leur potential petrolier et de leur degre d'evolution: Inst. Francais Petrole Rev., v. 32, p. 23-42.

Evans, C. R., D. K. McIvor, and K. Magara, 1975, Organic matter, compaction history and hydrocarbon occurrence, Mackenzie Delta, Canada: Proc. 9th World Petroleum Cong., v. 2, p. 149-157.

Gardett, P. H., 1971, Petroleum potential of Los Angeles basin, California, *in* Future petroleum provinces of the United States: AAPG Memoir 15, v. 1, p. 298-308.

George, A. E., et al, 1977, Simulated geothermal maturation of Athabasca bitumen: Bull. Canadian Petroleum Geology, v. 25, p. 1085-1096.

Hedberg, H. D., 1964, Geologic aspects of origin or petroleum: AAPG Bull., v. 48, p. 1755-1803.

——— 1978, Methane generation and petroleum migration: (this volume).

Hinch, H. H., 1978, The nature of shales and the dynamics of hydrocarbon expulsion in the Gulf Coast Tertiary section: (this volume).

Hitchon, B., 1968, Rock volume and pore volume data for plains region of Western Canada sedimentary basin between latitudes 49° and 60°N: AAPG Bull., v. 52, p. 2318-2323.

——— 1969, Fluid flow in the western Canada sedimentary basin: 1. Effect of Topography: Water Resources Research, v. 5, no. 1, p. 186-195.

——— 1974, Application of geochemistry to the search for crude oil and natural gas *in*, A. A. Levinson, ed., Introduction to exploration geochemistry: Applied Publishing, Ltd., p. 509-545.

Hobson, G. D., 1961, Problems associated with the migration of oil in "solution": Inst. Petroleum Jour., v. 47, no. 449, p. 170-173.

Hodgson, G. W., 1978, Origin of petroleum—in-transit conversion of organic compounds in water (abs.): AAPG Bull., v. 62, p. 522.

——— B. Hitchon, and K. Taguchi, 1964, The water and hydrocarbon cycles in the formation of oil accumulations: Recent Research in the Field of Hydrosphere, Atmosphere, and Nuclear Geochemistry, p. 217-242.

Hoots, H. W., A. L. Blount, and P. H. Jones, 1935, Marine oil shale, source of oil in Playa del Rey Field, California: AAPG Bull., v. 19, p. 172-205.

Hunt, J. M., 1961, Distribution of hydrocarbons in sedimentary rocks: Geochem. et Cosmochim. Acta, v. 22, p. 37-49.

——— 1963, Composition and origin of the bitumens of the Uinta basin: Oil and gas possibilities of Utah, reevaluated: Utah Geol. and Mineralog. Survey Bull. 54, p. 249-274.

——— 1977, Ratio of petroleum to water during primary migration in western Canada basin: AAPG Bull., v. 61, no. 3, p. 434-435.

Kidwell, A. L., and J. M. Hunt, 1958, Migration of oil in recent sediments of Pedernals, Venezuela: *in* L. G. Weeks, ed., Habitat of oil: AAPG, p. 790-817.

Lucas, P. T., and J. M. Drexler, 1975, Altamont-Bluebell: a major and overpressured stratigraphic trap, Uinta basin, Utah: Rocky Mountain Assoc. Geologists, Symposium (1975), p. 265-273.

Magara, K., 1978a, The significance of the expulsion of water in oil-phase primary migration: Bull. Canadian Petroleum Geology, v. 26, p. 123-131.

——— 1978b, Primary migration agents (abs.): AAPG Bull., v. 2, p. 538.

Mallory, W. W., 1977, Fractured shale hydrocarbon reservoirs in southern Rocky Mountain basins: Exploration Frontiers of the Central and Southern Rockies (Rocky Mountain Assoc. Geologists Symposium), p. 89-94.

Mason, B. B., 1971, Summary of possible future petroleum potential of Region 6, Western Gulf basin: AAPG Memoir 15, p. 805-812.

McAuliffe, C. D., 1979, Oil and gas migration—chemical and physical constraints: AAPG Bull., v. 63, p. 761-781.

McIver, R. D., 1967, Composition of kerogen—clue to its role in the origin of petroleum: Proc. 7th World Petroleum Cong., v. 2, p. 25-36.

Meinchein, W. G., 1959, Origin of petroleum: AAPG Bull., v. 43, p. 925-943.

Meissner, F. F., 1976, Abnormal electrical resistivity and fluid pressure in Bakken formation, Williston basin, and its relation to petroleum generation, migration and accumulation (abs.): AAPG Bull., v. 60, p. 1403-1404.

——— 1978, Petroleum Geology of the Bakken Formation, Williston basin, North Dakota and Montana: Williston Basin Symposium, Montana Geological Society, p. 207-227.

Miller, R. J., and P. F. Low, 1963, Threshold gradient for water flow in clay: Soil Sci. Soc. America Proc., v. 27, p. 605-609.

Momper, J. A., 1978, Oil migration limitations suggested by geological and geochemical considerations: AAPG Course Note Series 8, p. B1-B60.

Montgomery, D. S., R. C. Banerjee, and H. Sawatzky, 1974a, Optical activity of the saturated hydrocarbons from the Alberta heavy Cretaceous oils and its relation to thermal maturation: Bull. Canadian Petroleum Geology, v. 22, p. 357-360.

——— 1974b, Investigation of oils in western Canada tar belt, *in* Oil sands, fuel of the future: Canadian Soc. Petroleum Geologists, Memoir 3, p. 168-183.

Murray, G. H., 1968, Quantitative fracture study—Sanish pool, McKenzie County, North Dakota: AAPG Bull., v. 52, p. 57-65.

Philippi, G. T., 1965, On the depth, time and mechanism of petroleum generation: Geochim. et Cosmochim. Acta, v. 29, p. 1021-1049.

Poulet, M., and J. Roucaché, 1970, Etude geochemique des buits du bassin Parisien: Inst. Francais Petrole Rev., v. 25, p. 128-148.

Price, L. C., 1976, Aqueous solubility of petroleum as applied to its origin and primary migration: AAPG Bull., v. 60, p. 213-244.

——— 1977, Aqueous solubility of petroleum as applied to its origin and primary migration; reply: AAPG Bull., v. 61, p. 2149-2156.

——— 1978, New evidence for a hot, deep origin and migration of petroleum: AAPG Research Syposium on Problems of Petroleum Migration (Oklahoma City), 1978.

Roberts, W. H., III, 1978, Design and function of oil and gas traps as clues to migration (abs.): AAPG Bull., v. 62, p. 558.

Robinson, W. E., and G. L. Cook, 1975, Compositional variations of organic material from Green River oil shale—WOSCO Ex-1 Core (Utah): U.S. Bur. Mines Rept. Inv., 8017, p. 1-40.

Smith, J. E., J. G. Erdman, and D. A. Morris, 1971, Migration, accumulation and retention of petroleum in the earth: Proc. 8th World Petroleum Cong., v. 2, p. 13-26.

Smith, P. V., 1954, Studies on origin of petroleum: occurrence of hydrocarbons in sediments of Gulf of Mexico: AAPG Bull., v. 38, p. 377-404.

Snarskii, A. N., 1970, The nature of primary oil migration: Neft Gag 13, no. 8, p. 11-15 (Translated by Assoc. Technical Serv., Inc.).

Spillers, J. P., 1965, Distribution of hydrocarbons in south Louisiana by types of traps and trends—Frio and younger sediments (abs.) AAPG Bull., v. 49, p. 1749-1751.

Stach, E., 1953, The "Coalification Jump" in the Ruhr Carboniferous: Brennstoff-Chemie, v. 34, p. 353-355.

Teichmüller, M., 1975, Generation of petroleum-like substances as seen under the microscrope: Advances in Organic Geochemistry, 1973, p. 378-395.

——— and K. Ottenjann, 1977, Liptinites and lipoid materials in an oil source rock: type and diagenesis of liptinites and lipoid substances in an oil source rock on the basis of fluoresence microscopic studies: Erdol and Kohle, Erdgas, Petrochemie, v. 30, p. 387-398.

Tissót, B., and D. Welte, 1978, Petroleum formation and occurrence: New York, Springer-Verlag, 530 p.

——— et al, 1971, Origin and evolution of hydrocarbons in early Toarcian shales, Paris basin: AAPG Bull., v. 55, p. 2177-2193.

——— et al, 1974, Influence of nature and diagenesis of organic matter in formation of petroleum: AAPG Bull., v. 58, no. 3, p. 499-506.

——— G. Deroo, and A. Hood, 1978, Geochemical study of the Uinta basin: formation of petroleum from the Green River formation: Geochim. et Cosmochim. Acta, v. 42, p. 1469-1486.

Tóth, J., 1978, Gravity induced cross-formational water flow— possible mechanisms for transport and accumulation of petroleum (abs.): AAPG Bull., v. 62, p. 567-568.

Trofimuk, A. A., 1974, The fractionation of bitumoids in the course of their migration: Geologiya, Geofizika, v. 15, no. 5, p. 122-129.

Van Krevelen, D. W., 1961, Coal: Amsterdam, Elsevier Pub. Co.

Vigrass, L. W., 1968, Geology of Canadian heavy oil sands: AAPG Bull., v. 52, no. 10, p. 1984-1999.

Weeks, L. G., 1961, Origin, migration and occurrence of petroleum, *in* Petroleum Exploration Handbook, Chapter 5: New York, McGraw-Hill, p. 1-50.

Williams, J. A., 1974, Characterization of oil types in the Williston basin: AAPG Bull., v. 58, p. 1243-1252.

Wilson, H. H., 1975, Time of hydrocarbon expulsion, paradox for geologists and geochemists: AAPG Bull., v. 59, p. 69-84.

Zhuze, T. P., and E. G. Bourova, 1977, Influence des different processes de la migration primaire des hydrocarbons sur la compositon des petroles dans les gisements: Advances in Organic Geochemistry (1975), p. 493-500.

——— G. N. Jushkevich, and G. S. Vshakova, 1963, On the general rules of behavior of gas-oil systems at great depths (in Russian): Dokl. Adad. Nauk. SSSR, v. 152, p. 713-716.

Migration of Hydrocarbons in Compacting Basins [1]

By L. C. Bonham[2]

Abstract During migration of petroleum, either as phases separate from water or in water solution, an important role is assigned to pore fluid movements which result from compaction. The common view is that progressive burial of the sediments results in their compaction with consequent expulsion of pore fluids. These fluids are pictured as moving upward toward the depositional surface, even though the pathways (in actual detail) may include some lateral and downward movement. This commonly accepted view is often incorrect.

In the early stages of basin subsidence and sedimentation, the flux of water with reference to the depositional surface is downward, even though the flux of fluids expelled by compaction is upward across stratigraphic units. In later stages, a deep, subsiding basin contains a more or less constant volume of water. As subsidence, sedimentation, burial, and compaction continue, the sediments can be visualized as slowly moving downward through a fixed volume of water. Relative to a stratigraphic marker, the fluids move upward, but for the most part, they do not move to shallower positions relative to the surface of deposition.

When source sediments move downward into the "thermal window" for hydrocarbon generation, some of the hydrocarbons formed go into water solution. Subsequent migration and release of hydrocarbons from solution depend on the fluid flux and the positions of the isotherms. Exsolution occurs where the temperature of the solution falls below the saturation temperature (geologists have proposed that large volumes of deep, hot water are physically transported to shallower, cooler zones); however, exsolution occurs if the waters retain their position relative to the depositional surface while the isotherms are depressed.

Results of a Gulf Coast migration study indicate a formation temperature drop of about 50°F since early Pliocene time. In this case, a few thousand feet of section below the 200°F or 250°F isotherms could have exsolved hydrocarbons equivalent to the total known oil and gas in the area.

Quantitative modeling shows that some upward movement probably occurred when fluids from high pressure shales leaked through thin sands or along faults; however, migration by this mechanism is small in areas where the section retains abnormally high pressure and above-normal porosities.

INTRODUCTION

The conventional view of sediment compaction is that it provides a means for mass transport of fluids upwards through the sedimentary column. In most published papers the assumption is specifically or tacitly made that fluid movement is upward towards the ground surface or the depositional sediment-water interface in case of an offshore basin (Athy, 1930; Levorson, 1967; Dickey et al, 1968; Dott and Reynolds, 1969; Rieke and Chilingarian, 1974; Wolf and Chilingarian, 1976; Pirson, 1977). Local, lateral, or even downward movement of fluids may be described, but that the net flux of fluids is upward is implied if not stated. This compaction-induced upward fluid flow, if true, influences petroleum migration either in water solution or as a separate phase. Pore waters in source beds become saturated with hydrocarbons in the generative zone and move out of the source layers and upward as a result of expulsion by compaction. The upward movement carries the petroleum-saturated waters into areas of lower temperature, and hydrocarbon is exsolved as the waters cool off. The net result would be the transfer of some hydrocarbons by a solution mechanism from deep hot areas to shallower, cooler areas. Price (1976) considered this transfer of hydrocarbons in molecular solution to be an important mechanism of primary petroleum migration. However, if we consider migration of petroleum as a separate phase, then the buoyancy of the hydrocarbon phase is important in moving it upwards, particularly in porous carrier beds. Any upward compaction-

[1]Manuscript received, August 30, 1978; accepted, July 31, 1979.

[2]Chevron Oil Field Research Company, La Habra, California 90631.

Published with permission of the Chevron Oil Field Research Company. The writer acknowledges many colleagues who contributed data or ideas. Special thanks are given to P. D. Baranyai, H. O. Woodbury, H. N. Peterson, D. C. Barnum, R. W. Jones, G. B. Vockroth, and H. E. Province for geological and geochemical data; and to W. J. Plumley and L. P. Stephenson for consultation on compaction processes and geothermal restoration; and to W. T. Miller, D. G. Lyddon, G. W. Starke, and M. E. Osborne for developing the computer programs.

Article Identification Number: 0149-1377/79/SG10-0005/$03.00/0.

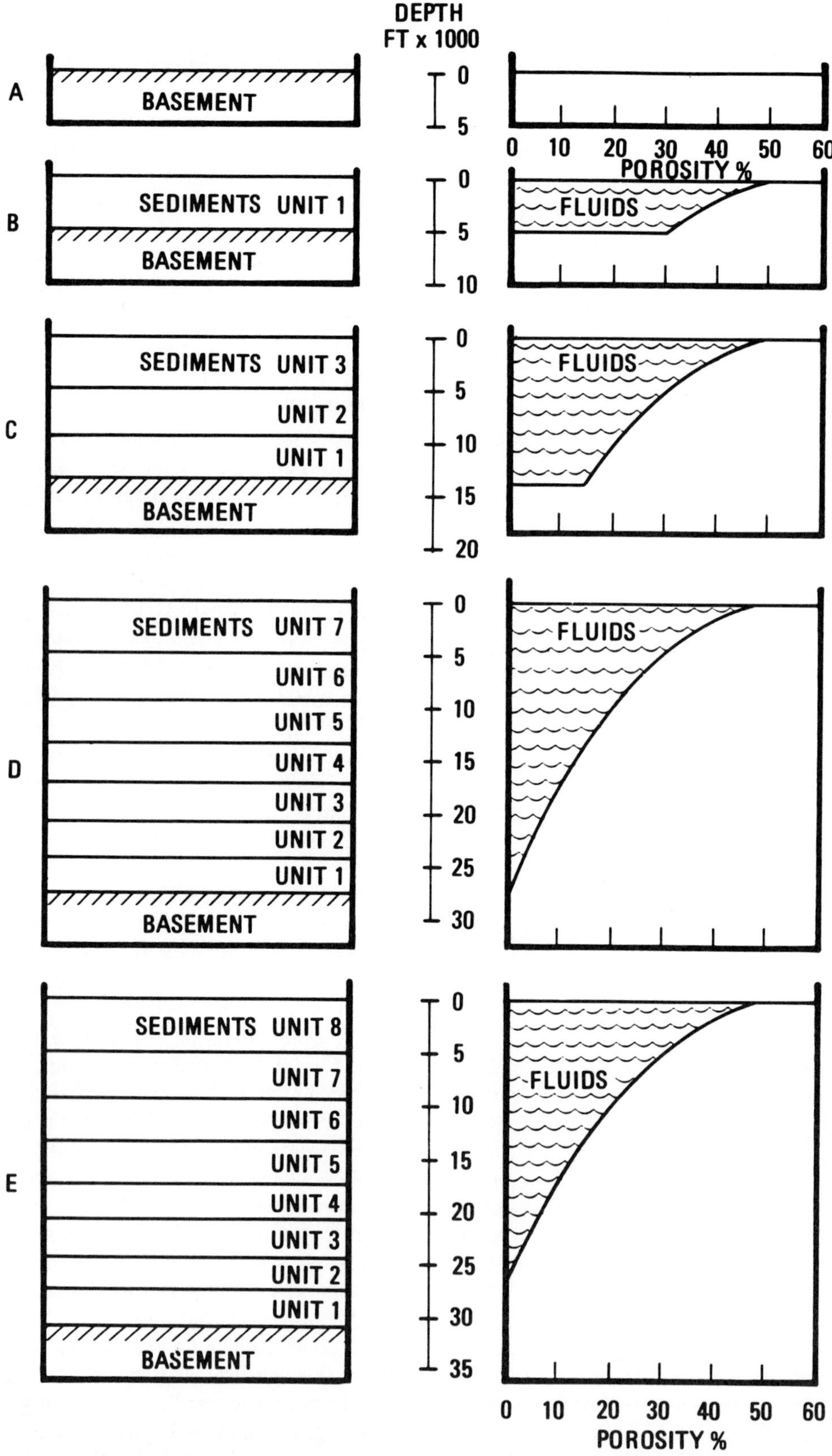

FIG. 1—Basin development model, schematic cross sections on the left represent sequential steps in basin subsidence, sedimentation, and compaction. An equivalent depth-porosity plot is shown on the right for each step.

induced flow of waters would assist in separate phase migration by helping to overcome capillary pressures. This study considers the conventional views of compaction-induced fluid flow to be inadequate. In petroleum migration the fluid movement must be considered both with respect to geologic horizons and with respect to a fixed datum such as sea level or the sediment-water interface. The following sections demonstrate the use of a conceptual model and computer modeling to evaluate two mechanisms of petroleum migration in water solution.

CONCEPTUAL MODEL OF FLUID MOVEMENT DURING COMPACTION

Figure 1 represents the development of a sedimentary basin through subsidence, sedimentation, and compaction. Figure 1A shows a basement which has little or no porosity or permeability. In Figure 1B, the basement has subsided and sediment unit 1 was deposited. A depth-porosity curve for the basin sediments is shown on the right. In Figure 1C, more subsidence has occurred and sedimentary units 2 and 3 were deposited. Unit 1 was compacted by the weight of the overburden and the expelled pore fluids passed upward stratigraphically into unit 2. If we look at the corresponding depth-porosity curves, the porosity of unit 1 is less in 1C than it was in 1B. But the total amount of fluid in the basin has increased. The total basin fluids can be represented by integrating the basin pore volume (see FLUIDS graph on right half of Fig. 1). This volume is zero in Figure 1A and increases as the basin develops in Figures 1B and 1C. In other words, the net movement of water is into the basin and downward to greater depths with time. The flux of water with reference to sea level is downward, even though the relative flux of fluids expelled by compaction is upward across stratigraphic units.

This condition will continue until basin subsidence and filling have progressed to the stage illustrated in Figure 1D. The sediment column has reached a thickness where the deepest layer (unit 1) has been compacted to zero porosity. From this point on, the total volume of water in the basin sediments is fixed; no new water enters with new subsidence and sedimentation, no pore water is expelled at the surface. Instead, the pore fluids of each new layer deposited and each underlying layer are displaced by the fluids expelled from the next deeper layer.

Figure 1D illustrates an equilibrium condition. We can visualize it as a fixed volume of water in the basin as represented by the fluid-zone area of the depth-porosity curve. As subsidence and sedimentation continue (Fig. 1E), the sediments that comprise the basin filling slowly move downward through the fixed volume of pore water.

Of course, this is a simplistic model. In actual basins, a variety of perturbations can and does occur. If high pressure develops due to rapid sediment loading or other causes, then the potential exists for fluid leaks to develop. Fluid bypassing can occur along a fault so that deep, high pressure pore waters are injected into shallower, low pressure zones or even expelled from the surface. However, the volumes of bypassing fluids must be limited. Only enough fluid flows to bleed off the excessive pressure. This mechanism will be discussed later in detail.

As a first order approximation, we consider the total pore fluids in a thick sedimentary basin as a fixed volume of water. Little or no movement occurs up or down with respect to a datum plane such as sea level. However, a relative upward movement occurs with respect to stratigraphic markers because these strata are sinking relative to sea level.

Relative Positions of Isotherms During Compaction

Where the geothermal gradient is constant throughout the sedimentary column from surface to basement (see the thermal gradient line in Fig. 2A), the temperature at a given depth is constant during subsidence, deposition, and compaction. In Figure 2A, point T (a given distance below sea level) maintains a constant temperature through time. Therefore, the isothermal surface on which point T lies is always located at a constant absolute depth below sea level.

We established that the movement of pore fluids during the pre-equilibrium stage of basin development is downward with respect to sea level. Therefore, we conclude that in a developing sedimentary basin with a uniform geothermal gradient, the movement of pore waters is downward across the isotherms into warmer and warmer regions. This occurs even though compaction and expulsion of fluids from sedimentary layers is going on at the same time. The relative movement of pore fluids with respect to stratigraphic markers is upward, but the net movement is downward, relative to a sea level datum. Thus, pore fluids in the pre-equilibrium stage undergo continuous heating.

The situation changes after the basin reaches the equilibrium stage (Fig. 1D). At this stage, the basin pore system contains the maximum amount of fluids it will ever have, and the fluids are static with respect to sea level. Because of the constant geothermal gradient, a given isothermal surface remains at a specific depth. Thus fluids do not cross the isotherm upward or downward even though subsidence, sedimentation, and compaction continue.

In some basins, the geothermal gradient is not constant through time. Figure 2B shows the case where the early (deeper) sediments have a geothermal gradient, g, the same as Figure 2A. Later sediments (possibly due to a change in lithology) have a lower gradient, g/2 (rate of change of temperature with depth).

The result is that the isothermal surface, temperature T, has moved from its original depth d_0, in 2A to

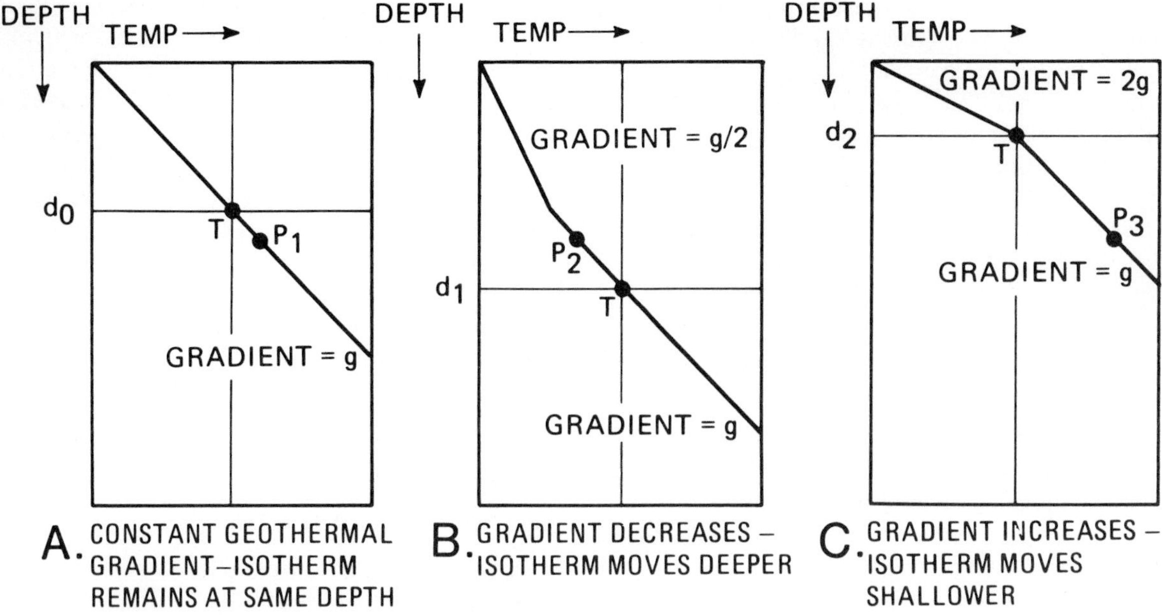

FIG. 2—Movement of an isothermal surface as the geothermal gradient changes with time.

a deeper depth, d_1, in 2B. This change in gradient may occur in the early phase of basin development before an equilibrium water volume is established. In this case, both the absolute fluid flux and the movement of the isotherms are downward. The relative flux of fluid may be up, down, or zero with respect to an isotherm depending on the rate of movement of the fluid relative to the isotherms. Where the basin has developed far enough for an equilibrium water volume to be established, with a decrease in thermal gradient, a given isotherm moves deeper (Fig. 2B). Because the water volume is fixed and static, a relative movement of water occurs upward across the isotherms. Thus, a volume of water at a given depth and temperature (point T in Figure 2A) becomes cooler with time even though it remains at the same depth (point T in Fig. 2B).

A similar set of conclusions and consequences can be developed for the situation illustrated in Figure 2C where the geothermal gradient increases to a higher value after early deposition at a lower value. In this case the relative movement of fluid is always downward with respect to an isotherm regardless of the stage of basin development.

Significance to Migration of Petroleum in Water Solution

Migration of petroleum in water solution is permissible or favored in some situations and prohibited in others. Hydrocarbons have a finite solubility in water which increases with temperature (This is true, even for the gaseous hydrocarbons which go through a minimum solubility point with increasing temperature at a constant pressure). The essence of the solution hypothesis of petroleum migration is that pore fluids become saturated with hydrocarbons deep in the basin. Mass transfer of the saturated solutions to shallower, cooler areas is usually called on for some of the hydrocarbons to come out of solution to form separate petroleum phases (gas and oil). In other words, the saturated solutions must move upward relative to isothermal surfaces.

Figure 3 summarizes the temperature effects on pore fluids caused by the thermal gradient behavior through time. There only is one condition under which we can be sure that the fluids will cool. This is in a basin which has reached the equilibrium stage and which has had a decrease in thermal gradient (Fig. 2B). Cooling is possible in basins which have a decrease in gradient but which have not yet reached equilibrium. As noted earlier, this will occur if the isotherms are moving downward faster than the downward movement of water into the developing basin.

Based on these models, migration in solution with exsolution to form a separate hydrocarbon phase should not occur in a basin with a constant (Fig. 2A) or an increasing (Fig. 2C) geothermal gradient through time. On the other hand, transfer of dissolved hydrocarbons to younger stratigraphic units with exsolution to form separate phase petroleum is at least technically feasible if the geothermal gradient changes from higher to lower values with time (Fig. 2B).

COMPUTER MODELING OF GULF COAST AREA

To evaluate the conceptual model, a representative Gulf Coast area was selected for computer modeling. The area is delimited by faults, salt domes, or synclinal axes, so that pore fluids (both water and petroleum) could reasonably be expected to remain within

		TEMPERATURE EFFECTS ON PORE FLUIDS	
(RATE OF CHANGE OF TEMPERATURE WITH DEPTH)	ISOTHERM MOVEMENT	PRE–EQUILIBRIUM STAGE	EQUILIBRIUM STAGE
CONSTANT	NO MOVEMENT	HEAT	CONSTANT TEMPERATURES
HIGH, THEN LOW	MOVES DOWN	HEAT OR COOL*	COOL
LOW, THEN HIGH	MOVES UP	HEAT	HEAT

*MOVEMENT OF H_2O > ISOTHERMS ⟶ HEAT
MOVEMENT OF ISOTHERMS > H_2O ⟶ COOL

FIG. 3—Effects on temperatures of pore fluids as the geothermal gradient changes through time.

the natural boundaries through time. The petroleum drainage area is about 150 sq mi (390 sq km; see Fig. 4). The area contains one field which has production from several sandstones in both normally pressured and geopressured zones. Ultimate recovery is estimated to be about 60 million bbl of oil plus oil equivalent of gas calculated on a BTU basis of 5,000 cu ft of gas equal to one barrel of oil.

The geothermal profile for this area suggests a geothermal history favorable for solution transfer of hydrocarbon during burial and compaction. The computer modeling procedure was to establish a 3-dimensional geometric replica of the present geology from well and geophysical data. Then successive units were removed from the top of the model while underlying units were moved upward and expanded by the amounts they were believed to have compacted during original burial. Results of this geological restoration are best illustrated by successive time restorations of the geological column at an arbitrarily selected point in the area (Fig. 5).

Input data for computer compaction and geological restoration modeling were obtained from over 60 wells in the area. The data consisted of subsea elevations of each of the horizons to be mapped, the isopach thickness of each unit, and the lithologic makeup of each unit (percent sandstone and shale). The depth to top of salt (Fig. 5) was estimated from geophysical measurements.

Establishing Depth-Porosity History Curves

For geological restoration modeling, it is necessary to establish the history of porosity reduction through time for each geologic unit to be modeled. Figure 6 represents the depth versus porosity of shales in several Gulf Coast wells as estimated from density log values. This change in porosity with depth might be approximated by the curve in Figure 7. The shale porosities show an offset to higher porosity values near the top of the geopressured zone. However, the curve in Figure 7 does not represent the porosity history of each shale interval through geologic time. This is more accurately represented by the curves in Figure 8. The burial pathway for shales above the top of the geopressured zone is indicated by the curve labeled normal compaction. However, shales below the top of the geopressured zone were inhibited from compaction almost from the time of deposition. Data from many Gulf Coast wells suggest that the thermal dehydration of mixed-layer clays, which commonly occurs near the top of the geopressured zone, does not result in an increase in porosity, but instead arrests the reduction in shale porosity over a depth interval. Thus the depth-porosity history curve for shales below the top of the geopressured zone has followed the pathway of the curve labeled *inhibited compaction.* In a similar manner, porosity history curves can be developed for sandstones with separate normal compaction and inhibited compaction curves but without the effects of clay alteration. The normal and inhibited compaction shale porosity curves for modeling the test Gulf Coast area are shown in Figure 9. They were derived from density log values of wells in the test area. Values below 18,000 ft (5,486 m) are extrapolated.

Geological Restoration and Uncompaction

The computer modeling program accepts structural tops, unit thicknesses, lithologic compositions, and

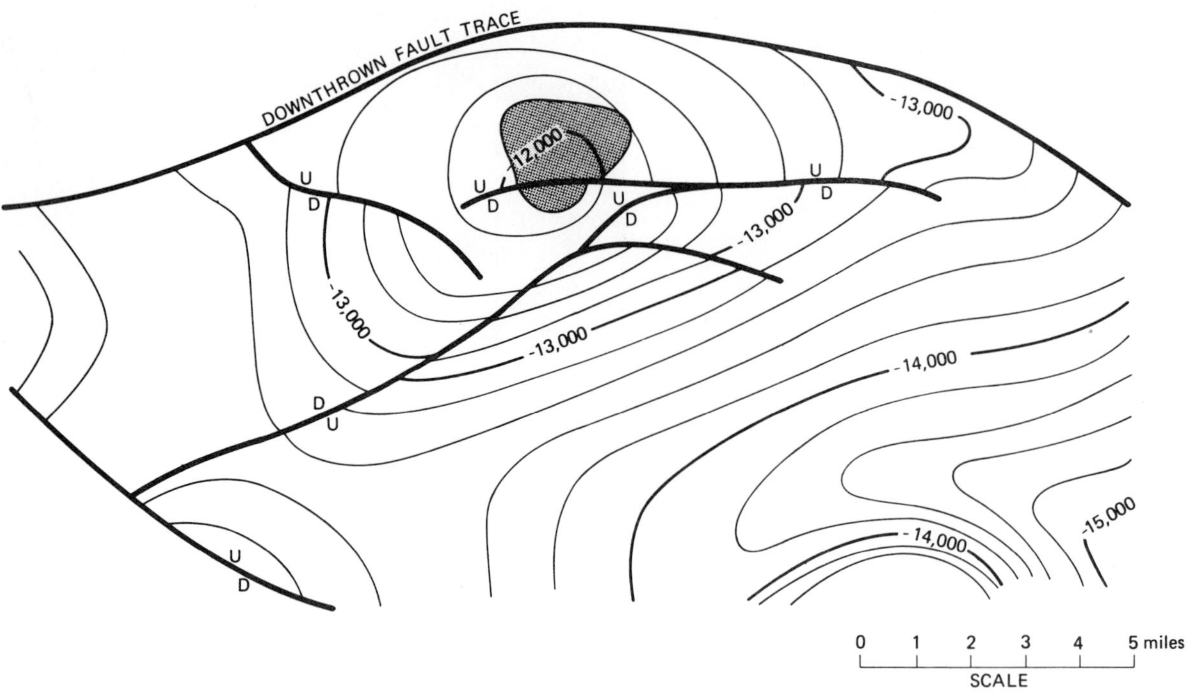

FIG. 4—Gulf Coast test area, structure is on top of Unit 4 (approx. top of Miocene).

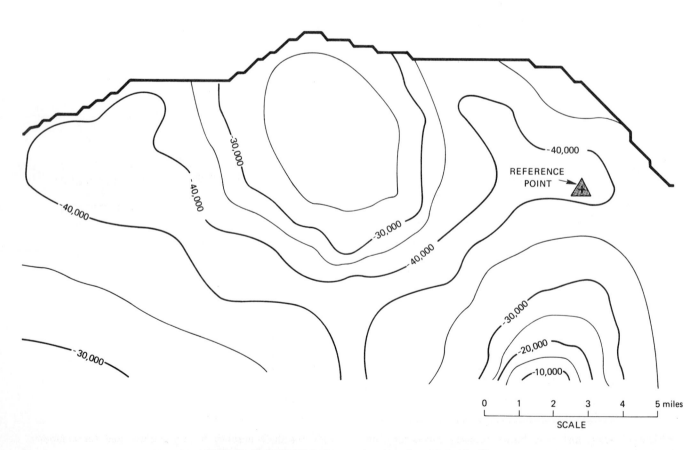

FIG. 5—Gulf Coast test area, structure is on top of salt.

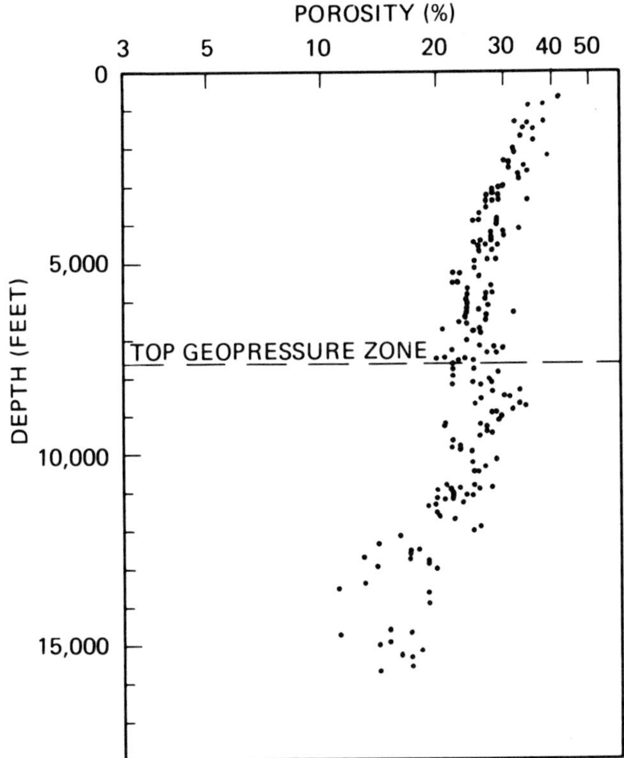

FIG. 6—Depth versus porosity of Gulf Coast shales from density log values.

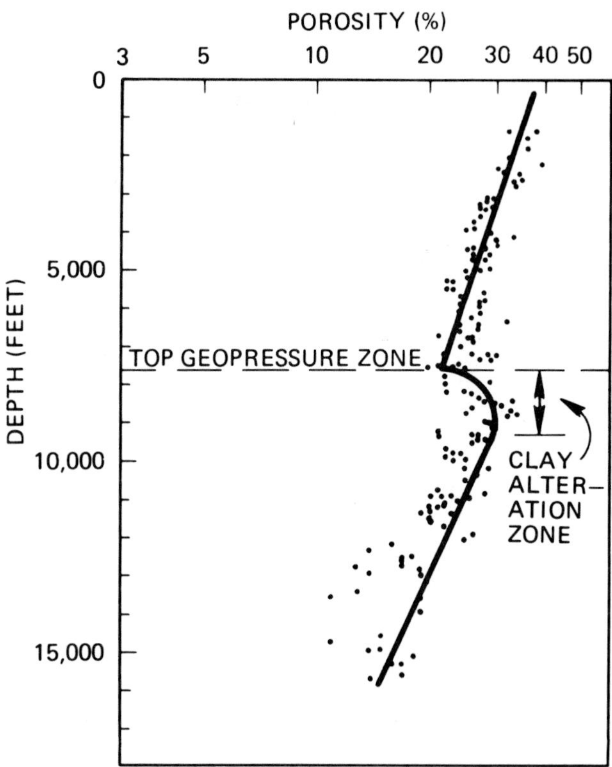

FIG. 7—Depth-porosity curve for Gulf Coast shales, this curve does not represent the porosity history of the shales in the geopressured zone.

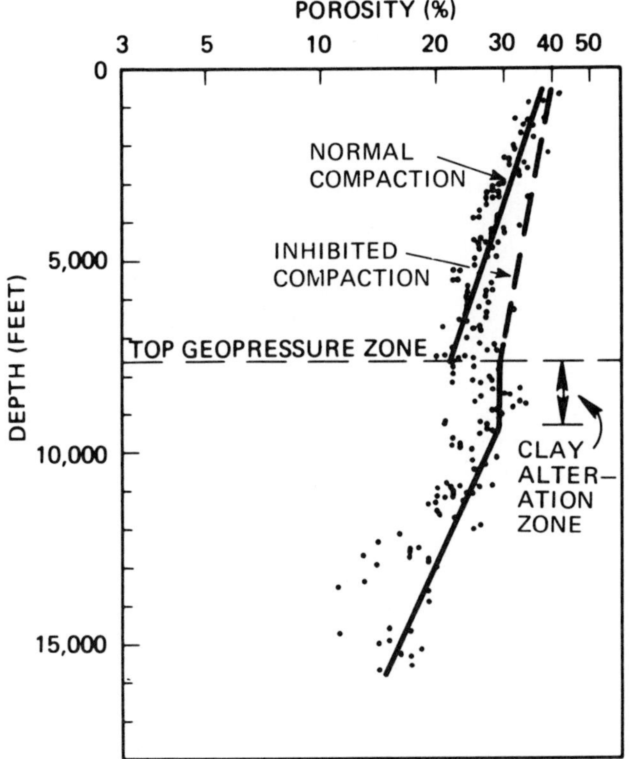

FIG. 8—Porosity history or burial pathway curves for Gulf Coast shales.

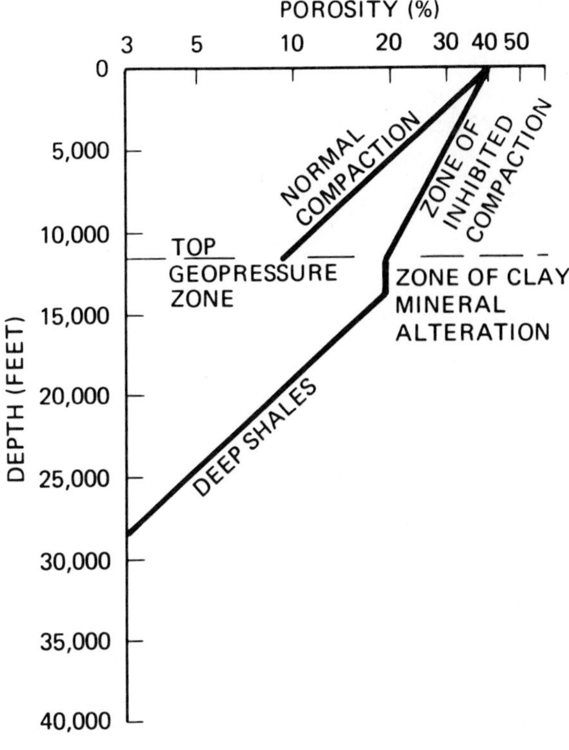

FIG. 9—Shale porosity history curves used for computer modeling of the Gulf Coast test area.

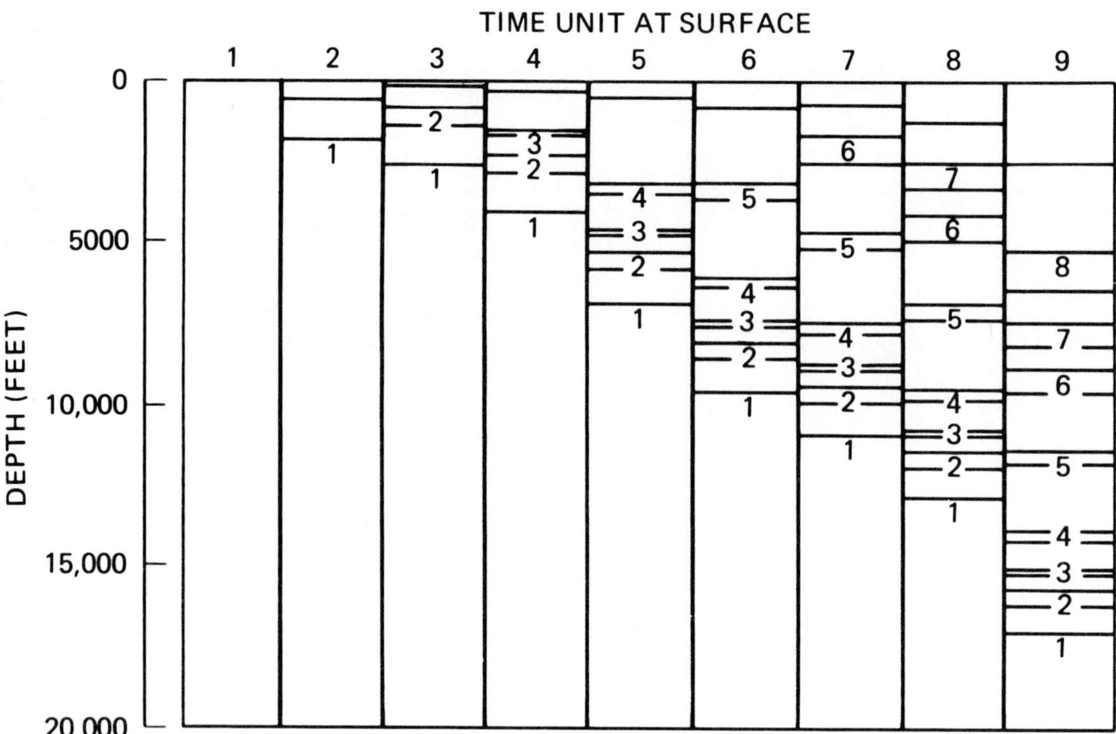

FIG. 10—Step-by-step geologic restoration of the sedimentary column at an arbitrary reference point in the test area (see Fig. 5). Depths and thicknesses of each unit are restored with uncompaction.

sediment porosities as they exist at present. Then individual units are removed from the top, one at a time, to restore the geology successively to earlier times. During this restoration, calculations are made of the new structural elevations of each geologic top, the expanded thickness of each unit caused by moving the specified lithologies up appropriate depth-porosity curves, the average porosity of each restored unit, and the fluid flux caused by the change of vertical thickness of each unit. This type of sequential restoration was performed on grids or arrays of 3-dimensional data which represent the geology of the Gulf Coast test area from the surface to the top of the salt. The geology at present or for any time step in the restoration can be displayed on maps or sections. In addition, any grid or array which represents sediment thickness or porosity can be integrated to calculate the area covered and the volumes of sediments or fluids. The computer modeling package provides output data on the restored elevations of each geologic unit in the model for each time step specified. Figure 10 illustrates the step-by-step restoration at the arbitrary reference point in the basin. It shows the depths and thicknesses for each unit modeled and for each time step from the present where unit 9 is at the surface back through time to where unit 1 was at the surface. The thicknesses for each unit in Figure 10 represent the expanded or uncompacted thicknesses derived from the computer model output. Figure 11 shows the lithologic makeup of the sediments. Unit 1 is a massive shale; units 2 through 5 are shales with interfingered sandstones; and units 6 through 9 contain massive sandstones which account for up to 50% of the total column.

Geothermal Restoration

This study assumes the generation of enough petroleum hydrocarbons from organic source material to saturate the pore fluids in the source zones is primarily temperature dependent. Many investigators have shown that the level of transformation depends both on temperature and time (Teichmüller and Teichmüller, 1966; Lopatin, 1971; Bostick, 1973; Demaison, 1974; Hood et al, 1975; and Zieglar and Spotts, 1978). The time factor is important if estimates are to be made of the total amount of hydrocarbons generated from organic source material because the rate of conversion depends on the temperature. A common assumption is that the reaction rate doubles with each 10°C increase in temperature (Lopatin, 1971; Laplante, 1974; Momper, 1972). In this study we were concerned not with the total amount of petroleum generated, but with the amount that could be contained in solution in the pore fluids. Consequently, it was assumed that any of the Tertiary shales in the test area which reached a temperature of at least 200°F remained long enough to generate sufficient hydrocarbons to saturate the pore fluids.

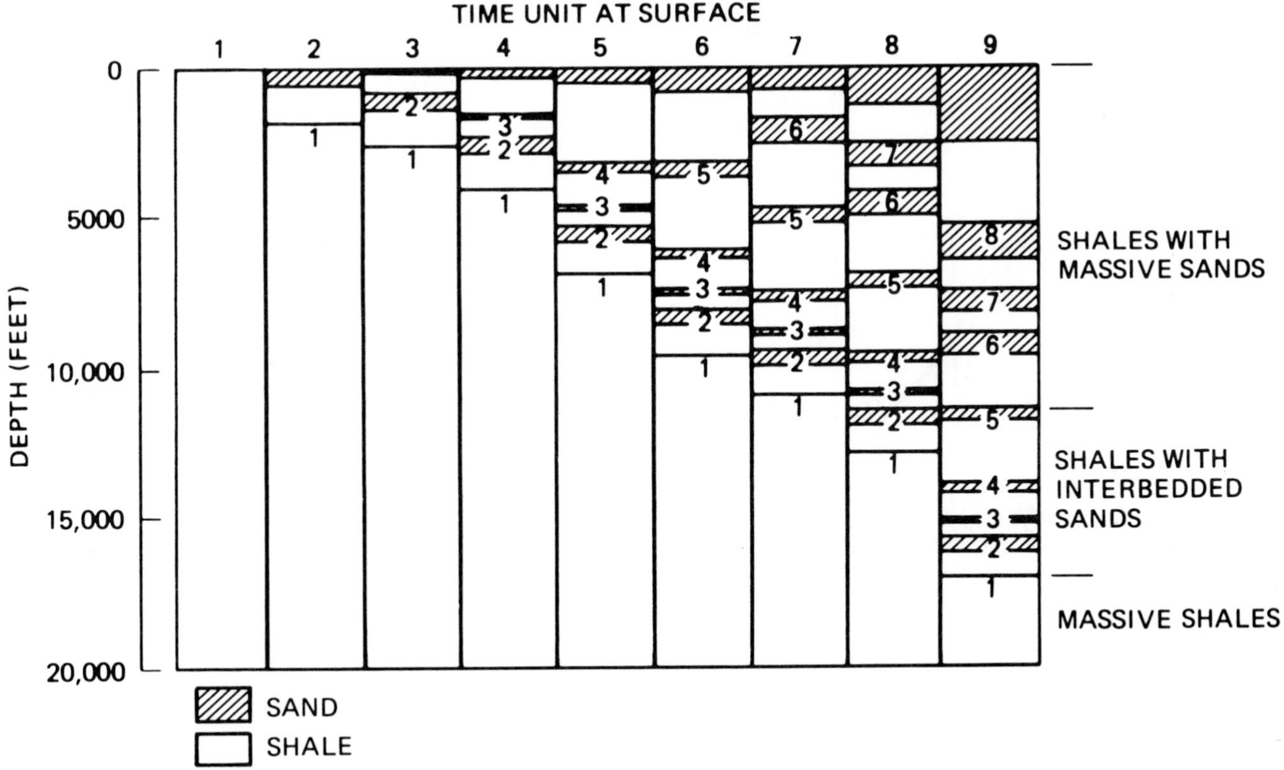

FIG. 11—Lithologic makeup of sediments in the test area.

Once hydrocarbons are generated, they are mobile; they can move with formation fluids in water solution, or they can move as a separate fluid phase under proper conditions. It is necessary first to identify the geologic units which have been in a favorable thermal range for petroleum generation and then to determine the thermal history and movement of the fluids which were in contact with the hydrocarbons. The approach used here is to designate an isothermal surface below which it is believed the temperature was high enough to generate significant quantities of hydrocarbons. Then the movement of that isothermal surface and the direction of fluid flux relative to the isotherms throughout the time represented by the model are charted. This is complicated by the fact that the geothermal gradient did not remain constant throughout the time interval of interest. Figure 12 is a depth-temperature profile from a deep well in the test area. The temperature data are from a wireline survey made after the well was shut in for three months. Also shown on Figure 12 are the present depths to the geologic units included in the model in the vicinity of the well. In Figure 12 the present day geothermal gradient is about 1.25°F per 100 ft (30 m) from the surface to a depth of about 11,000 ft (3,353 m). Below 11,000 ft (3,353 m) the gradient is about 1.8°F per 100 ft (30 m) to the maximum drilled depth. This change in geothermal gradient occurs near the top of unit 5 which also is near the top of the geopressured zone. The change in thermal and pressure gradients is lithologically related in that the deeper section is massive shale or predominantly shale with some interbedded sandstones, whereas the shallower section contains massive sandstones.

There is no complete agreement by various investigators about the specific temperatures at which generation of significant hydrocarbons begins. However, the concensus is that for Gulf Coast Tertiary sediments, such as the ones in this test area, the onset of generation occurs somewhere in the vicinity of 200°F and peak generation occurs somewhere near 250°F. The approximate depth and stratigraphic position of these two key isotherms at present is indicated in Figure 12. We shall follow the positions of these two isotherms through time in the test area by determining their depth and stratigraphic position at specific restoration times in the past. In the first restoration step Figure 10, we removed unit 9 from the top of the section and have restored unit 8 to the surface. All of the deeper units are moved upward an appropriate amount. We assume that the change or "knee" in the geothermal gradient curve in the past was related to the change in lithology at or near the top of unit 5, as it is today. Thus we can move the point of change of geothermal gradient upward by an amount equal to the upward movement of the top of unit 5. We assume that the geothermal gradient in the massive shales was about 1.8°F per 100 ft (30 m) in the past as it is today, and

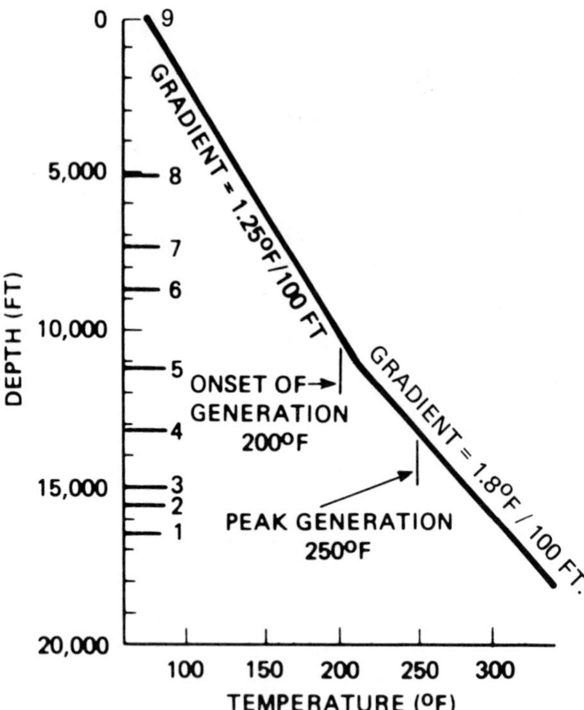

FIG. 12—Present depth-temperature profile for a well in the test area. Present depths to geologic units 1 through 9 also are shown.

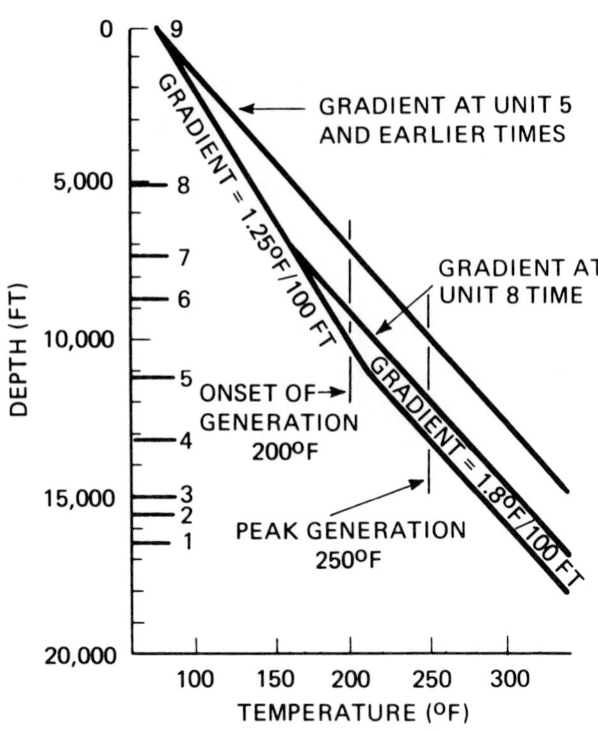

FIG. 13—Restored depth-temperature profiles, shown here are geothermal gradient curves for the present, for unit 8, and for unit 5 and earlier times.

that the gradient in the shallower sandy section was 1.25°F per 100 ft (30 m) as it is today. The result is a restored depth-temperature profile for the time when unit 8 was at the surface (Fig. 13). Note that at this time the depth to the 200° or 250° isotherm is less than it is at present. In a similar manner, step-by-step geothermal restoration can proceed until unit 5 is at the surface (Fig. 13). Prior to the time of deposition of unit 5, the entire section from surface to top of salt is predominantly shale and was thought to have a single gradient of 1.8°F per 100 ft (30 m). Thus this single gradient is used for unit 5 and all older units. Again, the depth to the 200°F and 250°F isotherms was less at the time of deposition of unit 5 than it has been at any time since. The restored depths of the key isotherms also are plotted on Figure 14. The line 00′ represents the depth and stratigraphic position through time of the onset of petroleum generation, and the line PP′ shows the depth and stratigraphic position of peak generation. Any other desired isotherms can be restored in a similar manner.

Identification and Timing of Fluid Sources

Figure 14 shows not only restoration snapshots at specific instants of geologic time, but also charts the depths and stratigraphic positions of two key isotherms, which are related to the upper boundary of the thermal window favorable for petroleum generation. Next it is necessary to consider the relationship of the pore fluids in compacting sediments as they pass through this thermal window. Where hydrocarbons are thermally generated from kerogen in a fine-grained source bed, the pore water in that layer should become saturated with hydrocarbons at the local temperature and pressure. Once in solution the hydrocarbons can move with the pore water in response to pressure gradients. Most of this movement is upward with respect to the sedimentary strata and is due to continued sedimentation, burial, and compaction. Any time that the hydrocarbon-saturated pore fluids are cooled some of the dissolved hydrocarbons may come out of solution. If this happens in a porous and permeable rock, the globules of separate phase hydrocarbons can agglomerate to form a thin film or slug of oil or gas which can migrate updip due to buoyancy. Thus it is important to specifically look for the periods of time when the fluids were cooling after having been heated at least to 200°F and preferably to 250°F or more in contact with organic source rocks.

The conditions for cooling of pore fluids were described previously. Net flow of fluids is into a basin and downward during the basin development stage, and continues until an equilibrium distribution of porosity versus depth is reached. Afterward, the total fluid in the basin is constant; no fluid enters or leaves. Thus, the only way for the pore fluids to cool is for the isotherms to move downward. Before the equilibrium stage is reached, in order to cool pore fluids, the

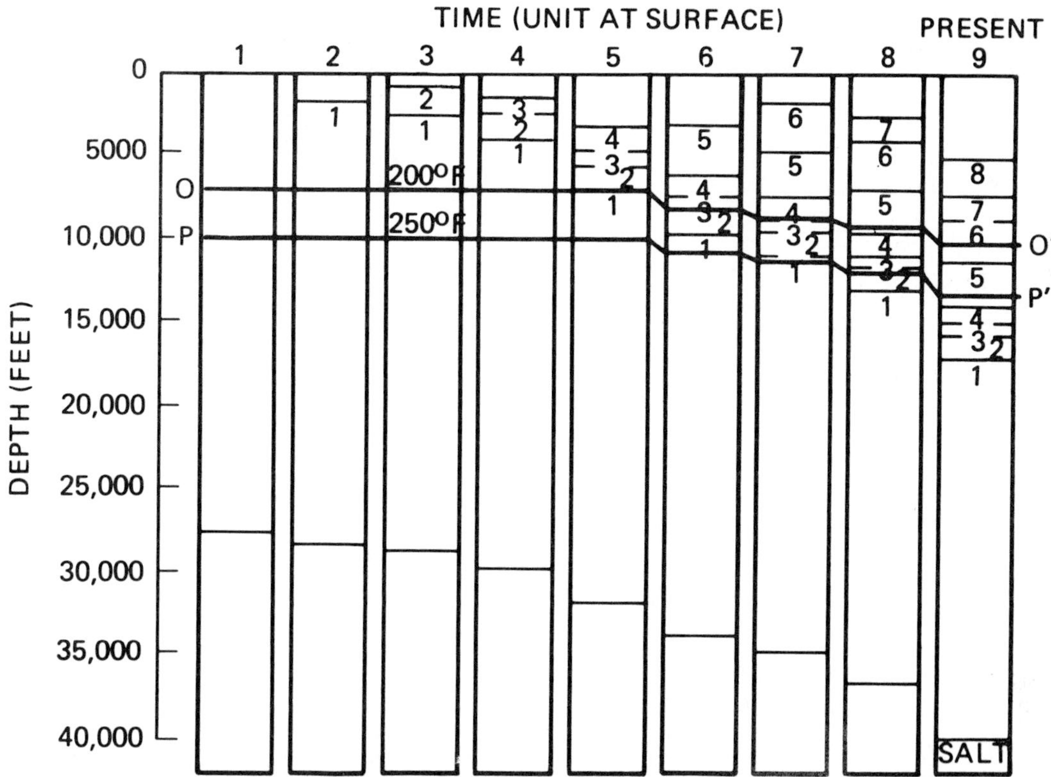

FIG. 14—Restored isotherm depths superimposed on restored sedimentary column.

isotherm must move downward faster than the absolute downward movement of water into the developing basin. After equilibrium is reached, any downward movement of the isotherm causes cooling of pore fluids.

Movement of the isotherms implies a change in geothermal gradient. An increase in gradient with time causes isotherms to move shallower. A decrease in gradient with time causes the isotherms to move deeper, which is the condition under which migration and separation of hydrocarbons from solution can occur.

In Figure 13, the geothermal gradient in the test area decreased from a value of 1.8°F per 100 ft (30 m) during the time of deposition of unit 5 and earlier, to a value of 1.25°F per 100 ft (30 m) since unit 5 time. This also is reflected in Figure 14 by the depression of the 200° and 250°F isotherms since the time of deposition of unit 5. Also in Figure 14, by the time of unit 5 deposition the total sediment thickness to top of salt was over 30,000 ft (9,144 m). From this it can be estimated that the sediments reached the equilibrium stage by the time unit 5 was deposited; that is, the lower part of the unit 1 massive shales was at, or near, zero porosity and no fluids were entering or leaving the basin.

With all these considerations in mind, it can be concluded that petroleum migration in water solution by the mechanism described could have been significant in this test area only since the time when unit 5 was deposited, and only by fluids which occupied the pore spaces of stratigraphic units between the 200° or 250°F isotherms and the top of salt.

Fluid Volume Calculations

Figure 14 shows that the 200° and 250°F isotherms were at constant depths in the test area from the time unit 1 was deposited until unit 5 was deposited. Since then however, both have been depressed. The 200°F isotherm dropped 3,100 ft (945 m) (from 7,000 ft [2,133 m] at unit 1 time to 10,100 ft [3,078 m] at present), and the 250°F isotherm dropped 3,400 ft (1,036 m) (9,800 ft [2,987 m] to 13,200 ft [4,023 m]) during the same time interval. Figures 13 and 14 show that the entire volume of pore fluid from the surface to salt has cooled since unit 5 was deposited. In Figure 14, the present depth to 200°F is close to the depth of the 250°F isotherm at the time unit 5 was deposited and earlier. In other words, the pore fluids have cooled about 50°F since the deposition of unit 5. To obtain a rough estimate of how much hydrocarbon could come out of solution in this process, it is possible to calculate the exsolution that would result from two cases.

1. If all pore waters below the 200°F isotherm (at the time unit 5 was deposited) were cooled 50°F; and
2. If all pore waters below the 250°F isotherm (at the time unit 5 was deposited) were cooled 50°F.

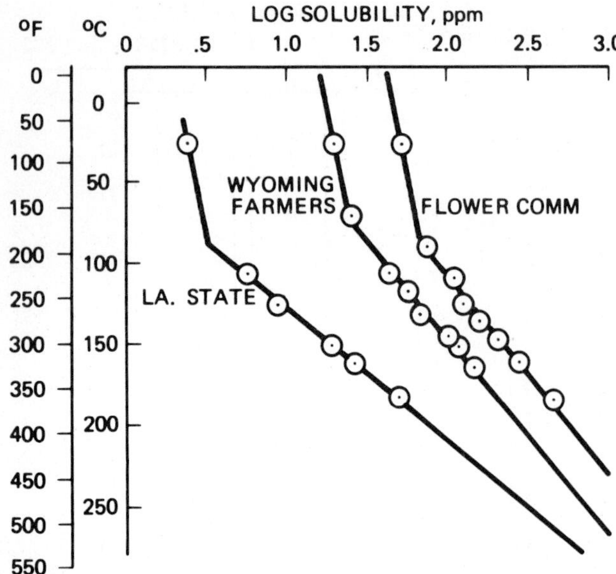

FIG. 15—Solubilities of 3 whole crude oils (from Price, 1976).

Several assumptions are made for these calculations. First, because it is necessary to have data on hydrocarbon solubilities, it is assumed that Price's (1976) data for three whole crude oils are representative and we will use his values for the most soluble and least soluble oils. Figure 15 shows that the log solubility versus temperature for each of these oils is a straight line at temperatures above 175°F.

Second, because of the need for values of the pore volumes of sediments from 7,000 ft (2,133 m) (200°F at unit 5 time) and deeper, it is necessary to obtain these by integrating the depth-porosity curve (Fig. 9). At the depths of interest, the section is almost all shale, so we use the shale depth-porosity curve in Figure 9 which best represents the compaction history pathway for the deep shales in the test area.

Results of the calculations are plotted in Figures 16 and 17; calculations were made on the assumption that all the petroleum in the test area originated and migrated within the 150 sq mi (388 sq km) downthrown fault block as outlined in Figure 4. It is possible that the gathering area for the test area extended beyond these boundaries during some intervals since unit 5 was deposited.

DISCUSSION OF RESULTS—MIGRATION IMPLICATIONS

The total recoverable petroleum in the test area is estimated at 55 to 65 million bbl of oil plus oil equivalent of gas. The total oil and gas in place is perhaps twice as much. From Figures 16 and 17 one can determine how much of the sedimentary column below the 200°F or 250°F isotherm must be included to account for these volumes.

If the petroleum has a solubility similar to that of Price's Flower Community crude, then 6,000 ft (1,828 m) of sediments directly below the 200°F isotherm would have exsolved enough hydrocarbons after a 50°F temperature drop to account for the recoverable oil and gas. If we use 250°F as the top of the generative zone, then only 4,500 ft (1,372 m) of sediment are needed to account for the recoverable oil and gas. Equivalent intervals to account for total oil and gas in place would be 9,500 ft (2,895 m) and 8,000 ft (2,438 m).

The second oil used for calculations (Louisiana State crude) has a much lower solubility. It would require about 16,000 ft (4,876 m) of section below the 200°F isotherm or 14,000 ft (4,267 m) below the 250°F isotherm to account for the recoverable volumes in the test area.

On the basis of this study, it is reasonable to hypothesize a solution transfer mechanism to contribute to the first stage migration of petroleum (Fig. 18). The hydrocarbons formed from kerogen in the source layers dissolve in the adjacent pore waters. With time, these solutions pass into successively overlying strata. This does not occur by upward movement of the solutions relative to sea level, because the pore fluids at this stage in basin development are more or less static. Instead, the solutions will be in younger and younger sediments because the sediments are passing downward through a fixed body of water. This also means that great thicknesses of sediments may have passed through a limited volume of water. Even though the sediments may be relatively poor in organic content, this hot water, "zone extraction" mechanism builds up the concentration of hydrocarbons in solution, to saturation.

Once saturation has been achieved in the extraction zone, any subsequent cooling of the solutions will cause exsolution of separate phase hydrocarbons. If the exsolution occurs in fine-grained shales, further migration is inhibited, unless enough separate phase hydrocarbon is exsolved to form a continuous hydrocarbon network through the pore structure. If the exsolution occurs in the coarser grained siltstones and sandstones, then the continual downward translation of the sediments through the water body will sweep the dispersed droplets to the top of the bedding unit. The droplets will coalesce to form discontinuous thin films of oil or gas.

From this point on, accumulation of the thin film of oil or gas into reservoirs depends on the geometry and fluid connectivity ("plumbing") of the local geology. If the hydrocarbon film forms in a layer with any appreciable dip, then it will migrate updip due to buoyancy as soon as the vertical extent of the film is adequate to overcome capillary effects. Levorsen (1967, p. 554) reported that a continuous oil phase with vertical continuity of 1 to 10 m (3 to 33 ft) provides the necessary conditions for oil to migrate updip.

These separate phase films or slugs of petroleum

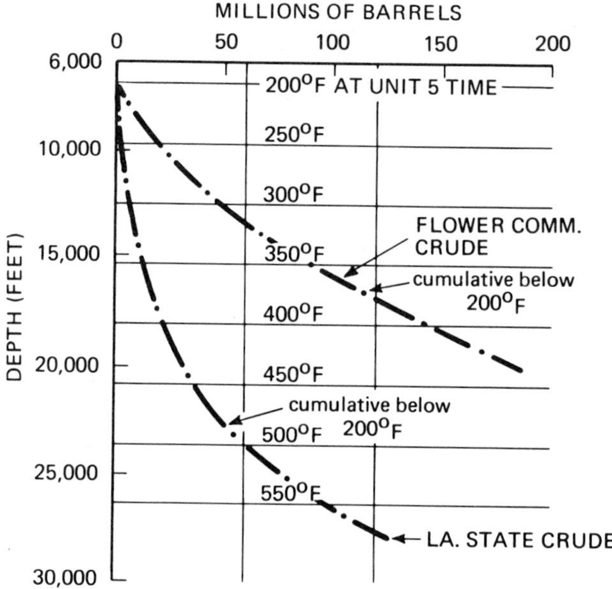

FIG. 16—Cumulative volumes of hydrocarbons released from solution below the 200°F isotherm in the test area by a 50°F drop in temperature.

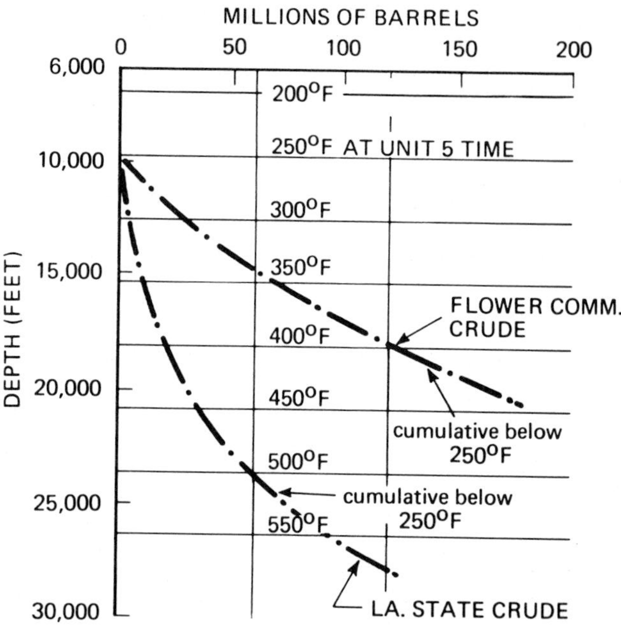

FIG. 17—Cumulative volumes of hydrocarbons released from solution below the 250°F isotherm in the test area by a 50°F drop in temperature.

will continue migrating updip until they reach some structural or stratigraphic barrier. At this stage it is likely that many "mini-accumulations" form, most of which are not commercial.

In the Gulf Coast area, these mini-accumulations would be favored where growth faults cut the numerous sandstones and offset them against shales in the interfingered zone between the massive sandstones and massive shales. Commonly these faults are not open conduits for upward flow of fluid. If they were, we would not expect to find abnormally high pressure in any sandstone adjacent to a fault.

Hubbert and Ruby (1959) postulated the formation of many Gulf Coast growth faults saying that abnormally high fluid pressures reduce the friction along detachment or glide zones so that large blocks of sediments can move down slopes as gentle as 1°. During the movement and for some time afterwards, some fluid migration could occur up the fault zone, particularly in the upper part where the tensional, pull-apart effect is much greater than in the deeper zone where the fault soles out in bedding planes.

At these times of intermittent fluid flow up the faults, petroleum from the mini-accumulations in the deep, thin sand stringers also can migrate up the fault. The flow mechanism would involve both the hydraulic pressure gradient and buoyancy of the petroleum. As petroleum migrates up the fault, it will pass into any reservoir sandstone where the potential gradient is favorable. Normally pressured sandstones just above the top of a geopressured zone would be especially favorable for accumulating large quantities of petroleum in this manner.

In brief, it is postulated that most of the petroleum which migrates up fault zones during periods of structural movement is oil or gas that already existed in small reservoirs as a phase separate from water. This migration mechanism differs from those proposed by Price (1976) and others in that the bulk of the petroleum does not move up the faults in water solution.

The water volumes that would be necessary to carry the known reserves in the test area up the faults in solution are excessive unless we are willing to accept a very great depth for hydrocarbon generation. For example, at a depth of 14,000 ft (4,267 m), temperature equal to 326°F at the time unit 5 was deposited, about 9,000 bbl of water would be required to move up the fault and cool down 50°F to release one bbl of oil, if its solubility were about the same as the Flower Community crude. At a depth of 9,800 ft (2,987 m), which is equivalent to the 250°F which we accept as a peak generation temperature, about 19,000 bbl of water per bbl of oil would be required. The equivalent water volumes required to release a barrel of oil of Louisiana State solubility would be 60,000 bbl at 14,000 ft (4,267 m) or 147,000 bbl at 9,800 ft (2,987 m).

It is because of these unreasonable volumes of water that Price (1976) proposed a deep (14,000 to 40,000 ft; 4,267 to 12,192 m), hot (275°F to 1,030°F) source for petroleum. At these temperatures the higher hydrocarbon solubilities would require less water for transport in solution. The consensus of current geochemical thinking is that the bulk of Gulf Coast petroleum did not generate at these excessively high temperatures. As

GEOLOGICAL EVENT	MIGRATION EFFECT
EARLY BASIN DEVELOPMENT	NET DOWNWARD FLUID FLUX
MATURE BASIN	STATIC WATER BODY – SEDIMENTS MOVE DOWNWARD
HYDROCARBON GENERATION	SOURCE SEDIMENTS MOVE DOWN THROUGH THERMAL WINDOW
HYDROCARBONS DISSOLVE	PORE FLUIDS BECOME SATURATED
GEOTHERMAL GRADIENT CHANGES	ISOTHERMS DEPRESSED
PORE FLUIDS COOL	HYDROCARBONS EXSOLVE
SEPARATE PHASE HYDROCARBONS	SWEPT TO TOP OF CARRIER BEDS
UPDIP MIGRATION	BUOYANCY EFFECT
INTERMITTANT FAULTING	PETROLEUM MIGRATES TO SHALLOWER TRAPS

FIG. 18—A petroleum migration mechanism.

noted earlier, a temperature of 200°F for initial significant generation and 250°F for peak generation is thought to be reasonable.

As we review the proposed migration mechanism for the test area, we see that the timing of geologic events is critical. First, the mini-accumulations could form only if sandstones pass down through a static volume of pore fluids at the time the isotherms are being depressed. Second, the migration of petroleum from scattered small accumulations up fault zones to form large commercial accumulations is favored by intermittent opening of the faults to fluid flow during periods of growth.

Figure 19 shows a structural growth curve superimposed on the stratigraphic and isotherm restoration diagram (Fig. 14) for the test area. The growth curve represents the percent of stratigraphic thickening versus time for a selected well on the downthrown side of the master growth fault as compared to a well on the upthrown side. In Figure 19, the structural movement culminated between the time unit 4 was deposited and unit 5 was deposited. But this was before the time of isotherm depression, so we would not expect much separate phase petroleum to be in a position to migrate up the faults. If the migration mechanism proposed here is valid, it is more likely that migration up the fault occurred during the period of secondary growth which culminated during the unit 7–unit 8 depositional time interval.

The test area reserves of about 55 to 65 million bbl are small compared to the capacity of the traps. This could be due to the fact that only minor structural growth occurred during the pore-fluid cooling period.

Work Needed

There are two aspects of this hypothesis where additional data will help in accepting or rejecting the mechanism:

1. Measurements of petroleum solubilities in water at elevated temperatures and pressures. Price's measurements on hydrocarbon solubilities were made in closed vessels which were temperature controlled, but not pressure controlled. Neither were the measurements made in the presence of light, normally gaseous, hydrocarbons. Careful measurements are needed to determine the solubility in water of selected whole crude oils and associated gases.

2. Measurements of the actual dissolved hydrocarbon content of deep formation waters. If the proposed migration mechanism is valid, then most of the deep pore waters in petroliferous basins should be saturated with hydrocarbons. Very few actual data exist on this point. The most widely cited reference is Buckley et al (1958) which includes hundreds of special water samples from drillstem tests of water-bearing formations. Significant quantities of dissolved hydrocarbons were found in nearly all formations sampled, although they apparently sampled only hydrostatically pressured aquifers. Multiple samples from the Woodbine Sandstone in East Texas and the Frio Formation from the upper Gulf Coast area of Texas were at or near saturation in every well sampled. Their work was mostly on the gaseous hydrocarbons; analytical procedures at the time were not adequate for quantitative determination of the heavier hydrocarbons in solution. A project to sample and analyze deep, supernormally pressured, Gulf Coast formation waters for dissolved hydrocar-

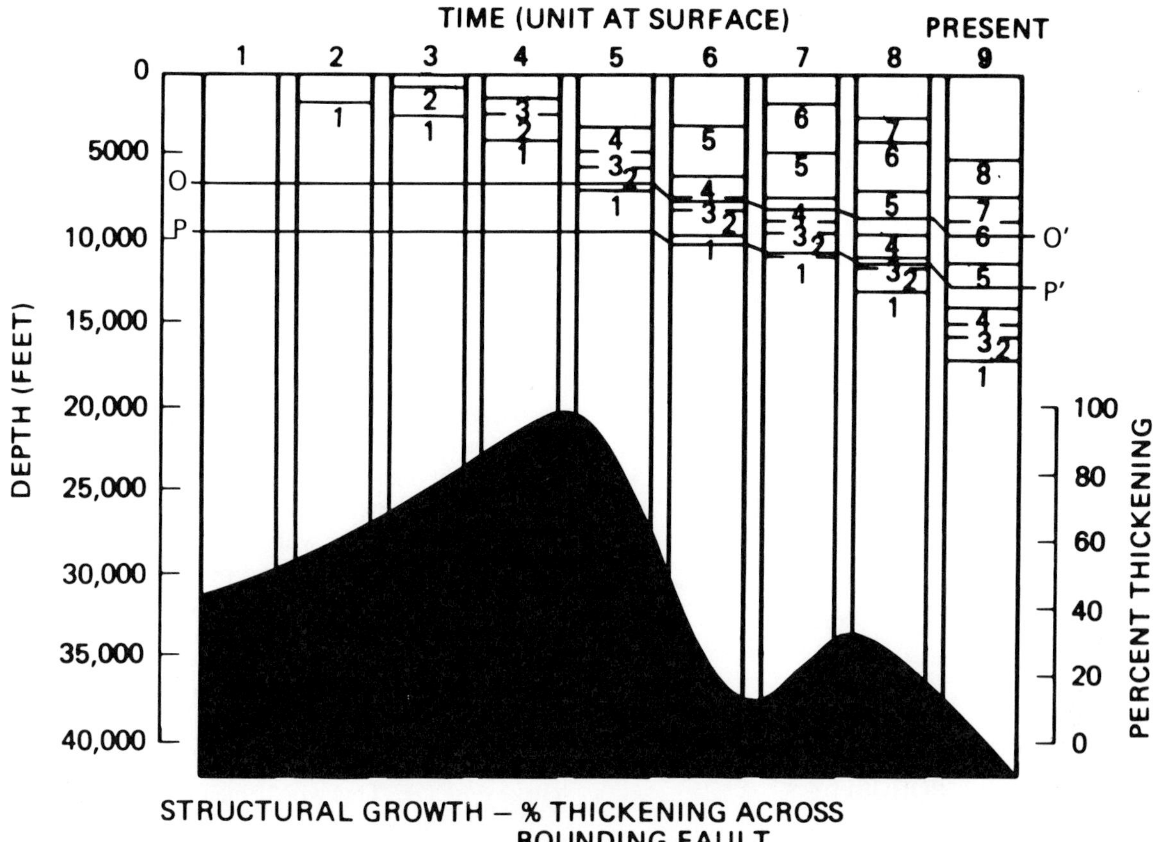

FIG. 19—Structure growth curve superimposed on stratigraphic and isotherm restoration diagram.

bons could add to our knowledge about generation and migration of petroleum.

COMPACTION OF ABNORMALLY HIGH PRESSURED SHALES AND THEIR CONTRIBUTION TO PETROLEUM MIGRATION

The concept was developed here earlier that most of the pore fluids in compacting basins move downward with respect to sea level in the early stages of basin development or remain static in later stages. The question that arises is when, if ever, do fluids move upward with respect to a datum such as sea level. If significant volumes can move from deep, hot zones to shallower, cooler zones, then the moving water can carry hydrocarbons in solution. If the water were saturated with hydrocarbons at a given depth-temperature condition, then some of the hydrocarbon would come out of solution as the water moved upward and cooled off. The purpose of this study is to evaluate this mechanism under geologic conditions representative of actual sedimentary basins.

In Figure 20, the curve ABCDE represents the present depth versus porosity for shales and is typical of Gulf Coast sediments. We believe that the shales above the geopresured zone were normally compacted along curve AB. However, the writer believes the high pressure shales below 12,000 ft were inhibited in compaction from the time of deposition and followed a compaction curve ABCDE. Regardless of history, a shale at point C has abnormally high fluid pressure and a porosity of 20%.

Consider what would happen to a unit volume of shale at point C (a cubic foot of shale can be visualized as 0.8 cu ft of solids and 0.2 cu ft of water). If the abnormally high pressure were to bleed off, the cubic foot of shale at C would lose fluid and move to the left towards B. If all the excess pressure were relieved, then the sediment could be represented by point B on the normal compaction curve at 11.1% porosity (0.1 cu ft of fluid in 0.9 cu ft bulk volume). The cubic foot of shale at C would have lost a volume of water 1 ft square by 0.1 ft deep. Also, the entire stratigraphic column from A to B would have moved downward 0.1 ft with respect to sea level. At the same time, a layer of water 1 ft square by 0.1 ft deep (0.1 cu ft) would be expelled from the sediments at A.

In many instances the 0.1 ft subsidence of the surface would be accompanied by additional sedimentation. But the newly deposited sediments would have about 40% porosity, so 40% of the 0.1 ft deep water volume (0.04 cu ft) would remain in the new sediments and 60% (0.06 cu ft) would be expelled at the surface.

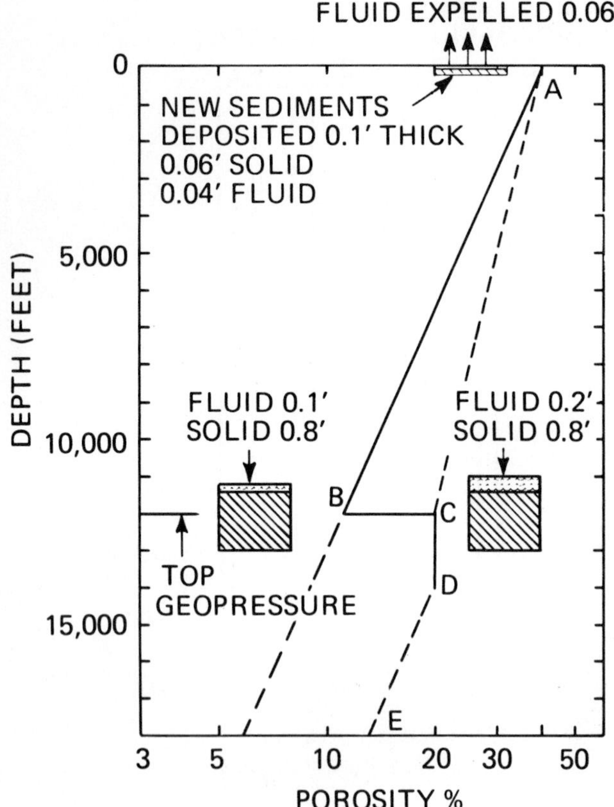

FIG. 20—Fluid bypassing model of volume relationships in compaction of geopressured shales.

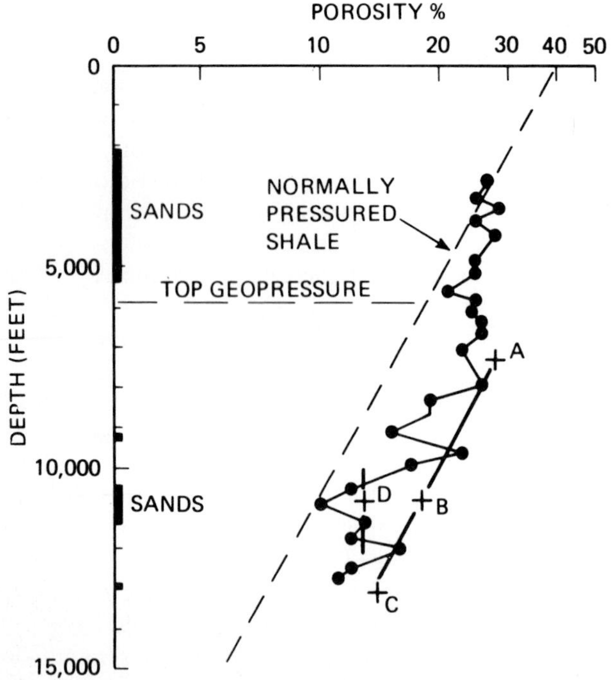

FIG. 21—Measured porosities of shales from a Gulf Coast well.

The point is that geopressured shales with higher than normal porosity will cause an absolute upward flux of water if they move toward the normal compaction curve. This flux need not be directly upward; it will be in the direction of least resistance (i.e., in response to the potential gradient) which can be upward across layers or along faults, laterally along permeable zones, or even downward to permeable sandstones providing there is some hydraulic connection with the surface.

Accepting 200°F as the temperature where significant generation of hydrocarbons from kerogen begins, then the generative zone in most Gulf Coast areas is restricted to the sediments below the top of geopressure. But most of the shales at these depths still have abnormally high porosities. Thus they have not compacted the sediment in a manner to cause an absolute upward flow of fluid which could contribute to petroleum migration. Yet we find petroleum accumulations under these conditions with reservoirs in both the normal and supernormal pressure zones.

One possibility exists under which absolute upward fluid flux could have occurred. This would be if any significant thickness of shale had porosities at an earlier time that lay to the right (ex., were higher than) the curve ABCDE. This would mean that they already compacted some and moved toward the normal compaction curve AB.

For the shales above the top of geopressure (Point B) this could have little or no effect on migration because the temperature in this zone is not high enough for petroleum generation. For geopressured shales, we see no evidence that the porosities have been any higher than those represented by a curve similar to CDE.

In a few places, evidence exists for fluid expulsion from high pressure shales directly above or below a sandstone interval. Figure 21 is an example which shows a deviation of the shale depth-porosity curve toward the normal compaction curve adjacent to sandstone intervals in the 9,000 to 13,000 ft (2,743 to 3,962 m) depth interval. These examples show that these sandstones had some hydraulic continuity with a lower pressured or perhaps a normal hydrostatically pressured zone and presumably to the surface for a limited period of time. These "leaks" are limited either in time or magnitude because the section still has abnormally high pressure and above-normal porosities. In the Gulf Coast area at least, it is likely that these leaks were intermittent and associated with periods of movement along the listric faults in the area. This type of fluid leakage can be called "bypassing," because a limited volume of fluid bypasses the main body of static pore fluids and is transported physically along a pressure gradient, commonly to a shallower zone.

To estimate the effect of "bypassing" on migration of petroleum in water solution, one can calculate the amount of hydrocarbons that might be transported as a result of the "leak" at about 11,000 ft (3,352 m; Fig. 21). Line ABC represents the average porosity curve for the high-pressure shales; D represents the average porosity of the shales near the leaky zone. Thus shales near the leak have been compacted from an original porosity of about 18% at a point B to 12% at point D.

It is estimated that the leak affected shales with about 1,000 ft (305 m) aggregate thickness. Thus 68.2 cu ft of fluid were squeezed out per square foot of area in the 1,000 ft (305 m) of shales.

Assume that these fluids were transported from the leak at 11,000 ft (3,352 m) to the normally pressured zone at about 6,000 ft (1,829 m). In a representative Gulf Coast area, this means that the fluid would move from a depth where the temperature was about 212°F to a depth where it was about 149°F. In other words, the 68.2 cu ft of fluid would be cooled about 63°F. If these fluids were saturated with hydrocarbons at the higher temperature, how much would come out of solution as they moved up into the cooler area? If the petroleum had the solubility of Price's (1976) most soluble crude oil (Flower Community crude), then the saturation value at 212°F is about 87 ppm and the value at 149°F is about 63 ppm. Thus 0.00164 cu ft of petroleum per square foot of area would come out of solution as the fluids moved from a depth of 11,000 (3,352 m) to a depth of 6,000 ft (1,828 m). This is equivalent to a little over 8,000 bbl of oil per sq mi of source area.

We can compare this volme to that which must have been transported to form known accumulations. In the test area, for example, the writer calculated that about 400,000 bbl of hydrocarbon per square mile of source area migrated to form the field. The "bypassing" example calculated above, therefore, could account for only about 2% of the known migration in the test area.

THE BYPASS COMPUTER MODEL

The previous section was based on a single example thought to be representative of Gulf Coast conditions. It seems appropriate to expand the study to a general model; that is, to determine the maximum amount of petroleum that could be transported by bypassing for a broad range of depths, temperatures, and degrees of compaction.

Conditions of the Bypass Model

The important factors to consider to construct the bypass model are the changes in petroleum solubility with temperature, the changes in temperature with depth, and the reduction of porosity with depth, which controls the volume of fluids available for bypassing.

This study used Price's (1976) data on solubilities of whole crude oils in water at elevated temperatures (see Fig. 15); and to establish an upper limit of petroleum migration by bypassing, used his data on the most soluble oil, the Flower Community crude. The highest temperature at which Price determined solubility for this oil was 359°F. It was assumed that solubilities follow the extrapolated linear segment to higher temperatures.

For a depth-temperature curve, the representative data cited previously for the Gulf Coast test area was used. This curve (Fig. 12) has a gradient of 1.25°F per 100 ft (30.5 m) to a temperature of 212°F at about 11, 000 ft (3,352 m). Below that the gradient is 1.8°F per 100 ft (30.5 m).

Next it is necessary to consider what volume of fluid could be expelled from deep shales and bypassed to a shallower zone. For the depth-porosity curves of normally pressured and geopressured shales, the curves in Figure 22 were used and are believed to be representative for Gulf Coast sediments. The curves are extrapolated beyond drilled depths for estimation of normal and geopressured porosities at great depths. In this model it is assumed that no high pressure shale has a porosity greater than the geopressured shale curve in Figure 22, and that if bypassing is permitted by faults, fractures, etc., the shale would compact, but the low porosity limit would be defined by the normally pressured shale curve in Figure 22. The difference in porosity at any depth, then, can be used to

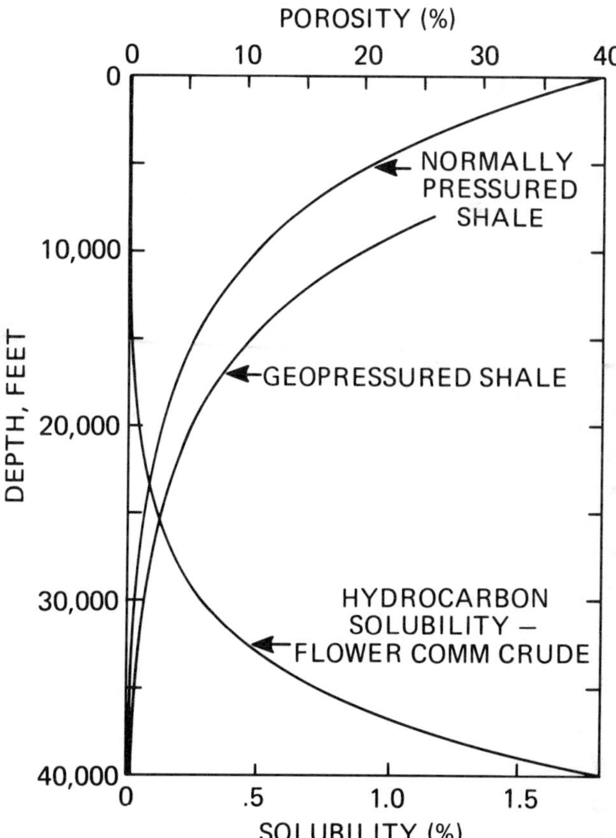

FIG. 22—Porosities of Gulf Coast shales and solubility of crude oil in pore fluids.

calculate the volume of fluid available through bypassing as follows:

$$\frac{V_{bn}}{V_{bg}} = \frac{(1 - \phi_g)}{(1 - \phi_n)} \tag{1}$$

When V_b is the bulk, water-saturated volume of the sediment, ϕ is the decimal porosity, and the subscripts (n and g) represent the normal and geopressured conditions, respectively. For example, the calculation is as follows for the maximum amount of water per square foot of area that could be expelled and bypassed upward from a 100-ft (30.5 m) thick geopressured shale at 10,000 ft (3,048 m): The geopressured shale porosity at 10,000 ft (3,048 m) is 20%; the normal porosity is 11%. From equation (1),

$$\frac{V_{bn}}{100} = \frac{(1 - 0.2)}{(1 - 0.11)} \tag{2}$$

$$V_{bn} = 89.89 \text{ cu ft}$$

$$V_{bg} - V_{bn} = 10.11 \text{ cu ft/sq ft of area}$$

The depth-porosity curves in Figure 23 show that the volume of fluid available for bypassing falls off rapidly with depth. For example, at a depth of 28,000 ft (8,534 m) the water available for bypassing from a 1,000 ft (305 m) geopressured shale is only 8.79 cu ft/sq ft of area.

Changes in pressure are important in bypassing if we are dealing with gas or oil-gas mixtures. Adequate data on solubility of oil-gas systems in water is not available at present. The best solubility data (Price, 1976) are for oils that were exposed to the atmosphere with consequent loss of light components, so this study is restricted to oil-water systems.

Another factor which affects solubility is the salinity of the formation fluids. If salt is added to a hydrocarbon-saturated water solution, some of the hydrocarbon comes out of solution. In the Gulf Coast area, a zone of hypersalinity (concentrations several times that of sea water) commonly overlies a zone of reduced salinity (concentrations about equal to sea water or less). Many geologists have speculated that compaction causes deep, low salinity waters, which become saturated with hydrocarbons, to be injected into shallower zones where the high salinity causes exsolution. However, extensive waterflooding experience tells us that water injection rarely leads to extensive fluid mixing, which would be required for the salting-out effect. Instead, fluid injection commonly results in a piston-like displacement of contained fluids. After injection, diffusion will slowly blur sharp concentration boundaries and eventually homogenize the salinities but the time required is of the order of available geologic time. Salting-out effects should not be ignored, but they are not included in this study because there was not a specified quantitative model.

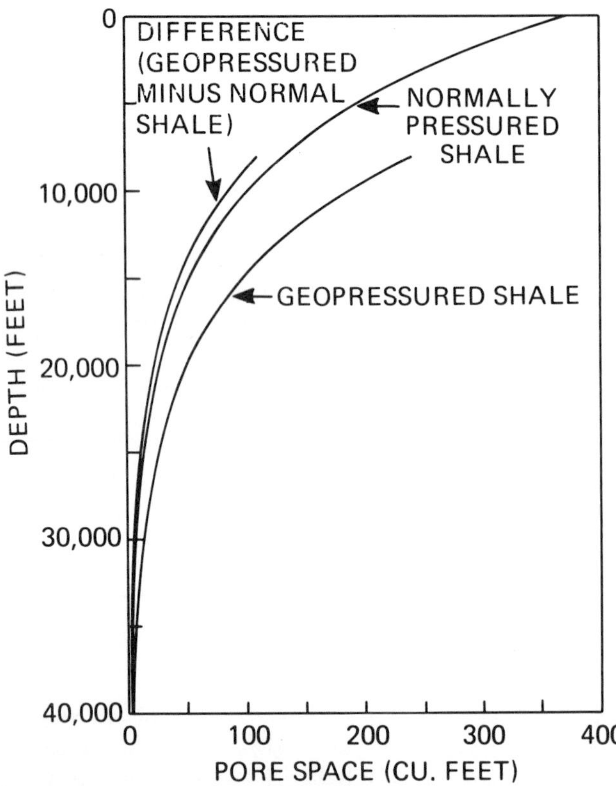

FIG. 23—Volume of pore space in a column of shale 1 ft square by 1,000 ft thick.

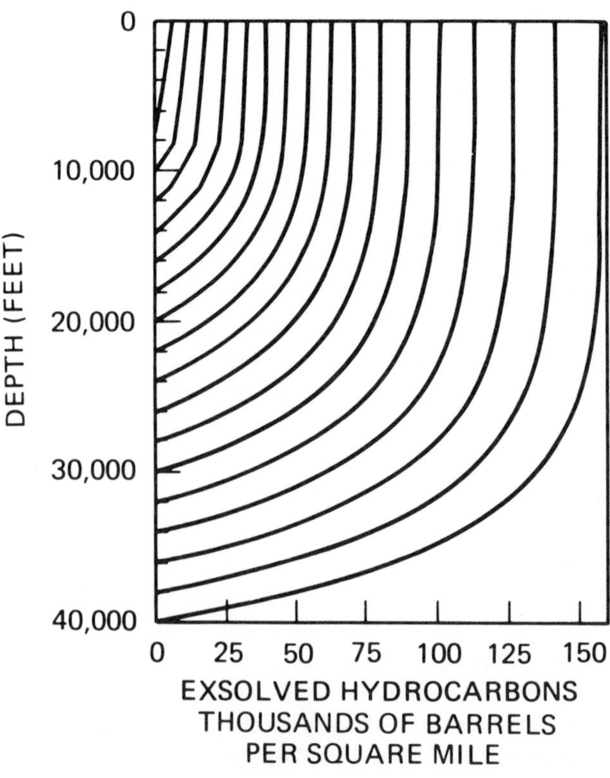

FIG. 24—Petroleum transport by bypassing from 1,000 ft of geopressured shale.

AREA	DRAINAGE AREA (MI^2)	ULTIMATE RECOVERY (BBLS)	$BBLS/MI^2$
1	154	65×10^6	422,000
2	90	360×10^6	4,000,000
3	2806	$2,639 \times 10^6$	940,000

FIG. 25—Petroleum migration requirements for three Gulf Coast areas.

Bypass Model Results

Results of the bypass model calculations are summarized in Figure 24. The model incorporates the conditions and assumptions of temperature, petroleum solubility, and shale porosity described previously. The study started with a 1,000 ft (305 m) section of geopressured shale at a given depth (say, 24,000 ft [7,315 m]). Its porosity was about 3.2% (Fig. 22), its temperature about 446°F, and pore water at these conditions could contain a maximum of about 956 ppm dissolved petroleum (Fig. 15). However, a normally compacted shale at this depth has about 1.73% porosity (Fig. 22). Thus, a column of shale one foot square by 1,000 ft (305 m) thick under these conditions could expel about 15 cu ft of water if it compacted from high pressure to normal conditions. This 15 cu ft of water could contain up to 0.01434 cu ft of dissolved petroleum. If this 15 cu ft were bypassed to some shallower zone (to 8,000 ft [2,438 m]), it would be cooled to a temperature of about 174°F. At this temperature it could hold no more than 63 ppm of petroleum in solution, so up to 893 ppm could come out of solution as a result of the bypassing. This is equivalent to about 65,000 bbls of oil per sq mi. Figure 24 provides results calculated in this manner for fluid expelled from a 1,000-ft (305 m) thick shale at any depth up to 40,000 ft (12,192 m) and bypassed to any shallower depth. Results are given in maximum barrels of oil per square mile that could come out of solution and thus contribute to accumulations.

DISCUSSION OF RESULTS

With the results shown in Figure 24, a better evaluation of bypassing as a petroleum migration mechanism is possible. In the Gulf Coast area, several examples have been found that may be fluid "leaks" in the high pressure section. Most of those are associated with sandstone intervals, which suggests the sandstones at some time have had hydraulic connection with shallower, lower pressured zones. Maximum bypassing would have occurred if the leak developed after the shales reached their present depths, and calculations in the study were made on this basis to establish an upper limit on migration by bypassing. However, it should be kept in mind that leaky zones evidenced by reduced porosities in shales above or below sandy zones could have been active throughout the history of burial and compaction. In this case, migration by bypassing would have been ineffectual.

Figure 24 shows that the maximum amount of hydrocarbons that could be transported and exsolved from a thousand feet of shale even as deep as 40,000 ft (12,192 m) is about 160,000 bbl per sq mi. For comparison, Figure 25 provides some estimates of the volume of petroleum per square mile of source drainage area in three Gulf Coast areas. And this requires that no exsolution occurs during the transport upwards. However, Figure 24 shows that more than 80% of the dissolved hydrocarbons would be exsolved in the first few thousand feet of ascent (below 17,000 ft [5,181 m] for a source at 25,000 ft)[7,620 m]; below 13,000 ft [3,962 m] for a source at 20,000 ft [6,096 m]).

Figure 24 is compiled from the maximum cumulative exsolution values which implies that hydrocarbons dissolved at the source temperature would remain in solution during bypassing and then come out of solution all at once in response to the lower temperature at a given shallower depth. Actually, there would be continuous exsolution as the fluids were bypassed upwards.

These data indicate that the water from several thousand feet of high pressure shale would have to be released upward from deep burial in order for bypassing to contribute significantly to migration of known accumulations. The evidence that this has not been a major migration mechanism is best provided by the fact that the bulk of the deep Gulf Coast shale retains high pressures and high porosities.

However, in the discussion of bypassing as a mechanism for transporting hydrocarbons in solution, it is important not to overlook the role of bypassing as a means of pushing along hydrocarbons that may already exist as a separate phase. This may be an important mechanism for movement of dispersed droplets and slugs or films along carrier beds or up faults.

SUMMARY AND CONCLUSIONS

A new conceptual model of primary petroleum migration was developed from consideration of the mass balance of fluids in a sedimentary basin. In the early stages of basin development, the fluids move downward and become warmer. Deep basins develop a more or less static body of water with sediments moving downward through the water. This is a dynamic extraction system whereby heated pore waters can become saturated with hydrocarbons as they are thermally generated in the source beds.

Continued basin subsidence, sedimentation, and compaction cause the pore waters with dissolved hydrocarbons to pass out of the source layer, usually upward stratigraphically but not necessarily upward in relation to sea level. Any subsequent cooling causes some hydrocarbons to come out of solution. If the exsolution occurs in porous and permeable carrier beds, the separate-phase hydrocarbons can then migrate to traps in response to buoyancy.

This migration concept was tested in a Gulf Coast area by computer modeling. The known accumulation can be accounted for by this mechanism, but this does not prove that is the way it happened. Most likely this solution-exsolution mechanism is one of several that contribute to primary petroleum migration.

A bypassing mechanism also was modeled wherein deep, hot, high-pressured pore fluids are injected into shallower zones. With reasonable geologic assumptions, this mechanism can account for only a small fraction of known accumulations.

REFERENCES CITED

Athy, L. F., 1930, Compaction and oil migration: AAPG Bull., vol. 14, p. 32.

Bostick, N., 1973, Time as a factor in thermal metamorphism of phytoclasts, *in* 7th International Congress of Carboniferous Stratigraphy and Geology, v. 2, Maastricht: Netherlands, E. Van Aelst, p. 183-193.

Buckley, S. E., C. L. Hocott, and M. S. Taggart, Jr., 1958, Distribution of dissolved hydrocarbons in subsurface waters, *in* Habitat of oil: AAPG, p. 850-882.

Demaison, G. J., 1974, Relationships of coal rank to paleotemperatures in sedimentary rocks: Colloque Internat. Petrographie de la matiere organique des sediments, Centre National de la Recherche Scientifique (Paris), Sept., 1973.

Dickey, P. A., C. R. Sheram, and W. R. Paine, 1968, Abnormal pressures in deep wells of southwestern Louisiana: Science 160, p. 614.

Dott, R. H., and M. J. Reynolds, 1969, Sourcebook for petroleum geology: AAPG Memoir 5, p. 189.

Hood, A., C. C. M. Gutjahr, and R. L. Heacock, 1975, Organic metamorphism and the generation of petroleum: AAPG Bull., v. 59, p. 986-96.

Hubbert, M. K., and W. W. Rubey, 1959, Role of fluid pressure in mechanics of overthrust faulting: Geol. Soc. America Bull., v. 70, p. 115-116.

Laplante, R. E., 1974, Hydrocarbon generation in Gulf Coast Tertiary sediments: AAPG Bull., vol. 58, p. 1281-1289.

Levorsen, A. I., 1967, Geology of petroleum: San Francisco, W. H. Freeman Co., Inc.

Lopatin, N. V., 1971, Temperature and geologic time as factors in coalification: Akad. Nauk. SSSR Izv. Ser. Geol., no. 3, p. 95-106 (English translation by N. H. Bostick, Illinois Geol. Survey, 1972).

Momper, J. A., 1972, Evaluating sourse beds for petroleum (abs.): AAPG Bull., v. 56, p. 640.

Pirson, S. J., 1977, Geological well log analysis: Houston, Gulf Publishing Co., p. 261.

Price, L. C., 1976, Aqueous solubility of petroleum as applied to its origin and primary migration: AAPG Bull., vol. 60, p. 213-224.

Rieke, H. H., and G. V. Chilingarian, 1974, Compaction of argillaceous sediments: New York, Elsevier, p. 7.

Teichmüller, M. and R. Teichmüller, 1966, Geological causes of coalification, *in* Coal science, advances in chemistry, Ser. 55: Am. Chemical Soc., p. 133-155.

Wolf, K. H., and G. V. Chilingarian, 1976, Compaction of coarse-grained sediments: New York, Elsevier, p. 371.

Zieglar, D. L., and J. H. Spotts, 1978, Reservoir and source-bed history of Great Valley, California: AAPG Bull., vol. 62, p. 813-826.

Oil and Gas Migration: Chemical and Physical Constraints[1]

By Clayton D. McAuliffe[2]

Abstract Primary migration of oil in aqueous solution is not possible because the composition of dissolved hydrocarbons is vastly different from that of crude oils. Migration of oil solubilized in surfactant micelles is also rejected because of the large amount of surfactant required, and because there has been no demonstration that micelles are formed in source rocks. Migration by oil-droplet expulsion also is not feasible, because of the high interfacial forces of small droplets within fine-grained source rock; in addition, at least 7.5% organic matter by volume would need to be converted to oil to attain 30% oil saturation required for separate-phase flow; even higher oil saturations would be required for "squeezing" oil from pores.

It is proposed that oil and gas are generated in, and flow from, source rock in a three-dimensional organic-matter (kerogen) network. Oil or gas flowing in this hydrophobic network would not be subject to interfacial forces until it entered the much larger water-filled pores in the reservoir rock. Oil saturation in the kerogen for oil flow to occur is indicated to be from 4 to 20%.

Secondary migration of separate-phase oil and gas should occur by buoyancy, when their saturations attain 20 to 30% along the upper or lower surfaces of the reservoir rock. Oil or gas entering at the lower surface would intermittently cross the rock when the buoyancy head became sufficient. Efficient migration from source to trap could then occur as rivulets along the upper few centimeters in the reservoir rock. The volume of conducting reservoir rock attaining oil or gas saturation during secondary migration should be small, with most of the pores remaining water-filled.

In contrast, secondary migration of gas or oil in solution would be very inefficient and require large volumes of water. Unless all pores in the reservoir rock attained 20 to 30% gas or oil saturation, separate-phase flow could not occur, and oil and gas would remain locked in the pores and would not form reservoirs in trap positions. Attaining a 30% pore volume (PV) gas or oil saturation would require a flow of about 90 to 200 PV of gas-saturated water, and 15,000 to 200,000 PV of oil-saturated water. Residual gas and oil in cores taken along suspected secondary-migration pathways should show this residual gas or oil saturation, and recovered water should always contain equilibrium concentration of dissolved hydrocarbons, but this has seldom been observed.

The proposed mechanisms of primary migration of oil and gas through a kerogen network, and secondary migration by separate-phase buoyant flow do not require the flow of water. Water flow probably disperses water-soluble constituents instead of concentrating them in reservoir traps.

INTRODUCTION

It is generally accepted that crude oil is generated from organic matter deposited with fine-grained sediments, and little, if any, is generated in reservoir rock. It is a further consensus that most oil is formed at temperatures between 60 and 150°C (140 and 302°F) corresponding to burial depths of about 1,500 to 4,500 m (4,921 to 14,760 ft) in areas with normal geothermal gradients (Philippi, 1965; Tissot et al, 1971; Cordell, 1972; Dickey, 1975; Hunt, 1975). During this generation period, primary oil migrates from the source rock to the reservoir rock, which is followed by secondary migration through the reservoir rocks to trap positions.

At depths below 1,500 m (4,921 ft), the porosity of fine-grained clastic sediments becomes 25% or less, with pores having diameters of 50 to 100 Å (Hobson and Tiratsoo, 1975). Such small pores inhibit the movement of individual oil droplets because of high interfacial tensions (Hobson and Tiratsoo, 1975). It is also assumed that oil and gas must move from the source rock to reservoir rock with water, and this has led to suggested migration mechanisms such as aqueous solubility of petroleum (Meinschein, 1959, 1961; Kartsev et al, 1971; Price, 1976) and petroleum solubilized in surfactant micelles (Baker, 1959, 1960, 1962, 1967; Cordell, 1973). These mechanisms depend upon complex sequences of temperature, salinity, pH, and ionic composition, all associated within a geologic framework. This assumption has led to numerous studies (summarized by Cordell, 1972; Price, 1976) on the origin of water flow out of fine-grained rocks by compaction, and from montmorillonite to illite clay

[1]Manuscript received, June 6, 1978; accepted, November 20, 1978. This paper also has been published in the AAPG Bulletin (v. 63, p. 761-781).

[2]Chevron Oil Field Research Company, La Habra, California 906 31.

Article Identification Number:
0149-1377/80/SG10-0006/$03.00/0.

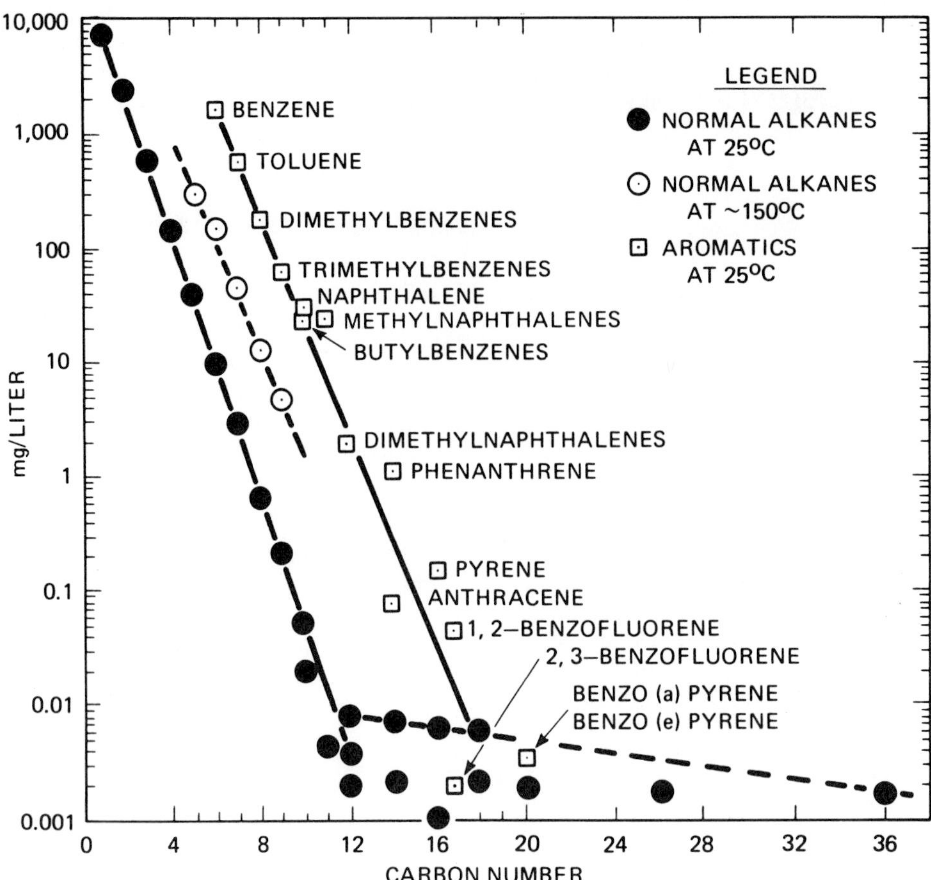

FIG. 1.— Solubilities of normal alkane and aromatic hydrocarbons in water (data from McAuliffe, 1963, 1966, 1969; Baker, 1967; Wauchope and Getzen, 1972; Sutton and Calder, 1974, 1975; Button, 1976; Eganhouse and Calder, 1976; Price, 1976; Schwartz and Wasik, 1976; and Mackay and Shiu, 1977).

transformation. The emphasis has been on water flow, and comparatively little attention has been given to the movement of oil and gas from source rock to reservoir rock by other migration mechanisms.

In this paper various migration mechanisms are reviewed, with emphasis on certain aspects that have not previously been considered.

Secondary migrations by separate-phase buoyant flow, and in water solution, are also discussed in order to evaluate adequately the overall movement of oil and gas from source to trap. Hydrocarbon solubilities are reviewed because of their role in several of the proposed migration processes. For a comprehensive review of petroleum generation and migration, the reader is referred to Cordell (1972).

SOLUBILITIES OF HYDROCARBONS IN WATER

Pure Hydrocarbon Solubilities

In recent years the solubilities in water of a relatively large number of individual pure hydrocarbons have been determined (McAuliffe, 1963, 1966, 1969; Baker, 1967; Wauchope and Getzen, 1972; Sutton and Calder, 1974; Button, 1976; Price, 1976; Schwartz and Wasik, 1976; Mackay and Shiu, 1977). Figure 1 summarizes many of these solubilities.

There is a marked decrease in solubility with increase in molecular weight (carbon number) for each class of hydrocarbons present in petroleum (alkanes, cycloalkanes, and aromatics). For normal alkanes, the solubility decreases six to seven orders of magnitude between carbon numbers 1 and 12. For aromatics, the solubility decreases similarly between carbon numbers 6 to 24 (Fig. 1). The line for cycloalkane hydrocarbons, when plotted, is parallel with that for aromatics, but is slightly to the right of the normal alkane line (McAuliffe, 1966). Values derived from the lines in Figure 1 show that addition of one carbon atom to the molecule (for example, pentane versus hexane; toluene versus benzene) decreases solubility by 75% for normal alkanes and 70% for aromatics.

The data in Figure 1 also demonstrate that aromatics are much more soluble than alkanes for a given carbon number. For example, hexane, cyclohexane, and benzene, each with six carbon atoms in the molecule, have respective solubilities of 9.5, 60, and 1,750 mg/L (McAuliffe, 1966; Price, 1976). The values for the hydrocarbons with seven carbon atoms, heptane, methycyclohexane, and toluene, are 2.5, 15, and 530 mg/L, respectively. Thus, benzene is 185 times more soluble than hexane, toluene 210 times more than heptane. The aromatic to n-alkane solubility ratio increases (Fig. 1) so that dimethylnaphthalenes are over 600 times more soluble than n-C_{12}.

Figure 1 shows an apparent change from true solubility (individual hydrocarbon molecules dispersed in the aqueous phase) to one of accommodation when the number of carbon atoms in the n-alkanes exceeds 12. Baker (1967) attributed this accommodation to individual alkane hydrocarbons associating in the water phase to form dimers or higher molecule number aggregates. Thus, n-alkanes having from 12 to 36 carbon atoms are accommodated in water to about the same degree. The n-alkane solubility data shown in Figure 1 were measured at 25°C. It is not known if this accommodation exists at higher temperatures where Brownian motion is much greater and should be more disruptive of aggregates. Such disaggregation has been observed for surfactant micelles, inasmuch as micelle formation requires higher concentrations of surfactant at higher temperatures.

Other factors influencing hydrocarbon solubilities in water are temperature, salinity, and pressure. Increasing temperature increases hydrocarbon solubilities if pressure is maintained on the system (Price, 1976). Solubility increases are moderate up to 125°C and become significant only above 150°C. Price showed that the increase is somewhat more rapid for the higher carbon numbered normal alkanes than for the lower. This relative increase in solubility with increase in carbon number is, however, relatively insignificant, as shown by the dashed line for C_5 through C_9 n-alkanes in Figure 1. The plotted values at 150°C (136°C for nonane) are from Price (1976). The slope is changed slightly, but it is still apparent that hydrocarbon solubilities decrease very rapidly with increase in carbon number.

Price (1976) also concluded that increasing temperature greatly increased the relative solubilities of aromatic hydrocarbons compared with the corresponding carbon number cycloalkanes and n-alkanes. Cycloalkane solubilities increased slightly more than alkane. However, with limited data, more studies are required to confirm this hypothesis. If additional investigations confirm these preliminary data, the aromatic dominance shown at 25°C would be further accentuated.

Hydrocarbon solubilities decrease with increase in water salinities. McAuliffe (1971) found solubilities of toluene in 3.5 and 20% NaCl to be 70 and 16% of that in distilled water. For 12 aromatic hydrocarbons (C_7 to C_{10}), Sutton and Calder (1975) found that compared with distilled water, the mean reduction in solubilities at 25°C in seawater was 68 ±4.4%. Price (1976) measured the following solubilities in 20% NaCl solution compared with distilled water: pentane, 15%; benzene, 22%; toluene, 19%; methylcyclopentane, 14%.

Increasing pressure decreases solubilities of liquid hydrocarbons in water but to a much lesser extent than the changes induced by temperature and salinity (Price, 1977).

Hydrocarbons Dissolved from Crude Oil

The preceding discussion deals only with the solubilities in water of pure, individual hydrocarbons. This section will consider hydrocarbons dissolved from crude oils, as related to crude oil compositions.

If water is equilibrated with crude oils, hydrocarbons will be found in the water phase, in concentrations related to the water solubilitiy of each pure hydrocarbon and its mole fraction concentration in the crude oil. Table 1 summarizes the measured concentrations of individual C_1 to C_{10} hydrocarbons in waters equilibrated with representative crude oils (Anderson et al, 1974; Rice et al, 1976; McAuliffe, 1977). The crude oils used in these experiments had most of the gas separated. Thus the concentrations of methane through pentanes in water are relatively low, particularly the lower molecular weight hydrocarbons such as methane and ethane.

Table 2 presents the solubilities in water of the higher molecular weight hydrocarbons from two of the crude oils (Anderson et al, 1974). The data in Tables 1 and 2 reflect the general observation drawn from Figure 1 as to the solubilities of pure hydrocarbons. Although most of the gas had been separated from the crude oils, the remaining C_1 to C_5 hydrocarbons in the oil produced higher C_1 to C_5 hydrocarbon concentrations in water than the sum of the C_6 and C_7 saturate fraction which then exceeded the higher molecular weight hydrocarbons. For example, concentrations of hydrocarbons in 2.0% NaCl solution equilibrated with Kuwait crude oil (Table 1) decreased as follows: C_1 to C_5, 10.4 mg/L; C_6 and C_7, 1.2 mg/L; C_{12} to C_{24}, 0.004 mg/L. The latter value is from Anderson et al (1974). If the C_1 to C_5 hydrocarbons were present as in reservoir crude oil, their concentration in water would be very much higher, reflecting both their relatively high concentrations in crude oils and their relatively high solubilities in water.

A similar solubility decrease with increase in carbon number is shown for the aromatic hydrocarbons. The concentrations (in mg/L) for south Louisiana crude oil (Table 1) decrease in the following order: benzene plus toluene, 10.9; dimethylbenzenes, 1.96; trimethylbenzenes, 0.76; naphthalenes, 0.30; biphenyls, fluorenes, and phenanthrenes, 0.1. These decreases are very pronounced, even though the number of hydrocarbons within the groupings (for both saturated and aromatic hydrocarbons) increases with molecular weight; that is, two hydrocarbons (benzene plus toluene) versus 20 to 30 hydrocarbons for the last group of aromatics.

Table 1. Hydrocarbons Dissolved in Water Equilibrated with Oil Samples (a)

Crude	Murban	La Rosa	South Louisiana	Kuwait	Cook Inlet	Prudhoe Bay
Water Salinity	Sea	Sea	2% NaCl	2% NaCl	Sea	Sea
Published Source	(McAuliffe, 1977)		(Anderson et al, 1974)		(Rice et al, 1976)	
Hydrocarbons	Concentrations in mg/L					
Methane	––	.26	––	––	.81	––
Ethane	.23	2.01	.54	.23	3.63	.01
Propane	2.15	3.63	3.01	3.30	10.70	.79
Isobutane	.80	.76	1.69	.90	1.90	.33
n-Butane	2.88	1.88	2.36	3.66	5.02	1.50
Isopentane	1.03	.60	.70	.98	1.02	.40
n-Pentane	1.34	.60	.49	1.31	1.11	.56
Hexanes + cyclopentane	.85	.50	.38	.59	.96	.50
n-Hexane	.50	.15	.09	.29	.21	.13
Methylcylopentane	.35	.27	.23	.19	.63	.32
Benzene	6.08	3.30	6.75	3.36	8.10	9.30
Cyclohexane	.41	.19	(b)	(b)	.80	.24
n-Heptane	.33	.10	.06	.09	.16	.11
Methylcyclohexane	.23	.16	.22	.08	.42	.21
Toluene	6.16	2.80	4.13	3.62	4.10	6.58
Ethylbenzene	.82	.27	(c)	(c)	.35	.49
m-, p-Xylene	1.94	.84	1.56	1.58	1.06	1.58
o-Xylene	1.01	.35	.40	.67	.45	.52
Trimethylbenzenes	.75	.30	.76	.73	.29	.34
C_1-C_5 Alkanes	8.43	9.48	8.79	10.38	24.19	3.59
C_6 + C_7 Saturates (d)	2.67	1.37	.98	1.24	3.18	1.51
Total Saturates	11.10	11.11	9.77	11.62	27.37	5.10
Benzene + Toluene	12.24	6.10	10.88	6.98	12.20	15.88
Dimethylbenzenes (C_8)	3.77	1.46	1.96	2.25	1.86	2.59
Trimethylbenzenes (C_9)	.75	.30	.76	.73	.29	.34
Naphthalenes (Table 2)	––	––	.30	.07	––	––
Biphenyls, Fluorenes, and Phenanthrenes (Table 2)	––	––	.01	.01	––	––
Total Aromatics	16.76	7.86	13.60	9.96	14.35	18.81
Total Hydrocarbons	27.87	19.97	23.37	21.72	41.72	23.90
Benzene/Hexane	12	22	75	12	39	72
Toluene/Heptane	19	28	69	40	26	60
(Benzene + Toluene / Total Aromatics) x 100	73	78	80	70	85	84
(Benzene + Toluene / Total C_6+ Hydrocarbons) x 100	64	66	75	62	70	78

(a) All analysis performed at Chevron Oil Field Research Company
(b) Included with benzene
(c) Included with m-, p-Xylene
(d) Saturated (alkane plus cycloalkanes)

The pronounced dominance of benzene and toluene in waters equilibrated with oil are shown as percentage calculations (Table 1). Benzene plus toluene constitute 70 to 85% of the aromatic hydrocarbons, and 62 to 78% of the total C_6+ hydrocarbons (saturates plus aromatics).

It is apparent from Tables 1 and 2 that benzene and toluene are the predominant hydrocarbons dissolved in waters equilibrated with crude oils. They are present in about the same concentration, despite the approximately 3.3 times higher solubility of benzene compared with toluene (1,760 versus 530 ppm). This reflects the approximately three times higher concentration of toluene in crude oils compared with benzene for the oils shown. The group-type analysis for Murban and La Rosa crude oils (Table 3) confirms the higher toluene concentrations.

Hydrocarbons dissolve in water from crude oils in accordance with their pure solubility in water and mole fraction concentrations in the crude oil. Because we do not know the exact molecular weights of crude oils, we cannot calculate mole fractions but they may be estimated (or weight percent is sufficiently close for these discussions). In Table 1, the ratios of benzene and toluene to their corresponding carbon number n-alkanes are calculated. The data show that benzene is present in water concentrations from 12 to 75 times higher than hexane, and toluene is 19 to 69 times higher than heptane. These concentration ratios are high, but they are not as high as they would be if benzene and toluene were present in crude oils in the same concentrations as hexane and heptane. The pure hydrocarbon solubilities (McAuliffe, 1966; Price, 1976) have the following ratios: benzene/hexane, 185; tou-

Table 2. Higher Molecular Weight Aromatic Hydrocarbons Dissolved in 2% NaCl Solution Equilibrated with Two Crude Oils*

Hydrocarbon	South Louisiana Crude Oil	Kuwait Crude Oil
	Concentrations in mg/L	
Naphthalene	0.12	0.02
1-Methylnaphthalene	0.06	0.02
2-Methylnaphthalene	0.05	0.008
Dimethylnaphthalenes	0.06	0.02
Trimethylnaphthalenes	0.008	0.003
Biphenyl	0.001	0.001
Methylbiphenyls	0.001	0.001
Dimethylbiphenyls	0.001	0.001
Fluorene	0.001	0.001
Methylfluorenes	0.001	0.001
Dimethylfluorenes	0.001	0.001
Dibenzothiophene	0.001	0.001
Phenanthrene	0.001	0.001
Methylphenanthrenes	0.002	0.001
Dimethylphenanthrenes	0.001	0.001
Totals	0.309	0.081

* From Anderson et al (1974).

Table 3. Group-Type Analysis of the Naphtha Fraction of Two Crude Oils

Crude Oil	LaRosa		Murban	
API Gravity @ 15.6°C	**23.9°**		**39.0°**	
Sulfur (wt. %)	**1.73**		**0.82**	
204°C Minus Fraction	**11 vol. %**		**19 vol. %**	
	Weight %			
	In Naphtha Fraction	In Oil	In Naphtha Fraction	In Oil
Benzene	0.6	0.07	0.7	0.13
Toluene	2.0	0.22	2.6	0.49
C_8 Aromatic	3.4	0.37	4.6	0.87
C_9 Aromatic	2.7	0.30	3.9	0.74
C_{10} Aromatic	1.3	0.14	1.8	0.34
C_{11} Aromatic	0.5	0.06	0.7	0.13
C_{12} Aromatic	0.2	0.02	0.2	0.04
C_{13} Aromatic	0.1	0.01	0.0	0.00
Naphthalenes	0.0	0.00	0.0	0.00
Indans	0.5	0.06	0.4	0.08
Total Aromatics	11.3	1.25	14.9	2.82
Alkanes	46.7	5.14	65.8	12.50
Cycloalkanes	38.3	4.21	17.5	3.32
Dicycloalkanes	3.7	0.41	1.8	0.34
Total Hydrocarbons	100.0	9.76	100.0	16.16

lene/heptane, 210. Analyses in Table 1 indicate that benzene and toluene are each present in the crude oils at average concentrations of only 7% of those of hexane and heptane.

Crude Oil Composition

The hydrocarbon compositions of the representative crude oils are markedly different from the hydrocarbons found in water in equilibrium with them, both as to molecular-weight distribution and amounts of the three classes of hydrocarbons.

Crude oil is a complex mixture, with hydrocarbon molecular sizes ranging from 1 to more than 40 carbon atoms. Crude oils contain the three principal hydrocarbon classes (alkanes, cycloalkanes, and aromatics) in approximately equal concentrations. Mixed molecules such as cycloalkane-aromatic are present in the higher molecular weight fraction. In addition, crude oils contain small amounts of oxygen, sulfur, and nitrogen as heteroatoms generally in large organic molecules. Table 4 shows the composition of a typical crude oil, although it should be realized that the compositions vary widely.

Each hydrocarbon class contains a range of molecules with different carbon numbers best exemplified by the normal alkanes (n-paraffins or waxes). Excluding C_1 to C_5 gases, n-alkanes often show n-C_6 or n-C_7 as slightly predominant in concentration, with decreasing amounts of n-alkanes as the carbon number increases to n-C_{40} and beyond. Individual branched alkanes decrease more rapidly with carbon number because of the increased number of isomers per increased carbon number. The same holds true for cycloalkane and aromatic hydrocarbons.

Table 3 shows the group-type analysis for two of the crude oils reported in the water equilibrium experiments in Table 1. Toluene is the predominant aromatic, followed by the four C_8 aromatics (o-, m-, p-xylene, and ethylbenzene). Benzene is next in concentration, followed by decreasing amounts of the other aromatic hydrocarbons with increasing carbon numbers. The relative concentrations of benzene and toluene in the oils are as predicted from the amounts found in the water. Toluene is present in the oils in concentrations about 3.3 times that of benzene (within experiment error).

Table 4. Typical Crude Oil Composition

Molecular Size	Weight Percent
C_4-C_{10}, gasoline	31
C_{11}-C_{12}, kerosene	10
C_{13}-C_{20}, gas oil	15
C_{21}-C_{40}, lube oil	24
>C_{40}, residuum	24
Molecular Type	**Weight Percent**
Alkanes	30
Cycloalkanes	49
Aromatics	15
Asphaltics	6

Tables 4 and 5 show that 70 or 80% of crude oils consist of hydrocarbons larger than C_{10}. Many other analyses confirm this, including those of Martin and Winters (1959), Martin et al (1963), Bailey et al (1973), and Milner et al (1977). The low-molecular-weight aromatic hydrocarbons are generally present in amounts equal to or lower than the corresponding carbon number cycloalkanes and alkanes. Thus, hydrocarbons in crude oils are vastly different from those in water solution, as related to both molecular size and molecular type.

Table 5. Gasoline-Range Hydrocarbons in Stoughton Crude Oil*

Hydrocarbon	Wt. %
Benzene	0.891
Toluene	0.887
Cyclopentane	0.151
Cyclohexane	0.783
Methylcyclopentane	0.580
Methylcyclohexane	0.610
1, C, 2-Dimethylcyclopentane	0.060
2, 3-Dimethylbutane	0.072
3-Methylpentane	0.547
2-Methylpentane	0.628
n-Pentane	1.016
n-Hexane	1.009
n-Heptane	0.945

*After Bailey et al (1973).

DISSOLVED HYDROCARBONS AS POSSIBLE PRIMARY MIGRATION MECHANISM

Crude Oils

The preceding discussions on the composition of hydrocarbons dissolved in water show that solution is unlikely to be a mechanism for petroleum migration from source rock to reservioir rock.

Several studies (Philippi, 1965, 1974, 1975; Tissot et al, 1971; Dow, 1974; Hunt, 1977; and others) have shown a correlation between petroleum in source rocks and in reservoired oil, so that there appears to be little question that oil found in a given reservoir originated in a specific source rock. If solution were a significant migration mechanism, we should be able to contact petroleum with water in a source rock and move it relatively unchanged to a reservoir trap.

The water in equilibrium with the crude oils shown in Table 1 would be representative of the process. If this water were moved to a new trap and the hydrocarbons brought out of solution (mechanism unspecified), the resulting "oil" would be dominated by hydrocarbons of lower molecular weight. Benzene and toluene would comprise 65 to 80% of the C_6+ hydrocarbons. In many oils this dominance would be even greater. Bailey et al (1973) found a Stoughton crude oil (Table 5) to contain a higher percentage of benzene and toluene than in Murban and La Rosa crude oils (Table 3). Milner et al (1977) gave the weight percents of benzene plus toluene in crude oils as follows: Malmo, 1.1; Leduc, 0.8; Westerose, 1.2; Innisfail, 1.1; and Simonette, 1.4.

Although the concentrations of hydrocarbons in solution will be higher than at 25°C (77°F), temperatures from 60 to 150°C (140 to 302°F), the range for most petroleum generation, are unlikely to change significantly the composition of dissolved hydrocarbons.

Natural Gas

Natural gas consists principally of methane, with decreasing amounts of ethane-through-pentane hydrocarbons (usually these are a few percent). Gas may migrate principally by separate-phase buoyant flow, as discussed later, but it may also migrate in true solution. Methane and other gaseous hydrocarbons have relatively high solubilities in water as shown in Figure 1 (McAuliffe, 1969). Conybeare (1970) presented the solubility of natural gas in waters of different salinities, and as a function of temperature and pressure on the basis of the data of McKetta and Wehe (1962).

The solubility of natural gas decreases with increasing temperature (to about 80°C and then increases), increases with increasing pressure, and decreases with increasing salinity. Table 6, adapted from Conybeare (1970), shows a few of the data points. Temperature and salinity are shown to have smaller effects on natural gas solubility than that produced by pressure change.

Methane is produced throughout the petroleum-generation process, becoming the principally produced hydrocarbon as temperatures become high with deep burial of organic matter (time-temperature effects). Despite the relatively high solubility of methane in water, and the production of methane throughout petroleum generation, migration in solution may not be the principal mechanism for natural gas, which is discussed in more detail in a later section on secondary migration.

OTHER SUGGESTED MECHANISMS FOR OIL MIGRATION IN WATER FROM SOURCE ROCKS

Several other methods have been proposed for the migration of crude oil from source to reservoir rocks in association with water.

Oil Solubilized in Surfactant Micelles

Baker (1959, 1960, 1962, 1967) proposed the solubilization of hydrocarbons in source rocks within surfactant micelles. More recently Cordell (1973) has supported this concept in a detailed analysis of the proposed process.

Micelles are molecular aggregates formed in solutions of surfactants. Surfactants or surface-active agents (also called detergents, dispersants) are compounds that orient at an interface such as between oil and water. The molecules have a nonpolar (hydrophobic) end (usually an n-alkyl hydrocarbon chain containing 8 to 18 carbon atoms) attached to a polar, water-soluble group. For crude oils, the principal group is carboxylic, that is, organic acids (R-COOH, where R can be a saturated linear or branched chain, a cyclic saturated ring [cycloalkane], or aromatic ring; Seifert, 1975).

In general, micelles are capable of solubilizing an amount of oil approximately equivalent to the amount of surfactant in the micelle (Rehfeld, 1970). Micelles of ionic surfactants (the type that may form or be present in organic matter of source rocks) have aggregation numbers ranging from 10 to 100. They are slightly flattened spheres with interiors resembling those of liquid hydrocarbon droplets (Fisher and Oakenfull, 1977). Baker (1967) gave their diameters as about 60 to 100 Å. Prince (1977) identified microemulsion micelles as having diameters of 2,000 Å (0.2 μm) down to 100 Å (0.01 μm).

Surfactant molecules in the micelles are in constant

Table 6. Solubility of Natural Gas in Fresh Water and Seawater at Various Pressures and Temperatures*

Hydrostatic Pressure, psi (1 psi=0.07 kg/cm²)	Equivalent Depth in Feet (1 ft=0.3048 m)	Temperature, °F of Equivalent Depth (2°F/100 ft)	Solubility in cu ft/bbl at 100°F (37.8°C)	150°F (65.6°C)	200°F (93.3°C)
Fresh Water (Hydrostatic Pressure Gradient = 43.3 psi/100 ft)					
1,000	2,308	106	8.7	7.6	7.2
3,000	6,924	199	17.6	15.2	15.4
5,000	11,540	291	22.5	19.3	20.1
Brine: 34,00 mg/L - Seawater (Hydrostatic Pressure Gradient = 44.3 psi/100 ft)					
1,000	2,258	105	6.8	6.3	6.0
3,000	6,774	195	14.2	12.6	13.2
5,000	11,290	286	18.3	16.0	17.6

*Adapted from Conybeare (1970).

equilibrium with those in true solution. The exchange is very rapid, with the surfactant molecule having a residence time in the micelle of about 10^{-5} sec with a range from 10^{-2} to 10^{-9} sec (Fisher and Oakenfull, 1977).

Micelles do not form until a critical micelle concentration (CMC) is reached. The CMC is a rather narrow concentration range, below which the surfactant molecules are in true solution (molecular dispersion) and above which the excess surfactant molecules form more and more micelles. Table 7 summarizes the CMC values for several surfactants (Mukerjee and Mysels, 1971). The CMC is shown to decrease with increased length of the hydrophobic hydrocarbon tail. However, the larger the surfactant molecule, the larger the micelle.

The likelihood of micellar migration of petroleum has been questioned by different workers, including Hobson (1961), Hedberg (1964), Erdman (1965), Hunt (1967), Dickey (1975), Hobson and Tiratsoo (1975), and Price (1976). Their objections are based upon (1) the requirement of many times more surfactant than hydrocarbon for hydrocarbon solubilization; (2) explaining the origin and fate of the large quantities of surfactant required; (3) the lack of micelle detection in the sedimentary section; (4) the fact that micelles appear to be larger than the pore throat constrictions in source rocks; (5) the electrical barrier caused by negative charges on both micelle and mineral surfaces; (6) adsorption of surfactant onto minerals as evidenced by failure of surfactant flooding in petroleum reservoirs; (7) the need for the sodium salt of organic acids to be produced by alkalinity in source rocks to form micelles; (8) the apparent lack of a mechanism for breaking down the micelles in traps; and (9) the apparent lack of sufficient water flow to carry micelles from the source rock.

All of these problems must be overcome, but the need for large amounts of surfactant and the large size of micelles relative to pore-throat constrictions in source rock appear to be the major constraints on this mechanism.

Price (1976) pointed out that at least 500 ppm of surfactant is needed to attain micelle formation at the CMC. He further quoted Kennedy (1963) as showing that the CMC increases markedly with increase in temperature, and that surfactant adsorption on clays would consume 1% surfactant solution.

For example, consider a source rock with 1% by weight of uniformly distributed organic matter and a rock density of 2.2 (25% porosity). The organic matter content would be 2.2% of the bulk volume and 8.8% of the pore volume, leaving 91% water. If the source rock contained an estimated 600 ppm oil (300 ppm $C_{15}+$ fraction; Hunt, 1961, 1967), and an estimated 300 ppm C_1 to C_{14} fraction, the oil would be 0.53% pore volume. If we assume (1) 1% adsorption loss to mineral surfaces, (2) that 1,000 ppm (0.1%) of surfactant is needed (Table 7) to form micelles, and (3) that as much surfactant as oil is required to solubilize the oil, the following minimum amounts of surfactant would be needed in each pore, as percent of pore volume: adsorption on minerals = 1.0%; to attain CMC in water = 0.1%; to solubilize 0.5% oil = 0.5%; total = 1.6%. Thus, 18% of the organic matter in the rock would need to be surfactant, either preserved or formed from the originally deposited organic matter. Even if the source rock contained a much higher percentage of organic matter, the likelihood of attaining sufficient surfactant to attain micelle formation does not appear probable.

Table 7, Critical Micelle Concentrations for Several Surfactant Solutions*

Surfactant	Length of Hydrocarbon Tail	Type of Polar Group	CMC, %
Sodium decanoate	C_{11}	Carboxylic	0.56
Sodium hexadecanoate	C_{15}	Carboxylic	0.096
Sodium octadecanoate	C_{17}	Carboxylic	0.055
Sodium undecyl l-sulfate	C_{11}	Sulfate	0.41
Sodium undecly sulfonate	C_{11}	Sulfonate	0.18

* From Mukerjee and Mysels (1971).

There are two items not listed previously that should also be considered: (1) oil-in-water microemulsions versus water-in-oil microemulsions, and (2) capability of micelles to produce a crude oil of given hydrocarbon composition from oil of different composition in source rock.

The original proposal of micelle transport of solubilized oil assumed that an oil-in-water system would develop. However, it is easier to prepare water-in-oil microemulsions than the oil-in-water type (Shinoda and Kunieda, 1977). Thus, water might just as logically be incorporated into oil in the source rock. Organic acids have been found in waters associated with crude oils in reservoirs; this has been used to imply that these materials are present in source rocks. They probably are present in relatively low concentrations. However, if emulsions are formed during production of crude oils (not in the reservoir, but as the oil flows into the wellbore or through the pumping or production system), they are apparently always the water-in-oil type, which would indicate that unless the surfactant composition in oil reservoirs is very different from that in source rocks, oil-in-water dispersions are not likely to form there.

Baker (1959) found the micellar solubility of toluene to be between 2.5 and 5.7 times that of benzene. In(1960) he stated, "In an analogous manner, the characteristic distribution of the heavier n-paraffin hydrocarbons may also be explained. Since the micellar solubilities of the n-paraffin hydrocarbons show no periodicity, but decrease smoothly with increasing molecular weight, it is apparent from the solubility measurements reported by Baker (1958) that the formation waters would transport evenly distributed, decreasing relative amounts of n-paraffin hydrocarbons such as characterize crude oils. All that would be required is that the sediments contain an excess of n-paraffin hydrocarbons over that needed to saturate the formation waters."

This statement is incorrect, because hydrocarbons in water would equilibrate with oil in the source rock according to their individual pure solubilities in water, and their mole fractions in the oil. If the n-alkanes in source rock oil have an odd-over-even predominance of 2.0, the odd-over-even predominance in water associated with this oil would remain 2.0 (assuming for this discussion that solubilities are unchanged with increase in molecular weight, and the mole fractions are the same).

Oil Droplet Expulsion

Solution and micelle-migration mechanisms were evolved in attempts to overcome the difficulties encountered when very small oil droplets flowing with water meet even smaller pore-throat constrictions in source rocks. Very high interfacial forces of deforming small droplets through small pore-throat constrictions have been discussed (Levorsen, 1954; Welte, 1965; Baker, 1967; Hobson and Tiratsoo, 1975; Cartmill, 1976). Cartmill, for example, stated, "Extremely small droplets of oil in a water-wet medium would be like balls of steel when deformation became necessary for passage through constrictions of very small pore throats."

Another important constraint is the amount of oil available for movement through a water-wet source rock. On the basis of laboratory core studies of reservoir rocks (much larger pores and pore-throat constrictions), oil concentrations of 20 to 30 vol. % are required for oil to flow with water. Assuming that interfacial tensions can be reduced in fine-grained source rock so that oil would flow at 30% pore volume (PV), the required oil saturation would not be produced by the conversion of all organic matter to crude oil in the example given previously (2.2% volume organic matter, or 8.8% of the pore volume). To obtain sufficient oil would require an organic-matter content very much higher, and/or heterogeneous rock with portions having high organic matter contents. In the latter condition, the oil from pores with high oil contents would have to escape without appreciably leaking oil into pores with little or no oil.

If this were a valid mechanism, a source rock that had generated and flowed oil at 30% PV would retain this degree of oil saturation. Further, all source rocks would retain this residual oil saturation (e.g., 6% by weight for 20% porosity). Source rocks with this high oil content are not common, if they exist at all (Hunt and Jamieson, 1956; Philippi, 1965; Schrayer and Zarrella, 1966).

Hobson and Tiratsoo (1975) discussed the suggestion that oil droplets can be driven from pore to pore by compaction, but concluded that this does not seem to be a particularly efficient mechanism. As compaction occured, porosity reduction would first squeeze out water, which would move more easily through a water-wet system. Squeezing of oil from pores would require essentially complete filling with oil which would then require a very high percentage of the rock to be organic matter, and a very high percentage conversion of it to oil. Further, as for 20 to 30% oil saturation, all source rocks that had generated and migrated oil by this mechanism should still have pores filled with oil, unless they were thermally matured beyond the oil-preservation limit.

Conceivably, rich source rocks that are heterogeneous could have a water-free system in parts of the bedding planes. The oil could be squeezed out along bedding planes subsequently cut by fractures, allowing the oil to enter reservoir rock (Momper, 1978).

Migration of Soluble Nonhydrocarbon Organic Compounds

Hunt (1968) suggested the possibility that some more soluble parental substance, rather than hydrocarbons, undergoes primary migration. Such compounds might be ketones, acids, and esters. However, these organic compounds are present in rather low concentrations in source rocks, and are subject to adsorption to mineral surfaces (Cordell, 1972). Of equal importance is the problem of why these organic compounds would be preferentially retained in the reservoir traps: that is, if they were in true solution in water and migrated from source rocks, why wouldn't they flow with water through the reservoir-rock trap? Water may actually disperse the soluble compounds, including hydrocarbons, and be destructive of petroleum accumulations (with the possible exception of natural gas).

CRUDE OIL MOVED IN GAS PHASE

Migration of liquid hydrocarbons dissolved in gases has been proposed. The occurrences of gas condensates at considerable burial depths illustrate the solubility of liquid hydrocarbons in gas. However, it is doubtful that gas can contain the heavier hydrocarbons and N-S-O compounds in high-molecular-weight crude oil fractions. Certainly the ratios of crude oil to gas in many reservoirs rule out migration in a gas phase, unless one postulates that most of the gas escaped from the reservoir, leaving the crude oil behind.

Separate-phase flow of gas through the aqueous phase in pores of fine-grained source rocks probably does not occur. The small gas bubbles would incur the same resistance as for oil to deformation in pore-throat constrictions. The gas-water interfacial tension

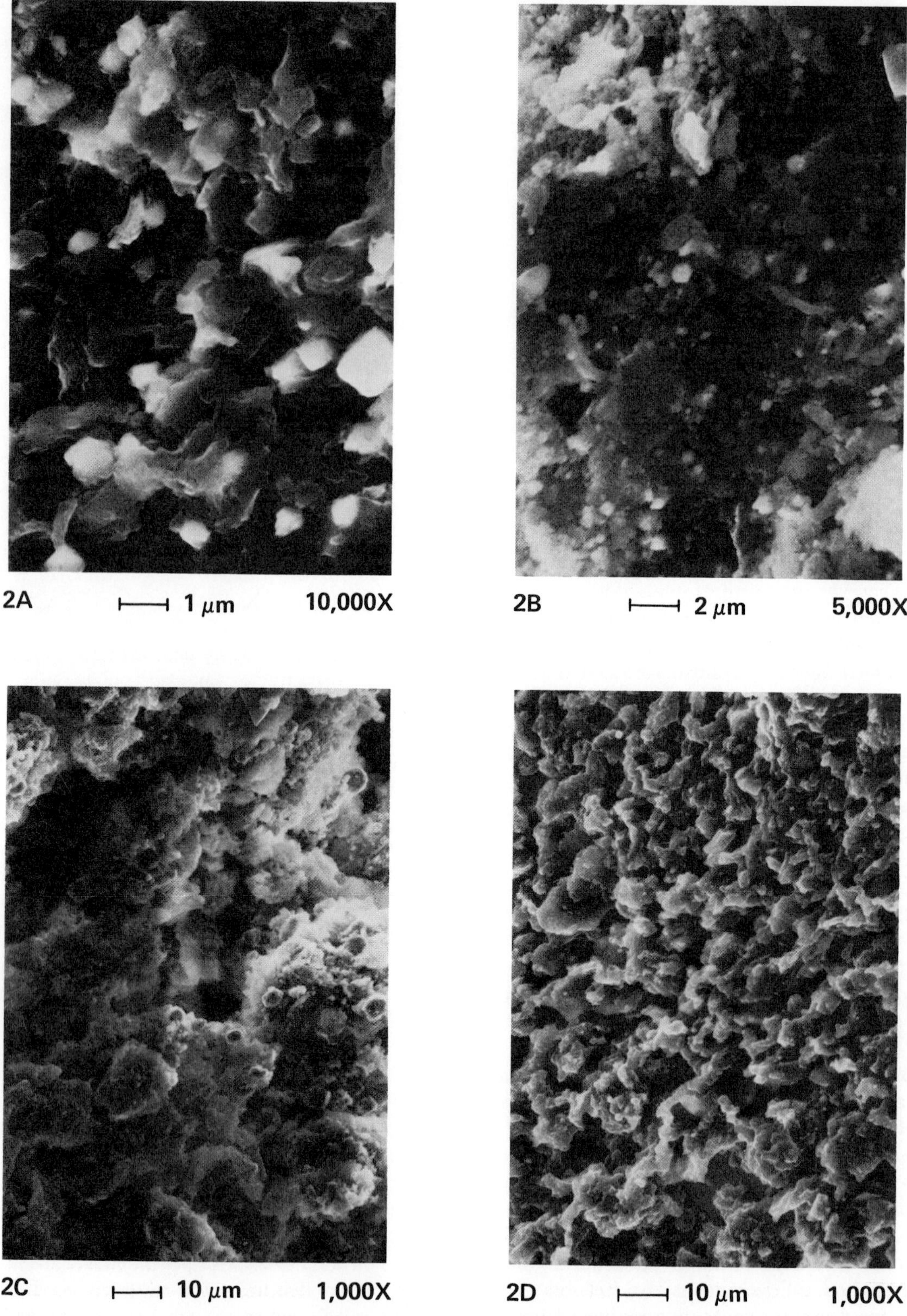

FIG. 2. —Scanning electron micrographs of kerogen in shale samples. Samples were extracted with carbon disulfide to remove hydrocarbons and permit good contact of acids with minerals, then were subjected to acid treatment (hydrochloric acid followed by hydrofluoric) to remove mineral matter. **A.** 1,980 m (6,500 ft), 1% organic matter; **B.** 2,400 m (7,900 ft), 4.3% organic matter; **C, D,** about 2,680 m (8,800 ft), 5.7% organic matter. White cubic structures in **A, B** are pyrite.

is somewhat higher than crude oil–water interfacial tension.

PETROLEUM MIGRATION FROM SOURCE ROCKS THROUGH THREE-DIMENSIONAL ORGANIC MATTER (KEROGEN) MATRIX

The chemical and physical constraints previously discussed illustrate that petroleum (with the possible exception of natural gas) cannot migrate from source rocks to reservoir rocks by entering into and moving with water by processes proposed to date. I suggest that oil and gas are generated in an organic matrix and flow through it to reservoir rock, where droplets of crude oil or gas bubbles form and attain sufficient pore volume (20 to 30% to coalesce and move, by buoyancy, toward a trap position. The flow of water and petroleum are probably independent processes. This concept has apparently been given little consideration because of two assumptions: a three-dimensional kerogen network does not exist; and reservoir rock along the migration pathway to a trap does not show 20 to 30% residual oil saturation.

Organic Matter Network

Erdman (1965) suggested that oil migration may be along fine strands of kerogen, and if the kerogen content were too low, no vehicle for migration would exist. More recently Dickey (1975) suggested that primary migration of oil from source rocks is in a liquid phase. He suggested that some of the interior surfaces of the rock are oil-wet, and that a large fraction of the water in the pores is structured and effectively solid. If this is true, the saturation of oil, in terms of the total pore liquid, may become very large (>50%). Under these conditions, he proposed that oil may flow more easily than water. The mechanism for creating a continuous thread of oil through the pore system was not addressed, and he apparently did not consider petroleum moving through kerogen.

It is proposed here that oil may flow through organic matter or kerogen when the petroleum concentration in the organic phase is sufficiently high. Figure 2 shows scanning electron micrographs of kerogen in four shale samples, which might be considered source rock on the basis of their organic-matter contents of 1.0 to 5.7%. Organic matter equals the extractable organics plus 1.22 times the organic carbon. Organic carbon is determined by combustion of samples following solvent extraction of hydrocarbons and treatment with acid to remove carbonate. All four micrographs clearly show a three-dimensional network.

If the proposed mechanism is to work, petroleum must be able to flow through the kerogen network. Even if we assumed that generated oil enters the aqueous phase, it must first migrate from the interior of the kerogen shown in Figure 2. The cross-sectional dimensions of the kerogen appear to approach several micrometers, although there may be microstructure not shown in the electron micrographs. Momper (1978) described the very large sizes of kerogen particles, compared with the mineral particles, in source shales. The large size difference suggests that it would be difficult to visualize nonconnection of the organic matter, particularly when it approaches or exceeds 5% by weight (10 to 12% by volume). An analogy might be straw in a bucket of marbles. As compaction occurs and the organic matter is altered, its large particles fuse and become continuous and essentially hydrophobic. High interfacial forces would prevent water flow in the organic matter (kerogen) network. Water would flow through the portions of the rock consisting of the closely packed small mineral particles with water-wet surfaces.

Petroleum, however, may be able to flow through this three-dimensional network with ease, much like oil through a wick. Even a very low permeability would be adequate because of the long time available for petroleum to flow from source to reservoir rock. Differential pressure to cause flow can result from the relatively greater compaction of source shales compared with the more structured reservoir rock; by expansion caused by gas formation; by volume expansion associated with oil generation; or by thermal expansion of oil with increasing temperature resulting from deeper burial. If the kerogen were hydrophobic, interfacial tension between oil and water would not exist. Only when oil or gas entered larger, water-filled pores, as in the reservoir rock or in fractures, would the oil contact water. As the oil and/or gas flowed to these larger pores, droplets or bubbles would grow until the oil or gas saturations became sufficiently high to coalesce and move by buoyancy.

Oil Concentration in Kerogen Required for Oil Flow

The petroleum saturation in the organic matter required for flow may be relatively low considering the organic matter and hydrocarbon content of source rocks reported by Hunt and Jamieson (1956), Philippi (1965, 1975), Schrayer and Zarrella (1966), and Hunt (1967).

Table 8, adapted from Hunt (1967), shows typical values for organic matter and hydrocarbon contents of source rocks reported by many investigators. There are wide variations in the contents of both organic matter (0.1 to 13%) and hydrocarbons (70 to 3,800 ppm) in these various suspected source rocks, but correlation between the two is relatively good. The correlation is reflected in a much narrower range in percent hydrocarbon in the organic matter, from about 2 to 17% with a mean of 5.4%. If only shales are considered, the range is only 2 to 6%, with a mean of 3.1%.

Philippi (1975) found a similar wide range in organic carbon and hydrocarbon contents. Hydrocarbons in the organic matter of possible source rocks ranged

Table 8. Distribution of Hydrocarbons and Organic Matter in Nonreservoir Rocks*

Rock Type	Hydrocarbons (ppm)	Organic Matter (weight %)	Hydrocarbons in Organic Matter, %
Shales			
Wilcox, La.	180	1.0	1.80
Frontier, Wyo.	300	1.5	2.00
Springer, Okla.	400	1.7	2.35
Monterey, Calif.	500	2.2	2.27
Woodford, Okla.	3,000	5.4	5.56
Limestones and Dolomites			
Mission Cyn. Limestone, Mont.	67	0.11	6.09
Ireton Limestone, Alta.	106	0.28	3.79
Madison Dolomite, Mont.	243	0.13	16.7
Charles Limestone, Mont.	271	0.32	8.47
Zechstein Dolomite, Denmark	310	0.47	6.60
Banff Limestone, N.D.	530	0.47	11.3
Calcareous Shales			
Niobrara, Wyo.	1,100	3.6	3.06
Antrim, Mich.	2,400	6.7	3.58
Duvernay, Alta.	3,300	7.9	4.18
Nordegg, Alta.	3,800	12.6	3.02
	Mean		5.4
	Mean Less Two High Samples		4.1

*After Hunt (1967).

from 2 to 11% with a mean of 4.8%. In 1965 Philippi reported the average percentages of hydrocarbons in organic matter in some ancient sediments from California, Alberta, Colorado, Venezuela, Colombia, Texas, and Montana to range from 3 to 8% with a mean of 5.0%. A factor of 1.22 was used to correct organic carbon to organic matter (Hunt and Jamieson, 1956).

Philippi (1965) hypothesized that organic matter can dissolve and hold hydrocarbons because of its hydrophobic nature. From available data he suggested that the sorptive capacity of the organic matter is normally exceeded when hydrocarbons are 3 to 12% (2.5 to 10% based on organic matter content) of the noncarbonate carbon.

Hunt and Jamieson (1956) discussed possible petroleum generation in the Frontier formation in the Powder River basin in Wyoming. This formation is fairly uniform in thickness, and contains about 274 m (900 ft) of shale. About 61 m (200 ft) of sandstone, in the form of lenses, is distributed through the formation. Because the sandstone bodies are completely surrounded by shale, it is probable that the oil trapped in them is from the Frontier shale. Hunt and Jamieson's calculations indicate that the oil in the shale is about 2.5% of the total organic matter in the formation. All these data suggest that oil is able to flow through a kerogen network when its content is between about 2.5 and 10%. The hydrocarbons extracted and measured in these studies were the $>C_{15}$ fraction. If the $<C_{15}$ fraction is added (see Table 4) the oil contents would approximately double (5 to 20%).

Richer source rocks have greater capacity to generate larger amounts of oil as shown by Barker's (1977) correlation of increasing number of giant oil fields with increasing organic carbon content of shale. However, source rocks with lower content of organic matter may still have the capacity of furnishing appreciable amounts of petroleum.

The degree of oil saturation required for oil and gas to flow through the kerogen network may depend upon the type of organic matter and the degree of alteration during petroleum generation. The flow of oil through kerogen allows migration throughout the petroleum-generation period. As oil is formed it can flow continuously to reservoir rocks independent of water flow. The two may be correlated, because they are subjected to the same pressure differential produced by shale compaction.

Petroleum generation probably is slow, thereby allowing oil and gas to flow at a slow rate. The kerogen network need not be so great as to have many connections in all directions. Conceivably the network in shale may be rather complete in two dimensions (along bedding planes), but with only a few interconnections in the third dimension. Because of the slow flow rates, these sparse cross-connections are sufficient for oil and gas flow to reservoir rocks or to fractures.

The amount of organic matter may become too low,

however, to produce the necessary interconnection. About 0.5 to 1.0% organic matter (usually the estimated lower limit for a shale to be a source rock) may not be due to a lack of petroleum generation but, rather, to an insufficient three-dimensional kerogen network to provide continuous flow for primary oil migration.

Residual Oil Saturation During Secondary Migration

Secondary petroleum migration by separate-phase oil flow has been discounted owing to the absence of high residual-oil saturations (estimated 20 to 30 % in the reservoir rock along the path to a trap. Cartmill (1976) suggested that oil migration as a continuous phase through reservoir rock to a trap would be very inefficient. Most of the oil might be lost as residual-oil saturation along the migration path, with little oil arriving at the trap. He also suggested that oil saturation should easily be found in wells drilled off structure in the migration pathway. The general assumption has been that much of the reservoir rock along migration pathways should contain residual oil.

To the contrary, secondary migration of oil as a continuous phase is probably the most efficient mechanism for moving oil and gas from source rocks to reservoir traps, leaving a minimum of residual oil along the migration pathway. Hobson and Tiratsoo (1975) have described how oil emerging from source rocks would accumulate along the interface with the reservoir rock until saturation was sufficient for oil droplets to coalesce and flow updip by buoyancy. Oil entering from the bottom of the reservoir rock would collect along that interface until buoyancy was sufficient to allow it to cross to the top of the reservoir rock.

The height of the vertical column of oil required for oil to flow by buoyancy is determined by the rock's pore-throat constrictions (ratios of curvature of the leading and following portions of an oil drop or oil mass extending through several pores) and the density difference between the oil and water (Hobson and Tiratsoo, 1975; Cartmill, 1976). Cartmill (1976) indicated that a fine-grained sandstone conducting low-gravity oil would require a buoyant head of 3.0 m (10 ft). Aschenbrenner and Achauer (1960) estimated the buoyant head needed to be about 35 cm (1 ft) for a medium-coarse sandstone.

It is generally believed that the separate-phase oil migrates updip in the uppermost few centimeters of the carrier rock. The thickness depends upon the dip of the reservoir rock, which determines the length of continuous-phase oil along the top or bottom of the reservoir rock required to attain a buoyant head for flow. For a 9° dip, the length would be approximately 30 m (100 ft) for a 3-m head, and 3.5 m (12 ft) for a 35-cm head. Oil accumulation on the bottom side of the reservoir rock from an underlying source would therefore attain sufficient buoyancy to start across the formation every 30 m or 3.5 m for the examples given. If the dip were less, cross-formational flow would occur at greater distances. Conceivably this flow could have a small cross-sectional area, leaving most of the rock water-saturated.

Because the top of the reservoir is unlikely to be a perfect plane, the movement of oil may be restricted to only a part of the upper surface as rivulets. Likewise, a nonuniform plane would result in small pools where buoyant depressions occurred.

Thus, oil moving by buoyancy from source rock to reservoir trap can leave a residual oil saturation only in a relatively small volume of the conductor reservoir rock. Because of small pore-throat constrictions of the smaller pores, oil would not enter. The small pores through which the oil moved (also those in the trap) would remain water saturated.

With only a small volume of reservoir rock possessing residual oil saturation, it is unlikely that it would have been noticed in wells drilled outside the reservoir limits if, indeed, attempts have been made to find this residual oil in seemingly water-saturated cores. Even if attempts were made, a well conceivably could be cored from top to bottom of the reservoir rock and not encounter residual oil saturation. Other cores might show a small amount of oil which would be attributed to a small reservoir (or show) rather than to oil migrating as a separate phase.

Hobson and Tiratsoo (1975) suggested that residual oil saturation may be lower when fluid-flow rates are much lower than those used in laboratory core studies and oil field waterfloods. High rates would favor disruption of stringers of oil at pore-throat constrictions that might not occur with slow flow.

The rate of buoyant flow of oil would depend on the dip of the conducting reservoir rock, but may not be appreciably slower than oil field waterfloods. In terms of geologic time required for petroleum generation and migration from source rock, separate-phase buoyant flow of oil and gas would be almost instantaneous. Therefore, buoyant flow would be by slugs (interrupted flow). Oil and gas would flow once buoyancy were attained, but replenishment from the source rock would be too slow to maintain continuous separate-phase oil from source to trap.

Secondary oil migration as separate-phase flow can occur by buoyancy alone. Water flow is not required other than that due to displacement by oil. Water, of course, does flow through reservoir rock and, depending upon its direction, would facilitate or impede the buoyant flow of oil.

The flow of oil by separate phase from source rock to reservoir trap minimizes exposure of oil to water, thereby, the loss of water-soluble constituents. Most of the water in the conducting reservoir rock may be well undersaturated with these water-soluble hydrocarbons.

By analogy, the flow of natural gas would be the

same as for crude oil, with consideration for differences in interfacial tensions, densities, and solubilities.

SECONDARY PETROLEUM MIGRATION IN SOLUTION OR BY MICELLES IS INEFFICIENT

In this section, secondary migration of gas, oil, and micelles in solution are discussed. Although oil-in-water solution and solubilization in surfactant micelles have been shown to be highly improbable, if not impossible, their secondary migration also is considered to illustrate the additional constraints imposed on these proposed mechanisms.

Natural Gas Flow in Solution

Figure 3 is a schematic diagram of a sandstone reservoir rock with a trap at position B. Water enters at point A that is just saturated with natural gas (principally methane). As water flows updip through the pore system of the rock, gas will be released as a separate phase as temperature and pressure drop. Water, containing residual dissolved gas at a given temperature and pressure, will flow throughout the conducting reservoir rock (assuming relatively uniform permeability) from points A to C and beyond.

The initial gas release in each pore will be small, but as more and more gas-saturated water arrives at any given point, additional gas, corresponding to its solubility at that temperature and pressure, will add to the separate-phase gas in each pore. Given sufficient flow of gas-saturated water, ultimately all pores (large and small) in the rock will attain higher and higher gas saturations until there is sufficient gas to coalesce and flow from the larger pores by buoyancy. A column of gas then moves to the upper surface of the reservoir rock and on to the trap at B. The separate flow of gas will occur from the larger pores (with larger pore-throat constrictions) first, to give a continuous gas column. Then as gas saturation in the pores increases, flow will be from progressively smaller pores.

Some estimates can be made for this process. For purposes of calculation assume point A (Fig. 3) is at 4,572 m (15,000 ft) and the trap is at 1,524 m (5,000 ft). Bonham (1978), using the data of Culberson and McKetta (1952) and Sultanov et al (1972), plotted the solubility of methane in water for a geothermal gradient of 2.7°C/100 m (1.5°F/100 ft) to the top of supernormal pressure at 2,896 m (9,500 ft), as for the U. S. Gulf Coast, and 3.6°C/100 m (2.0°F/100 ft) below the top of supernormal pressure (Fig. 4). A hydrostatic gradient was assumed to be 100 g/cm^2/m (0.433 psi/ft). If water saturated with methane at 4,572 m (15,000 ft) and 162°C (point A, Fig. 3) moved to 4,267 m (14,000 ft) and 151°C, 650 ppm of dissolved methane would be released (4,850 minus 4,200 ppm), the methane contents at 4,572 and 4,267 m respectively, (Fig. 4). Gas-saturated water flowing from 1,524 m (5,000 ft) to 1,219 m (4,000 ft) and from

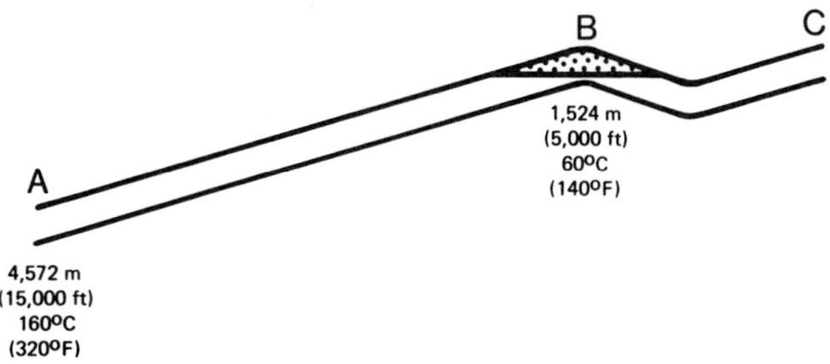

FIG. 3. — Schematic diagram of reservoir rock with trap.

63 to 55°C, would release 125 ppm of methane (1,600 minus 1,475 ppm). Thus, 5.2 times as much gas is released near point A compared with that at the trap.

If sufficient natural-gas-saturated water flows through the system, the question arises as to where the first gas saturation attains 20 to 30% for separate-phase flow to the trap. The preceding calculation shows that 5.2 times as much gas is released at 4,572 m (15,000 ft) as at 1,524 m (5,000 ft). However, the greater amount of gas in the pores at the greater depth would be more compressed than at the shallower depth.

If we assume an ideal gas law, the pore volumes of water flow required to attain separate-phase gas flow can be calculated. The 650 ppm of gas released from 4,572 to 4,267 m would be by weight, 0.65g of methane per liter of water. As one mole equals 22,400 mL of gas, 1 L of water would release 910 mL of gas at standard temperature and pressure (STP). Corrected for temperature and pressure at 4,572 m, the volume would be 910 × (14.7 psi × 435°K/6,495 psi × 273°K) = 3.3 mL/L or 0.33%.

On average, each pore from 4,572 to 4,267 m would contain 0.33% pore volume (PV) of gas phase for each pore volume of natural-gas-saturated water that flowed through the rock. Because the gas volume released is not quite linear (Fig. 4), the pores at 4,572 m would contain slightly more gas phase than those at 4,267 m. To attain 30% PV gas saturation would require 91 PV of water to flow through the reservoir rocks at 4,572 m.

In a similar manner at 1,524 m (5,000 ft), the 125-ppm release of gas would be 0.125 g/L or 175 mL at STP. 175 × (14.7 psi × 336°K/2,165 psi × 273°K) = 1.5 mL/L or 0.15% (200 PV to attain 30% gas saturation). The amount of gas released from 610 m (2,000 ft) to 305 m (1,000 ft) would be 300 ppm; the pressure, respectively, 866 and 433 psi; the temperatures, 37 and 29°C. The 300-ppm release of gas would be 0.30 g/L or 420 mL/L at STP. At 610 m the volume would be 8.1 mL/L or 8.1% (3.7 PV).

These data show that in Figure 3, the first separate-phase gas flow will be updip from the trap. Because buoyant gas flow is probably rapid, compared with the flow of water, the first gas to reach the trap

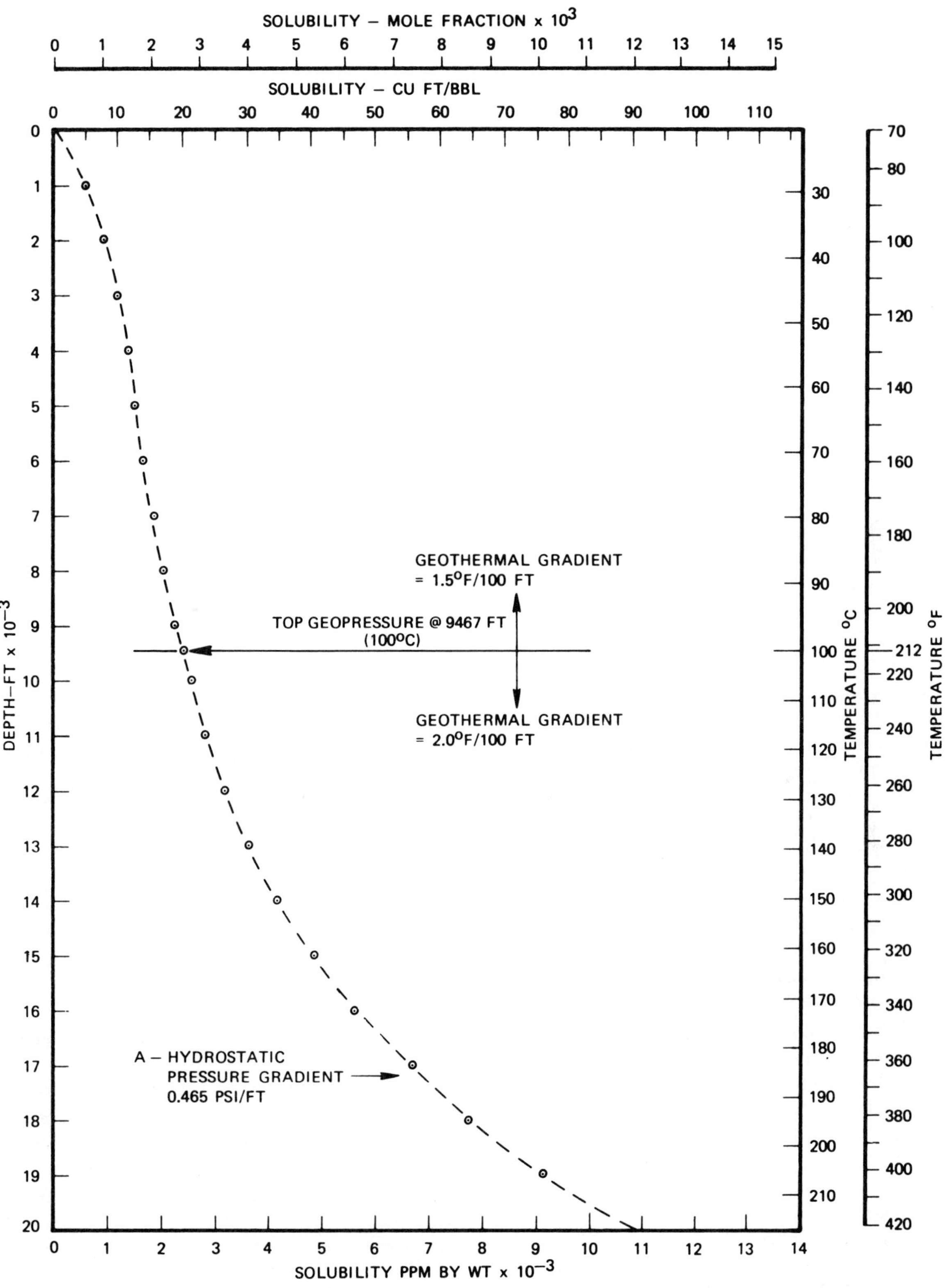

FIG. 4. — Solubility of methane in water at elevated temperatures and pressures (data of Culberson and McKetta, 1962, and Sultanov et al, 1972; modified from Bonham, 1978).

may be from near point A (the first reservoir rock below the trap to attain sufficient gas saturation for separate-phase flow).

Until sufficient gas saturation was attained for separate-phase flow, all the natural gas would be locked in the reservoir rock. Therefore, gas migration in solution would be an inefficient mechanism, compared with gas flowing as a separate phase along the top surface of the reservoir rock. The amount of natural gas-saturated water flowing through the rock determines whether a gas reservoir will form and, if so, its size. When the volume of reservoir rock in three dimensions is compared with the volume of the trap, the inefficiencies become very apparent for flow of gas in solution.

Because of the relatively high solubility of methane in water compared to other hydrocarbons, some gas would be lost to the water by solution during separate-phase gas flow from source rock to the trap. This loss, however, should be minor in comparison to a solution-flow mechanism.

If flow of natural gas as solution in water is a valid mechanism, all of the reservoir rock through which water flowed should contain 20% or more residual separate-phase gas, which would be found in all wells drilled at any point into the reservoir rock. All recovered waters (no gas loss during recovery) should be saturated with gas for the respective temperature, pressure, and salinity of water at its point of collection.

Hydrocarbons Dissolved in Water

Previous discussion shows that migration of hydrocarbons by solution in water is highly unlikely, if not impossible. This conclusion is based on (1) the lack of solubility of high-molecular-weight hydrocarbons compared with low-molecular-weight hydrocarbons, and (2) the markedly greater solubility of aromatic hydrocarbons, compared with the corresponding-carbon-number saturate hydrocarbons.

Without knowing the chemical composition of the dissolved fraction, the solubility versus temperature data from Price (1976) can be used for discussion. Solubility data for a Wyoming Farmers whole crude (gas removed) has been selected because the 20-ppm solubility reported at 25°C is similar to that reported in Table 1.

Using the same temperature and pressure gradients as for gas (Fig. 3) the temperature at 4,572 m (15,000 ft) at point A is 162°C (324°F) and 63°C (145°F) at 1,524 m (5,000 ft) near the trap. The release of separate-phase oil with a decrease in temperature from 160 to 150°C (320°F to 302°F) is about 20 ppm; the release from 60 to 50°C (140°F to 122°F) is approximately 1.5 ppm. The volume of release per 10°C (18°F) at the higher temperature is about 13 times greater. The difference in the release rates for point A and trap positions is greater than for natural gas release, previously discussed. Therefore separate-phase oil flow (if attained) would occur first at point A and advance more slowly toward the reservoir than would natural gas in solution.

Oil, like gas, would be slowly released in all pores of the reservoir rock. If there were a sufficient volume of hydrocarbon-saturated water, the smaller pores would attain oil saturation corresponding to those at which water ceases to flow.

The "oil" released near point A would not only be larger in amount than at the trap, but the composition would be slightly different. Because the hydrocarbons of higher molecular weight are relatively more soluble at high than low temperatures, they likewise will come out of solution preferentially as temperatures decrease. The composition of oil near point A would have slightly higher relative concentrations of hydrocarbons with higher molecular weight. The composition would continuously change slowly until the hydrocarbons in solution in the reservoir rock near the trap and on to point C (25°C) were similar to those in water in equilibrium with crude oils, shown in Table 1. However, the composition of hydrocarbons at point A would closely resemble those at the trap, and not those of a typical crude oil. The hydrocarbons of low molecular weight would predominate over high, and the aromatics (particularly benzene and toluene) would greatly exceed the C_6+ saturated hydrocarbon fraction.

To attain 30% separate-phase oil near point A and at the trap would require respectively 15,000 (300,000 ppm ÷ 20 ppm) and 200,000 (300,000 ppm ÷ 1.5 ppm) PV of hydrocarbon-saturated water, disregarding oil and water volume changes with temperature and pressure. There are no reasonable sources for such large amounts of water below 4,572 m (15,000 ft).

Only one value has been published for hydrocarbons in solution at high temperature (Price, 1977). This is for a crude-oil-distillation fraction (C_{14} to C_{20}) at 272°C (522°F). If one assumes a temperature gradient of 3.6°C/100 m (2°F/100 ft) below 4,572 m (15,000 ft), 270°C (518°F) would be attained at 7,925 m (26,000 ft) and a pressure of 791 kg/cm^2 (11,250 psi). Under these conditions the C_{14} to C_{20} distillation fraction has a solubility of 950 ppm (Price, 1977). If the low-temperature oil solubility data curve (about 140 ppm) at 160°C (320°F) is extrapolated to the 950-ppm value, the slope indicates that about 70 ppm of disolved hydrocarbons would be released for a 10°C (18°F) temperature decrease (272 to 262°C or 522 to (505°F). At 7,925 m (26,000 ft) about 4,300 PV of water would be needed to attain 30% oil saturation (300,000 ppm ÷ by 70 ppm). Again, this amount of hydrocarbon-saturated water below 7,925 m (26,000 ft) does not seem reasonable.

These evaluations indicate that a solution mechanism for crude oil migration is unrealistic. Any oil moving in solution would be trapped in the reservoir-rock pores, below the oil saturation required for separate-phase flow, and a reservoir would not form in the trap.

The preceding calculations for PV water flows of gas or hydrocarbon-saturated waters used freshwater data. If the waters were saline, as is usual, solubilities of hydrocarbons would be decreased (carrying capacities of waters decreased) and the number of PV of water required to attain separate-phase gas or oil flow would be even larger.

Micellar Migration

Secondary migration of oil solubilized in surfactant micelles would also be an inefficient mechanism. Micelles, although particulate in the water phase, are so small that Brownian motion keeps them completely dispersed. Thus, as for dissolved hydrocarbons, there is no gravity segregation, even under high fields of the ultracentrifuge (Prince, 1977). Therefore micelles, if able to escape from the small pores of the source rock, would travel with the water and could pass through all pores in the reservoir rock (it is assumed that the smallest pore-throat constrictions in the reservoir rocks exceed those in the source rock through which the micelles passed).

Different conditions can be postulated for breaking micelles to release the solubilized oil, but a likely mechanism is adsorption of the surfactant molecules on the mineral surfaces of the resevoir rock. The CMC is a narrow concentration range, and the exchange of surfactant molecules in micelle with molecules in water is rapid (10^{-5} sec). Thus, only a slight reduction in the surfactant concentration in the water would cause the micelles to break, releasing the solubilized oil.

Micelles moving into the reservoir rock, for example at point A (Fig. 3), would start to break and continue to do so as surfactant was adsorbed. The amount of oil released would depend upon the adsorptive capacity of the reservoir rock for surfactant. Although adsorption would release oil by breaking micelles, it is questionable whether sufficient adsorption of surfactant would occur to release 20 to 30% oil saturation required for separate-phase oil flow by buoyancy. Because micelles are approximately half surfactant and half oil, the pores would contain 20 to 30% surfactant to attain 20 to 30% oil. This does not seem reasonable. More likely, an adsorption of 1 to 5% surfactant would release 1 to 5% oil. If sufficient flow were available, the water and micelles would flow throughout the reservoir rock and on through the trap. However, buoyant separate-phase oil flow would not occur, and the oil and surfactant would be trapped in the reservoir-rock pores. Cores taken from the migration pathway would reveal both oil and surfactant.

In addition to the other constraints on solution mechanisms, the potential for large loss of natural gas and crude oil to the pores of reservoir rock during secondary migration indicates that solution is unlikely to play a significant role in petroleum migration from source rocks to reservoir traps.

ROLE OF WATER

The evidence indicates that the much more efficient separate-phase flow of gas and oil from source rock to trap is the most likely migration mechanism. Petroleum generated in an organic three-dimensional system would flow through the organic matter to the reservoir rock and then move by separate-phase flow to the trap. For this mechanism, water is involved only in providing buoyancy during secondary migration. Oil, moving always as a separate phase, would be independent of water properties such as salinity, ionic composition, and as solvent for hydrocarbons and organic compounds.

It is then not necessary to account for the large water flows required for solution mechanisms. One should consider water as a dispersant for water-soluble constituents, rather than a method of concentrating gas and oil into reservoir traps. Waters that are in close proximity to gas and oil in source rocks probably do become saturated with hydrocarbons and soluble organic compounds. However, as this water moves into the reservoir rock, it is probably diluted by other waters.

The water in contact with gas or oil along the migration path of separate-phase oil or gas flow from source to trap also would attain equilibrium concentrations, but would represent a small volume. Hydrocarbons and organic compounds probably stay well below saturation as they move with water upward through the sedimentary section. The soluble organic materials may be discharged at the surface or more likely destroyed by bacteria (Winters and Williams, 1969; Bailey et al, 1973, 1974; Milner et al, 1977; Philippi, 1977). These authors have shown how dissolving, "waterwashing" hydrocarbons from oil in reservoirs changes crude oil compositions.

CONCLUSIONS

Chemical and physical constraints make unlikely such primary migration mechanisms for oil as (1) hydrocarbons dissolved in water, (2) oil-droplet explusion with water, or (3) surfactant micelles.

The compositions of hydrocarbons dissolved in water from crude oils are vastly different from those of crude oils, both as to molecular size and molecular structure, which rules out solution as a primary migration mechanism over the usually accepted temperature range of 60 to 150°C for petroleum generation. Very high temperatures are needed to get hydrocarbons into solution in significant amounts. However, there is insufficient water available for secondary migration of oil dissolved in water. Also, all hydrocarbons except methane are thermally unstable at very high temperatures and will break down.

Oil-droplet flow with water is prevented by high interfacial tensions of small droplets attempting to pass through smaller pore throats. Even if this were possible, large amounts of organic matter would be required, and the residual oil saturation in each source rock would be 20 to 30%. Forcing oil out by compaction or volume expansion would require even higher organic-matter contents and even higher oil saturation.

Migration of oil solubilized in surfactant micelles is ruled out by lack of evidence of their formation and, if formed, their ability to flow through fine-grained source rock.

The concept of primary migration by flow of gas and oil through the organic-matter phase has been largely neglected, yet it may be the answer to this perplexing problem. Deposited organic matter has larger dimensions than the fine-grained mineral matter. With compaction the organic matter may fuse and provide a three-dimensional flow network for petroleum as it is generated within the organic phase. Flow of oil or gas through the organic matter (kerogen) to the reservoir rock would not encounter interfacial restrictions.

Once out of the source rock, the now-separate-phase oil and gas would be most efficiently moved (secondary migration) through the reservoir rock by buoyancy. Most of the conducting reservoir rock would remain water wet, and much of the water would remain undersaturated with soluble hydrocarbons. Water flow would not be required for either primary or secondary migration.

In contrast, secondary migration of oil or gas by proposed solution mechanisms would be inefficient and require more water than could be accounted for reasonably. Secondary migration by solution would leave all pores (large and small) in the conducting reservoir rock with 20% or higher oil or gas saturation. All water in the reservoir rock would have equilibrium concentrations of soluble hydrocarbons. These oil and gas saturations and dissolved hydrocarbon concentrations have not been observed in reservoir rock through which petroleum has migrated, which rules out solution as a principal petroleum-migration mechanism for either oil or gas.

Water appears to disperse soluble oil and gas constituents instead of concentrating them in reservoir traps.

SELECTED REFERENCES

Anderson, J. W., et al, 1974, Characteristics of dispersion and water-soluble extracts of crude and refined oils and their toxicity to estuarine crustaceans and fish: Mar. Biology, v. 27, p. 75-88.

Aschenbrenner, B. C., and C. W. Achauer, 1960, Minimum conditions for migration of oil in water-wet carbonate rocks: AAPG Bull., v. 44, p. 235-243.

Bailey, N. J. L., C. R. Evans, and C.W. D. Milner, 1974, Applying petroleum geochemistry to search for oil: examples from Western Canada basin: AAPG Bull., v. 58, p. 2284-2294.

——— et al, 1973. Alteration of cude oil by waters and bacteria — evidence from geochemical and isotope studies: AAPG Bull., v. 57, p. 1276-1290.

Baker, E. G., 1959, Origin and migration of oil: Science, v. 129, p. 871-874

——— 1960, A hypothesis concerning the accumulation of sediment hydrocarbons to form crude oil: Geochim. et Cosmochim. Acta, v. 19, p. 309-317.

——— 1962, Distribution of hydrocarbons in petroleum: AAPG Bull., v. 46, p. 76-84.

——— 1967, A geochemical evaluation of petroleum migration and accumulation, *in* Fundamental aspects of petroleum geochemistry: New York, Elsevier, p. 299-329.

Barker, C., 1977, Aqueous solubility of petroleum as applied to its origin and primary migration: discussion: AAPG Bull., v. 61, p. 2146-2149.

Bonham, L. C., 1978, Solubility of methane in water at elevated temperatures and pressures: AAPG Bull., v. 62, p. 2478-2481.

Button, D. K., 1976, The influence of clay and bacteria on the concentration of dissolved hydrocarbon in saline solution: Geochim. et Cosmochim. Acta, v. 40, p. 435-440.

Cartmill, J. C., 1976, Obscure nature of petroleum migration and entrapment: AAPG Bull., v. 60, p. 1520-1530.

Conybeare, C. E. B., 1970, Solubility and mobility of petroleum under hydrodynamic conditions, Surat basin, Queensland: Geol. Soc. Australia Jour., v. 16, p. 667-681.

Cordell, R. J., 1972, Depths of oil origin and primary migration: a review and critique: AAPG Bull., v. 56, p. 2029-2067.

——— 1973, Colloidal soap as proposed primary migration medium for hydrocarbons: AAPG Bull., v. 57, p. 1618-1643.

Culberson, O. L., and J. J. McKetta, 1952, Phase equilibrium in hydrocarbon-water systems: AIME Petroleum Trans., v. 192, p. 223-226.

Dickey, P. A., 1975, Possible primary migration of oil from source rock in oil phase: AAPG Bull., v. 59, p. 337-345.

Dow, W. G., 1974, Application of oil-correlation and source-rock data to exploration in Williston basin: AAPG Bull., v. 58, p. 1253-1262.

Eganhouse, R. P., and J. A. Calder, 1976, The solubility of medium molecular weight aromatic hydrocarbons and the effects of hydrocarbon co-solutes and salinity: Geochim. et Cosmochim. Acta, v. 40, p. 555-561.

Erdman, J. G., 1965, Petroleum — its origin in the earth, *in* Fluids in subsurface environments: AAPG Mem. 4, p. 20-22.

Fisher, L. R., and D. G. Oakenfull, 1977, Micelles in aqueous solution: Chem. Soc. Rev., v. 6, p. 25-42.

Hedberg, H. D., 1964, Geologic aspects of origin of petroleum: AAPG Bull., v. 48, p. 1755-1803.

Hobson, G. D., 1961, Problems associated with the migration of oil in "solution": Inst. Petroleum Jour., v. 47, p. 170-173.

——— and E. N. Tiratsoo, 1975, Introduction to petroleum geology: Beaconsfield, England, Scientific Press Ltd., 300 p.

Hunt, J. M., 1961, Distribution of hydrocarbons in sedimentary rocks: Geochim. et Cosmochim. Acta, v. 22, p. 37-49.

——— 1967, The origin of petroleum in carbonate rocks, *in* G. V. Chillingar et al, eds., Carbonate rocks: New York, Elsevier, p. 225-251.

——— 1968, How gas and oil form and migrate: World Oil, v. 167, no. 4, p. 140, 145, 148-150.

——— 1975, Is there a geochemical depth limit for hydro-

carbons?: Petroleum Eng., March, p. 112-124.
——— 1977, Distribution of carbon as hydrocarbons and asphaltic compounds in sedimentary rocks: AAPG Bull., v. 61, p. 100-104.
——— and G. W. Jamieson, 1956, Oil and organic matter in source rocks: AAPG Bull., v. 40, p. 477-488.
Kartsev, A. A., et al, 1971, The principal stage in the formation of petroleum: 8th World Petroleum Cong. Preprints, Moscow, PD 1, paper 1, 17 p.
Kennedy, W. A., 1963, Solubilization of hydrocarbons as a process of formation of petroleum deposits: PhD thesis, Univ. Texas at Austin.
Levorsen, A. I., 1954, Geology of petroleum: San Francisco, California, W. H. Freeman, 703 p.
Mackay, D., and W. Y. Shiu, 1977, Aqueous solubility of polynuclear aromatic hydrocarbons: Jour. Chem. Eng. Data, v. 22, p. 399-402.
Martin, R. L., and J. C. Winters, 1959, Composition of crude oil through seven carbons as determined by gas chromatography: Anal. Chemistry, v. 31, p. 1954-1960.
——— ——— and J. A. Williams, 1963, Distribution of n-paraffins in crude oils and their implications to origin of petroleum: Nature, v. 199, p. 110-113.
——— ——— and ——— 1964, Composition of crude oils by gas chromatography: geological significance of hydrocarbon distributions: 6th World Petroleum Cong., 1963, Frankfurt/Main, Proc., sec. 5, p. 231-260.
McAuliffe, C. D., 1963, Solubility in water of C_1-C_9 hydrocarbons: Nature, v. 200, p. 1092-1093.
——— 1966, Solubility in water of paraffin, cycloparaffin, olefin, acetylene, cyclo-olefin, and aromatic hydrocarbons: Jour. Phys. Chemistry, v. 70, p. 1267-1275.
———1969, Solubility in water of normal C_9 and C_{10} alkane hydrocarbons: Science, v. 163, p. 478-479.
——— 1971, GC determination of solutes by multiple phase equilibrium: Chem. Technology, v. 2, p. 46-51.
——— 1977, Evaporation and solution of C_2 to C_{10} hydrocarbons from crude oils on the sea surface, *in* D.A. Wolfe, ed., Fate and effects of petroleum hydrocarbons in marine ecosystems and organisms: New York, Pergamon Press, p. 363-372.
McKetta, J. J., and A. H. Wehe, 1962, Hydrocarbon-water and formation water correlations, *in* T. C. Frick and R. W. Taylor, eds., Petroleum production handbook, II, chap. 22: New York, McGraw-Hill, p. 22-1–22-26.
Meinschein, W. G., 1959, Origin of petroleum: AAPG Bull., v. 43, p. 925-943.
——— 1961, Significance of hydrocarbons in sediments and petroleum: Geochim. et Cosmochim. Acta, v. 22, p. 58-64.
Milner, C. W. D., M. A. Rogers, and C. R. Evans, 1977, Petroleum transformations in reservoirs: Jour. Geochem. Exploration, v. 7, p. 101-153.
Momper, J. A., 1978, Oil migration limitations suggested by geological and geochemical considerations, *in* Physical and chemical constraints on petroleum migration: AAPG Continuing Education Course Notes Ser. 8, p. B1-B60.
Mukerjee, P., and K. J. Mysels, 1971, Critical micelle concentrations of aqueous surfactant systems: Natl. Bur. Standards Natl. Stand. Ref. Data Ser., NSRDS-NBS 36, 222 p.
Philippi, G. T., 1965, On the depth, time and mechanism of petroleum generation: Geochim. et Cosmochim. Acta, v. 29, p. 1021-1049.
——— 1974, Depth of oil origin and primary migration: a review and critique: discussion: AAPG Bull., v. 58, p. 149-150.
——— 1975, The deep subsurface temperature controlled origin of the gaseous and gasoline-range hydrocarbons of petroleum: Geochim. et Cosmochim. Acta, v. 39, p. 1353-1374.
——— 1977, On the depth, time and mechanism of origin of the heavy to medium-gravity naphthenic crude oils: Geochim. et Cosmochim. Acta, v. 41, p. 33-52.
Price, L. C., 1976, Aqueous solubility of petroleum as applied to its origin and primary migration: AAPG Bull., v. 60, p. 213-244.
——— 1977, Aqueous solubility of petroleum as applied to its origin and primary migration: reply: AAPG Bull., v. 61, p. 2149-2156.
Prince, L. M., 1977, Schulman's microemulsions, *in* L. M. Prince, ed., Microemulsions — theory and practice: New York, Academic Press, Inc., p. 1-20.
Rehfeld, S. J., 1970, Solubilization of benzene in aqueous sodium dodecyl sulfate solutions measured by differential spectroscopy: Jour. Phys. Chemistry, v. 74, p 117-122.
Rice, S. D., et al, 1976, Acute toxicity and uptake-depuration studies with Cook Inlet crude oil, Prudhoe Bay crude oil, no. 2 fuel oil and several subarctic marine organisms: Northwest Fisheries Center Auke Bay Fisheries Lab. Processed Rept.
Saxby, J. D., and M. Shibaoka, 1975, Depth of oil origin and primary migration: discussion: AAPG Bull., v. 59, p. 721-723.
Schrayer, G. J., and W. M. Zarrella, 1963, Organic geochemistry of shales, I. Distribution of organic matter in the siliceous Mowry shale in Wyoming: Geochim. et Cosmochim. Acta, v. 27, p. 1033-1046.
——— ——— 1966, Organic geochemistry of shales, II. Distribution of extractable organic matter in the siliceous Mowry Shale of Wyoming: Geochim. et Cosmochim. Acta, v. 30, p. 415-434.
Schwartz, F. P., and S. P. Wasik, 1976, Fluorescence measurement of benzene, naphthalene, anthracene, pyrene, fluoranthene, and benzo(e)pyrene in water: Anal. Chemistry, v. 48, p. 524-527.
Seifert, W. K., 1975, Carboxylic acids in petroleum and sediments: Progress in Chemistry of Organic Natural Products, v. 32, p. 1-49.
Shinoda, K., and H. Kunieda, 1977, How to formulate microemulsions with less surfactants, *in* L. M. Prince, ed., Microemulsions — theory and practice: New York, Academic Press, Inc., p. 57-89.
Sultanov, R. G., V. G. Skripka, and A. Yu. Namiot, 1972, Solubility of methane in water at high temperatures and pressures: Gasovaia Promyshlennost, v. 17, p. 6-7.
Sutton C., and J. A. Calder, 1974, Solubility of higher-molecular-weight n-paraffins in distilled water and seawater: Environmental Sci. and Technology, v. 8, p. 654-657.
——— ——— 1975, Solubility of alkylbenzenes in distilled water and seawater at 25.0°C: Jour. Chem. Eng. Data, v. 20, p. 320-322.
Tissot, B., et al, 1971, Origin and evolution of hydrocarbons in early Toarcian shales, Paris basin, France: AAPG Bull., v. 55, p. 2177-2193.
Wauchope, R. D., and F. W. Getzen, 1972, Temperature dependence of solubilities in water and heats of fusion of solid aromatic hydrocarbons: Jour. Chem. Eng. Data, v. 17, p. 38-42.
Welte, D. H., 1965, Relation between petroleum and source rocks: AAPG Bull., v. 49, p. 2246-2268.
Winters, J. C., and J. A. Williams, 1969, Microbiological alteration of crude oil in the reservoir: Am. Chem. Soc. Natl. Mtg., New York, Petroleum Div. Preprints, v. 14, no. 4, p. E22-E31.

Some Economic Aspects of Water-Rock Interaction[1]

By Brian Hitchon[2]

Abstract Water, through its unique and extreme properties, is the fundamental fluid genetically relating all mineral deposits in sedimentary rocks. Economically important mineral deposits in sedimentary rocks which are the result of natural water-rock interaction include petroleum and Mississippi-type lead-zinc deposits. Understanding of the origin of these deposits through water-rock interaction requires knowledge of the relations between hydrochemistry and hydrodynamics. The recovery of some of these mineral deposits involves man-imposed water-rock interactions, for example, during water flooding of petroleum reservoirs, in-situ steam injection into oil sand deposits and underground coal gasification. These man-imposed water-rock interactions may result in subsurface reactions which can reduce permeability, produce toxic or deleterious substances which require removal before reuse of the produced water, contaminate local potable groundwater, or cause problems in waste injection wells because of subsequent water-rock reactions. Although we understand some of the principles involved, it is clear that considerably more thought and additional research effort needs to be directed to these and other economic aspects of water-rock interaction.

INTRODUCTION

Water, through its unique and extreme properties, is the fundamental fluid genetically relating all mineral deposits in sedimentary rocks. It is the vehicle for the transport of materials in solution and suspension throughout the hydrologic cycle, and it takes part in reactions during the dissolution of minerals in chemical weathering, diagenesis and metamorphism of sediments and sedimentary rocks, and the remelting and crystallization of igneous and metamorphic rocks. The intimate interplay between the hydrologic and the geochemical cycles is illustrated in Figure 1. The geochemical cycle is not closed, either energetically or materially, and the cycle may be temporarily halted, short-circuited, or reversed. Nevertheless, it is a convenient concept and framework within which to evaluate and discuss the role that water plays in the accumulation of mineral deposits in sedimentary rocks.

Although the geochemical cycle comprises two interlinked environments, a primary environment (involving magmas, and igneous and metamorphic rocks) and a secondary environment, it is only the latter with which we are concerned in this paper. This should not be construed to mean that there are no examples of economically important aspects of water-rock interaction in the primary environment; one has only to cite the variety of hydrothermal mineral deposits in igneous and metamorphic rocks to realize the importance of water-rock interaction in the primary environment to the economic activities of man.

The secondary environment encompasses the relatively shallow processes of weathering, sedimentation, diagenesis and low-grade metamorphism, that occur in the environment of effective circulation of meteoric water; it includes the two most important short-circuits within the geochemical cycle between sedimentary and metamorphic rocks and soils and sediments, as illustrated by the broken arrows in Figure 1. The term meteoric water is here used in the same sense as by Clayton et al (1966); that is, there is an

[1]Manuscript received, October 3, 1978; accepted, July 31, 1979.

[2]Alberta Research Council, Edmonton, Alberta, T6G 2C2, Canada. Contribution Number 931, Alberta Research Council.

The writer acknowledges with appreciation the critical and valuable reviews of the manuscript by J. A. Boon and E. I. Wallick of the Alberta Research Council, and A. A. Levinson of the Department of Geology, University of Calgary.

Funding for some of the research reported on in-situ steam injection into oil sand deposits came from the Alberta Oil Sands Technology and Research Authority; funding for the research reported on the underground coal gasification field test at Forestburg, Alberta, was supplied through a contract between the Research Council of Alberta and a consortium of eleven private companies and four departments and agencies representing the governments of Canada, Alberta, British Columbia, and Saskatchewan, with major funding coming from Alberta Energy and Natural Resources. This funding, and permission to publish the information cited in this paper, is gratefully acknowledged.

Finally, the writer would like to record his sincere thanks and appreciation to G. W. Hodgson, Director, Environmental Sciences Centre (Kananaskis), The University of Calgary, who so kindly and ably presented this paper on the writer's behalf at the AAPG Research Symposium on the Problems of Petroleum Migration (Oklahoma City, April, 1978).

Article Identification Number:
1049-1377/79/SG10-0007/$03.00/0.

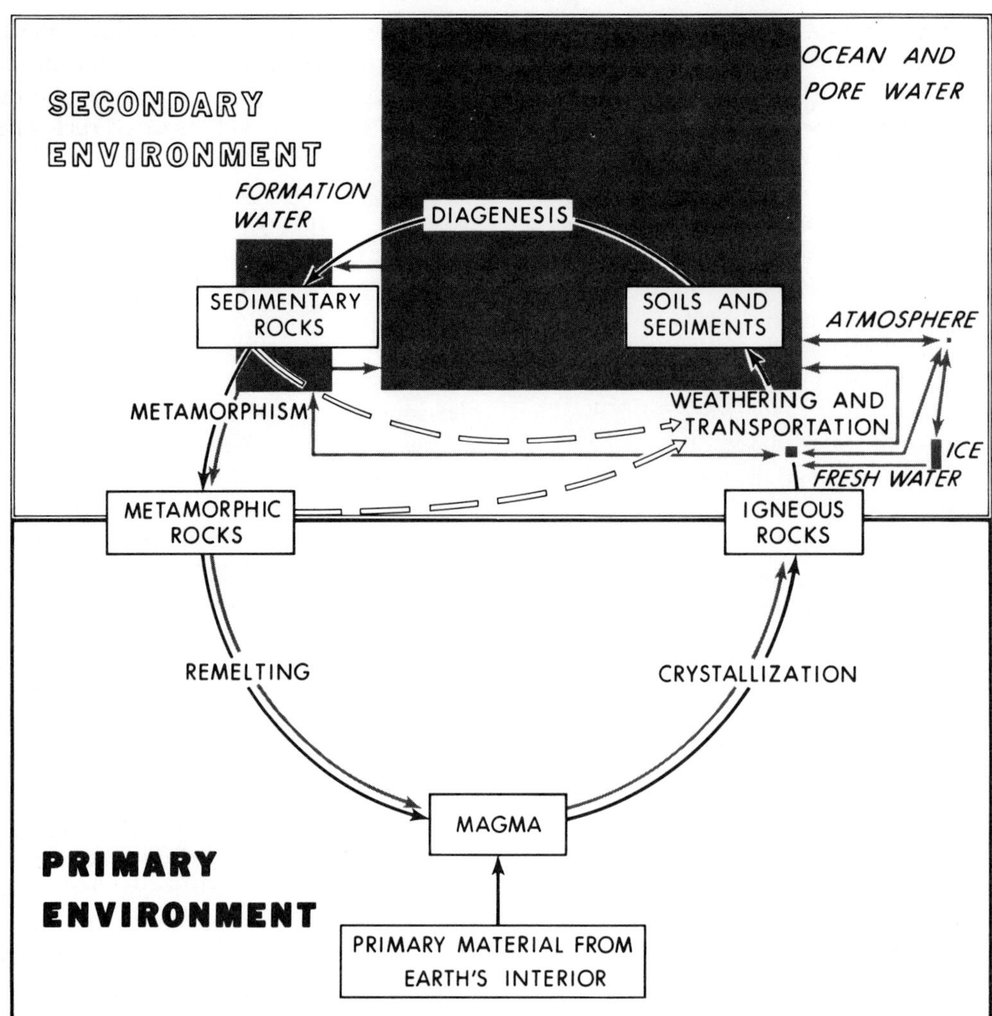

FIG. 1—Relation between the geochemical cycle (modified from Saxby, 1969) and the hydrologic cycle (screened arrows). The area of the rectangles in the hydrologic cycle are proportional to the volumes of water in the respective reservoirs, based on data from Horn and Adams (1966) and Penman (1970). Reproduced with permission of Elsevier Scientific Publishing Co., Amsterdam, (from Hitchon, 1976).

isotopic component in all water in the secondary environment which originates in the atmosphere and this component is absent in water from the primary environment. Additional aspects concerning this meteoric isotopic component as they relate to the origin of formation waters found in sedimentary rocks are discussed by Hitchon and Friedman (1969). However, there is no clearcut demarcation between the primary environment and the secondary environment. The nuances of the chosen limit may be appreciated by considering that circuit of the hydrologic cycle which follows directly in the path of the geochemical cycle (explained further by Hitchon, 1976).

It is the intention of the writer to emphasize that there are two extremely important factors to bear in mind when considering the role played by water in sedimentary basins. These factors are the nature of water-rock interaction and the relations between hydrochemistry and hydrodynamics. Specific examples will be cited in which a knowledge of these factors has assisted in appreciating the economic aspects of man-imposed water-rock interaction. All examples are taken from the work of the writer and his colleagues at the Alberta Research Council.

NATURE OF WATER-ROCK INTERACTION

In the strictest sense we can only obtain a comprehensive knowledge of sedimentary basins by including in our considerations all interactions between the rocks and the contained fluids. This implies study of the complex interactions and inter-relations between the individual minerals of the original sediment, and any subsequent diagenetic minerals, and the aqueous and organic fluids in the rocks. We need a wide knowledge of water-mineral interactions as well as information on reactions between these same minerals and bitumen, crude oil, and natural gas. It is suspected that the presence of certain organic materials may interfere with water-mineral interactions but the degree of the interference and its economic implications are almost unknown. The authors of many of the papers in this volume have dealt at length with various aspects of the origin, migration, and accumulation of hydrocarbons and some have stressed the importance of formation waters in the processes involved. It is my opinion that a greater research effort needs to be devoted to the relations and interactions among the minerals and the various organic phases, especially reactions with clay minerals, not only with regard to possible interference with water-mineral interactions but also to attempt to understand, at the molecular level, how hydrocarbons originate, migrate, and accumulate in what is primarily an aqueous environment; and it should be borne in mind that not only is the aqueous environment commonly saline but the aqueous fluid itself is continually changing because it is in motion and therefore continually adjusting its composition toward equilibrium with the adjacent minerals. The writer's thoughts on various aspects of this theme have been presented previously (Hitchon, 1971, 1974, 1977b), but a comprehensive evaluation must await further research. It is sufficient, here, to emphasize the complexity of the situation and to confine attention to the nature of water-rock interaction *sensu stricto*.

The chemical and physical attributes of each formation water sample may be thought of as the result of static equilibrium with the rock from which it was obtained, and may be expressed in terms of the degree of saturation with respect to the specific minerals in that rock. The properties of the system result in constraints which must be considered in water-rock interaction studies. All are well known to geochemists and, fundamentally, are concerned with energy and mass transfers among solid, liquid, and gaseous phases. They include the laws of conservation of mass and energy, the law of mass action, Gibbs' phase rule, diffusional transfer, oxidation-reduction reactions, stoichiometric and nonstoichiometric reactions, and solid solution phenomena. In addition to these major constraints, there exist the possibilities of reactions involving irreversible changes, such as may occur in weathering, diagenesis and evaporative concentration, osmotic and ion-exchange phenomena with clay minerals, the nonideal behavior of the gas phase, incongruent reactions, chromatographic effects, volume changes either in the fluid phases or in the solid phases due to the precipitation or solution of reactants, and of course, the possibility of partial equilibrium. It is beyond the scope of this paper to discuss any of these constraints in detail, and the interested reader is referred to standard textbooks on physical chemistry and the many geochemically oriented papers in the literature on this topic. As far as this writer knows, there are no definitive studies in the geochemical literature dealing with organic, stable isotope, and radioactive isotope interactions with minerals such as exists for the "simple" water-mineral system.

Although it is possible to write and program (for computer calculation) steps in the calculations for determining the solution-mineral equilibria for most individual minerals (provided the fundamental physical and chemical data are available), we found a general purpose computer program (Kharaka and Barnes, 1973) of great value in preliminary studies of the complex solution-mineral situations such as are found in sedimentary basins. The program computes the equilibrium distribution of 162 inorganic aqueous species generally present in natural waters over the temperature range of 0°C to 350°C from the reported chemical analysis, temperature, pH, and Eh. Interpolated dissociation constants of the aqueous complexes and the computed activity coefficients also are used in these calculations. States of reactions of the aqueous solutions with respect to 158 solid phases (minerals) are computed from the distribution of aqueous species and an internally consistent set of thermodynamic data. Using this program it is possible to examine the degree of saturation with respect to specific minerals. The program calculates the deviation from equilibrium (ΔG_{DIFF}) rather than equilibrium. When supersaturation is found (ΔG_{DIFF} is positive) the minerals that supersaturate the solution will eventually precipitate. Minerals with respect to which the aqueous phase is undersaturated (ΔG_{DIFF} is negative) will keep dissolving either until they are exhausted or until they saturate the aqueous phase. When kinetics of all the reactions considered by Kharaka and Barnes are known, it should be possible to link hydrochemistry and hydrodynamics by developing a partial equilibrium simulation model of the chemical state of a regional formation water flow system that incorporates spatial mass transfer rates and reaction kinetics. Having described the nature of water-rock interactions it is obvious that rigid and precise techniques for the collection, preparation and preservation of the water samples is mandatory if the results of the solution-mineral equilibria computations are to be meaningful.

RELATIONS BETWEEN HYDROCHEMISTRY AND HYDRODYNAMICS

Water is the dominant fluid in sedimentary basins, and if the movement of water in the basin were to cease, chemical and physical equilibration between it and the rocks would ultimately occur. At that time there would be no further opportunities to generate new mineral deposits (both hydrocarbon and nonhydrocarbon) in the rocks. This hypothetical and (as we shall see) effectively impossible situation stresses the significance of hydrodynamics to the genesis of mineral deposits in sedimentary basins. It should be noted here that this same situation applies whether petroleum migrates dissolved in formation waters or as separate oil and gas phases; the dominant fluid is water and if it were to cease movement probably only buoyancy forces would be available to assist petroleum migration if it were in separate phases.

The flow of fluids through porous media is governed by a well recognized system of energy potential fields (fluid, thermal, electric, and chemical). Of these, the fluid potential is defined as the amount of work required to transport a unit mass of fluid from an arbitrary chosen datum (usually sea level) and state, to the position and state of point considered. Flow may be considered as a transient (unsteady-state) or steady-state phenomenon. Flow from a borehole is an example of transient flow and probably best analyzed using unsteady-state techniques. When consideration is given to studies of small drainage basins or to flow regimes across large sedimentary basins, it can be assumed that a relatively steady state has been reached. In short, we have a case of dynamic equilibrium.

Hitchon (1976) briefly reviewed the history of development of fluid flow models from Hubbert's (1940) classic work, through Toth's (1962, 1963) earlier analytical methods, and to the more complex models by Freeze (1966, 1969) and Freeze and Witherspoon (1966, 1967, 1968). The latter developed both analytical and numerical methods of analysis for a wide variety of different models, with up to three continuous or discontinuous layers of differing permeabilities, at various attitudes within the model, with a sloping basement, and with a variety of topographies including a general configuration. Because the parameters they used are dimensionless, the resulting flow nets apply to sedimentary basins of all sizes. The Toth-Freeze-Witherspoon model has been applied to the western Canada sedimentary basin (Hitchon, 1969a, b). These studies show that the main variables affecting the fluid potential distribution are topography and geology (lithology and permeability). The fluid potential in any part of the basin corresponds closely to the fluid potential at the topographic surface in that part of the basin. Major recharge areas correspond to major upland areas, and major lowland regions are major discharge regions. Large river valleys commonly exert a drawdown effect on the fluid potential distribution, which may be observed to depths of up to 1,500 m (Hitchon, 1971, Fig. 4). Beds of relatively high permeability, if sufficiently thick, can significantly affect the regional fluid potential distribution by acting as channels that draw down the fluid potentials, although the dominant features are topographically controlled.

Recently, Toth (1980) provided additional evidence from Alberta and around the world that gravity-induced cross-formational flow of formation waters is a reality and an important aspect of hydrodynamics in sedimentary basins. Indeed, it is this writer's opinion that a sedimentary basin can be considered as analagous to a giant sponge comprising layers of sedimentary rocks with varying degrees of porosity and permeability through which fluids are more or less free to move from regions of high energy to regions of lower energy. The degree of movement is related to the degree of difference in porosity and permeability between the high energy and low energy regions. While this analogy may not be strictly correct or commonly acceptable, it does at least have the merit of being simple and easy to understand. Energy potential fields other than the dominant fluid potential field, as well as such phenomena as compaction, the aquathermal effect and the results of montmorillonite-illite conversion may all play a role in fluid movement in sedimentary basins, but a return to considering hydrodynamics in sedimentary basins as largely due to gravity-induced cross-formational flow within a giant sedimentary "sponge" might be helpful in putting hydrocarbon accumulation into its proper perspective. Once that perspective is attained it becomes apparent that the probability of cessation of fluid movement in a sedimentary basin is effectively zero, hence water-rock interaction as a continuous phenomenon must always be thought of in terms of the continuous flow of fluids, and the potential for the generation of mineral deposits is persistently present. The close relations between hydrochemistry and hydrodynamics then becomes apparent.

What has been explained so far concerning fluid flow models applies more properly to older sedimentary basins with recognizable topographic relief. In many onshore coastal basins, and certainly in offshore basins, compaction phenomena and other processes dominate the regional hydrodynamic pattern.

RELATIONS BETWEEN HYDROCARBON AND NONHYDROCARBON DEPOSITS IN SEDIMENTARY BASINS

Before discussing economic aspects of man-imposed water-rock interactions, the writer cites a well documented example (from the Devonian strata in the western Canada sedimentary basin) of the close association of hydrocarbon occurrences and lead-zinc deposits of the Mississippi-type. This digression represents a condensation of the concepts presented by Hitchon (1977a) in which an attempt was made to demonstrate the geochemical links between oil fields and ore deposits in sedimentary rocks. Comparison of the composition of geothermal brines, believed to be ore-bearing fluids, with that of formation waters shows that the content of effectively all components in geothermal brines falls within the concentration limits of formation waters. The features of high salinity and high metal content of the geothermal brines are matched by the deeper, hotter, more saline formation waters, which have leached evaporites to produce the solutes (possibly assisted by membrane filtration) and leached shales to produce the enhanced content of metals. An analogy is apparent between the generation and leaching of hydrocarbons from shale source rocks (and their subsequent migration and accumulation as oil fields) and the leaching of metals from shales by hot saline formation waters (with their subsequent migration and accumulation as ore deposits). Thus the source-migration-accumulation scenario for oil fields and ore deposits in sedimentary rocks is controlled by the temperature, pressure, hydrodynamic, and fluid-rock interaction continuum that exists in all sedimentary basins. The most important requirement for migration is the presence of a conduit for channel flow, and one of the best candidates for future channel-flow conduits for hydrocarbon- and metal-laden formation waters are reef or shelf carbonate fronts, close to both evaporites and organic-rich muds. The length and attitude of the carbonate front with respect to the future sedimentary basin are crucial. Accumulation of hydrocarbons and metals results from their removal from the moving saline formation waters through decrease in temperature, pressure or salinity, and by changes in P_{H_2S}, P_{CO_2}, brought about through a variety of processes, including water-rock interaction. Because it is conceptually difficult, at this time, to formulate the exact deposition mechanism that can result in concurrent deposition of both metallic minerals and petroleum, it is possible that their presence at the same site may imply multiple pulses of deposition. Subsequently, both the oil fields and ore deposits may be brought into the zone of near-surface local groundwater flow resulting in weathering features, and in the case of oil fields, to the effects of waterwashing and biodegradation. The conceptual model that has been developed for the history of both hydrocarbon and nonhydrocarbon deposits in sedimentary basins thus comprises five stages—deposition, burial and leaching, migration, accumulation, and uplift and dissipation.

The Devonian strata of the western Canada sedimentary basin will serve as an example of the conceptual model developed by Hitchon (1977a) and of the close association of hydrocarbon and nonhydrocarbon deposits in a specific sedimentary basin. At this time, approximately two-thirds of the reserves of conventional crude oil and somewhat less than one-third of the reserves of recoverable natural gas from the western Canada sedimentary basin are found in Devonian strata. The only commercial lead-zinc deposit in the basin also is found in Devonian rocks. The hydrocarbons are widely distributed in numerous carbonate reefs and shelves, with the exception of southern Alberta (a groundwater recharge region) and the shallow strata adjacent to the Devonian outcrop. What is perhaps less well known than the distribution of hydrocarbons is the presence in the same carbonate reefs and shelves of a sufficient number of occurrences of lead-zinc mineralization to suggest that the conceptual model cited above is well exemplified by the Devonian of western Canada.

The distribution of lead-zinc occurrences in Devonian strata of Alberta, British Columbia, and the southern North West Territories is shown in Figure 2 in relation to the main carbonate fronts and the presence of reservoir bitumens. There are no occurrences of mineralization without the presence of crude oil or reservoir bitumens, although the converse is not true. With the possible exception of the Oldman River deposit, all lead-zinc occurrences are associated with carbonate-shale facies boundaries, including Rainbow, which is a pinnacle reef. Evaporites are adjacent to some of the occurrences. The reservoir bitumens are clearly of two origins, those which occur in the deeper parts of the basin and are the result of the deasphalting of crude oil, and those which occur in the outcrop and shallow parts of the basin and have been subjected to the influence of near-surface groundwater flow. Although sphalerite is the most abundant sulfide (with only minor amounts of galena), the reverse is sometimes found. Gangue minerals include smithsonite and cerussite, which in the surface deposits probably result from the weathering of their respective sulfides. However, the presence of smithsonite at the Duhamel oil field, at a depth of 1,370 to 1,480 m, with an original P_{H_2S} of 62 kPa and P_{CO_2} of 270 kPa in the associated natural gas (Hitchon and Friedman, 1969), and formation waters with 1.65 mg/l zinc and a salinity of 210, 000 mg/l (Hitchon et al, 1971) suggest coprecipitation with the sphalerite.

Studies of the organic carbon and trace metal content of Devonian shales from the Pine Point region (Macqueen et al, 1975) and the trace metal content of some Devonian shales reported by Hitchon (1977a), together with the leaching experiments of H. H. Williams on outcrop samples of Alberta shales (see Billings et al, 1969), all point to the presence of high contents of leachable trace metals, particularly zinc, in the Devonian shales from western Canada. Williams' results are similar to those of Hathaway et al (1972) which suggest, although their conditions of leaching were different, that the leachability of trace metals from shales is probably a feature of many sedimentary basins.

Mention has already been made of the channel flow developed in the Devonian carbonates, and the extension of the latter into the deeper, western parts of the basin is well illustrated in Figure 2. In addition to the hydrodynamic studies of the writer (Hitchon, 1969a, b; Hitchon 1971, Figs. 5 and 6), Kesler et al (1972) used the horizontal asymmetry of single crystals and crystal aggregates in flat-floored vugs at Pine Point to indicate that fluid movement during sulfide deposition was the same as that shown by hydrodynamics. Re-

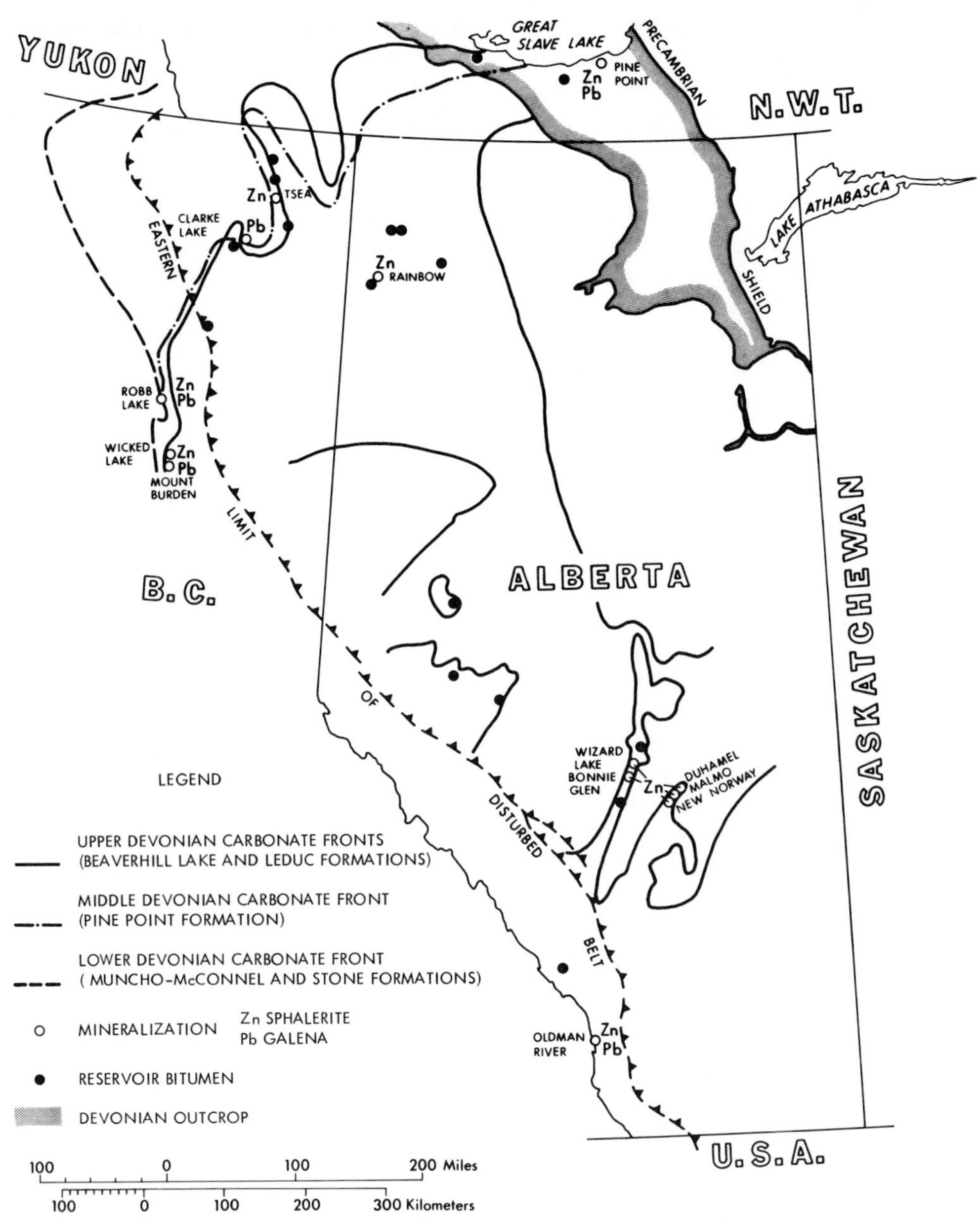

FIG. 2—Distribution of lead-zinc occurrences in Devonian strata of Alberta, British Columbia, and southern North West Territories, Canada. Reproduced with permission of Imperial College, London, (from Hitchon, 1977a).

gional salinity maps (Hitchon, 1964) show more saline formation waters in the deeper parts of all reef trends, compared to the relatively fresher formation waters in the shallower, eastern, updip parts of the reefs and shelves.

It is thus quite clear that the concepts developed by Hitchon (1977a) concerning the relations between hydrocarbon and nonhydrocarbon deposits in sedimentary basins can be demonstrated in the Devonian strata of the western Canada sedimentary basin, and almost certainly may be applicable to all sedimentary basins. They show, in an actual situation, the importance of the nature of water-rock interaction and the relations between hydrochemistry and hydrodynamics, an understanding of which may be of economic value as applied to natural situations.

MAN-IMPOSED WATER-ROCK INTERACTION—IN-SITU RECOVERY FROM OIL SAND DEPOSITS

In-situ technology for the recovery of crude bitumen from oil sand deposits is, in essence, the application of physical and/or chemical forces to the oil sand deposits in the subsurface environment to mobilize the crude bitumen. During the application of these forces, reactions may take place which change the composition of both the rocks and the fluids (original formation fluids and the injected fluids). Understanding these fluid-rock interactions may be an important part of the process of optimizing crude bitumen recovery, and specifically may provide information about the solution and precipitation of minerals during in-situ operations. In addition, a variety of geochemical parameters are important from the environmental point of view, and here one should bear in mind the possibility of polluting the subsurface environment as well as the surface environment. Some of the subsurface reactions may involve the production of toxic or deleterious substances which require removal before reuse of the produced water; monitoring may be necessary to avoid contamination of the local potable groundwater. If the produced water is not reused but is separated from the produced fluids (which may include bitumen, bitumen-in-water emulsions, and water-in-bitumen emulsions in addition to water and steam) and subsequently disposed of into waste injection wells, further water-rock interaction may take place, possibly causing problems in disposal. Previously, a close relation between hydrochemistry and hydrodynamics was pointed out. This relation remains important during in-situ recovery operations from oil sand deposits because water-rock interactions may take place anywhere along the flow path once the physical and chemical conditions are right for the reactions to occur. By analogy to the natural environment in a sedimentary basin, flow will be from the high energy recharge region (the injection well in an in-situ operation) to the low energy discharge region, which will be the production well, unless thief zones, fingering, or surface breakthrough occur. The reader is referred to the first paragraph of the section of this paper dealing with the nature of water-rock interaction, and specifically to the possible importance of organic-mineral interactions. Further, it should also be noted that there are large reserves of crude bitumen in carbonate rocks of the Upper Devonian Grosmont Formation in Alberta. The same remarks in this paragraph apply to in-situ recovery from carbonate rocks except that the reactions are likely to be more complex, especially if arenaceous or argillaceous impurities are present in the carbonates. There also may be production of large volumes of carbon dioxide, which may not be detrimental to the in-situ recovery economics because it is known that injection of carbon dioxide can enhance the recovery of conventional crude oil.

There are basically two different techniques that have been applied in field pilot experiments to in-situ recovery from oil sand deposits. The first essential step in both techniques is the creation of a communications path for the fluids between the injection and production wells if a natural communications path is not present. In one technique steam, sometimes with additives such as light hydrocarbons or reactive and nonreactive nonhydrocarbon gases, is injected into the oil sand deposits to raise the reservoir energy by increasing the temperature and pressure, and consequently mobilizing the crude bitumen so that it can flow to the production well. In the other technique the crude bitumen is ignited, by a variety of methods, thus similarly raising the reservoir temperature and pressure; ignition commonly is followed by steam injection to enhance flow of the mobilized bitumen to the production well. It is beyond the scope of this paper to explain further all the engineering aspects of these two techniques; the reader is referred to the volume, *The Oil Sands of Canada-Venezuela 1977*, published by The Canadian Institute of Mining and Metallurgy (CIM Special Volume 17).

In order to place the man-imposed water-rock interactions which take place during in-situ recovery from oil sand deposits in their proper perspective, it is instructive to consider the temperatures and pressures obtained during in-situ operations in relation to natural metamorphism. The pressure and temperature limits of the five main types of metamorphism (diagenesis, burial metamorphism, dynamothermal metamorphism, contact metamorphism, and anatexis) are shown in Figure 3. In terms of type of metamorphic environment, the maximum temperature (336°C) and pressure (13.8 Pa) reported by Winestock (1974) for some of Imperial Oil's Cold Lake steam injection operations, fall at the low pressure end of burial metamorphism. In-situ combustion recovery takes place at comparably low pressures ($<$20 Pa), but at temperatures a few hundred degrees higher. In effect, in-situ operations correspond to conditions of low pressure diagenesis, burial metamorphism, or contact metamorphism. Hitchon (1977b) provided further details of the mineral changes which take place in these natural metamorphic environments in arenaceous, argillaceous, and carbonate rocks, as they may relate to in-situ recovery operations. It should be clear from this metamorphic analogy that study of in-situ operations in terms of metamorphic reactions may lead to a better understanding of underground conditions during steam injection or combustion. However, it must be granted that many of the laboratory studies carried out in order to understand burial-metamorphic reactions took place in systems with little water present; nevertheless, the principles formulated are still relevant. The use of selected mineral transformations to monitor the progress of in-situ reactions has considerable merit and is especially pertinent to combustion recovery (and underground coal gasification) where the temperatures are higher and the reliability of other temperature-measuring devices less certain.

Understanding of the water-rock interactions during in-situ recovery from oil sand deposits can only come after a thorough appreciation of the theory of solution-mineral chemistry, realization of the relations of hydrochemistry and hydrodynamics, and careful labo-

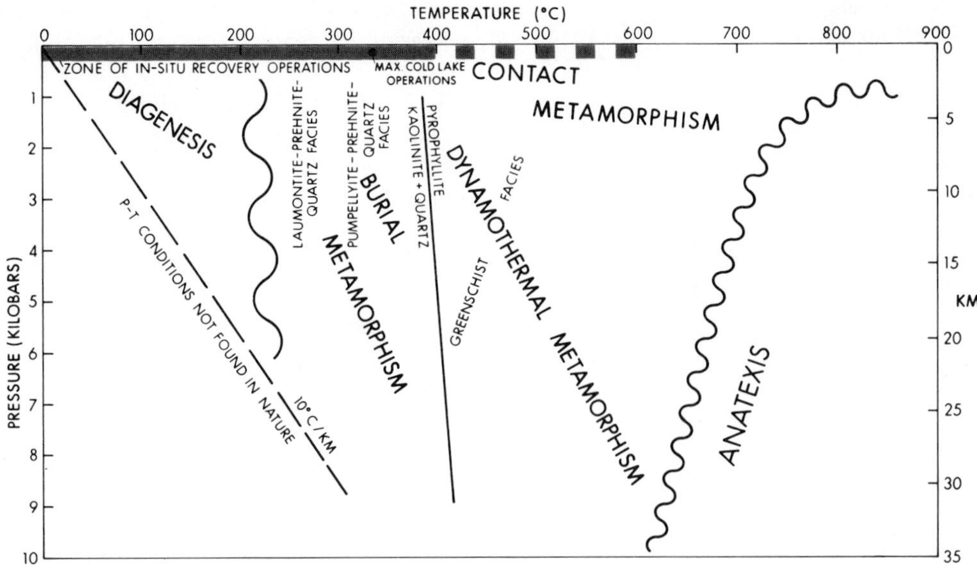

FIG. 3—Schematic pressure-temperature diagram for different types of natural metamorphism (after Winkler, 1967), showing the operative zone for in-situ recovery from oil sand deposits (screened area). Reproduced with permission of The Canadian Institute of Mining and Metallurgy, Montreal, (from Hitchon, 1977b).

ratory and field studies directly relevant to the specific oil sand deposit. Three examples of the work carried out by the writer and his colleagues at the Alberta Research Council are described to help in attaining this understanding.

The first specific information obtained which indicated the possibility that water-rock interactions might take place during in-situ recovery operations has been reported by Hitchon (1977b). Formation water from a reliable drillstem test run in the McMurray Formation in the southern part of the Athabasca oil sand deposit was collected, prepared, and preserved using appropriate procedures designed to retain components in their natural state. Solution-mineral equilibria data were obtained using a computer program (Kharaka and Barnes, 1973). As might be anticipated, the formation water at reservoir temperature was saturated with respect to those minerals which commonly occur in the McMurray Formation. If one were to heat that formation water (say to 200°C), for example during steam injections, then the ΔG_{DIFF} values for the new temperature can be computed by the program, with a minor adjustment to pH. The recalculated ΔG_{DIFF} values indicated that the hot formation water would be undersaturated with respect to all the common minerals in the McMurray Formation. Dilution of the formation water to one-tenth its original concentration (an arbitrary dilution factor), for example by mixing with condensed steam, and consideration of ΔG_{DIFF} at the same high temperaure again showed undersaturation with respect to all common minerals present. Recognizing that the Kharaka and Barnes program is for a static case, it was clear that in-situ steam injection might result in the heated underground water becoming undersaturated with respect to the most common minerals in the McMurray Formation and, consequently, dissolution of these minerals might take place. Reduction in temperature either along the flow path to the production well or elsewhere during production, waste disposal, or reinjection might cause the water to again become saturated or supersaturated with respect to these same minerals, with possible consequent scaling on equipment or reduction in porosity and permeability in the waste-disposal strata, respectively, as a result of precipitation of these minerals.

Evaluation of all information available to us suggested that because quartz is the major mineral in the oil sand deposits of Alberta, its dissolution in the high temperature regions of the formation during in-situ steam injection and its subsequent precipitation in cooler parts of the formation could lead to a significant decrease in permeability, and consequently the possibility of loss of communication between the injection and production wells. Although the reaction kinetics of pure quartz were known up to about 100°C, it was clear that the temperatures and pressures at which most experiments dealing with the dissolution, transportation, and precipitation of silica as reported in the literature were above those normally experienced during in-situ operations. Accordingly, it was decided to carry out a series of bomb runs at 150°C, 170°C, and 205°C using bitumen-extracted clean sand from the McMurray Formation. Boon (1977a) reported the initial results of these experiments and, after further work (Boon, 1977b) concluded that silica dissolution and precipitation can be expected to occur to a significant degree in the Athabasca oil sands under the conditions that prevail in the reservoir during in-situ recovery by steam injection; further, these processes are expected to affect significantly the reservoir permeability. These later experiments suggest that, in addition to temperature and pH, the presence of bitumen was an additional main factor affecting the rate of dissolution of quartz. All three factors interact with each other and with salinity. These observations further underscore earlier remarks concerning the importance of understanding organic-mineral interactions.

Levinson and his co-workers (Levinson and Vian, 1966; Levinson and Day, 1968) were the first to report the low temperature hydrothermal synthesis of montmorillonite. The conditions under which they obtained montmorillonite suggested to us that it should also form during in-situ steam injection into oil sand deposits. In later experiments (Boon, 1977b) considerable amounts of montmorillonite (a mineral not normally present in the McMurray Formation) were formed in a number of cases when illite- and kaolinite-rich shales associated with the oil sand deposits of the McMurray Formation were reacted in bombs at similar temperatures, but at a higher pH than those for the clean quartz runs. Again, initial results suggest that the presence of bitumen affects the reaction. The fact that montmorillonite has greater swelling properties than either illite or kaolinite implies a potential for permeability reduction through its production during in-situ steam injection.

Our laboratory studies clearly demonstrate the need for further research related to water-rock interaction during in-situ recovery from oil sand deposits by steam injection. More effort needs to be directed to the kinetic aspects of these reactions and the development of a predictive mathematical model. Very little has been done to evaluate the possibility of subsurface contamination during in-situ recovery and because most of the deleterious and/or toxic elements probably will originate from shales associated with the oil sand deposits, a series of experiments on the leaching of shales by appropriate solutions would be of interest—as would some similar carbonate leaching experiments. Related knowledge of potential reactions that might take place during waste disposal is required, although, because the availability of water for steam generation may be an economically limiting factor, it may be more useful to consider ways of treating the contaminated produced water for reuse as boiler feedstock. In this regard, it may ultimately be possible to control underground water-rock interactions to maximize reuse of the produced water for steam raising and the economics of in-situ recovery from oil sand deposits.

MAN-IMPOSED WATER-ROCK INTERACTION—UNDERGROUND COAL GASIFICATION

The problems which arise through man-imposed water-rock interaction during underground coal gasification are in many ways similar to those created in in-situ recovery operations from oil sand deposits. The pressures are of the same order of magnitude as those occurring during steam injection or in-situ combustion, but the temperatures may be higher than those during in-situ combustion. As with in-situ recovery from oil sand deposits, the first essential step is the creation of a communications path for the injected air and steam, and for the produced gas. After ignition of the coal by a variety of methods, air and steam are injected and sometimes with additional oxygen. It seems likely that the gases produced may arise through a combination of pyrolysis and gasification of the coal. The main chemical reaction involved is that between carbon and water to yield carbon monoxide and hydrogen, both of which are combustible, so that the mixture can serve as a low-BTU fuel gas. This carbon-steam reaction is endothermic, that is, it requires heat to proceed, and this heat is provided by combustion of some of the coal in the underground environment. The water can either be supplied with the air as steam, or it can be provided by heating the local groundwater.

The Alberta Research Council recently conducted an underground coal gasification field test in a shallow (20 m), 3m-thick coal seam at Forestburg, Alberta, some of the results of which have been published (Berkowitz and Brown, 1977). The 18×9 m, 4-well system chosen for the test installation was oriented with the long axis in the direction of the major fracture pattern (cleat) of the coal seam and was surrounded by a series of monitoring wells up to two hundred meters from the test installation. Groundwater samples were collected prior to the test from most of the twelve monitoring wells, and at both the monitoring wells and three of the test wells about two days, 170 days, and 270 days after completion of the burn. Details of the results of the water analyses are not available now, except to note that some toxic elements were found in water samples at the burn site at concentrations of more than four orders of magnitude above the baseline contents. Concentrations of these elements declined sharply in subsequent sampling and reached levels of about one order of magnitude above baseline contents 270 days after cessation of the burn. No samples were obtained later than this time because the site was excavated and one can only speculate that baseline contents would be reached eventually.

It is not known if the cause of the rapid decline in concentration resulted from adsorption of the trace elements onto the burnt and charred coal or dispersion in the natural groundwater flow system, but there can be little doubt that the initial high concentrations of at least some of the ions analyzed are due to the release of those ions from either the coal or associated shale partings as a result of high temperature water-rock interaction during the burn. This work points out the importance of water-rock interaction during underground coal gasification, especially with respect to possible pollution of local potable groundwater supplies. Because only gas effectively is produced during underground coal gasification, there is little water-disposal problem; the relatively minor amount of water found in the water-knockout system at the Forestburg site would not create the sort of water disposal problems cited for in-situ steam recovery operations in oil sand deposits. However, the close relation of hydrochemistry and hydrodynamics means that similar precautions must be taken during underground coal gasification as were suggested for in-situ recovery from oil sand deposits with respect to pollution of local potable groundwater. The test installation at the Forestburg site was mined out after the burn. The conditions found in the burnt and partly burnt coal seam indicate that although some ions found in the groundwater samples collected after the burn were at high concentrations, permeability reduction through mineral precipitation was not likely to create problems during underground coal gasification.

OVERVIEW

In this paper an attempt was made to point out the importance of water-rock interaction in both the natural environment of sedimentary basins and the artificial environment created during in-situ recovery from oil sand deposits and underground coal gasification. The economic aspects of water-rock interaction to exploration and recovery of mineral deposits in sedimentary rocks should be obvious and if this paper does nothing more than to draw the attention of the reader to the importance of water-rock interaction, it will have served its purpose.

REFERENCES CITED

Berkowitz, N., and R. A. S. Brown, 1977, In-situ coal gasification—The Forestburg (Alberta) field test: Canadian Mining and Metall. Bull., v. 70, p. 92-96.

Billings, G. K., S. E. Kesler, and S. A. Jackson, 1969, Relation of zinc-rich formation waters, northern Alberta, to the Pine Point ore deposit: Econ. Geology, v. 64, p. 385-391.

Boon, J. A., 1977a, Mass transfer of silica during steam injection: Proc. 2nd Internat. Symposium on Water-Rock Interaction, v. IV, p. 199-206.

——— 1977b, Fluid-rock interactions during steam injection, *in* D. A. Redford and A. G. Winestock, eds., The oil sands of Canada-Venezuela 1977: Toronto, Canadian Inst. Mining and Metallurgy, p. 133-138.

Clayton, R. N., et al, 1966, The origin of saline formation waters—I. Isotopic composition: Jour. Geophys. Research, v. 71, p. 3869-3882.

Freeze, R. A., 1966, Theoretical analysis of regional groundwater flow: PhD Thesis, Univ. California, Berkeley.

——— 1969, Theoretical analysis of regional groundwater flow: Canadian Inland Waters Branch, Sci. Ser. 3, 147 p.

——— and P. A. Witherspoon, 1966, Theoretical analysis of regional groundwater flow, 1. Analytical and numerical solutions to the mathematical model: a Water Resources Research, v. 2, p. 641-656.

——— ——— 1967, Theoretical analysis of regional groundwater flow, 2. Effect of water-table configuration on subsurface permeability variation: Water Resources Research, v. 3, p. 623-634.

——— ——— 1968, Theoretical analysis of regional groundwater flow, 3. Quantitative interpretations: Water Resources Research, v. 4, p. 581-590.

Hathaway, L. R., O. K. Galle, and T. Evans, 1972, Brine leaching of the Heebner Shale (Upper Pennsylvanian) of Kansas: Kansas Geol. Survey Bull. 204, p. 15-18.

Hitchon, B., 1964, Formation fluids, *in* R. G. McCrossan and R. P. Glaister, eds., Geological history of western Canada: Alberta Soc. Petroleum Geologists, p. 201-217.

——— 1969a, Fluid flow in the western Canada sedimentary basin, 1. Effect of topography: Water Resources Research, v. 5, p. 186-195.

——— 1969b, Fluid flow in the western Canada sedimentary basin, 2. Effect of geology: Water Resources Research, v. 5, p. 460-469.

——— 1971, Origin of oil: geological and geochemical constraints, *in* R. F. Gould, ed., Origin and refining of petroleum: American Chemical Society, Advances in Chemistry Series, v. 103, p. 30-66.

——— 1974, Application of geochemistry to the search for crude oil and natural gas, *in* A. A. Levinson, ed., Introduction to exploration geochemistry: Calgary, Applied Publishing Ltd., p. 509-545.

——— 1976, Hydrogeochemical aspects of mineral deposits in sedimentary rocks, *in* K. H. Wolf, ed., Handbook of strata-bound and stratiform ore deposits: New York, Elsevier, v. 2, p. 53-66.

——— 1977a, Geochemical links between oil fields and ore deposits in sedimentary rocks, *in* P. Garrard, ed., Proceedings of the forum on oil and ore in sediments: London, Imperial College, p. 1-37.

——— 1977b, Geochemical aspects of in-situ recovery, *in* D. A. Redford and A. G. Winestock, eds., The oil sands of Canada-Venezuela, 1977: Canadian Inst. Mining and Metallurgy, p. 80-86.

——— and I. Friedman, 1969, Geochemistry and origin of formation waters in the western Canada sedimentary basin, I. Stable isotopes of hydrogen and oxygen: Geochim. et Cosmochim. Acta, v. 33, p. 1321-1349.

——— G. K. Billings, and J. E. Klovan, 1971, Geochemistry and origin of formation waters in the western Canada sedimentary Basin, III. Factors controlling chemical composition: Geochim. et Cosmochim. Acta, v. 35, p. 567-598.

Horn, M. K., and J. A. S. Adams, 1966, Computer-derived geochemical balances and element abundances: Geochim. et Cosmochim. Acta, v. 30, p. 279-297.

Hubbert, M. K., 1940, The theory of groundwater motion: Jour. Geology, v. 48, p. 785-944.

Kesler, S. W., R. E. Stoiber, and G. K. Billings, 1972, Direction of flow of mineralizing solutions at Pine Point, N.W.T.: Econ. Geology, v. 67, p. 19-24.

Kharaka, Y. K., and I. Barnes, 1973, SOLMNEQ: Solution-mineral equilibrium computations: NTIS PB-215 899, 82 p.

Levinson, A. A., and J. J. Day, 1968, Low temperature hydrothermal synthesis of montmorillonite, ammonium-micas and ammonium zeolite: Earth and Planetary Sci. Letters, v. 5, p. 52-54.

——— and R. W. Vian, 1966, The hydrothermal synthesis of montmorillonite group minerals from kaolinite, quartz and various carbonates: Am. Mineralogist, v. 51, p. 495-498.

Macqueen, R. W., et al, 1975, Devonian metalliferous shales, Pine Point region, District of Mackenzie: Canada Geol. Survey Paper, 75-1, Pt. A, p. 553-556.

Penman, H. L., 1970, The water cycle: Sci. American, v. 223, no. 3, p. 98-108.

Saxby, J. D., 1969, Metal-organic chemistry of the geochemical cycle: Rev. Pure and Appl. Chemistry, v. 19, p. 131-150.

Toth, J., 1962, A theory of groundwater motion in small drainage basins in central Alberta, Canada: Jour. Geophys. Research, v. 67, p. 4375-4387.

——— 1963, A theoretical analysis of groundwater flow in small drainage basins: Jour. Geophys. Research, v. 68, p. 4795-4812.

——— 1980, Gravity-induced cross-formational flow: Possible mechanism for the transport and accumulation of petroleum: (this volume).

Winestock, A. G., 1974, Developing a steam recovery technology, *in* L. V. Hills, ed., Oil sands fuel of the future: Canadian Soc. Petroleum Geologists, p. 190-198.

Winkler, H. G. F., 1967, Petrogenesis of metamorphic rocks: New York, Springer-Verlag, Inc., 237 p.

Cross-Formational Gravity-Flow of Groundwater: A Mechanism of the Transport and Accumulation of Petroleum (The Generalized Hydraulic Theory of Petroleum Migration)[1]

By József Tóth[2]

"Unquestionably the people in the oil industry will find much to learn in familiarizing themselves with the work of and cooperating with those interested more particularly in hydrology where the scientific approach to the problem has been an axiom of long standing" (Muskat, 1932, p. 401).

Abstract Observations in deep sedimentary basins around the world confirm the theory of gravity-induced cross-formational flow of groundwater. Thus, geologically mature basins are hydraulically continuous environments in which the relief of the water table, commonly a subdued replica of the land surface, generates interdependent systems of groundwater flow with patterns modified by permeability differences. In these systems, meteoric waters infiltrate and move downward in upland recharge areas, migrate laterally under regions of medium elevations, and are discharged in topographic depressions. Where flow systems meet or part, relatively stagnant zones develop and flow directions change abruptly.

The theory is advanced that in geologically mature basins, gravity-induced cross-formational flow is the principal agent in the transport and accumulation of hydrocarbons. The mechanism becomes operative after compaction of sediments and the concomitant primary migration cease, and subaerial topographic relief develops. Hydrocarbons from source or carrier beds are then moved along well-defined migration paths toward discharge foci of converging flow systems, and may accumulate en route in hydraulic or hydrodynamic traps. Accordingly, deposits are expected and observed to be associated preferentially with ascending limbs and stagnant zones of flow systems and hence to be characterized by relative potentiometric minima, downward increase in hydraulic heads possibly reaching artesian conditions, reduced or zero lateral hydraulic gradients, and relatively high groundwater salinity. Continuous flow of meteoric waters imports hydrocarbons into such traps until the trap capacity is reached. The excess becomes source material for new accumulations. A temporal change in surface topography causes a proportionate but delayed readjustment of the flow pattern and redistribution of petroleum. Nevertheless, some hydrocarbons may remain in place, constituting residual deposits in discharge and stagnant regions of relict flow systems.

This study reconfirms previous versions of the hydraulic theory of petroleum migration as valid representations of component parts of the migration-accumulation process. However, by introducing the geometry of migration paths in the form of quantitatively defined groundwater flow patterns it integrates existing concepts and observations into a generalized hydraulic theory of petroleum migration.

INTRODUCTION

In view of a long line of classical contributions starting with Munn's (1909) treatise on "The anticlinal and hydraulic theories of oil and gas accumulation," including the papers by Shaw (1917), Dodd (1922), Rich (1921, 1923, 1931, 1934) and Illing (1933), and crowned by Hubbert's (1953) celebrated "Entrapment of petroleum under hydrodynamic conditions" it may appear audacious and misguided to make yet another attempt at explaining the transport and accumulation of hydrocarbons on the basis of gravity-induced groundwater flow. Nevertheless, there are at least two compelling reasons to do so. First, purely hydrogeological investigations during the last ten years have resulted in observations and conclusions (Tóth, 1968, 1970, 1978) that may be considered as independent corroboration of the principles of the hydraulic theory. Second, recent major advances in the theory of regional groundwater movement have produced concepts unknown to students of the migration problem prior to the 1960s and now offer a better understanding of the cause-and-effect relationships between the localization of petroleum deposits, on the one hand, and the regional flow of groundwater, on the

[1]Manuscript received, August 31, 1978; accepted for publication, August 20, 1979.

[2]Groundwater Division, Alberta Research Council, Edmonton, Alberta T6G 2C2, Canada.

The writer is indebted to his colleague, E. I. Wallick, for his encouraging interest and helpful suggestions during the latter phases of the study; to M. M. Madunicky for keeping the work material organized and drafting all the diagrams; to S. Cane for typing the manuscript; and to his wife, Erzsike, for being the patient companion of a man searching for answers.

Alberta Research Council Contribution No. 908.

Article Identification Number:
0149-1377/79/SG10-0008/$03.00/0.

other, than was possible before.

Specifically, a series of theoretical and empirical studies has shown that the energy distribution and traffic patterns of formation fluids in rigid rock framework are produced by elevation differences of the water table acting through a hydraulically continuous flow region. Theoretical considerations and field observations have indicated also that conditions for entrapment are more favorable in certain parts of the groundwater flow field than in others. However, as these parts also are genetically related to the configuration of the water table, their probable positions also are predictable on the basis of the theory of regional groundwater movement.

The above outlined concept combines most features of the various versions of the "hydraulic theory" including cross-formational flow of water offered by Munn and the hydrodynamic entrapment mechanism of Hubbert. In addition, it takes into account the effect of the relief of the water table. Hence it appears appropriate for brevity to refer to this theory of migration and accumulation of petroleum by gravity-induced cross-formational flow of groundwater as *the generalized hydraulic theory of petroleum migration.*

This paper presents this theory including its foundations, consequences, field evidences, and those particular features which are possibly useful in hydrocarbon exploration. As regional groundwater movement constitutes the basis of the migration theory, it also will be discussed in some detail.

REGIONAL FLOW OF GROUNDWATER IN LARGE DRAINAGE BASINS

Principles of Regional Groundwater Movement

Basic Laws of Groundwater Movement

A convenient starting point for an inquiry into the movement of subsurface fluids is the general equation of groundwater flow (Davis and DeWiest, 1966):

$$\nabla^2 h - 2g\beta\rho \frac{\delta h}{\delta z} = \frac{S_o}{K} \frac{\delta h}{\delta t} \qquad (1)$$

where $S_o = g\rho[(1 - n)\alpha + n\beta]$ = specific storage; K = hydraulic conductivity; ∇ = Laplacian operator =

$$\vec{i}\,\frac{\delta}{\delta x} + \vec{j}\,\frac{\delta}{\delta y} + \vec{k}\,\frac{\delta}{\delta z} =$$

gradient of a scalar function; h = hydraulic head; z = vertical co-ordinate (elevation), positive upward; g = acceleration due to gravity; t = time; ρ = density of water; n = porosity; α = compressibility of the rock framework; β = compressibility of water.

From the various special cases that can be derived from Equation (1) two are of interest presently, namely: (i) one-dimensional transient flow in a confined aquifer, and (ii) flow in unconfined regions.

The first may be derived from Equation (1) by integration (Davis and DeWiest, 1966) and is also known as the equation of diffusion:

$$\nabla^2 h = \frac{S_o}{K} \frac{\delta h}{\delta t} \qquad (2)$$

In an unconfined region the compressibilities of rock and water become unimportant rendering the terms containing α and β negligibly small. Thus, in this case Equation (1) reduces for both steady- and non-steady flow to the Laplace Equation:

$$\nabla^2 h = 0 \qquad (3)$$

In the above equations the hydraulic head, h, denotes the height above a standard datum to which the water rises from a point P in the flow region. It is related to the fluid potential, ϕ, and pore pressure, p, in point P by (Hubbert, 1940):

$$h = \frac{\phi}{g} = z + \frac{p - p_o}{\rho} \qquad (4)$$

where p_o = atmospheric pressure (the pressure at the water table), and z = elevation of the point of measurement above datum. Therefore, "h" is proportional to the energy per unit mass of water. The flow of fluid occurs from regions of high energy to regions of low energy and the relation between energy and flow is given by Darcy's equation:

$$\vec{q} = -\vec{K}\,\vec{\text{grad}}\,\phi \qquad (5)$$

where $\vec{q}$ = specific volume discharge indicating the direction and intensity of flow;

$$\vec{K} = \vec{k}\,\frac{\rho}{\eta} = \text{hydraulic conductivity};$$

$\vec{k}$ = intrinsic permeability dependent on the rock's texture; and η = viscosity of the fluid.

With Equations (1) through (5), spatial and temporal distributions of energy and flow can be calculated. Four energy-related parameters will be used in the present analysis, the distributions of which are diagnostic of the fluid-dynamic conditions in drainage basins. These parameters are: (i) hydraulic head which is directly proportional to the fluid potential:

$$h(x,z) = \frac{\phi}{g};$$

(ii) flow through a unit cross-sectional area per unit length of time: $\vec{q}(x, z)$; (iii) pressure versus depth: p(d); and (iv) dynamic pressure increment: Δp (z,d). The concepts and definitions of the first three parameters are commonly known. However, the dynamic pressure increment was defined recently (Tóth, 1978) as ". . . the difference between dynamic (real) and static (nominal) pressures at any given depth." The distribution characteristics of these parameters in typical drainage basins will be examined briefly in the following paragraphs.

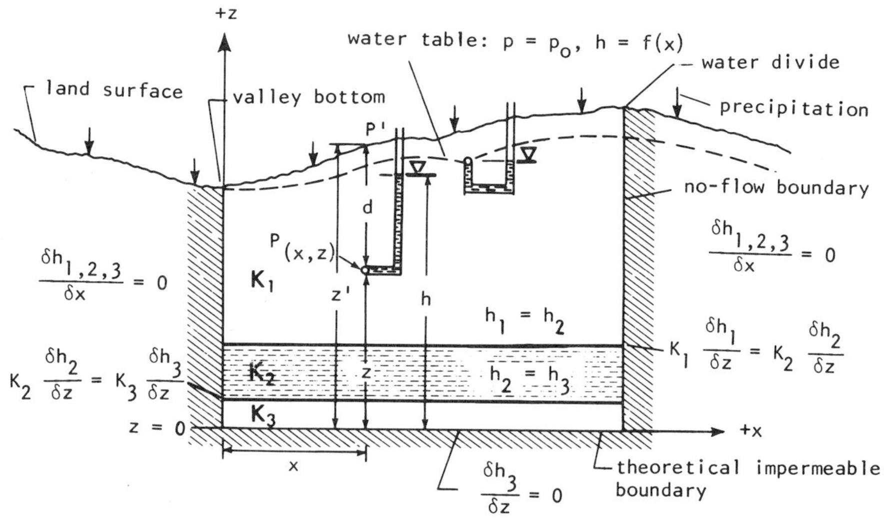

FIG. 1 Schematic representation of groundwater flow-region in drainage basin with stratified rock framework.

Steady-State Regional Flow of Groundwater

General Comments—Energy and flow are said to be steady state if their values (magnitude and direction) remain constant during the time interval considered. Averaged over a sufficiently long time, many natural processes that are subject to short-period changes can be treated as steady-state problems. Typical of this situation is the flow of groundwater in drainage basins whose patterns, when obtained as solutions to steady-state boundary value problems (Tóth, 1962, 1963; Freeze and Witherspoon, 1966, 1967, 1968) can be considered to represent long-term conditions.

Figure 1 is a schematic cross section of the subsurface flow region in a drainage basin. Owing to assumed symmetry the vertical planes beneath the crest and bottom of the basin may be considered as no-flow boundaries under natural conditions. The basin is underlain by a hypothetical impermeable boundary and is recharged by uniformly distributed precipitation. In areas with sufficient precipitation the water table is a subdued replica of the land surface and as a first approximation may be substituted by it.

The flow and energy conditions in such a region of subsurface space are defined by Equation (3) and the appropriate boundary conditions are shown in Figure 1. The solution of Equation (3) is a standard problem of potential-field theory and may be obtained analytically, numerically, or by analog models. The mathematical aspects are beyond this study and the reader is referred to standard texts on field theory or works mentioned in the following paragraphs.

Simple Basin Geometry, Homogeneous Rock Framework—A simple homogeneous drainage basin (Fig. 2a) contains, in general, three different groundwater regions, namely the areas of recharge, mid-line, and discharge. Relative to the water table, groundwater flow is descending, lateral, and ascending in the three regions, respectively. The intensity of movement decreases with increasing depth and away from the area of mid-line, and near-stagnant conditions exist in the lower corners of the flow field. Water levels decline with increasing depth in the recharge area, remain constant in the vicinity of the mid-line and rise in regions downslope from it, possibly to elevations above the land surface. Consequently, the potentiometric surface for a horizontal zone, s (Fig. 2a), of any depth across the basin is below, at, and above the water table in the recharge, mid-line, and discharge regions, respectively.

The pressure-depth relations (Fig. 2b) reflect these conditions by negative and positive deviations for the areas of recharge and discharge, respectively, from the normal (hydrostatic) pressure increase which is found in the region of lateral flow. The curves representing the vertical pressure distribution at sites 2 and 3 (Fig. 2a, b) are slightly arcuate: pressure gradient slopes are markedly different from hydrostatic slopes near the land surface but approach it at greater depths. As the flow region is homogeneous, thus hydraulically continuous, the departure from hydrostatic conditions can only be caused by water movement and is therefore termed here the "dynamic pressure increment", Δp. The Δp values are negative and minimum at the divide, increase gradually to zero as the mid-line region is approached, and become maximum in the valley bottom (Fig. 2a, b). Consequently, the slopes of the p(d) curves are inversely proportional to the topographic elevation in the basin (the curves rotate coun-

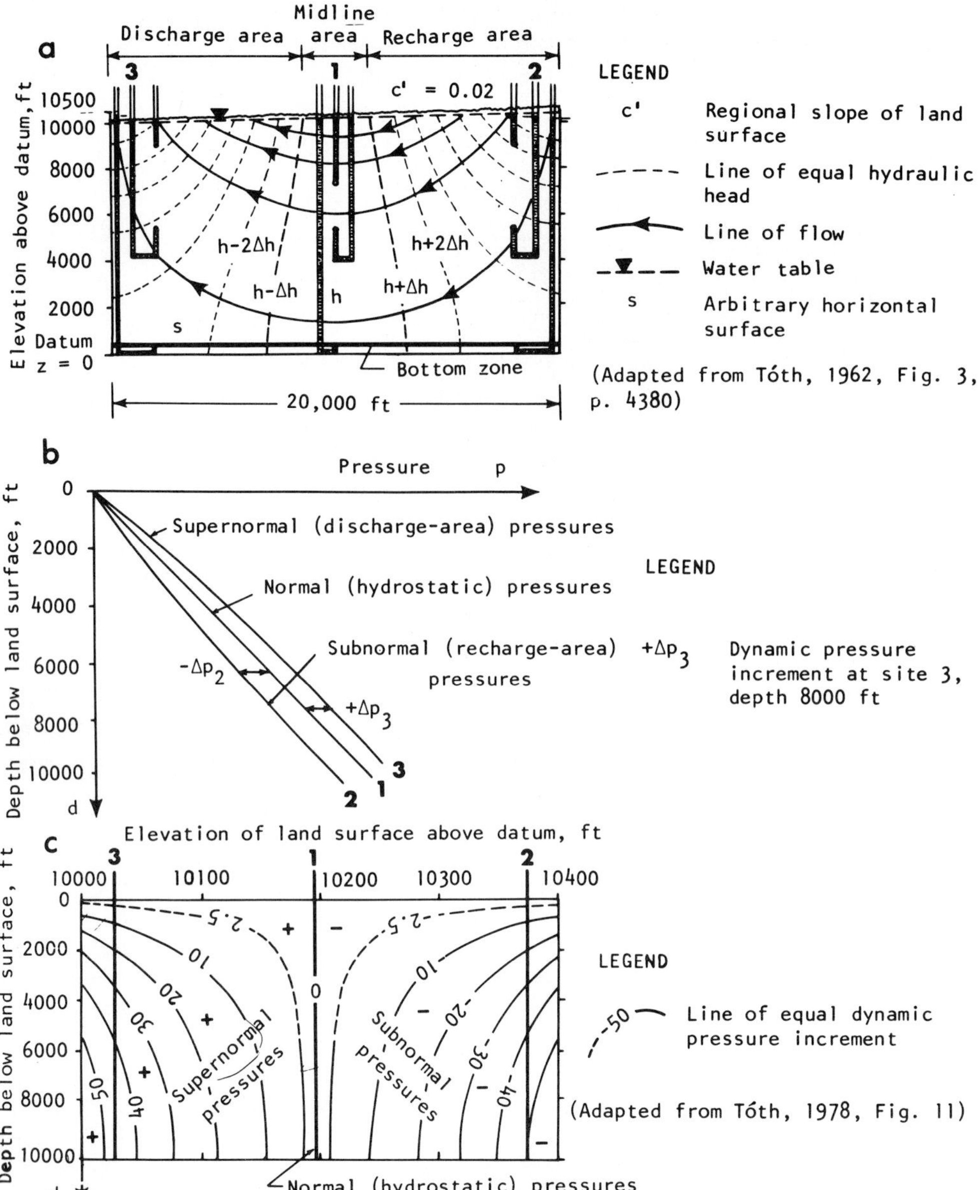

FIG. 2 Distributions of: (a) hydraulic head, h(x,z), and flow, $\vec{q}(x,z)$; (b) pore pressure, p(d); and (c) dynamic pressure increment, $\Delta p(z',d)$ in theoretical drainage basin with simple geometry and homogeneous rock framework.

ter-clockwise about the origin of the p-d coordinate system [Fig. 2b] as the elevation of the basin surface decreases).

These relations may be studied conveniently in a two-dimensional plot of Δp against topographic elevation and depth below land surface. The pattern of isolines of the dynamic pressure increment (Fig. 2c) also may be interpreted in terms of the subsurface dynamic conditions in the basin. Thus a vertical band of zero values indicates that no vertical force component exists between the water table and any depth beneath the mid-line region. Hence, any water present must either be static or move essentially horizontally. Negative Δp's indicate recharge conditions whereas upward directed forces, indeed flowing-well conditions, are reflected by positive pressure increments beneath elevations less than that of the mid-line region.

The Δp (z′,d) pattern in Figure 2c was computed from the theoretical potential distribution shown in Figure 2a. However, patterns of Δp also may be generated from field data. A method of interpretation of actual conditions consists of comparing the observed field patterns with hypothetical type patterns. The method has the advantage of concentrating pressure measurements scattered in three dimensions into a two-dimensional system while retaining a dynamically meaningful relation between Δp, position in the basin, d, and geometry of the basin, z′.

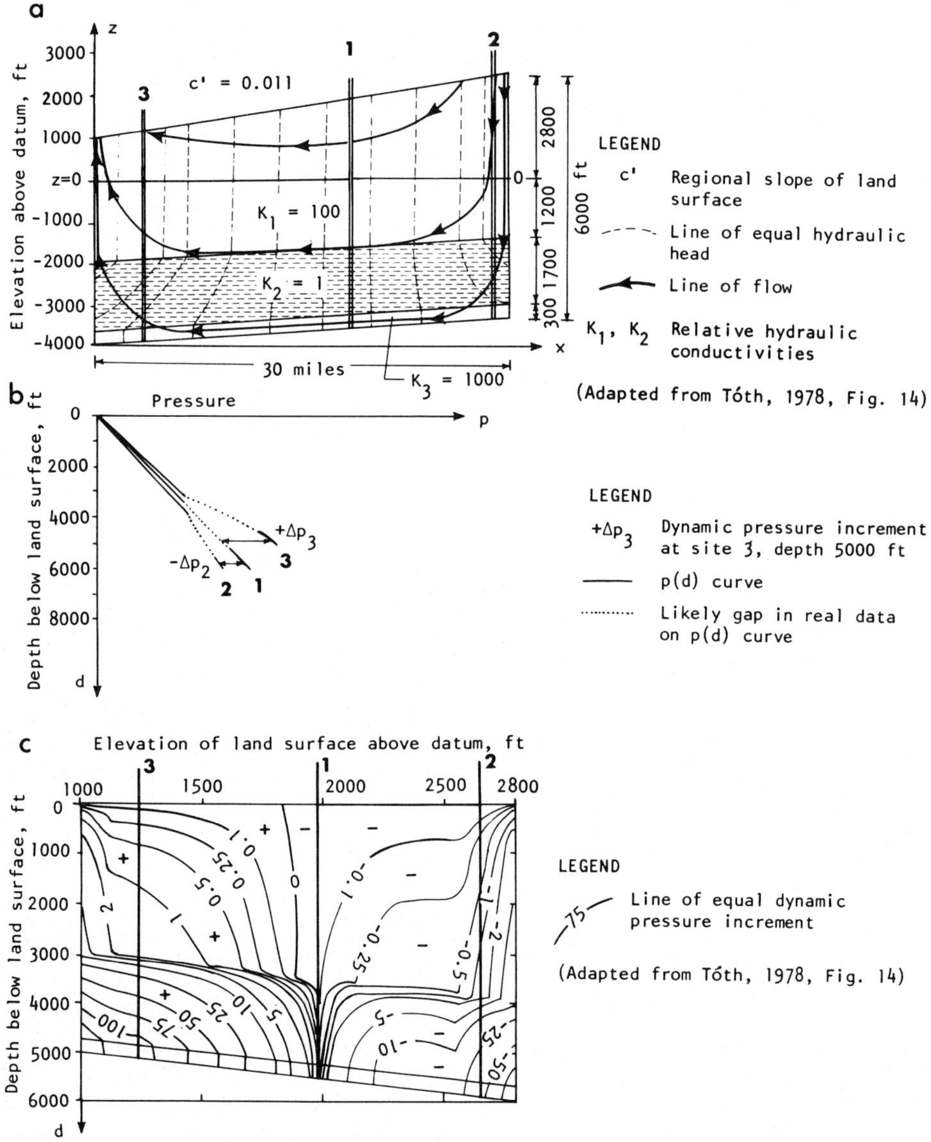

FIG. 3 Distributions of (a) hydraulic head, h(x,z) and flow, $\vec{q}$ (x,z); (b) pore pressure, p(d); and (c) dynamic pressure increment, $\Delta p(z',d)$ in theoretical drainage basin with simple geometry and extensive aquitard; also model for Paleozoic Hydrogeologic System, Red Earth region, Alberta, Canada.

Simple Basin Geometry, Areally Extensive Aquitard —The distribution of parameters h, q, p, and Δp in a drainage basin of simple geometry and containing an extensive aquitard (geologic formation, part, or group of formations having relatively low permeability as compared with adjacent rocks) are presented on a model (Fig. 3a, b, c) designed for the analysis of real conditions in the Red Earth area, northern Alberta (Tóth, 1978, 1979).

The distribution characteristics of the various parameters appear to be generally similar to those of the homogeneous basin, discussed above. However, a major modification, but no change in character, of the patterns results from the introduction of the massive (1,700 ft thick) aquitard extending across the entire basin. Lines of equal hydraulic head are nearly vertical throughout the upper and lower aquifers (Fig. 3a) resulting in essentially lateral flow and nearly hydrostatic rates of pressure changes vertically both above and below the aquitard (Fig. 3b, c). The principal areas of vertical flow are those of the divide and valley bottoms. However, vertical flow is evidently more widespread across the aquitard than in the aquifers as seen from the obliquely refracted lines of equal heads (Fig. 3a), rapid changes in pressure with depth (Fig. 3b), and high values combined with increased concentration of the iso-Δp lines (Fig. 3c) associated with the former.

Attention must be drawn to the possible misinterpretation of practical observations executed in a real

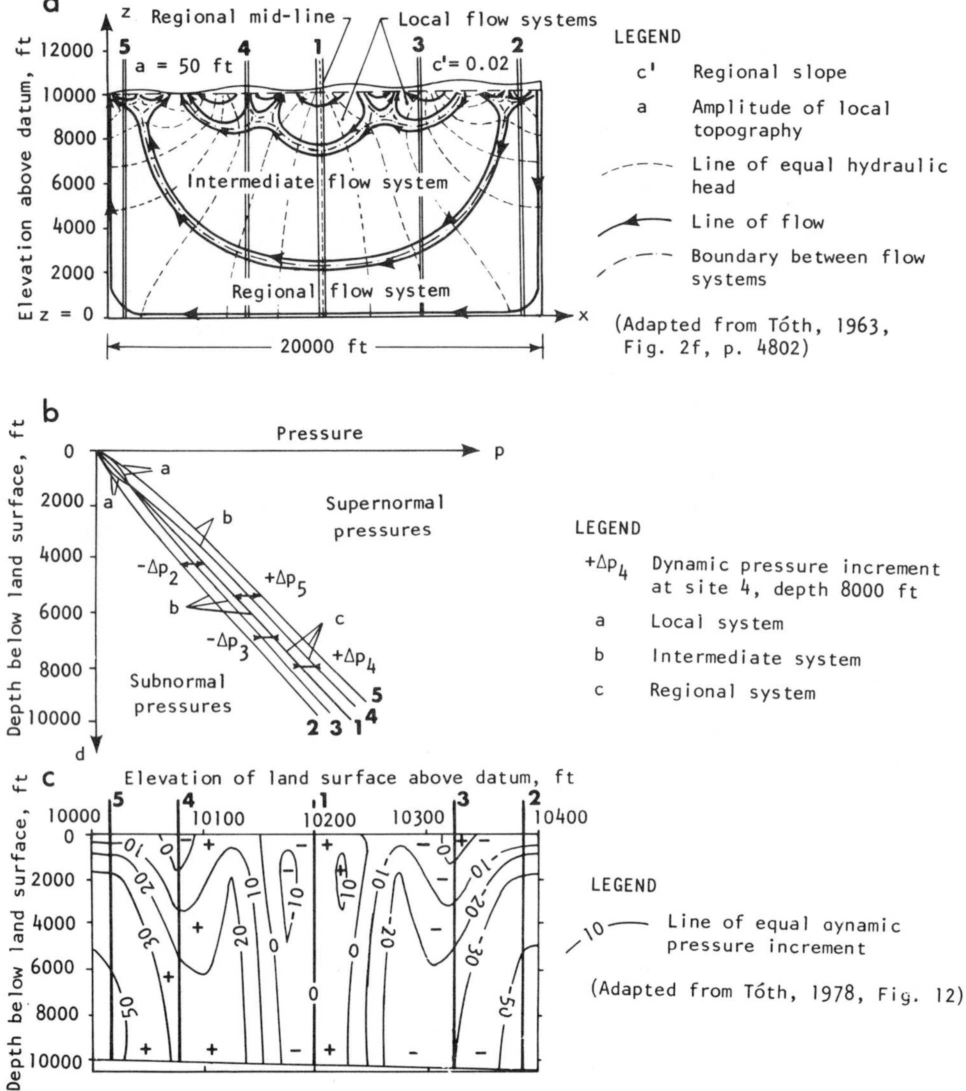

FIG. 4 Distributions of: (a) hydraulic head, h(x,z) and flow, $\vec{q}(x,z)$; (b) pore pressure, p(d); and (c) dynamic pressure increment, $\Delta p(z',d)$ in theoretical drainage basin with complex geometry and homogeneous rock framework.

situation that is similar to the one under discussion. First, most likely no pressures will be observed in the aquitard, thus the transitional parts of the measured p(d) curves will be missing (Fig. 3b). Second, the pressures in the upper permeable zone, to depths of 3,000 to 4,000 ft (914 to 1,219 m) will be observed to be largely hydrostatic, with a few and minor anomalies, both negative and positive. Third, major pressure anomalies will be recorded for the lower, highly permeable stratum. The conventional interpretation of the above observations would be in error, namely: that the upper and lower aquifers of the basin are hydraulically separated by an impermeable formation, with unconfined static conditions in the higher aquifer, and confined flow in the lower aquifer. In reality, all of the above phenomena are caused by a single, regionally confined, continuous-flow system crossing a massive slightly permeable formation.

Complex Basin Geometry, Homogeneous Rock Framework—A basin's geometry is considered to be complex where a local relief is superimposed on the regional slope. The local relief is represented by a sinusoidal curve in Figure 4a and its effect on the dynamic parameters is shown in Figure 4a, b, and c. Again, the basic features of the parameters are similar to those found in basins of simple geometry. Nevertheless, significant and, from the point of view of petroleum migration, fundamental modifications result from the superposition of the local topography.

The primary effect of the complexity of the basin's surface on groundwater movement is the generation of flow systems of different orders (Fig. 2a). A flow system has been defined (Tóth, 1963, p. 4,806) as ". . . a set of flow lines in which any two flow lines adjacent at one point of the flow region remain adjacent through the whole region; they can be intersected anywhere by an uninterrupted surface across which flow takes place in one direction only." Similar to those in the simple basin, each system has three segments: descending, lateral, and ascending. However, three different orders of flow systems may be distinguished in the present case, namely, local, intermediate, and regional (Fig. 4a). A system is termed *local* if its recharge and discharge areas are contiguous; *intermediate* if these areas are separated by one or more local systems but do not occupy the main divide and valley bottom; and *regional* if it links hydraulically the crest and bottom of the watershed.

The depth of penetration of the various flow systems is a function of the ratio of local relief and regional slope (Tóth, (1963) and may reach several thousand feet in a homogeneous basin under the effect of local topography of a few tens of feet. The important corollary to the complex pattern of flow systems is that different systems may be oriented in opposite directions. Stagnant or near-stagnant zones can develop where different systems meet or part, and *hydraulic traps* are formed in focal regions of flow converging from different directions.

Some pressure conditions which may be expected in topography-induced complex groundwater flow systems are shown in Figure 4b and c. Again beneath the regional mid-line area the vertical pressure gradient equals hydrostatic as indicated by the p(d) curve at site 1 (Fig. 4b) and by $\Delta p = 0$ at elevation range 10, 200 ft (3,108 m; Fig. 4c). The inverse relation of the vertical pressure gradient to the regional elevations is clearly indicated by the rotation of the p(d) curves with respect to the hydrostatic line. Changing, indeed reversed, trends of vertical pressure gradients across flow system boundaries are represented by p(d) curves crossing the line of normal hydrostatic pressure increase. The depth of the crossing point (less than 2,000 ft or 610 m in this case) is an approximate indication of the depth of local flow systems.

In a complex basin different subsurface points may have identical elevation or depth values or both, hence any point in the z′,d coordinate system may represent several subsurface points in the field. Consequently, the dynamic pressure increment also may have more than one value for any particular point on the Δp (z′, d) plot. Nevertheless, by virtue of accentuating the essential features of the pressure field with respect to the various elevation and depth ranges of the basin the Δp pattern reveals if there is any systematic relation between the pressures and topography. Thus the regional recharge and discharge character of the areas upslope and downslope of the mid-line are indicated respectively by the dominantly negative and positive Δp-fields in the Δp (z′,d) diagram (Fig. 4c). The alternating recharge and discharge zones of the shallow (local and intermediate) flow systems are reflected by the alternating sign of the dynamic pressure increments at shallow depths across the elevation range of the basin.

All these observations are useful in the interpretation of actual field situations. However, more important than the practical application is the conclusion derived from the above model that complex patterns of hydraulic head, flow, pressure, and pressure increments do not require the existence, hence are not necessarily indicative, of permeability differences or discontinuities in the flow. Instead, they can be caused in hydraulically continuous or even homogeneous rock framework simply by the relief of the local topography. A possible cause for misinterpretation can be demonstrated by considering that observations are available only for shallow and great (no intermediate) depths at site 4 (Figure 4a, b). The most likely interpretation would be a shallow underpressured and a deep overpressured reservoir with no communication. In reality the observed p(d) curves constitute parts only of one continuous p(d) relation rendered nonlinear by the effect of the local topography.

General Basin Geometry, Partial Aquitard—The patterns of dynamic parameters in a relatively general situation (Fig. 5), simulating conditions in the Upper Devonian and younger sediments in the Red Earth region, northern Alberta (Tóth, 1978), exhibit most major features of the previously discussed models. Thus, potential and flow distribution above the (approximately) 500-ft-thick (152 m) massive aquitard is noticeably sensitive to the local relief. Consequently, local recharge and discharge areas alternate across the basin (c.f. Fig. 4) and are reflected at site 3, for instance, by descending flow lines, sub-hydrostatic vertical pressure gradients, and negative Δp (Fig. 5a, b, c) and, respectively, by ascending flow, larger than hydrostatic pressure gradients, and a local field of positive dynamic pressure increments (site 2, Fig. 5a, b, c). The depths of the local systems appear to be controlled by, and equal to, that of the aquitard.

This also means that most of the flow is deflected against the surface of the aquitard and only a fraction of the total recharge penetrates the slightly permeable zone. The large losses of energy and the relatively small flow through this zone are indicated by a concentration of the lines of equal head (Fig. 5a), and by rapid drops in pressure and dynamic pressure increments (Figs. 3, 5b, c). Flow is slow and essentially lateral in the lower zone of high permeability as indicated by the vertical position of the lines of equal h and equal Δp as well as by the essentially hydrostatic rate of pressure increase at sites 1, 2, and 3. Slightly supernormal rates of pressure increase (site 4, Fig. 5b) and positive pressure increments under surface elevations of (approximately) 1,100 ft (335 m) or less (Figs. 5a, c) show clearly the regional discharge nature of the outcropping parts of the lower aquifer.

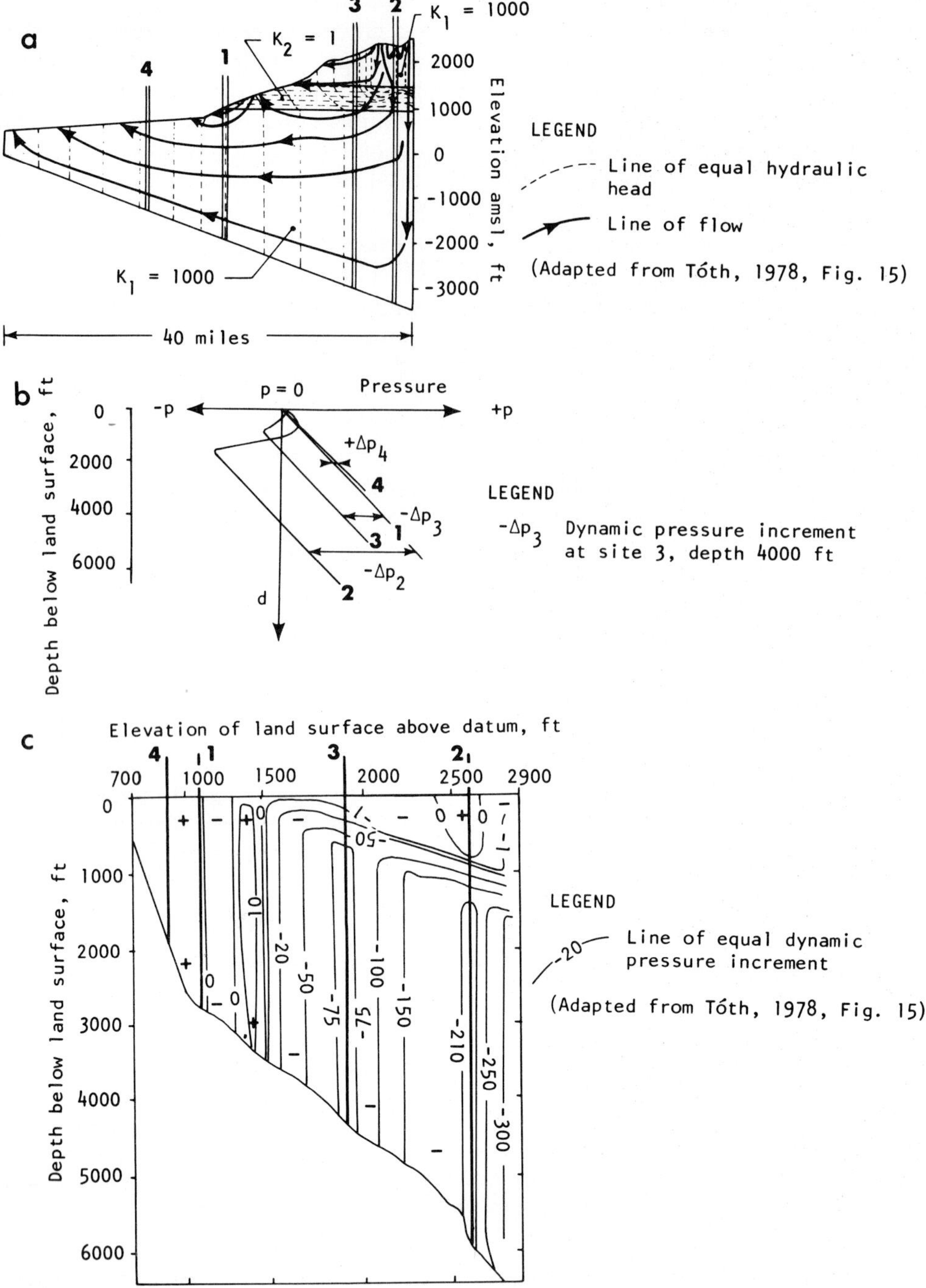

FIG. 5 Distributions of: (a) hydraulic head, h(x,z), and flow, $\vec{q}(x,z)$; (b) pore pressure, p(d); and (c) dynamic pressure increment, $\Delta p(z',d)$ in theoretical drainage basin with general geometry and extensive aquitard; also model for hydrogeologic units combined above the Devonian III Hydrogeologic Group, Red Earth region, Alberta, Canada.

Again it is evident that the combination of local relief and a low permeability stratum may cause in a hydraulically continuous flow field such lateral and vertical distributions of hydraulic head and pressure, including sub- and super-normal values, and abrupt changes that the effects could be suggestive of discontinuous conditions.

Highly Permeable Lens Completely Enclosed in Slightly Permeable Matrix— Contrary to common interpretation "anomalous" pressures (either sub- or super-normal) commonly observed in highly permeable lenses completely enclosed by low permeability matrix are not intrinsically indicative of hydraulic isolation of, and stagnant fluid conditions in, the lens. Indeed, they are necessary consequences of high flow intensity through the lens and reflect the accompanying distortion of the field of fluid potential.

The distorting effect of ellipsoidal lenses of different

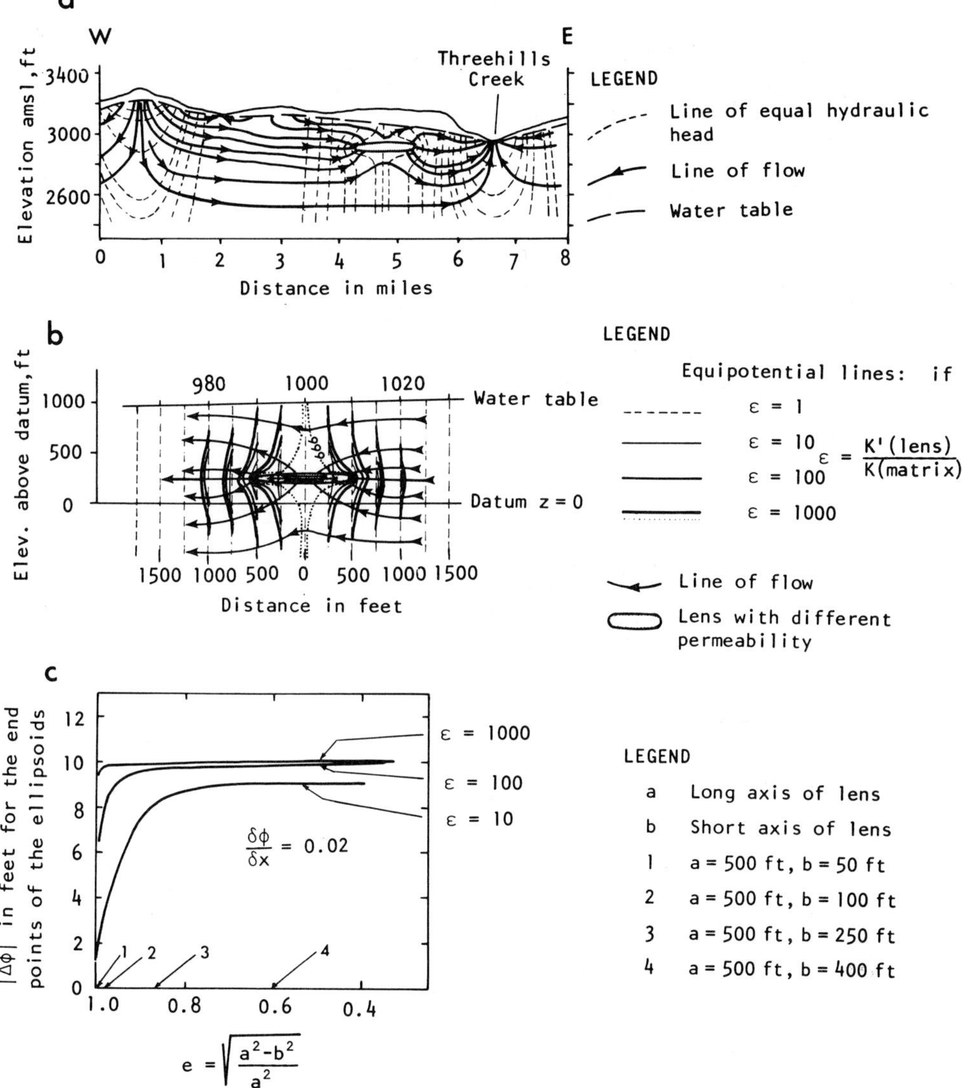

FIG. 6 Effect of a relatively highly permeable lens enclosed in slightly permeable matrix on homogeneous fields of hydraulic head and groundwater flow: (a) regional context; (b) geometry of lines of flow and equal head distorted by lenses of different permeability contrasts; (c) amount of hydraulic head-difference due to lenses of different permeability contrasts and geometry (from Tóth, 1962, Figs. 6, 7, 8).

sizes and relative permeabilities on a homogeneous potential field was examined earlier (Tóth, 1962, p. 4384). Figure 6a shows such a lens in its regional context while Figure 6b presents calculated distortions for different ratios of lens-to-matrix permeability ($\epsilon = K'/K$). It may be seen, for example, that while the topographic elevation is 990 ft (301 m) at the downslope extremity of the 1,000 ft (304 m) long and 100 ft (30 m) thick lens, the hydraulic head at that point is 999 ft (304 m) for a ratio of $\epsilon =$ 1,000, 995 ft (303 m) for $\epsilon =$ 100, and barely more than the original (undisturbed) head of 990 ft (302 m) if ϵ = 10.

Anywhere over the downslope half of the lens, flowing artesian conditions may exist, pressures will be "anomalously" high, and the vertical pressure gradient is greater than hydrostatic. Vertical gradients of hydraulic heads and pressures are greater near the lens than farther above or below it. No anomalous conditions exist at and near the middle of the lens, and changes in head are equal but opposite in sense (subnormal) at the upslope end. Figure 6c shows calculated head differences caused by 1,000 ft (304 m) long lenses of different thicknesses and permeabilities in a uniform potential field of a gradient of 0.02 (approximately 100 ft/mi or 19 m/km). It is apparent that the distortion of the field (anomaly of pressures) increases as a ratio of length to thickness decreases and as the permeability contrast, ϵ, increases. However, almost maximum distortion is reached already with even a relatively thin lens (thickness 100 ft or 30 m, length 1,000 ft or 304 m [case 1]) at a permeability contrast, ϵ, as low as 1,000.

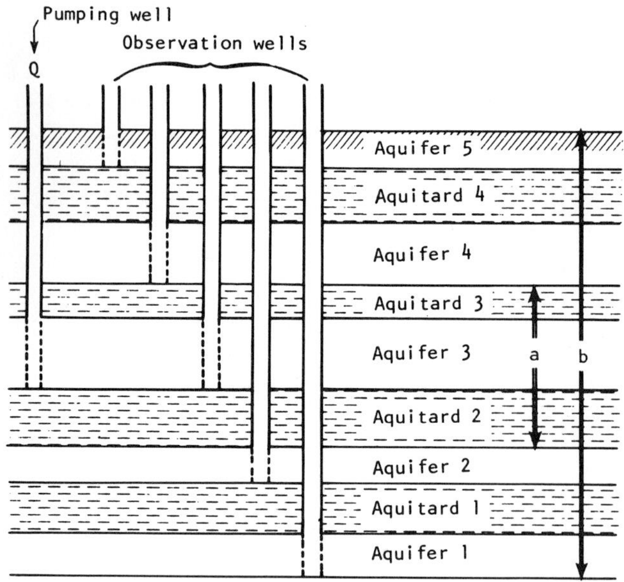

FIG. 7 Five-aquifer system (from Neuman and Witherspoon, 1969, Fig. IV-21, p. 143).

These calculations thus clearly show that significant pressure anomalies may be caused in a hydraulically continuous flow field by relatively small bodies of high permeability. Flow is increased through these bodies and converges from cross-sectional areas considerably larger than that of the lenses themselves.

Non-Steady State Regional Flow of Groundwater

General Comments—Changes in boundary conditions and the appearance of new or disappearance of previous sources of driving forces result in modifications of existing fields of force and flow. The state of change is termed non-steady state or transient condition. An insight in the causes, nature, and time rates of transient groundwater conditions is a prerequisite for the correct evaluation of the effects of different geologic events on the traffic patterns of subsurface fluids, including hydrocarbons, for the understanding of the time-dependent nature of permeability barriers (the concept of hydraulic continuity), and finally, for the appreciation of the lengths of time required for regional flow and energy fields to adjust to new boundary conditions.

The three main types of mechanisms resulting in transient conditions are: (1) active energy sources such as osmosis, phase changes of minerals, changes in fluid volumes due to heating and cooling, changes in rock-pore volumes, and so on; (2) deformation of the basin's rock framework as, for instance by tectonic compaction or sediment loading; and (3) changing boundaries in a geologically mature basin (one with a rigid rock framework). As the theory of petroleum migration to be developed in later sections of this paper is based on topography-induced groundwater flow, attention will be devoted here to transient conditions and their significance in drainage basins of consolidated (i.e. rigid) rock framework modified by erosion.

Selected Aspects of Transient Conditions in Drainage Basins—A detailed presentation of transient flow requires lengthy and very complex mathematical discussion which is beyond the scope of the present paper. Only three relevant points will be mentioned, based on published results, namely: (1) the relative nature of hydraulic continuity; (2) the time-rate at which changes in potential may propagate in a real situation; and (3) the evolution of potential- and flow-patterns in drainage basins. These are outlined here:

1. The relative nature of hydraulic continuity as reflected by transient flow conditions was clearly demonstrated by Neuman and Witherspoon (1969, p. 143-145) in their exacting analysis of multiple-aquifer systems. The gist of their findings is that the effect of pumping of any member of a system of several aquifers and aquitards (Fig. 7), in terms of disturbing the potential- thus flow-conditions in any other aquifer, depends on the hydraulic coefficients of the system's components and the length of time of pumping. Thus, if aquifer 3 is pumped, the effects of aquifers 2 and 4 (and more so of those of 1 and 5) will be negligible at short values of pumping time. System "a" (Aquifer 3–Aquitards 2 and 3) will behave independently of the rest and as a confined, isolated unit. However, for long values of time, leakage across the aquitards will take effect and have an influence on the drawdown in the pumped aquifer 3, as well as on the water levels in the unpumped aquifers.

In the words of Neuman and Witherspoon (p. 145) ". . . the definition of the boundaries of a multiple-aquifer system is a relative matter, and . . . it depends on the accuracy which one is willing to accept in applying our theory to the particular problem at hand." The caprocks and other confining strata that appear impermeable on short-term laboratory measurements or by the short-time reservoir-engineering tests and production histories may pass fluids easily during times significant on the geologic time scale.

2. With the above thoughts in mind a numerical example was produced for a real situation in northern Alberta (Figs. 20-23) by the solution of Equation (2) (Tóth, 1978). According to the results shown in Figure 8, the original steady-state head, h_o, at point P_I in the Devonian I Hydrogeologic Group, would remain unaffected for approximately 100,000 years following a stepwise drop, Δh_T, in head at the land surface caused presumably by erosion. Despite the two thick and very slightly permeable aquicludes between the surface and Group D_I one half of the total disturbance is transmitted to P_I in approximately 700,000 years, and hydraulic heads will be adjusted completely to the new boundary conditions after 4 million years. On the geologic time scale, the entire sequence is definitely hydraulically continuous.

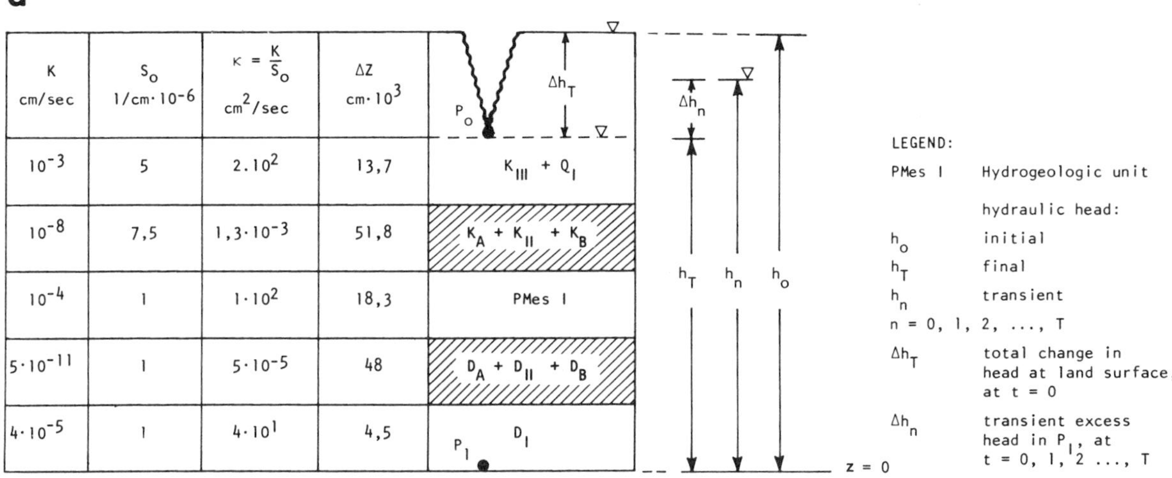

a
K cm/sec
So 1/cm·10-6
κ = K/So cm2/sec
ΔZ cm·103
10-3
5
2.102
13,7
10-8
7,5
1,3·10-3
51,8
10-4
1
1·102
18,3
5·10-11
1
5·10-5
48
4·10-5
1
4·101
4,5
Po
ΔhT
KIII + QI
KA + KII + KB
PMes I
DA + DII + DB
PI
DI
Δhn
hT
hn
ho
z = 0
LEGEND:
Time available for excess head in Paleozoic Hydrodynamic Zone:
2 to adjust to the land surface during Pliocene. (Thickness of line is qualitative indication of probability of particular duration).
PMes I
Hydrogeologic unit
hydraulic head:
ho
initial
hT
final
hn
transient
n = 0, 1, 2, ..., T
ΔhT
total change in head at land surface, at t = 0
Δhn
transient excess head in PI, at t = 0, 1, 2 ..., T

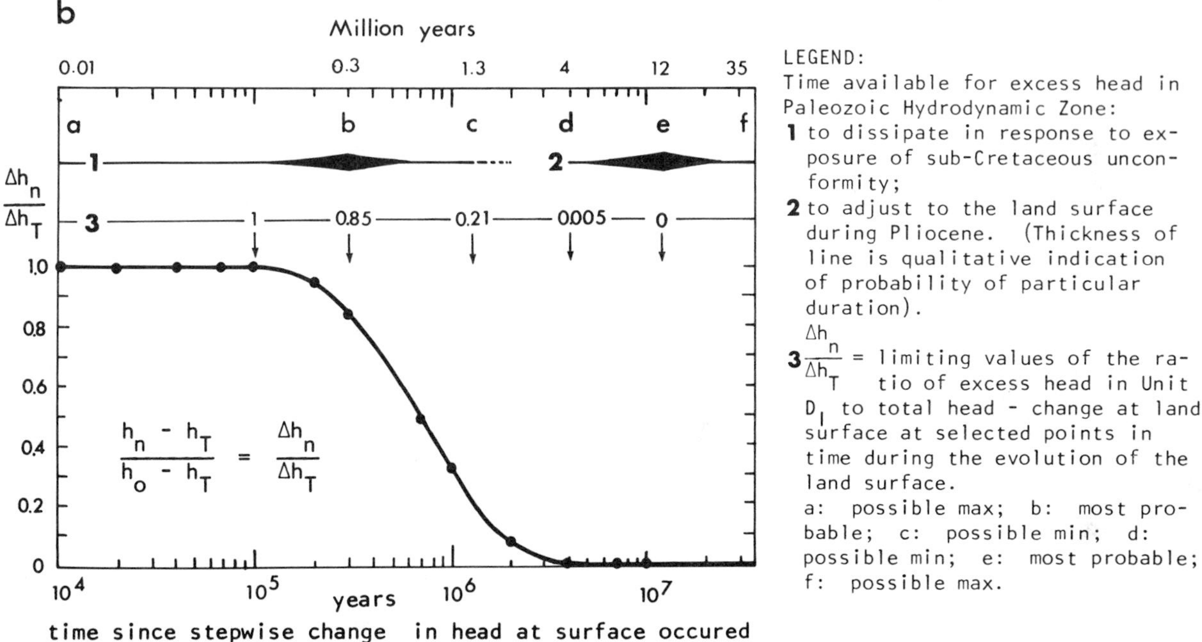

a: possible max; b: most probable; c: possible min; d: possible min; e: most probable; f: possible max.

FIG. 8 One-dimensional decay of hydraulic head in Group D_I in multiple aquifer system resulting from erosion-induced stepwise change in head at the land surface, northern Alberta, Canada (from Tóth, 1978, Figs. 41, 42).

3. Transient positions of the water table due to increased rainfall, the initial and final distributions of hydraulic heads and the corollary shift in positions of particular flow lines are shown in a hypothetical drainage basin (Fig. 9) computed by Freeze (1972, Fig. 7, p. 127). The different positions of the water table represent the changing boundary conditions. The shift in flow lines, influenced somewhat by a slightly permeable stratum, indicates the different routes that fluids originating at the same surface point can take under the influence of changing conditions.

Theoretically Expected Field Phenomena Associated with Regional Groundwater Flow

Due to water's ability to interact with chemical substances, to transport heat and suspended solids, and to support organic life, sustained patterns of its subsurface movement may be expected to cause characteristic distributions of physical, chemical, hydrological and biological processes and phenomena in space in addition to the previously outlined dynamic parameters. An understanding of the relation between the pattern of flow and its natural consequences is valuable both in the study of the genesis of certain natural phenomena and in recognizing the distribution of regional fluid movement.

A brief summary of these manifestations was presented before (Tóth, 1972), and only the ones directly relevant to the present problem are mentioned here (Fig. 10):

1. Contents of dissolved minerals are known to in-

crease with the water's subsurface residence time (along and in the direction of the flow lines). Consequently, the mineral content of groundwater is expected to be relatively low at shallow depths of the recharge areas and in short and active systems. Conversely, salinities will be high at great depths, in discharge areas and in extensive sluggish systems.

2. The amount and type of mineralization of ground water may vary significantly over short distances both laterally and vertically. Marked differences are expected across the boundaries of flow systems as a result of the possible juxtaposition of waters derived from different areas of recharge, or through different lengths of flow paths, or both.

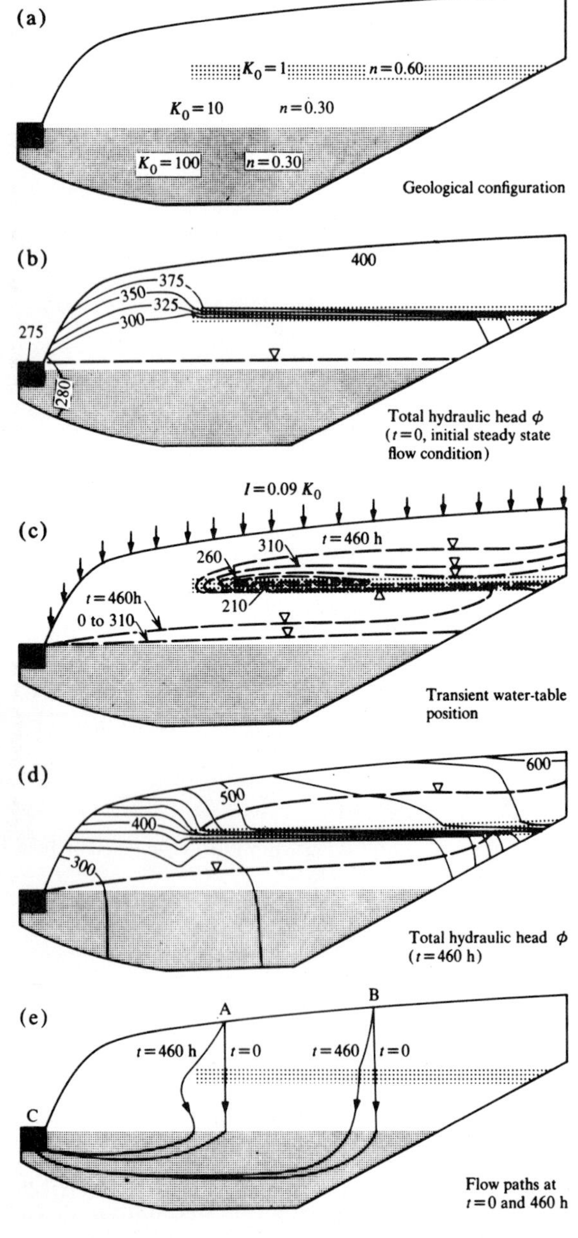

FIG. 9 Transient development of a perched regional flow system showing the time-dependent shift in the location of specific flow lines (after Freeze, 1972, Fig. 7, p. 127).

3. Accumulation of dissolved, suspended, emulsified, or colloidal matter is expected to take place at flow-pattern nodes due to a decrease in the transporting capacity of the flow. Substances capable of being transported by groundwater include: inorganic ions, compounds, and minerals; contaminants such as detergents, fertilizers, industrial and human wastes, natural hydrocarbons; and so on.

4. The amounts of moisture available to the areas of inflow and outflow are inherently different due to the difference in the average direction of regional groundwater movement. In areas of inflow a deficit of ground moisture may be expected as compared with the mid-line region taken as the basis of reference. However, in areas of discharge groundwater is added to local precipitation, causing the moisture content to exceed its reference value for the drainage basin and possibly resulting in marsh conditions. Also, the differences in moisture supply may produce observable contrasts in the vegetation, erosional features, soil types, and surface accumulation of salts. In arid regions extensive salt deposits may form in regional discharge areas.

5. Alternating or isolated areas of groundwater inflow and outflow may occur in composite drainage basins. Islands of discharge character may be generated by ascending limbs of local systems in regional recharge areas, whereas relatively arid conditions will arise where local flow is downward in an area of regional discharge.

6. Descending cool waters will lower the geothermal gradients in recharge areas while positive thermal anomalies may be expected over the ascending limbs of discharging deep flow systems.

Field Examples of Regional Groundwater Flow and Associated Natural Phenomena in Large Drainage Basins

General Comments

Geological literature is replete with references, observations, and data evidencing certain properties and manifestations of regional groundwater flow in extensive areas. It is nevertheless difficult to find or to compile even comprehensive flow patterns in large drainage basins, mainly because of the unfamiliarity of most hydrogeologists with the relatively recent concepts of cross-formational flow and the resulting failure to report all relevant information. Also, groundwater-related observations may only be incidental to principal objectives and therefore incomplete for the purposes of flow analysis. Even in those rare instances when flow across aquitards is allowed for, an attempt is seldom made to relate subsurface energy and flow conditions to the present or past relief of the land sur-

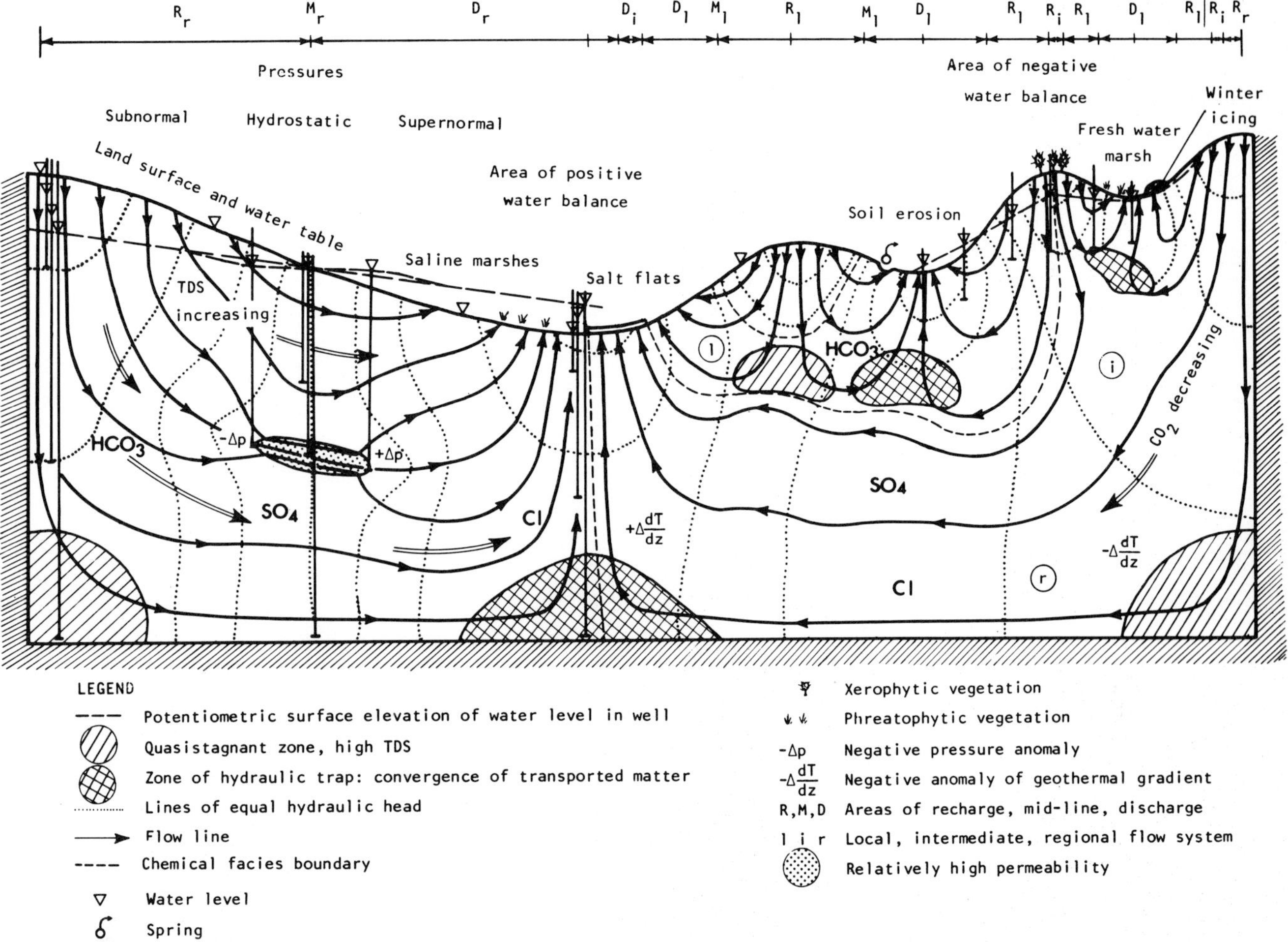

FIG. 10 Diagrammatic summary of properties and manifestations of regionally unconfined groundwater flow in simple and complex drainage basins.

face. This results in the usual omission of contour maps of the present- or paleotopography from publications.

The above problems notwithstanding, it was possible to evaluate regional groundwater conditions for several large basins, based on various or even unrelated sources and on theoretical concepts in most instances. Some examples are given in the following paragraphs.

Steady-State Conditions

Aquitaine, France—The Aquitaine region of France was one of the first areas where hydraulic communication across aquitards and the resulting control of the topographic relief on natural groundwater flow was conclusively documented (BRGM, 1969).

Figures 11 through 13 show the topography, water levels, and water level differences in various aquifers as compiled from a number of independently conducted studies (Astié et al, 1969; Coustau, et al, 1969; Margat, 1969; Martin, 1969; BRGM, 1970; Institut de Géographie, 1973). The two main features of the potential distribution are immediately obvious: the potentiometric surface remains similar to the land surface even in the deepest zones (Upper Cretaceous, Fig. 12d) though damped as depth increases, and hydraulic heads in the deeper Eocene exceed those in the shallower Tertiary formations ($\Delta p > 0$), in areas of low topographic elevation (Figs. 11, 13a, b). Invoking the concept of "leakage" developed earlier by Hantush and Jacob (1954) for the analysis of aquifers by pumping, the previous features of the potential distribution were interpreted by Margat (1969), p. 15) as " . . . a 'reflection' of the surface configuration of the phreatic systems on the confined aquifers which, on the final account, are drained by the former . . . " (writer's translation).

Studying water qualities in the western part of the area, Martin (1969 p. 85) stated explicitly that "With the exception of the sandstones of the Marl series the porous bodies of the Jurassic, Cretaceous, Tertiary and Quaternary constitute one single system. The contained fluids thus have one common hydraulic base" (writer's translation), asserting cross-formational hydraulic communications to depths greater than 5 km (16,000 ft). Based on the above considerations and

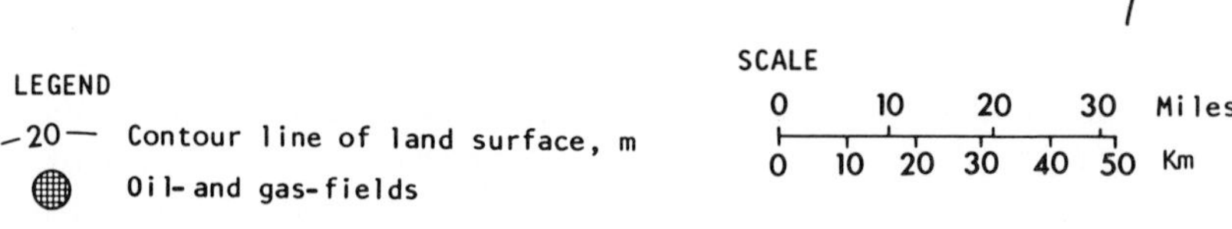

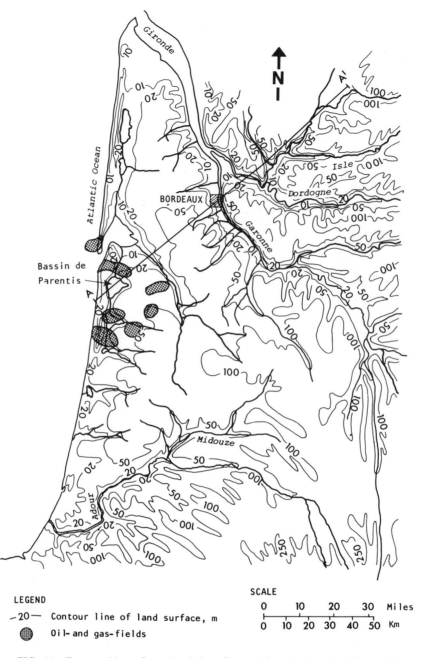

FIG. 11 Topographic surface, Aquitaine, France (from Institut de Géographie, 1973).

using the geologic cross section by Bourgeois (BRGM, 1970) the present pattern of regional groundwater flow for the northern part of Aquitaine was compiled by the present writer in Figure 14.[3] It is characterized by the control of the land surface and by flow across formation boundaries. All of its features are compatible with the distribution patterns presented for the various theoretical cases. The causal relation between the position of the oil fields and the areas of ascending water flow will be explained later.

[3]A very similar flow pattern published by Besbes et al (1978, p. 295, Fig. 1) came to the writer's attention after the completion of this manuscript.

The Great Lowland Artesian Basin, Hungary—The Great Lowland artesian basin of Hungary (approximately 50,000 sq km) is probably the world's (hydrogeologically) best known closed basin of comparable size. Over the past 100 years more than a million shallow wells and over ten thousand deep (up to 4 km) wells were drilled for the exploration and development of rich resources of cold and thermal waters as well as of oil and gas deposits and yielded abundant data on the geology, pressure, chemistry, and thermal conditions of the subsurface fluid environment (Máfi, 1958; Rónai, 1961). Nevertheless, in spite of the numerous explicit observations

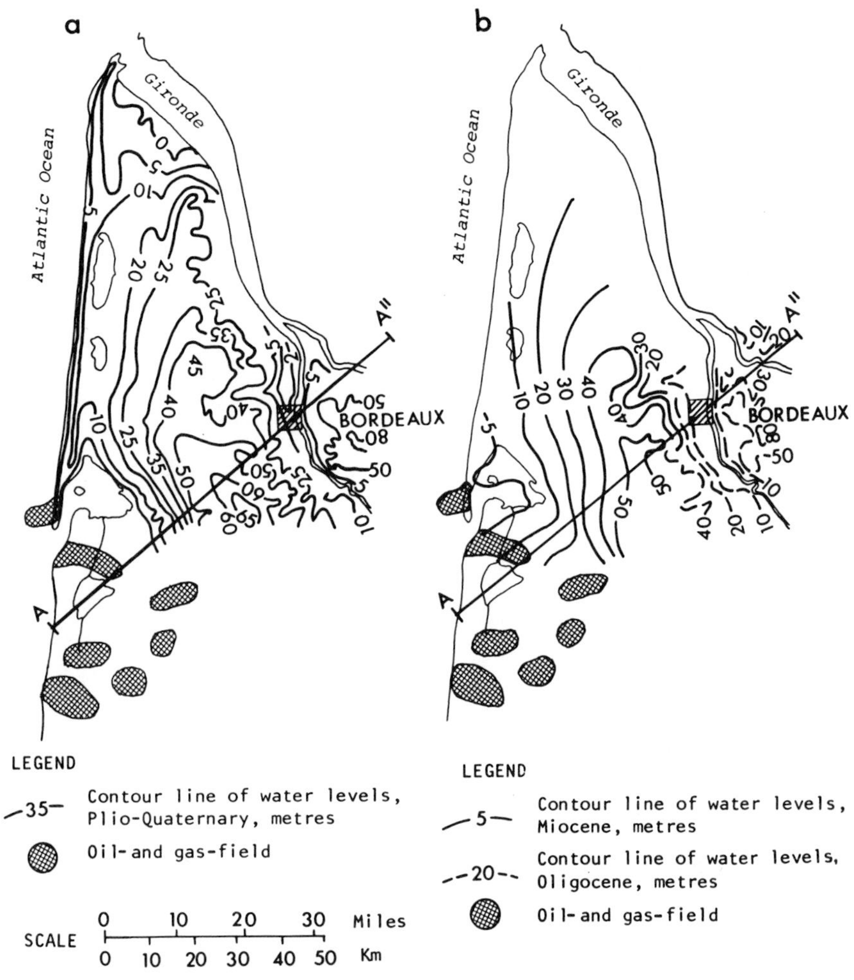

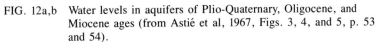

FIG. 12a,b Water levels in aquifers of Plio-Quaternary, Oligocene, and Miocene ages (from Astié et al, 1967, Figs. 3, 4, and 5, p. 53 and 54).

suggesting hydraulic communication between seemingly separated aquifers (Schmidt and Almássy, 1958; Szebényi, 1965, Figs. 1, 2, p. 126-127; Urbancsek, 1963, Figs. 9-25, p. 212-214) it was not until Erdélyi (1972) applied the principles of regional groundwater movement that the topography-dependent nature of the basin's fluid dynamics was established.

The correlation between the topographic relief and potentiometric surface of aquifers at depths of 150 to 300 m is illustrated by Figures 15 and 16. Clearly, water levels decrease from the hills on the northwest and may be expected by extrapolation to decrease also from the southeast toward the central lowland. Nevertheless, potentiometric highs coincide with the relatively low north-south ridge of the Duna-Tisza interfluve and the low mound northeast of Debrecen. An even more detailed agreement between the land surface and water levels is found at shallower depths (Rónai, 1961) and a progressive attenuation, but preserved similarity, at greater depths (Erdélyi, 1972, Fig. 29-4, p. 120). Examples for the relatively continuous vertical changes of hydraulic heads are given in Figure 17a, b and c, for three hydraulically different areas; namely recharge, mid-line, and discharge, respectively (Fig. 16). A significant, and theoretically anticipated, characteristic of the three conditions is their association with different average elevations of the land surface: decreasing from recharge to discharge, with the mid-line elevations in between.

The type and nature of the regional flow patterns were illustrated by Erdélyi (1972, Fig. 30-4, p. 121), as shown in Figure 18. He also stated that: ". . . the most important areas with energy potential decreasing with depth are the topographic highs inside the central basin." On the other hand and still after Erdélyi "The deepest parts of the Hungarian Basin are the areas of typical artesian flow where energy potential increases with depth . . . These discharge areas are covered by completely impervious alkali soils (solonetz and solonchak). . . (and) . . . occupy about half of the area of the basin.

"On the discharge areas, waters of higher salt

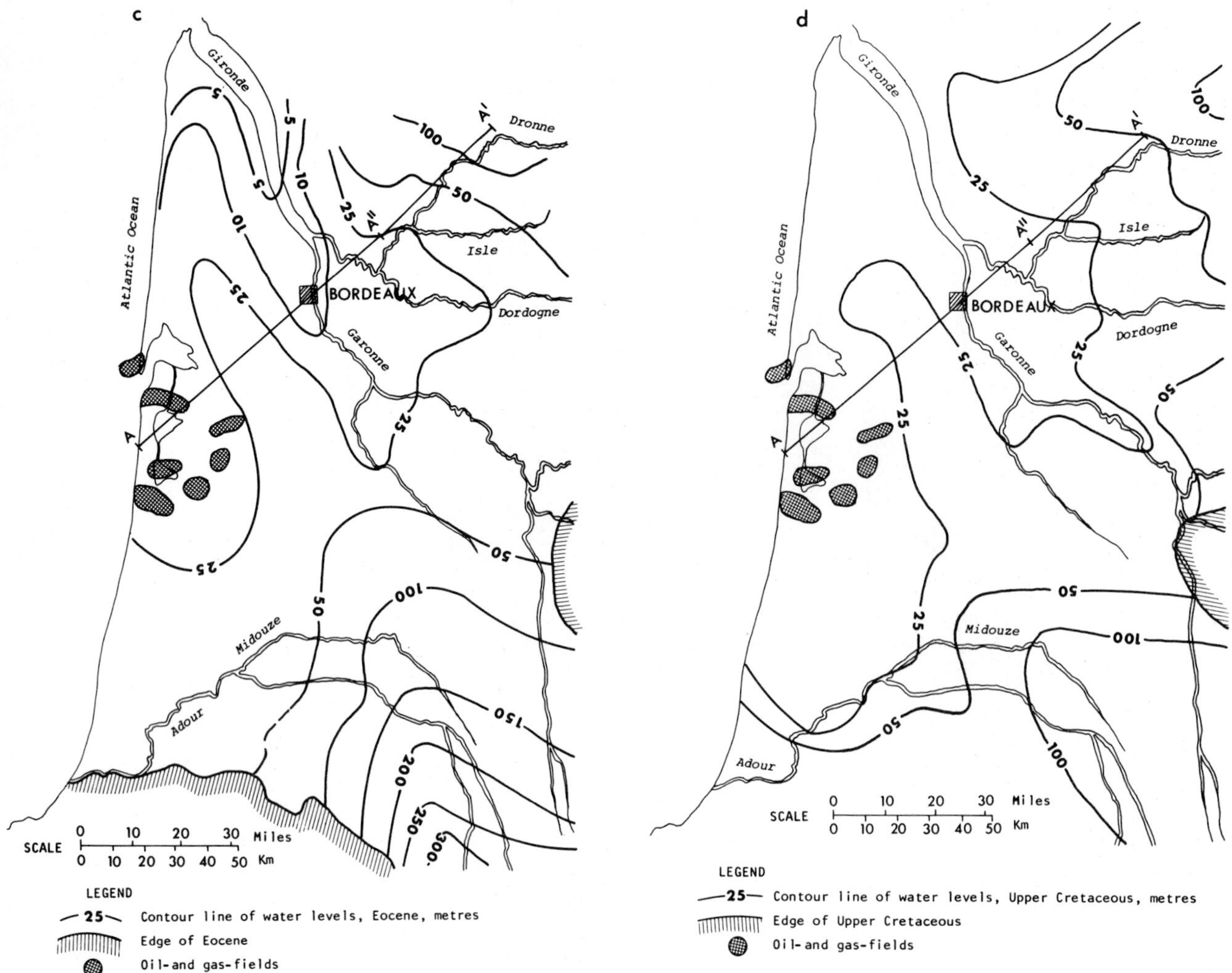

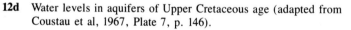

12c Water levels in aquifers of Eocene age (from Coustau et al, 1967, Plate 4, p. 143).

12d Water levels in aquifers of Upper Cretaceous age (adapted from Coustau et al, 1967, Plate 7, p. 146).

content move upward from deeper aquifers regionally by slow seepage, or locally along open fault lines, which localities are characterized by high salinity, marked geothermal anomaly and very high specific capacity of wells as well as by high gas content of the artesian waters." This is the general area where oil and gas deposits have been found. Therefore, it is reasonable to consider some causal relation between ascending water and the oil and gas deposits. If we add that the "Depression of Békés," coinciding with the valley of the Körös River north of that town (Fig. 15) and corresponding with the major discharge area east of the Tisza River (Figs 16, 17, 18), was a vast expanse of "marshy meadows" until extensive drainage and river training turned it into arable but locally strongly alkali dry land during the 19th century, it becomes evident that the Great Lowland artesian basin of Hungary presents nearly all of the diagnostic surface and subsurface features of regionally unconfined flow of groundwater.

The Artesian Basins of the Central Plateau, Iran—The Central Plateau of Iran is built of mountain ranges of poorly permeable volcanic, clastic, and evaporitic rocks containing closed sedimentary basins filled by coarse and fine alluvial material. Areal topographic relief can reach 1,200 m. The regionally unconfined character of the groundwater systems is clearly manifested by the configuration and position of the potentiometric surface and by several flow-related natural phenomena, including chemical type and total mineralization of groundwater, surface accumulation of salts, and the water balance (Issar and Rosenthal, 1968).

Figure 19 is a generalization of the various features described and presented graphically by Issar and Rosenthal (1968) for several basins of the Central Plateau. Typically, water levels in wells remain below the land surface near the mountains but rise above it

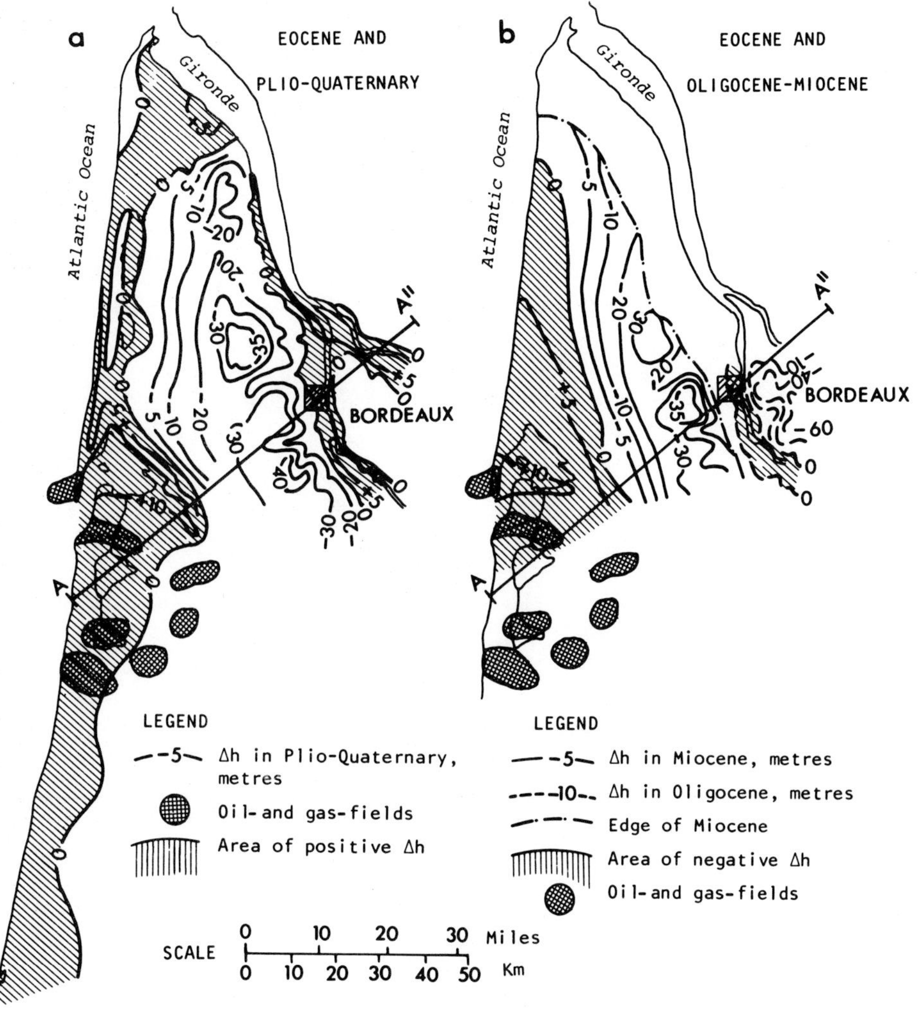

FIG. 13 Water-level differences between aquifers of different age (adapted from Astié et al, 1969, Figs. 6 and 7, p. 55; and estimated from Coustau et al, 1969, Plate 3, p. 142).

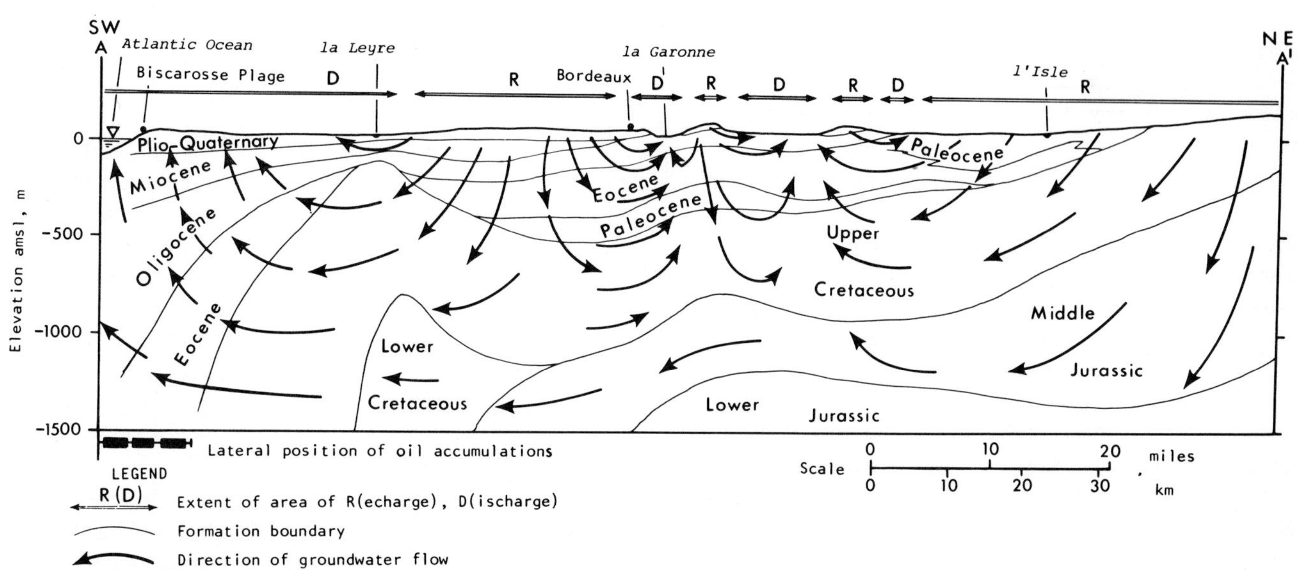

FIG. 14 Regional groundwater flow patterns along A-A[1], Northern Aquitaine, France (stratigraphy adapted from BRGM, 1970, cross section 1, Aquitaine Nord, by M. Bourgeois).

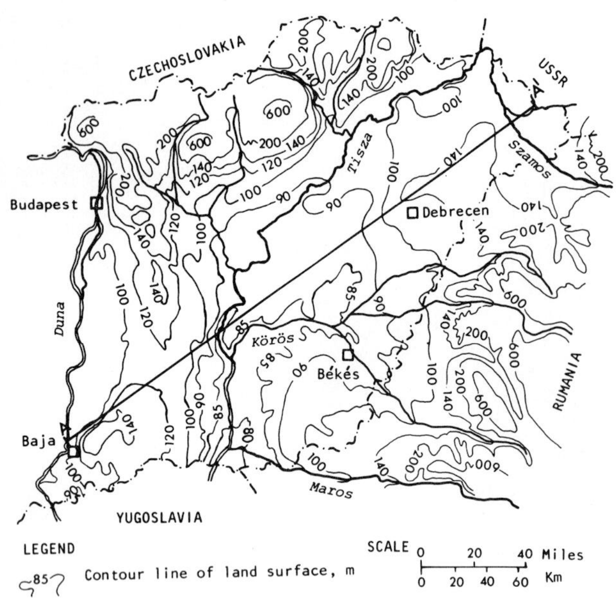

FIG. 15 Topographic surface, Great Lowland, Hungary (from MÁFI, 1958, map 1).

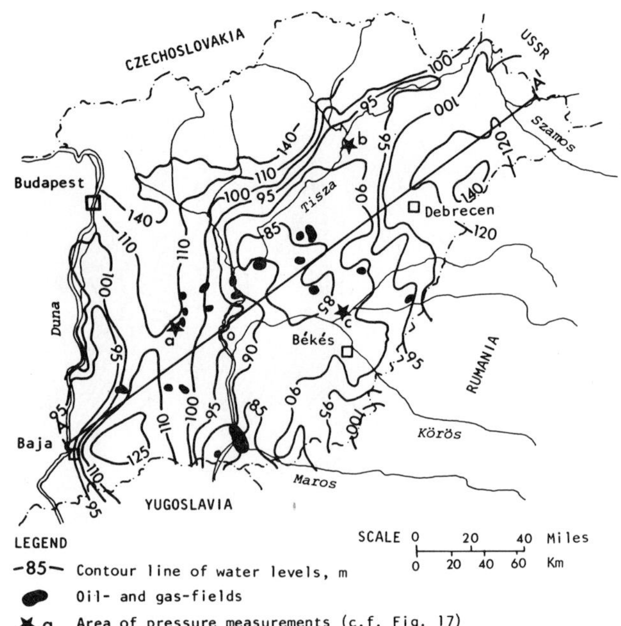

FIG. 16 Water levels in aquifers at depths of 150 to 300 m, Great Lowland, Hungary (from Erdélyi, 1972, Fig. 27-4, p. 118).

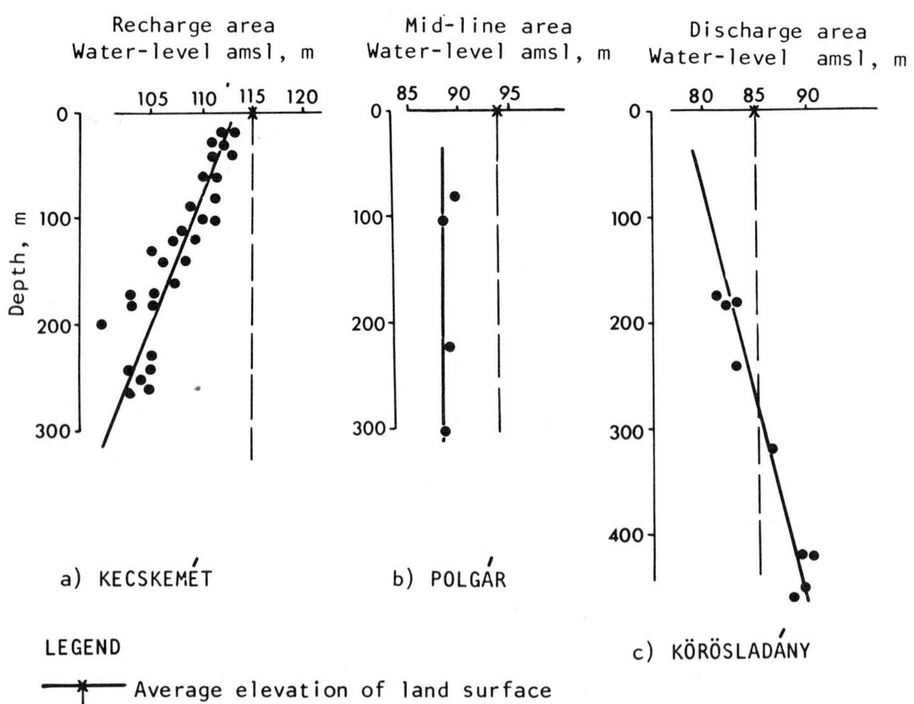

FIG. 17 Change in water levels with respect to depth in: (a) a recharge area, (b) a mid-line area, and (c) a discharge area in the Hungarian Great Lowland (after Urbancsek, 1963, Figs. 23-24, p. 214).

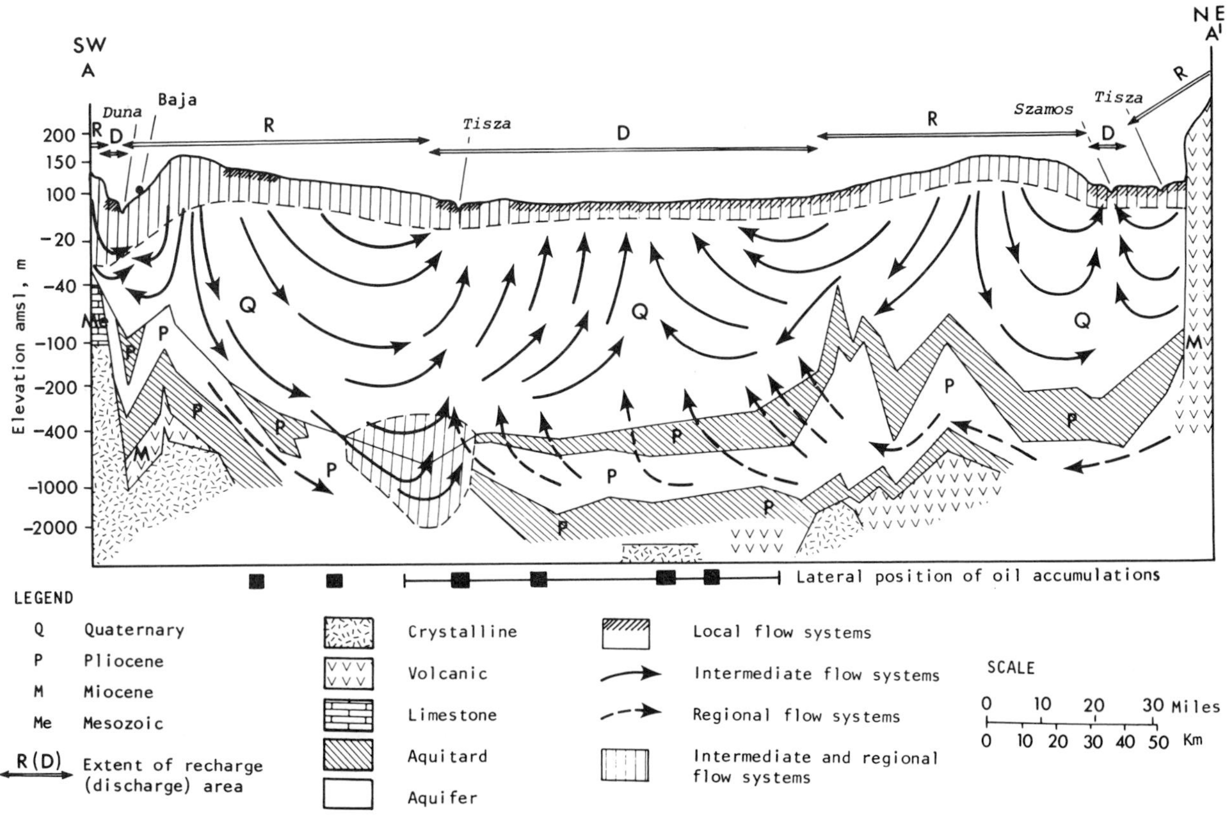

FIG. 18 Diagrammatic flow-pattern of the Great Lowland, Hungary (from Erdélyi, 1972, Fig. 30-4, p. 121).

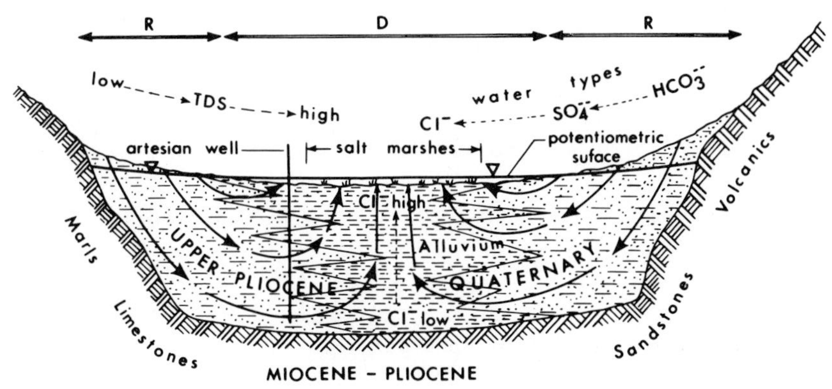

FIG. 19 Schematized groundwater conditions in the artesian basins of the Central Plateau of Iran (Generalized after Issar and Rosenthal, 1968, Figs. 1-4).

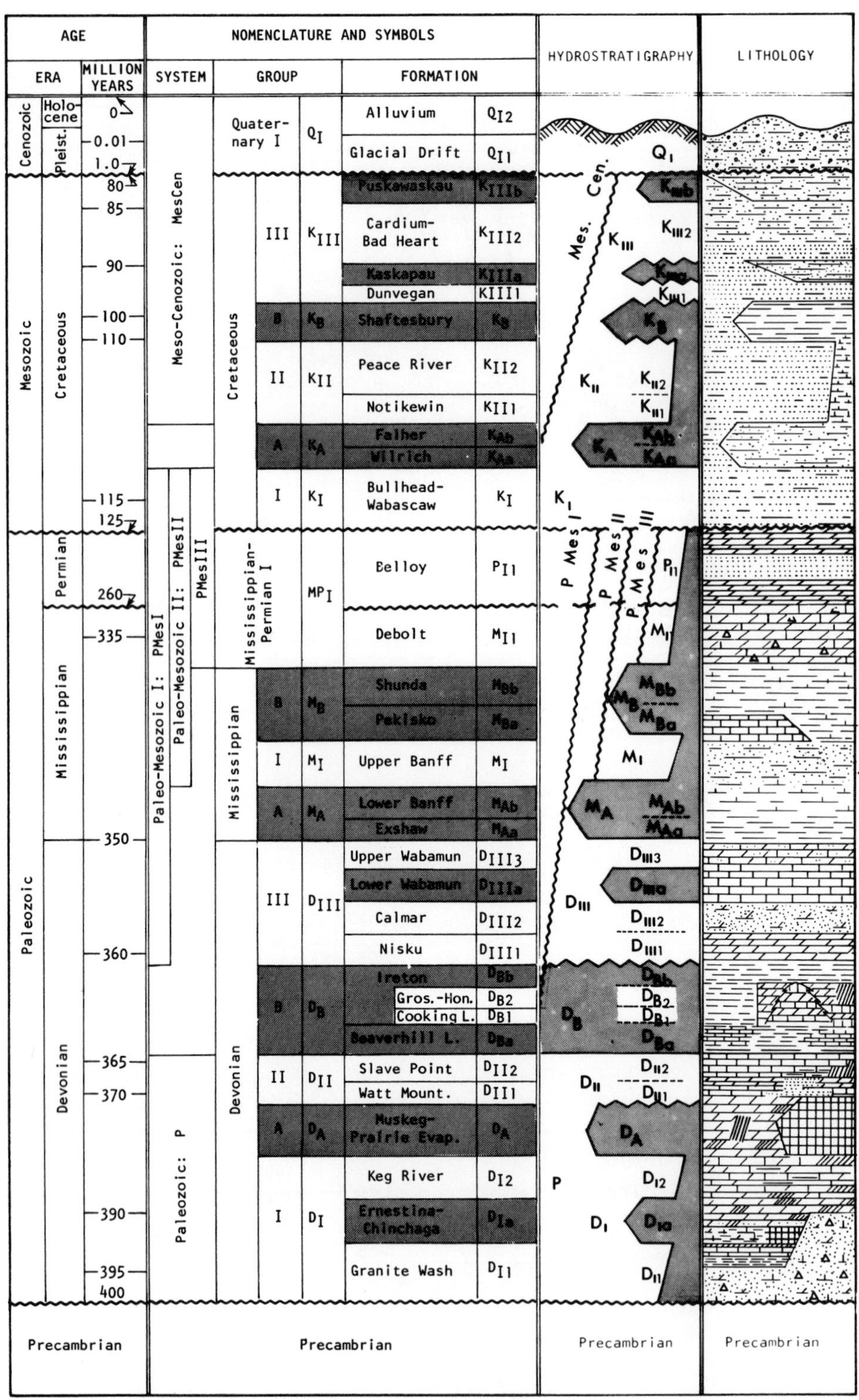

FIG. 20 Stratigraphy and lithology of hydrogeologic units, Red Earth region, northern Alberta, Canada (After Tóth, 1978, Fig. 3).

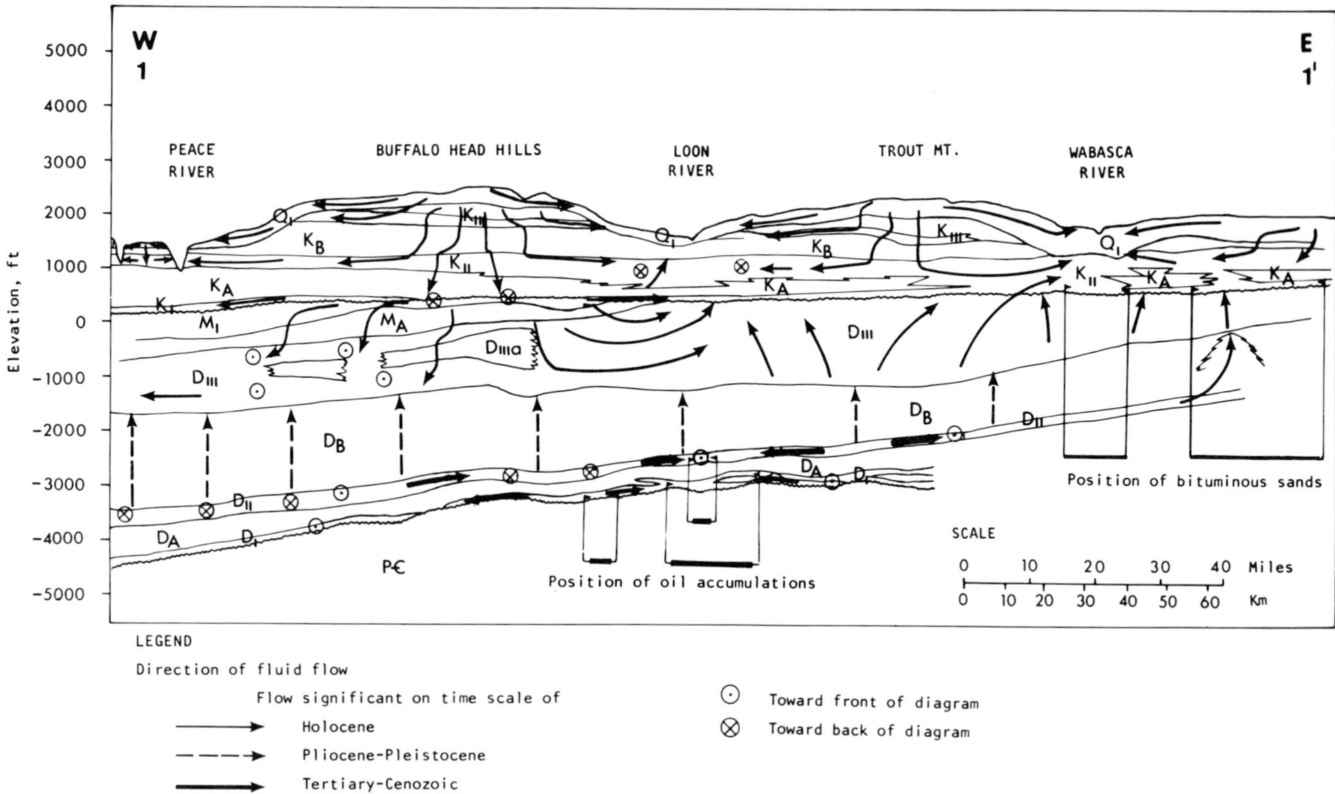

FIG. 21 Hydrogeologic cross section 1-1[1], Red Earth region, northern Alberta, Canada (Adapted from Tóth, 1978, Fig. 4).

near the center of the valleys, resulting in potentiometric surfaces intersecting the land surface in the area of mid-line or lateral flow. Total mineralization increases steadily from the areas of recharge towards the centers, with contents of chlorine ranging from 15 to over 3,000 ppm in the same direction. However, chlorine content increases from the base to the surface in the basins' center leading Issar and Rosenthal (1968, p. 5) to the conclusion that ". . . the upward seepage of water, under a certain hydrostatic pressure, causes a gradual leaching of the salts from the confined, and confining strata."

From the margins to the center of the basins the type of water changes from bicarbonate to sulfate to chlorine, which is the well-known sequence of Chebotarev (1955) indicating the direction of groundwater flow. The salts, imported by laterally moving and ascending groundwaters, accumulate in the central parts of the basins and form incrustations on dry surfaces or contribute to the marshes' salinity. In a hot and arid climate of barely 200 mm annual precipitation the existence of the marshes in itself is a manifestation of water rising from depth. Indeed, water-balance calculations indicated ". . . a clear discrepancy between the quantities of water lost from the marshes by evapotranspiration and the quantities of water which were found to flow into the marshes through the thin and near phreatic aquifer" (Issar and Rosenthal, 1968, p. 5).

Red Earth Region, Northern Alberta, Canada—Detailed studies of subsurface hydrodynamics (Tóth, 1978, 1979) of the Red Earth region indicated regional hydraulic communication in consolidated Mesozoic and Paleozoic rocks including a regionally extensive shaly aquitard with thicknesses ranging between 1,000 and 2,000 ft (304 and 609 m; Figs. 20, 21). The resulting effect of the land surface on the fluid potential distribution is observable to the Precambrian basement, i.e. to depths of over 6,000 ft (1,828m).

A genetic relation between the energy distribution in the basal or Devonian I (D_I) conducting unit and the topographic relief is suggested by the general similarity of the potentiometric surface in the D_I unit (Fig. 23) and the land surface (Fig. 22). Whereas there may be some lateral displacements in corresponding features of the two surfaces there is a general agreement in both the positions and orientation of all major rises and depressions. For obvious economic reasons, attention is called to the clustering of oil fields (Fig. 23) in or near the Cadotte, Loon, and Wabasca drainages. The common relationship to ascending water in these oil-productive drainage trends is to be explained later.

The rather well defined correlation between the D_I

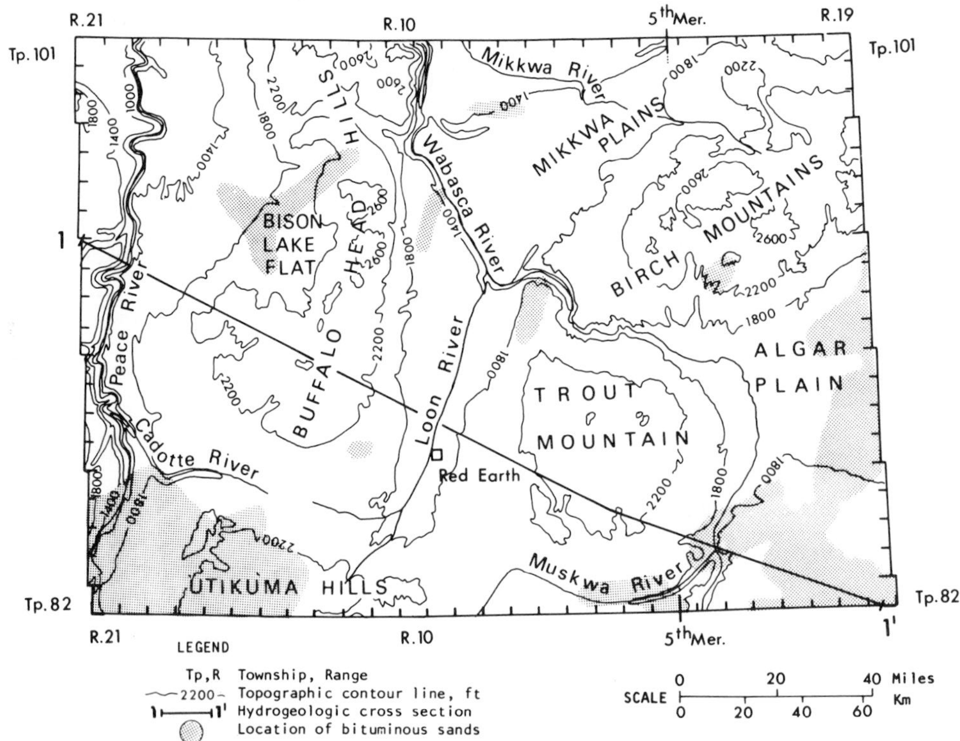

FIG. 22 Topography, location of hydrogeologic cross section 1-1[1], Red Earth region, northern Alberta, Canada (After Tóth, 1978, Fig. 1).

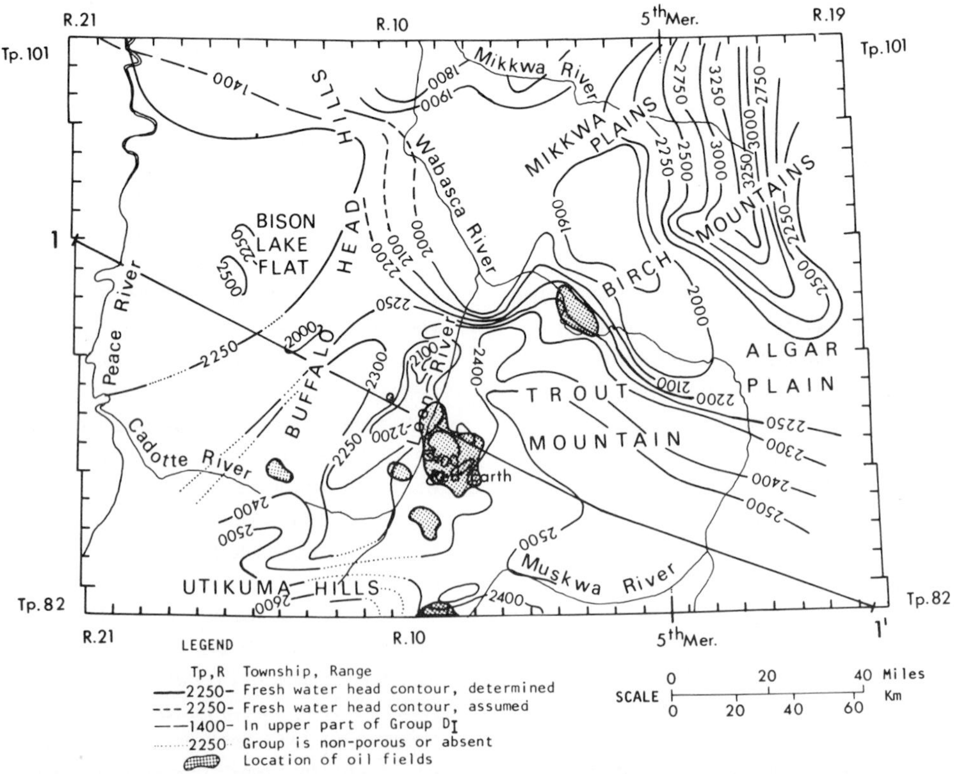

FIG. 23 Potentiometric surface and location of oil fields, Devonian I Hydrogeologic Group, Red Earth region, northern Alberta, Canada (Adapted from Tóth, 1978, Figs. 16 and 30).

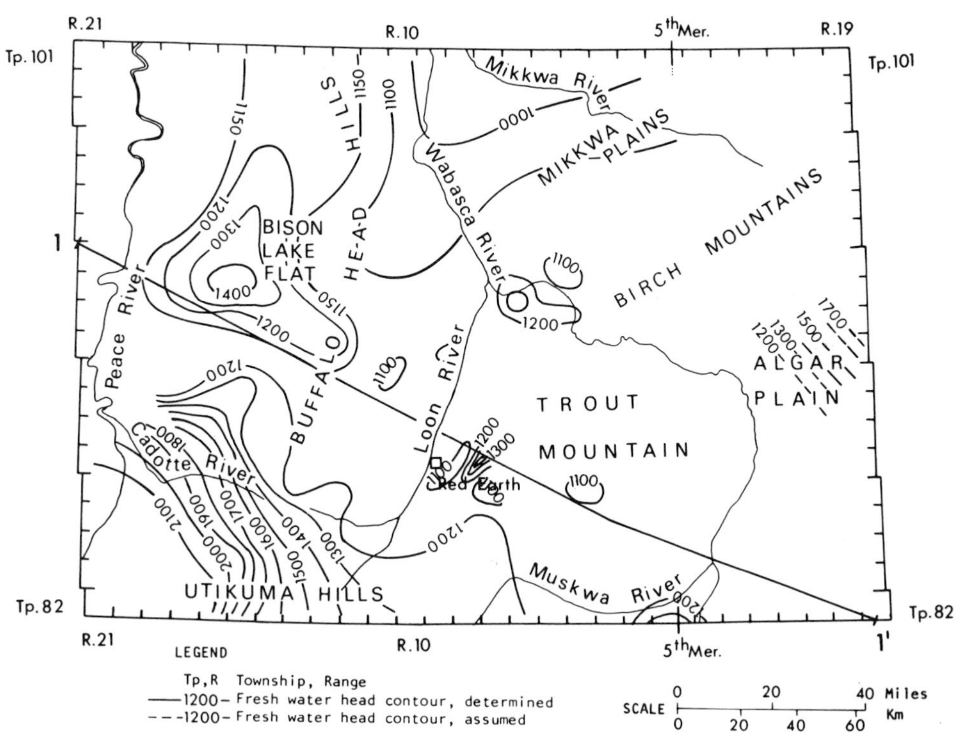

FIG. 24 Potentiometric surface, Devonian III Hydrogeologic Group, Red Earth region, northern Alberta, Canada (from Tóth, 1978, Fig. 26).

unit (Fig. 23) and the land surface (Fig. 22) becomes an enigma when it is realized that the general flatness and low value of the potentiometric surface of the intervening Devonian III unit (Fig. 24) indicates that no energy from above is being transferred across that zone to the D_{II} unit. The answer is found by an analysis of the p(d) and Δp (z′, d) distributions in the various conducting units.

The observed patterns of the pressures and dynamic pressure increments in the Devonian I Group (Figs. 25a, b) are comparable with those found for drainage basins with simple geometry (Figs. 2, 3) and hydraulically continuous geologic framework: the rate of vertical pressure increase is inversely proportional to the topographic elevation; pressures are evenly distributed about the hydrostatic values; dynamic pressure increments vary between extreme negative and positive values with an essentially vertical band of zero dividing the field into nearly equal regions corresponding to the recharge and discharge areas; sharp bends in the Δp = const lines reflect the presence of zones of reduced permeability. These features appear to reconfirm the generating influence of a land surface that is at least similar to that of today.

The pressure and dynamic pressure increment patterns of the Devonian III Group (Figs. 26a, b) are comparable with those parts of the theoretical cases (Figs. 5b, c) below the reach of the local systems. Here pressures are generally subnormal; the best fit p(d) curves are nearly parallel and arranged in order of increasing depth and decreasing altitude; the Δp = const lines are predominantly negative and essentially vertical, with abrupt breaks at the elevation range of 2,000 to 2,200 ft (609 to 670 m); and Δp values are slightly positive below elevations less than approximately 1,000 ft (304 m).

The observed features characterize the dynamic conditions in those parts of the continuous system modeled in Figure 5a, below the extensive aquitard (below Units $K_A + K_{II} + K_B$ in Figures 20 and 21). The effect of the aquitard, indicated by the sharp lateral turn of the Δp = const lines in Figure 5c, is represented by the Δp pattern observed in the Cretaceous I Hydrogeologic Group (Fig. 27b) which may be thought of as the top zone of the highly permeable complex under the $K_A + K_{II} + K_B$ water-retarding group. The lateral variations in the Δp pattern represent already the first noticeable effects of the overlying local topography which are clearer in the Δp pattern of shallower units. The increasing disorder in the arrangement of the p(d) curve may be attributed to the same cause. The effect of the local topography becomes quite evident in the distribution of the energy-related parameters of the highest water-conducting units, namely the Cretaceous II, III, and Quaternary I Groups (Figs. 28a, b). Whereas the generally negative values and distribution of Δp in the

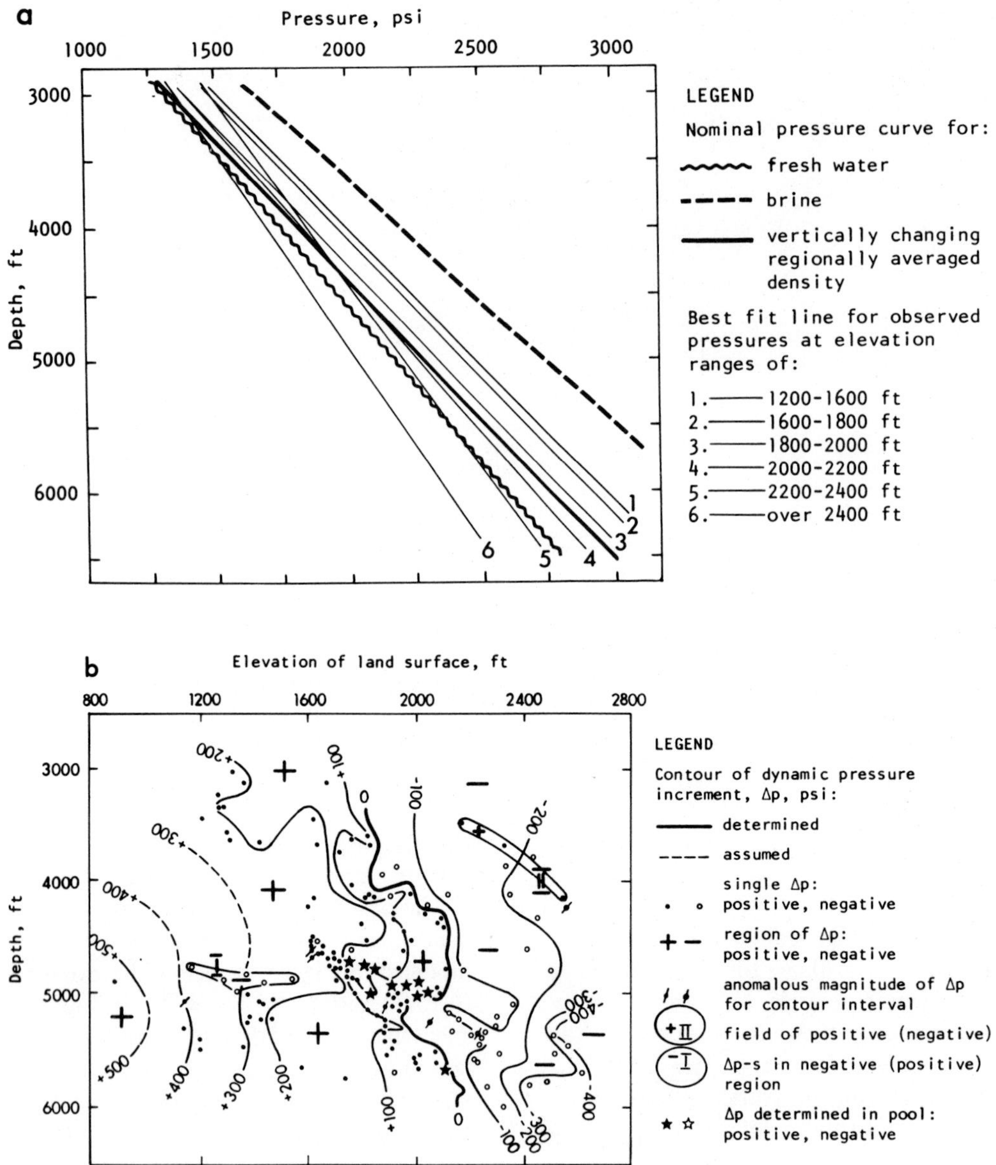

FIG. 25 Distribution of (a) p vs d, and (b) Δp vs. z′,d, in Devonian I Hydrogeologic Group, Red Earth region, northern Alberta, Canada (from Tóth, 1978, Figs. 18, 20).

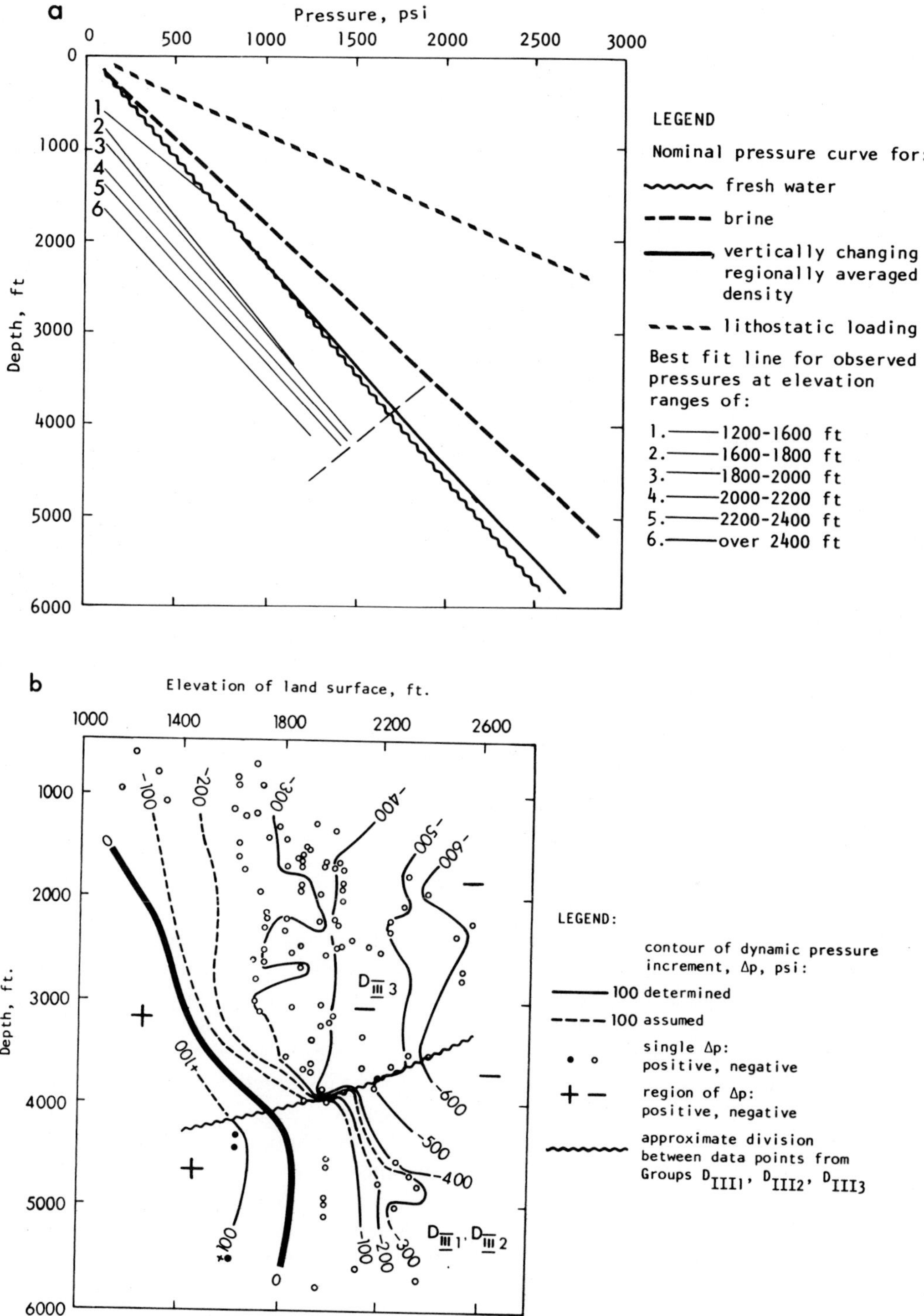

FIG. 26 Distribution of: (a) p vs. d, and (b) Δp vs. z′,d, in Devonian III Hydrogeologic Group, Red Earth region, northern Alberta, Canada (adapted from Tóth, 1978, Figs. 28, 29).

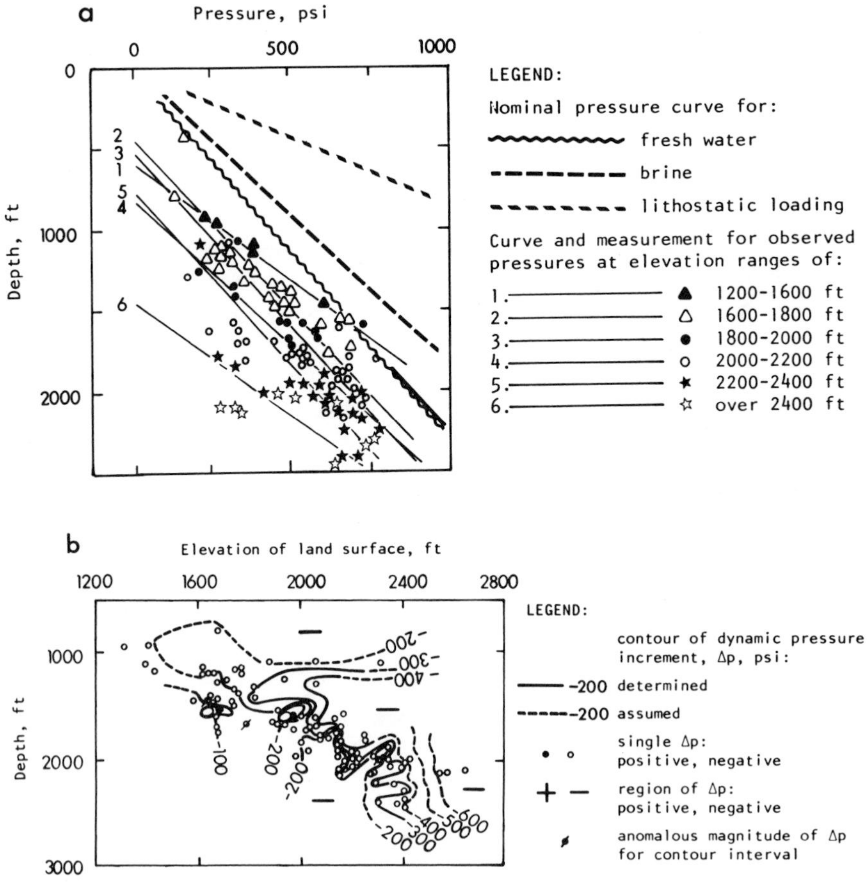

FIG. 27 Distribution of: (a) p vs. d, and (b) Δp vs. z′,d, in Cretaceous I Hydrogeologic Group, Red Earth region, northern Alberta, Canada (Adapted from Tóth, 1978, Figs. 34, 35).

lower units (K_{II} and K_{III}) are indicative of predominantly recharge conditions, the alternating negative and positive Δp regions for the shallow zones and the complete disorder in the p(d) values demonstrate flow in local systems with recharge and discharge areas occurring at any range of elevation, as is apparent from the shallow parts of the models computed for the general (Fig. 5c) and the complex homogeneous (Fig. 4c) cases.

From the comparison of the observed patterns with computed patterns it is therefore apparent that the main factor responsible for the energy distribution in the subsurface of the Red Earth region is the relief of the land surface. The greater the depth, the larger is the topographic feature whose effect counts. Uplands function as areas of recharge while both lateral and ascending flow converges towards topographic depressions. The combined units of Devonian III and Cretaceous I with the interposed sub-Cretaceous unconformity constitute a highly permeable zone. They function as a drain and channel both descending and ascending flow to the surface. The hydraulic continuity of the rock framework is demonstrated by the continuity of the energy parameters. The fact that in spite of the intervening energy drain the potentials in the basal aquifers still conform to the topography must be explained, tentatively at this point, by assuming that the time elapsed since the creation of the drain (because of the exposition of the sub-Cretaceous unconformity to the surface) is shorter than what is needed for the previously existing energy differences to decay beyond recognition. That this is a realistic assumption will be shown in connection with the non-steady state examples.

Non-Steady State Conditions

Non-steady state of the fluid-potential field may be caused by changing or changed boundary conditions. The rate of adjustment is a function of permeability and can be so slow as to make slightly permeable zones appear completely impervious. Based on the erroneous assumption of complete impermeability, recognition of the correct cause-and-effect relationships may be difficult, if not impossible. The following field examples will demonstrate the point as well as the actually pervious nature of slightly permeable rocks on three different time scales.

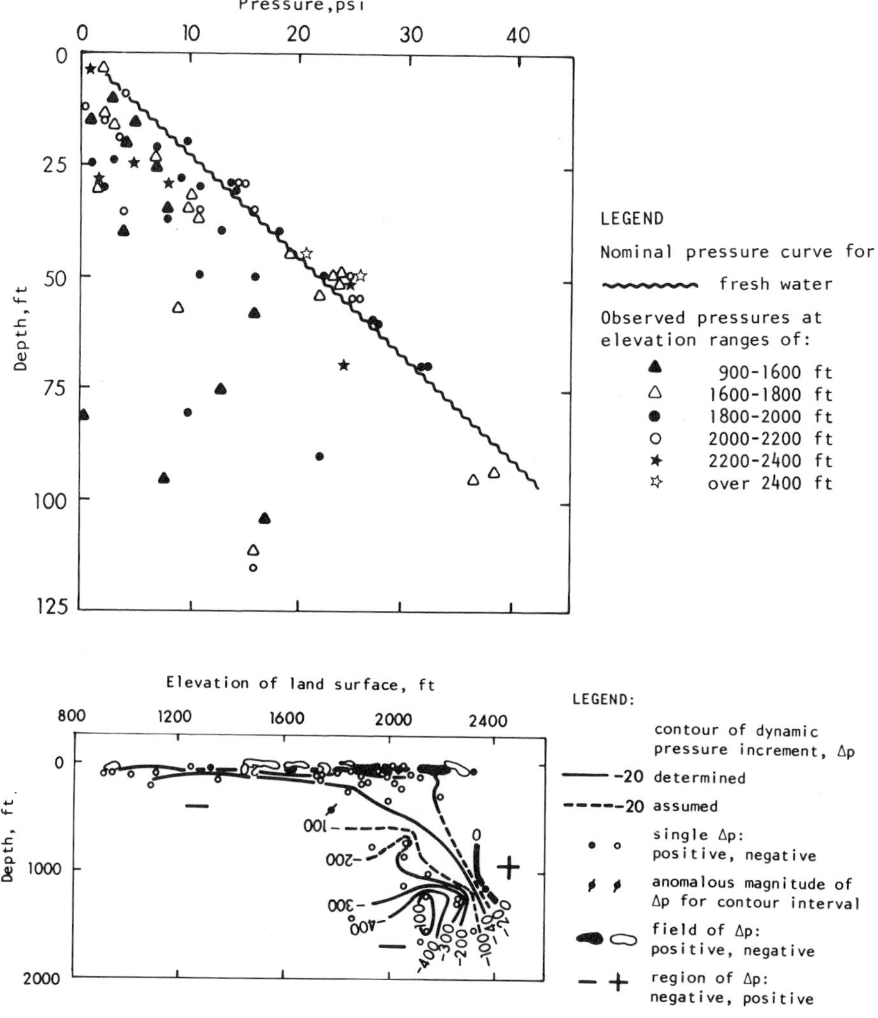

FIG. 28 Distribution of: (a) p vs. d, and (b) Δp vs. z',d, in Cretaceous II, III and Quaternary I Hydrogeologic Groups, northern Alberta, Canada (adapted from Tóth, 1978, Figs. 37, 38).

Delayed Response of Water Levels Through Caprock—In an attempt to determine the permeability of the caprock of an intended gas-storage aquifer, Witherspoon and Neuman (1967) pumped the Galesville sandstone and observed the water levels in the sub-caprock aquifer (Fig. 29a). The obviously delayed response of water levels in the observation well (Fig. 29b) was described by the authors as follows: "If a reference fluid level of 679.9 ft (207.23 m) above sea level was adopted, it was apparent that there is no evidence of drawdown until after about 30 days. An unmistakable downward trend began after 40 days and continued for 20 days after the water withdrawal stopped. About 45 days after the pumping ceased, the fluid levels began to recover." Had pumping and observation of water levels stopped before 30 days (few pump tests are conducted for more than a few days) the 16 ft (4.8 m) thick caprock would have appeared impermeable, and the two sandstones above and below it hydraulically unconnected. In fact, the authors concluded that the caprock's permeability could not be due to fracturing as the $k = 0.7 \times 10^{-4}$ md determined from the test was the same order of magnitude as that obtained from core analysis, i.e. 1.8×10^{-4} md.

Leakage Across Areally Extensive Aquitard—On the basis of production data, water levels, and flow net analysis, Walton (1960) determined the amount of water flowing vertically through the slightly permeable extensive Maquoketa aquitard of Illinois, shown on Figure 30 A, B, C. The original area of recharge was Area 1 (Fig. 30A) in which ". . . leakage downward through the Maquoketa Formation . . . was . . . about 1,000,000 gpd in 1864." At this time the piezometric surface was above the water table in Area 2 where the water was discharged ". . . by vertical leakage upward through the Maquoketa Formation and by leakage into the Illinois River valley." The regional

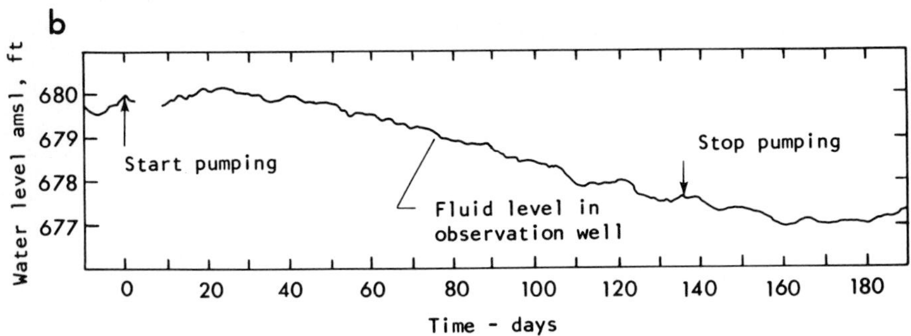

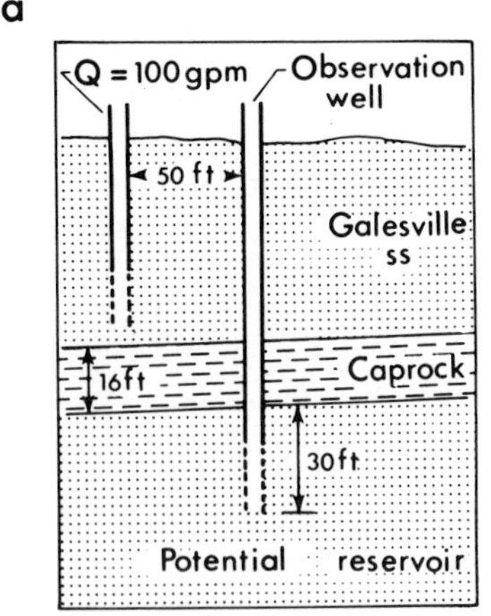

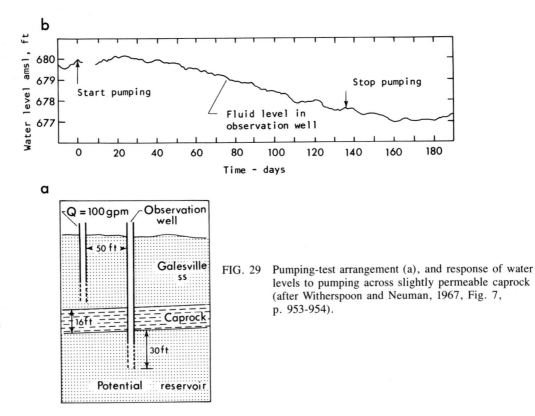

FIG. 29 Pumping-test arrangement (a), and response of water levels to pumping across slightly permeable caprock (after Witherspoon and Neuman, 1967, Fig. 7, p. 953-954).

average vertical permeabilities of the aquitard were computed to be 0.0001 gpd/sq ft in Area 1 and 0.00003 gpd/sq ft in Area 2. By 1958, as a result of heavy pumping (particularly in the Chicago area), leakage was vertically downward in both Areas 1 and 2 and amounted to approximately 8,400,000 gpd or 11% of the water pumped from deep wells. Without the above analysis of water balance (i.e. considering exclusively the declined position of water levels observed in 1958; Fig. 30B), the 200 ft (61 m) thick Maquoketa Formation would erroneously be evaluated as an effectively impermeable extensive aquiclude precluding hydraulic communication between the overlying and underlying water-conducting zones.

Relict Water Levels of Pliocene Times—One of the very few examples available to show water-level changes through geologic time was already discussed (Red Earth region). It will only be underscored here that the potential distribution observed today is not a result of the present topography despite the apparent correlation. Rather, it reflects the Pliocene topography which was similar to the present land surface (compare present land surface with pre-Quaternary surface, Fig. 21; also Tóth, 1978, Fig. 2: "Preglacial topography"). As pointed out earlier (Fig. 8), the time required for these potential differences to dissipate completely through the overlying rocks is approximately 4 million years. However, only about 100,000 years have elapsed since the previous boundaries changed; this was enough for the systems in Groups D_{III} and higher to adjust but not for the ones below D_B — which explains the apparent relation between land surface and potentiometric surface in D_I in spite of the intervening low-energy drain in D_{III}.

The above examples of nonsteady state flow demonstrate therefore two points that are germane to the question of regional groundwater movement: (1) even thick and very slightly permeable aquitards are conductive in terms of geologic time; and (2) delay in response to changes in fluid potential creates the appearance of hydraulic discontinuity, which however disappears on careful analysis.

Additional Observations of Regional Hydraulic Continuity, Flow Patterns and Associated Phenomena

At the risk of appearing to belabor the point, some further observations and evidences will be presented concerning the three basic aspects of regional groundwater flow, namely: hydraulic continuity, flow patterns, and associated phenomena. However, these points cannot be overemphasized, as is indicated by the fact that, despite countless observations and reports, the generating control of the topography on the flow distribution and its role in the transport and accumulation of petroleum has still not been realized.

Regional Hydraulic Continuity—Regional hydraulic continuity was already mentioned by Munn (1909, p. 524) who considered the effect of the increased friction of water through shales as compared with sandstones to be ". . . to decrease the rate of movement

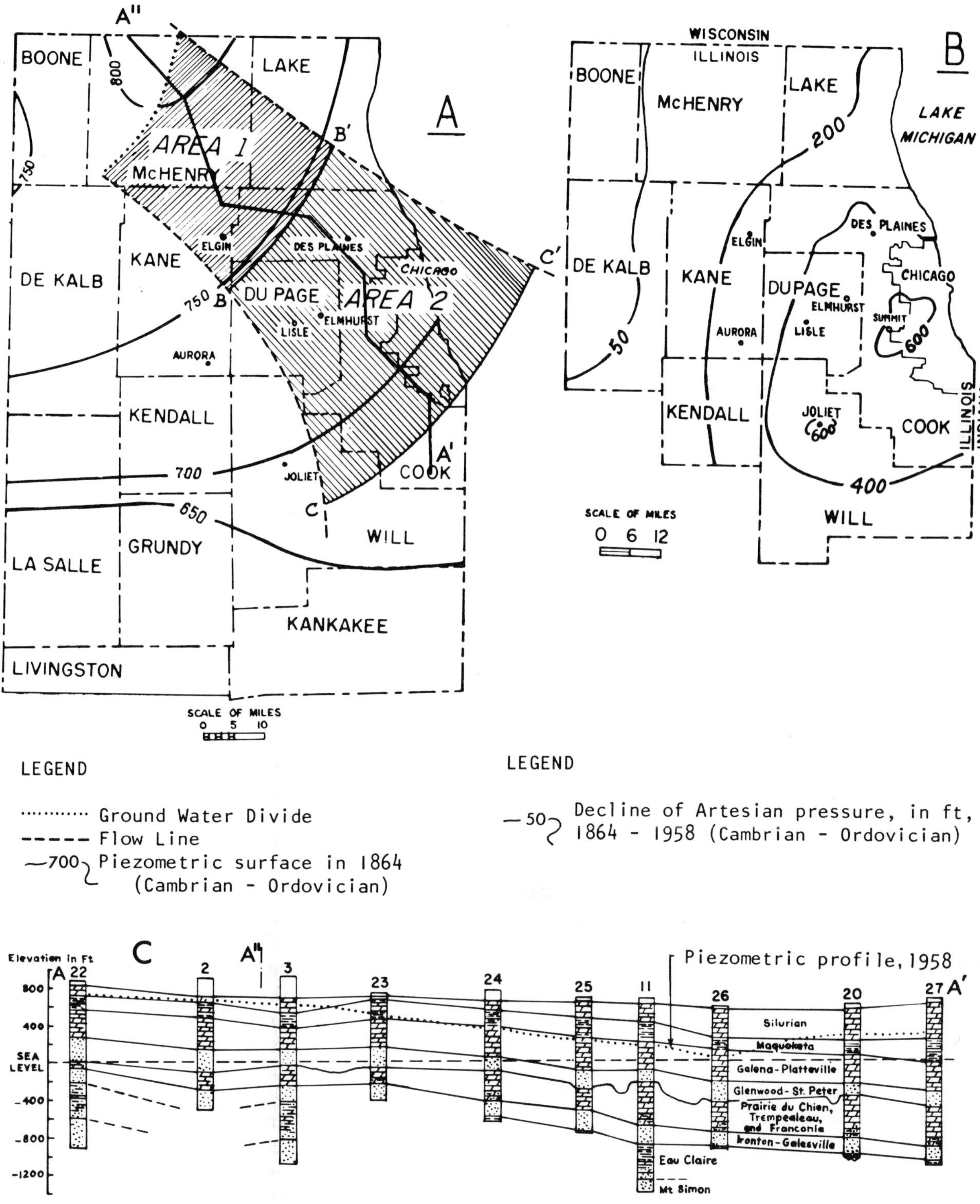

FIG. 30 (A) Piezometric surface about 1864; (B), decline in artesian pressure 1864-1958; (C), and geologic cross section, Cambrian-Ordovician aquifer, northeastern Illinois (after Walton, 1960, Figs. 10A, 11A, B, p. 21-22).

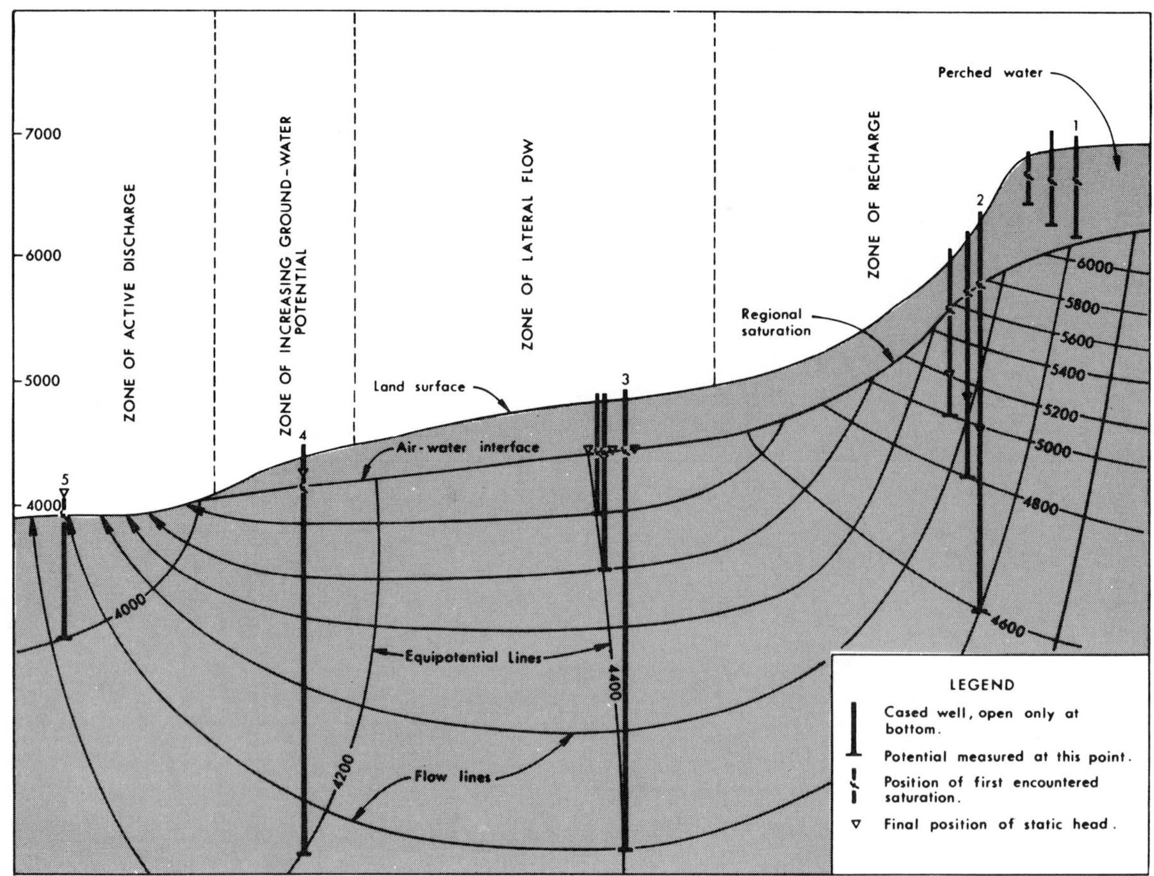

FIG. 31 Sketch of observed relationship in a typical Great Basin flow system (after Mifflin, 1968, Fig. 5, p. 12).

and active head at the edge of the saturated area as it (the water) spreads farther and farther away from the point of intake or the source of water supply." Muskat (1946), after observing that ". . . virgin-fluid pressures encountered in drilling are usually very near to the hydrostatic head corresponding to the depth, or approximately 433 lb/sq in per 1,000 ft. depth" (p. 28), stated that ". . . strata may be traversed by faults and joints through which the water may escape to overlying beds, or even by slow seepage through the covering strata an influx of water to a given formation may be dissipated via overlying formations or vertically to the surface" (p. 38).

In Hubbert's view (1953, p. 1,958), as ". . . the water table follows closely the earth's topographic variations . . . in a three dimensionally interconnected network of pore spaces the water is not in equilibrium and so must flow continuously, descending into the ground in regions where the topography is high and emerging in areas where the topography is low." Levorsen (1958) reaffirmed that (p. 386) "The average weight of oil field water is about 45 pounds per square inch per 100 feet, and most of the thousands of pools discovered have been found to have original pressures near this figure." He also explicitly recognized cross-formational flow (p. 398): "The fact that many sealed-reservoir pressures often approach what might be called a hydrostatic depth equilibrium must mean that there is or has been some connection — in other words, that there is some permeability, even in rocks that appear impermeable to oil, gas, and water."

On the basis of chemical similarities and potentiometric maps in the Four Corners area, Hanshaw and Hill (1969) observed that while ". . . the Cambrian-Devonian aquifers receive recharge primarily by cross-formational flow from overlying strata" (p. 270), also these higher zones, namely the Mississippian, Pennsylvanian, and Permian aquifers, are interconnected and are, ultimately, under the influence of the topography, thus forming one deep interconnected hydraulic system. Comparing waters from Oligocene sandstone aquifers with those of superjacent shale aquitards, Zatenatskaya (1969) found that in the Tobol-Ishim drainage basin in western Siberia the chemical composition of waters of the ". . . water-bearing horizon in most cases reflects the composition of the interstitial waters contained in the stratum of Aral clays overlying it" (p. 645), supplying one of the rare direct evidences of similar waters occurring in contiguous aquicludes and aquifers.

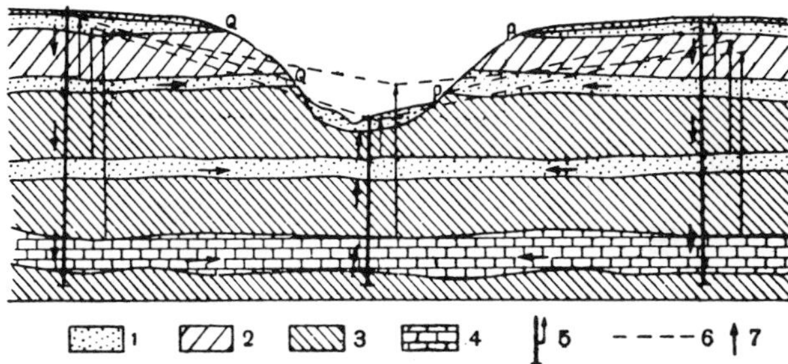

FIG. 32 Conceptual diagram of regional groundwater flow (after Kolesov, 1965, Fig. 2, p. 197).

The assumption of cross-formational flow is a necessity in many popular theories concerning subsurface aqueous processes. In their hypothesis of the origin of brines Bredehoeft et al (1963) stated (p. 257): "Under artesian conditions common in geologic basins, water enters aquifers near the outcrop area and is discharged slowly through confining layers in areas where the hydraulic head in the aquifer exceeds the head in the adjoining beds." Magara invoked both upward and downward flow through shales due to compaction (1969, Fig. 8, p. 30) and would allow the volumetric decrease of water due to cooling to induce flow from shales to more permeable beds during erosion and uplift (where there is no significant fluid movement from compaction).

Finally, Clayton et al (1966) found large variations of mean deuterium contents of oil field brines from the Illinois, Michigan, and Alberta basins and from the Gulf Coast. However, the δD isotopic compositions in the various basins correlate with those of local meteoric waters. This observation led the authors to the conclusion (p. 3,873) ". . . that the original water from the deposition and basin has been lost during compaction and subsequent flushing and that the formation water now encountered originated as precipitation over land, under climatic conditions not greatly different from those prevailing today," providing further independent indication of hydraulic continuity through deep sedimentary basins.

Regional Groundwater Flow Patterns—In the most comprehensive survey known to date Mifflin (1968) recognized 136 regional "flow systems" in the state of Nevada. The observed characteristics of the regional systems are summarized in Figure 31, and show all diagnostic features of a gravity-induced cross-formational system in a simple basin. A map accompanying Mifflin's report (1968, Plate I) indicates the outlines of the basins and shows that salt marshes or dry salt flats are associated with most areas of groundwater discharge.

The conceptual diagram (Fig. 32) of Kolesov (1965, Fig. 2, p. 197) is based on thousands of observations by a great number of investigators in Finnish Karelia, the Western Siberian Lowland, Russian Plateau, and the Dnieper-Donets artesian basin. While the distribution and cross-formation nature of the flow is self-evident from the diagram, the author stated explicitly (p. 197) that these features can be explained by considering the strata as ". . . one single complex entity hydraulically connected . . . and that "The watershed divide areas and elevations can be therefore considered as local areas of feeding of artesian waters, while river valleys and lake depressions as local areas of their discharge."

A conclusion virtually identical to that of Kolesov was reached by Hitchon (1969) who found for the Western Canada sedimentary basin that (p. 186): "The dominant fluid potential in any part of the basin corresponds closely to the fluid potential at the topographic surface in that part of the basin. Major recharge areas correspond to major upland areas, and major lowland regions are major discharge regions."

Albinet and Cottez (1967) established these same relations between the shallow water-table aquifers and the confined Cenomanian zone in the region of Touraine, France. Recognizing the importance of the process of leakage through slightly permeable strata between aquifers which concept ". . . has overturned the previous concepts regarding the replenishment of confined aquifers," they observed from potentiometric maps of good control that "Generally, the valley bottoms serve as discharge regions of the confined Cenoman aquifer. Conversely, the areas of recharge of this aquifer occupy the topographically high positions" (writer's translation).

And finally, a version of cross-formational gravity flow patterns was presented by Coustau et al (1975, Fig. 2, p. 108) which, although conceptualized, was no doubt based on extensive practical experience (c.f., Coustau et al, 1969; Chiarelli, 1973). Their diagram (Fig. 33) implies that, at the second stage of its development, a sedimentary basin becomes a drainage basin which is recharged at the elevated peripheries and allows meteoric water to move out by ascending through the retarding shale units into the low-lying central positions. With respect to the surface of the basin's center, vertical pressure gradients are larger than hydrostatic just as was found in the case of the theoretical basin with simple topography (Fig. 26).

Observed Field Phenomena Associated With Regional Groundwater Flow—Perhaps the most well documented phenomena associated with regional groundwater flow are the changes in the types and amounts of the water's constituents. On the basis of field studies, Chebotarev (1955) determined that salinity increases from

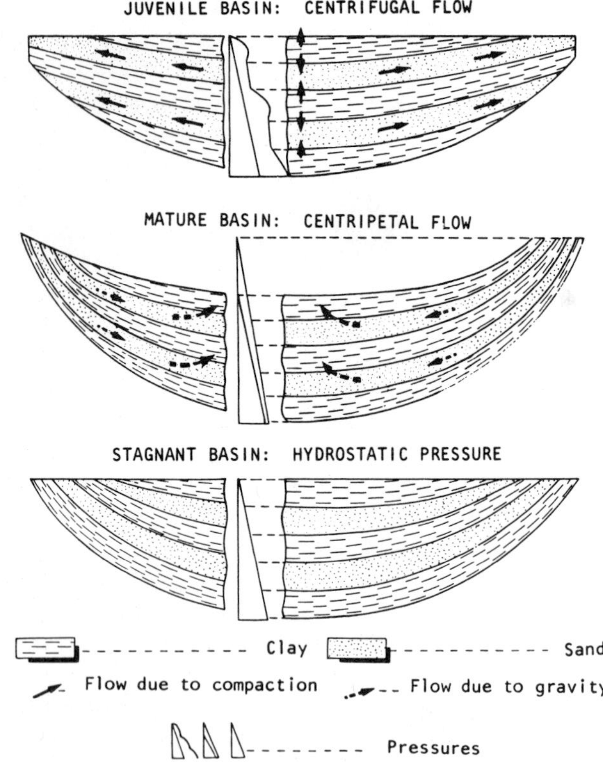

FIG. 33 The three hydrodynamic types of sedimentary basins (after Coustau et al, 1975, Fig. 2, p. 108).

areas of recharge to areas of discharge with the cominant anions changing according to the following sequence: $(HCO_3^-) \rightarrow (HCO_3^- + SO_4^{2+}) \rightarrow (SO_4^{2+}$ $\rightarrow (SO_4 + Cl^-) \rightarrow (Cl^-)$. From over 3,000 chemical analyses Back (1960) inferred a groundwater flow pattern in the Atlantic Coastal Plains of the United States (Fig 34). The pattern, subsequently confirmed by head measurements (Back, 1966), contained all major properties of regional groundwater movement, including the three orders of flow systems, recharge and discharge areas, and stagnant zones. In the Hungarian Great Lowland discussed earlier, groundwaters have total dissolved solids contents of less than 500 mg/l and are of bicarbonate type in recharge areas, whereas salinity increases to thousands of mg/l and becomes chlorine in nature toward the discharge areas (i.e. the central parts of the basin). Holmquest (1965, Figs. 11, 12, p. 270, 272) reported a clear correlation between the direction of water flow from the peripheries towards the centers of the Delaware and Midland basins and a change in the same direction of salinity of formation fluids from fresh (less than 5,000 ppm) to over 75,000 ppm. High salinities associated with stagnant zones (opposing flow directions) were observed by Martin (1969) in the Parentis basin, France.

The ultimate expression of the transport and distribution of dissolved chemical constituents by moving groundwater are the salt accumulations found commonly in the land surface in discharge areas. The possible examples are countless, and include most of the saline flats in the intermontane basins of Nevada (Mifflin, 1968), the "chotts" of North Africa (Guiraud, 1973), the sodium sulfate deposits of the Northern Great Plains of Canada and the United States (Grossman, 1968), and so on. An interpretive review and list of selected references on groundwater generated surface accumulation of salts was given by Williams (1970).

Along with the distribution of total salinity Holmquest (1965, Figs. 12, 13) attributes the regular downward decrease in the carbon dioxide contents in the Ellenburger and Siluro-Devonian formations also to the ". . . flushing and hydrodynamic action of meteoric waters" (p. 271) which, considering the topography of the Delaware and Midland basins and surrounding uplifts, requires topography-induced cross-formational flow.

The temperature modifying effects of descending cold meteoric waters in uplands and rising warm groundwaters in depressions has been observed in numerous places, including the Yellowstone Park geyser basins, and the thermal waters of the Hungarian Great Lowland (Erdélyi, 1972). Yerofeyev (1969) hypothesized that in the Volga-Ural oil-gas basin, 19 observed positive temperature anomalies are associated with ". . . abyssal waters rising through fracture zones . . . " while the 10 ". . . negative anomalies should occur in areas adjacent to regions of alimentation, where descending filtration is dominant."

A further manifestation of regional groundwater flow is the relative moisture deficiency and surplus in areas of recharge and discharge respectively. Documented cases are rare mainly because of a general lack of appreciation of the relation between the moisture balance of an area and groundwater flow. However, numerous lakes are known to be fed by discharging groundwater all over the world. A positive water balance maintained by groundwater is manifested by large marsh areas overlying artesian regions as was the case in the Hungarian Great Lowland and in les Landes southwest of Bordeaux, France, before drainage in both areas dried up the marshes in the 19th century, and by the oases of the deserts. Less obvious is the role of groundwater in areas where its surplus is expressed by secondary features such as vegetation contrasts and salt deposits, discussed in an earlier paper (Tóth, 1972, Fig. 7, p. 161). Much of the extra water in regional depressions rises to the surface through springs and seeps which are noticeably more abundant in areas of ascending flow. Of particular interest, it is in such regions that buried faults, fracture trends, and "lineaments" are commonly most visible on LANDSAT imagery of the earth's surface. Ascending waters seem to enhance the surface "signature" of the rock framework.

Discharge features deserving particular attention in the context of petroleum migration are the springs and seepages of oil found commonly in the sedimentary regions of the world. Levorsen (1958, p. 16) gave a concise overview of major occurrences. The association of the oil seeps with discharging groundwater and mud-volcanoes is obvious from those pages (15-19): they occur mostly in depressions and are mixed with water, some of them being mined and ". . . continually replaced by what slowly wells up."

Summary of the Principles and Manifestations of Regional Groundwater Flow in Large Drainage Basins

A drainage basin is a depression of the land surface partly or entirely surrounded by relatively high land and underlain at some depth by an effectively impermeable basement. In the present context a basin is considered large if it contains: several, or several sets of drainage channels; a rock framework that consists of various hydrogeologic units of differing permeabilities; and boundary uplands some tens to hundreds of kilometers distant. Geologic maturity denotes here a stable rock framework not subject to deformation either by tectonism or compaction; however, erosion may be active.

The two fundamental postulates of the theory of regional groundwater flow in this environment are that: (1) the rock framework is hydraulically continuous; and (2) the principal motive force of subsurface fluids is gravity. These postulates were proven by the agreement found between patterns of: flow, hydraulic head, vertical pressure gradients, and dynamic pressure increments evaluated mathematically, on the one hand, and those observed in real situations, on the other. Hydraulic continuity was thus established as being a generally valid principle and independent of the type of forces acting. However, the possible existence of energy sources other than gravity (compaction, osmosis, heat), is recognized but their importance in a mature basin must be temporal or local.

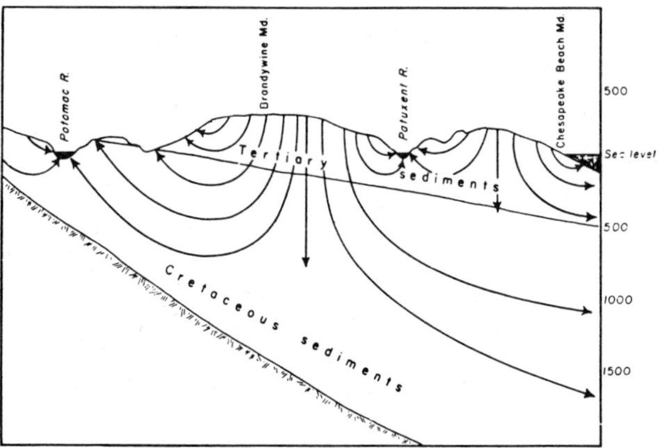

FIG. 34 Schematic flow pattern in the Atlantic Coastal Plain of the United States deduced from hydrochemical data (after Back, 1960, Fig. 5, p. 94).

Analyses of transient conditions indicated (see p. 22-24, Fig. 8) that the manifestations of hydraulic continuity are time-dependent: poorly permeable barriers do not arrest but only retard the propagation of energy disturbances and the movement of fluids. Depending therefore on the time scale used, a given deceleration may appear either as a complete stop, or a delay only, of the fluid movement. Consequently, parts of a flow field in heterogeneous media may be found at any given time to be out of phase with respect to existing boundary conditions, thus to one another.

The direct consequence of hydraulic continuity and gravity drive is that the relief of the water table, commonly a subdued replica of the land surface, generates interdependent systems of groundwater flow with well defined patterns which are modified by permeability differences (Fig. 10). Meteoric waters infiltrate and move downward in upland recharge areas, migrate laterally under regions of medium elevations, and are discharged in topographic depressions. Where flow systems meet or part, relatively stagnant zones, hydraulic traps, develop. Depending on the degree of continuity with the land surface, flow-system geometries and stagnant zones may change as a result of changes in the boundary conditions. In this type of environment the rational unit of fluid-flow analysis is the topographically defined drainage basin rather than one based on geologic criteria.

Important forms of permeability differences are the highly permeable lenses completely encased in a less permeable matrix. Placed in a homogeneous flow field these lenses were found to be preferential zones for fluid flow, and to cause local pressure anomalies depending on their size, geometry, and permeability contrast. Flow converges from cross-sectional areas that are much greater than those of the lenses; the pressure anomalies are negative at and near the upstream end and positive at and near the downstream end of the lenses.

Systematic redistribution of matter, and changes in physical and chemical properties along the flow systems result from the transporting ability of moving subsurface fluids. In general, regions of origin and termination of the flow paths are characterized by relative deficiencies and surpluses, respectively. Thus, as a rule salinity increases in the direction of the flow and high concentrations of salts may be found in areas where chemical and dynamic conditions are conducive for precipitation and deposition, such as in stagnant zones or at the land surface in discharge areas. Similarly, other matter including metallic, colloidal, and suspended material has the tendency to accumulate in discharge regions and hydraulic traps.

Removal of water by the descending limbs of flow systems may result in relatively deficient moisture conditions in upland regions whereas emerging groundwater may contribute to the maintenance of lakes and wet lands in discharge areas. Negative anomalies of the subsurface temperatures and geothermal gradients caused by the cooling effect of descending water are common in recharge regions as well as positive anomalies are in discharge regions.

In summary, groundwater-related accretions of mineral matter, heat, water, and so on, are not only the results of gravity-induced cross-formational flow but they may also be considered as manifestations thereof (as diagnostic features on the basis of which present and past systems can be recognized and reconstructed). Some seeps and accumulations of petroleum also can be viewed in this light.

TRANSPORT AND ACCUMULATION OF PETROLEUM BY GRAVITY-INDUCED CROSS-FORMATIONAL FLOW OF GROUNDWATER

Observed Relations between Petroleum Accumulations and Regional Groundwater Flow

General Comments

The generalized hydraulic theory of petroleum migration to be presented in this paper is based on the observation that accumulations of hydrocarbons are associated preferentially with particular parts of regional flow systems. The demonstration of this relation is rendered difficult by the paucity of published patterns of formation-fluid flow in large drainage basins or of data enabling their evaluation. For instance, few if any published potentiometric maps are accompanied by topographic maps; pressure-depth data commonly are expressed with reference to sea level, precluding the possibility of recognizing topographic effects; hydraulic data or hydraulic cross sections commonly are shown for selected horizons or strata only and do not normally cover the complete depth interval between land surface and petroleum accumulations; and so on. Nevertheless, it was possible for several areas to evaluate regional flow patterns and relate them to hydrocarbon occurrences, commonly based on different kinds of data and produced by different authors. The data used in the synthesis of observed relations range from original measurements processed into quantitative parameters by the writer, through maps and data sets adapted from publications, to qualitative reports of isolated observations placed in the present context on the basis of the groundwater flow theory.

Field Examples of Relations Between Petroleum Accumulations and Cross-Formational Gravity-Flow of Groundwater

Red Earth Region, Northern Alberta, Canada—There are two main zones with known significant accumulations of hydrocarbons in the Red Earth region: (1) liquid oil in the Devonian I and II Hydrogeologic Groups (Figs. 20, 21, 23); and (2) Cretaceous I and II Groups with heavy bitumen at the sub-Cretaceous unconformity (Figs. 20, 21, 22). Reference to Figure 23 shows that known accumulation of oil in Groups D_I and D_{II} tend to be concentrated between regional potentiometric mounds and in or near potentiometric depressions. In addition, with the exception of the field on the Wabasca River, all other occurrences are associated with low or zero lateral hydraulic gradients, and all oil pools are in areas of positive dynamic pressure increment (Fig. 25b).

The D_{III} Hydrogeologic Group is conspicuously devoid of commercial accumulations of liquid oil. However, areas of heavy bitumen occur in Groups K_I and K_{II} primarily along the sub-Cretaceous unconformity (Figs. 21, 22). At this level, flow distribution is closely adjusted to the present topography (Figs. 27, 28) and the general coincidence of the location of bituminous sandstones with topographic lowlands is equivalent to a coincidence with converging and/or ascending flow. Indeed, most of the known accumulations are clearly associated with foci of laterally, or laterally and vertically converging flow systems, or with low or zero lateral gradients. Inasmuch as the sub-Cretaceous unconformity functions as a drain collecting flow both from the underlying Devonian and overlying Cretaceous Groups, the bituminous sandstones are located in three-dimensional fluid-potential minima with outlets in only one lateral direction or only towards the land surface (Fig. 21).

Parentis Artesian Basin, Aquitaine, France—The gravity-induced and cross-formational nature of groundwater flow in the Aquitaine of southern France was discussed in a previous section. In the Parentis artesian basin of this region commercial oil fields occur in reservoirs of Late Jurassic and Early Cretaceous age. The relation between the position of the accumulations and the regional flow distribution may be summarized from Figures 11 through 14 as follows: all fields are located in areas of relatively low topographic elevations (Fig. 11) and at relatively low values of hydraulic head in all formations (Fig. 12a-d) which at the same time, are also areas of positive or slightly negative dynamic pressure increments (Fig. 13a, b). The exact depth of the deposits could not be ascertained from available literature. However, on the basis of explicitly stated hydraulic continuity, (Martin, 1969, p. 85), it appears highly probable that the dynamic nature of the flow pattern does not change significantly below the depths shown on Figures 14 (see also Footnote 3, this paper) and that therefore the oil accumulations are associated with the ascending (i.e. discharging) limbs of the oceanward systems. Due to the upward direction of the regional flow in this area the lateral hydraulic gradients are essentially zero, resulting in laterally stagnant conditions.

An expected manifestation of fluid stagnancy is high salinization which was indeed found by Martin

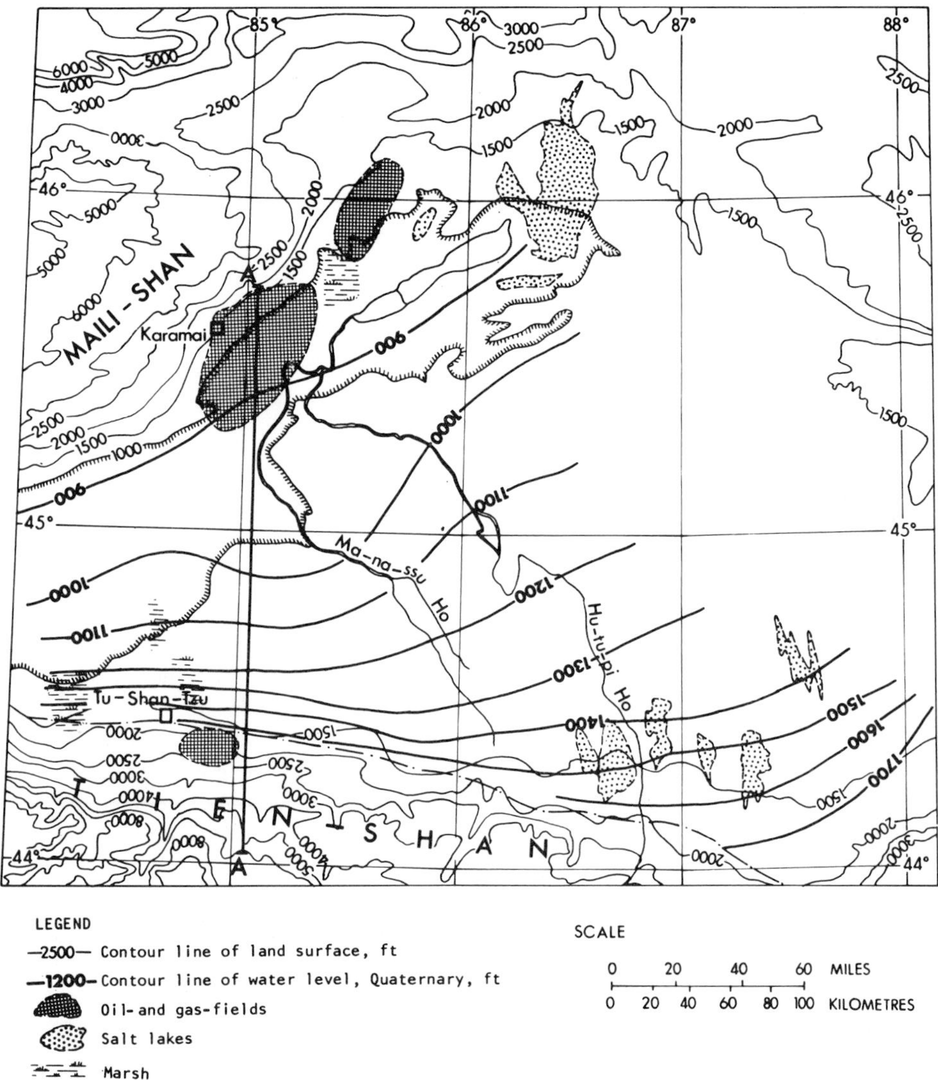

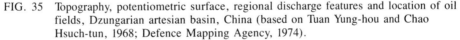

FIG. 35 Topography, potentiometric surface, regional discharge features and location of oil fields, Dzungarian artesian basin, China (based on Tuan Yung-hou and Chao Hsuch-tun, 1968; Defence Mapping Agency, 1974).

(1969, Figs. 12, 13, 14) in conjunction with the oil fields and in all higher zones. Coustau et al (1975, Fig. 7, p. 115) recognized the existence of gravity induced flow beyond the periphery of an area coinciding approximately with the 25 m hydraulic head contour in Figure 12c. However, they consider the presence of hydrocarbons and high salinity within the area, termed "réduit pétrolier" (petroleum shelter), as an indication that meteoric waters have not yet reached the deepest parts of the basin. Biodegradation, oxidization, and dilution have, therefore, not taken place. Contrary to this view, the theory presented in this paper asserts that gravity-induced cross-formational flow of meteoric waters has produced the accumulations of salts and oil in the Parentis basin, rather than having failed to destruct them.

Great Lowland Artesian Basin, Hungary—The majority of hydrocarbon accumulations in the Great Lowland of Hungary occur in or near areas of the lowest topographic positions and at the same time, the lowest potentiometric values (Figs. 15, 16). The topographic ridges are markedly devoid of known commercial-size deposits, with possible exceptions on the east side of the ridge of the Duna-Tisza interfluve, where some minor fields are located along the 100 and 110 m potentiometric contours. However, even these values are low as compared with those presumably prevailing under the topographic elevations exceeding 600 m both to the northwest and southeast. Therefore, the impression is strong that the accumulations are trapped by and between the discharging limbs of opposing regional systems originating in the northwest and southeast. The stagnant and discharging conditions are indicated by the low and even closed potentiometric surfaces (northwest and southeast of the town of Békés), and by the associated regional saline and naturally marshy

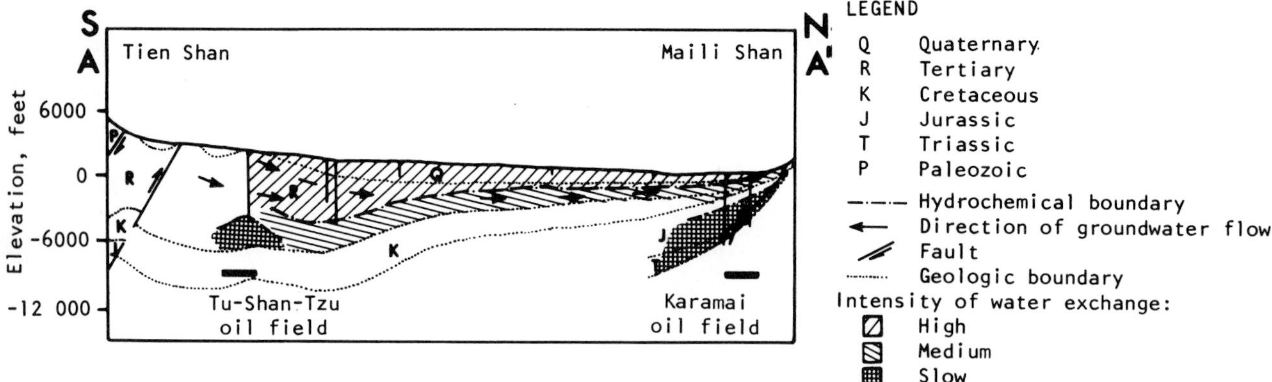

FIG. 36 Idealized hydrogeological cross section, Dzungarian artesian basin (after Tuan Yung-hou and Chao Hsuch-tun, 1968, Fig. 10, p. 931).

conditions (see earlier section on observed phenomena associated with regional groundwater flow). A secondary concentration of hydrocarbons appears to be effected by northeast and southwest directed systems descending between the Duna and Tisza rivers and near Debrecen (Fig. 16) and discharging in the aforementioned depressions (Fig. 18)[4].

Dzungarian Artesian Basin, Northwest China —The Dzungarian artesian basin of northwest China has been a closed inland basin since Permian time. Continental clastic sediments derived from the surrounding mountains fill the basin to depths exceeding 5,000 m and lie on igneous basement. The cross-formational gravity induced nature of the flow systems is evident both from the explicitly stated generally high permeability of the deposits and from the chemical and dynamic conditions of groundwater, including: total salinities of a few hundred to less than ten thousand mg/l from depths of 2,000 to 3,000 m; hydrogeological divisions (sic) which ". . . are adapted to modern geomorphology and do not correspond to the position of the tectonic axis of the basin" (Tuan Yung-hon and Chao Hsuch-tun, 1968, p. 921), i.e. the potentiometric surface reflects the topography (Fig. 35); and artesian conditions and other manifestations (salt flats, marshes) of regional groundwater discharge in the topographically deepest parts of the basin (Fig. 35).

These conditions have been summarized by the authors (op. cit., Fig. 10, p. 931) and reproduced in Figure 36, showing the present discharge center of groundwater to be located at Karamai and the Ma-na-szu Lake area. However, prior to the Quaternary (prior to the uplift of the Tien Shan), the discharge focus was in the Tu-shan-tzu area (Fig. 35). The clear coincidence of the two main oil producing areas of the basin with the pre- and post-Quaternary discharge centers of deep groundwaters prompted the authors to state (p. 931): "These two centers are exactly where the petroleum fields of Karamai and Tu-shan-tzu are now located." They thought therefore ". . . it appropriate to cite Altovski's deduction to the effect that 'ground water is the medium for the accumulation of primitive organic compounds and their transformation into petroleum.' The high pressure of the Karamai oil field may be caused by the driving force of a vast body of water in the southern part of the basin." Considering the complete lack in the basin of rock types and history related to marine sedimentation, and the striking coincidence of commercial hydrocarbon accumulations with regional discharge foci of groundwaters, Altovski's deduction appears, indeed, very plausible.

The Persian Gulf—Flow in the Cretaceous and Oligocene-Eocene series of the Persian Gulf is considered by Chiarelli (1973, p. 179) as an example of cross-formational ("leaky") discharge of gravity-driven formation fluids. This observation is based on the closed minimum of the Cretaceous potentiometric surface (Fig. 37) combined with upward decreasing hydraulic heads indicated by positive head differences between the Cretaceous and Eocene-Oligocene series (Fig. 38). In addition, it should be pointed out that the line of $\Delta h = O$ separating the areas of descending ($\Delta h < O$) and ascending ($\Delta h > O$) flow appears to divide the flow region in a manner seen before in the case of simple basin geometry.

The dynamic effect of moving groundwater on some petroleum accumulations in the Persian Gulf is clearly manifested by inclined oil-water interfaces which agree in magnitude and directions with the observed hydraulic gradients (Chiarelli, 1973, p. 179). However, beyond this agreement, it is also noteworthy that the majority of accumulations are in areas of low values and positive differences of hydraulic heads. In the Abu Dhabi region, where potentiometric information is lacking, a discharge character of the oil field area

[4] A similar interpretation (Erdélyi, 1976) was published and brought to the writer's attention in December, 1978, after completion of this manuscript.

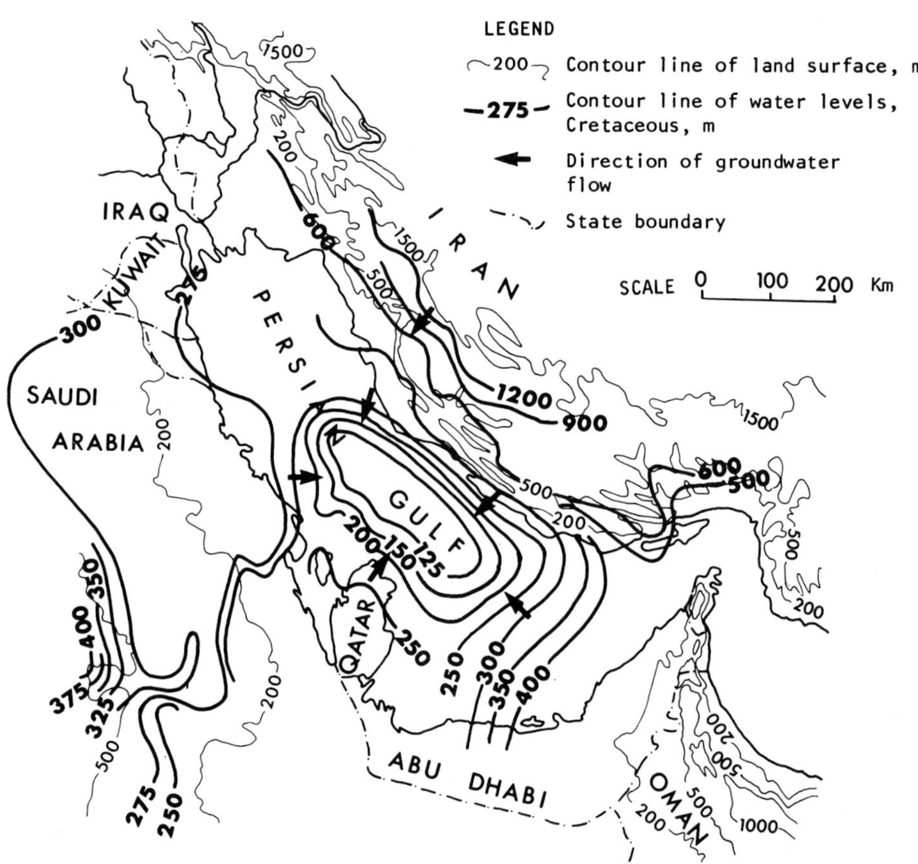

FIG. 37 Topography, potentiometric surface, and lateral flow directions in the Cretaceous, Persian Gulf (from Chiarelli, 1973, Fig. 101, p. 178; USSR, 1967, p. 143-144).

may be inferred from the low topographic position (Figs. 37, 38).

Etzikom Coulee Analogue Flow Pattern, Southern Alberta, Canada —Electrical analogue patterns of fluid flow and hydraulic heads were produced (Tóth, 1970, Fig. 8) along a 110 mi (176 km) cross section based on the assumption of hydraulic continuity in the modeled Jurassic and Cretaceous carbonate and lenticular clastic rocks. Figure 39 shows that 9 out of the 14 known occurrences of hydrocarbons along the section, namely those from 2 through 10, are located at or near areas of ascending flow or quasistagnant conditions of either ascending or descending direction. Only the five gas shows (1, 11, 12, 13 and 14) appear to be in areas of strong lateral flow, though in the case of 14 a shallow stagnant zone and changing flow directions are in evidence. It should be added that artesian conditions occur in all theoretically evaluated major discharge areas.

Additional Observations of Hydrocarbon Occurrences Related to Regional Groundwater Flow Patterns —Among the earliest published observations relating the location of petroleum fields to regional groundwater flow conditions were those made by Rich (1921). Exemplifying conclusions of his version of the hydraulic theory, he stated, ". . . the remarkable productivity of the Salt Creek field may be due to the fact that it is located near a point where the combined artesian flow from all artesian basins of central Wyoming drains eastward . . . " or, with reference to the mid-continent field of Oklahoma and Kansas: "Is it not significant that this region of supersaturation with oil and gas should lie in the natural zone of concentration on the outflow side of this great artesian basin?"

The surface occurrences of oil springs and seepages described by Levorsen (1958, p. 16) and mentioned earlier in this paper constitute evidence that liquid petroleum, or the ingredients thereof, may be transported to and emplaced in natural areas of relative energy minima by moving groundwater. Ryzkhov et al (1968) pointed out the relation between ascending waters and hydrocarbon transport in the northeast Fergana region of Uzbekistan by reporting the drying up of a ". . . spring of oily water in the Kul'mensay valley . . . " as a result of several years of pumping water of ". . . the sodium chloride type, showing an oil film" from a nearby test well. In addition, they emphasized the mobile nature of the accumulations by considering that the variability of the oil, gas, and chemical compositions of "many oil deposits (that) are located on both sides of the Mayli-Su fault" indicated ". . . a continuous integration and disintegration of the oil deposits."

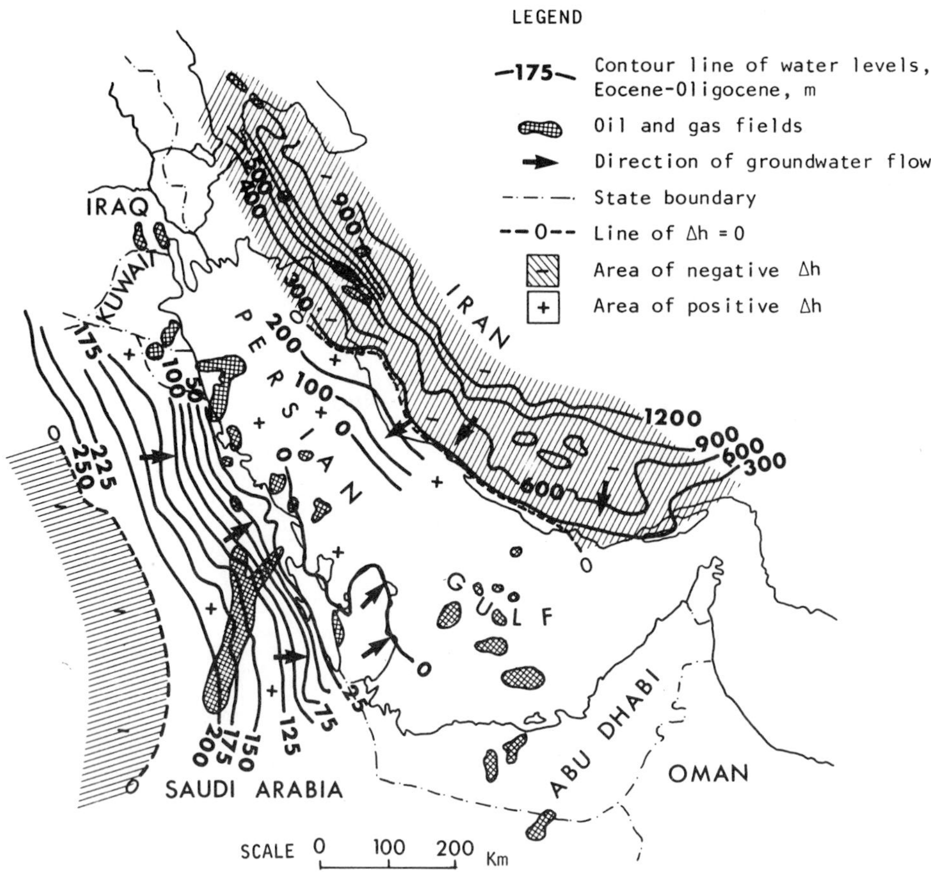

FIG. 38 Potentiometric surface and lateral flow directions in the Eocene-Oligocene, area of positive hydraulic head differences and location of hydrocarbon fields, Persian Gulf (from Chiarelli, 1973, Fig. 100, p. 177; Defence Mapping Agency, 1972/1973).

The results of detailed analyses of the frequency distributions of various hydrocarbon components, gas blowouts, and the presence of oil drops in salt were considered by Skrotskiy (1974, p. 201) as ". . . evidence of the upward flow of oil fluids through the actual salt," in the giant El'ton salt dome in the western Caspian basin.

Al-Shahristani and Al-Atyia (1972, p. 929) used vanadium and nickel concentrations in oil to study the history of migration and accumulation in Iraqi oil fields and find the results to indicate ". . . that oil in northern Iraq originated during the Middle/Late Cretaceous, and has migrated vertically to the Tertiary reservoirs where it now is found." Based on maps and cross sections of measured water levels Hitchon and Hays (1971, p. 658 and Fig. 8) concluded for the Surat basin of Australia that ". . . hydrocarbon accumulations are associated with quasistagnant portions of discharge areas in which flow is vertical with respect to the bedding, the result being increased salinity of the formation waters caused by membrane filtration and consequently conditions more favorable for the disaccommodation and subsequent accumulation of hydrocarbons."

Bars et al (1961) illustrated their interpretation of observed relations between hydrocarbon occurrences and groundwater flow conditions in the North, Middle, and South Caspian petroleum basins in an explicit diagram (Fig. 40) and summarized their findings in terms familiar from the theory of regional groundwater flow: ". . . an oil-gas basin is the interior part of an artesian basin . . . " (p. 585), within which ". . . favorable for exploration for oil and gas pools are areas of modern and ancient flow and discharge of saline artesian waters of a reducing nature. For example, the Cheleken and Nebit-Dag fields occur at open active foci of discharge of subsurface waters, and the Kotur-Tepe and Kum-Dag fields at closed and ancient foci" (p. 584), but "The oil gas basin cannot include within it an area of modern infiltration . . . " (p. 583).

Nine types of "piezometric minima" have been distinguished by Kudryakov (1974), eight of which involve cross-formational flow explicitly (Fig. 41). On the basis of observations in western Siberia, Kara-Kum arch, Ciscarpathian trough, and Don-Caspian ridge, he found that "Piezometric minima are one of the manifestations of hydrogeologic factors in the development and distribution of oil and gas pools (p.

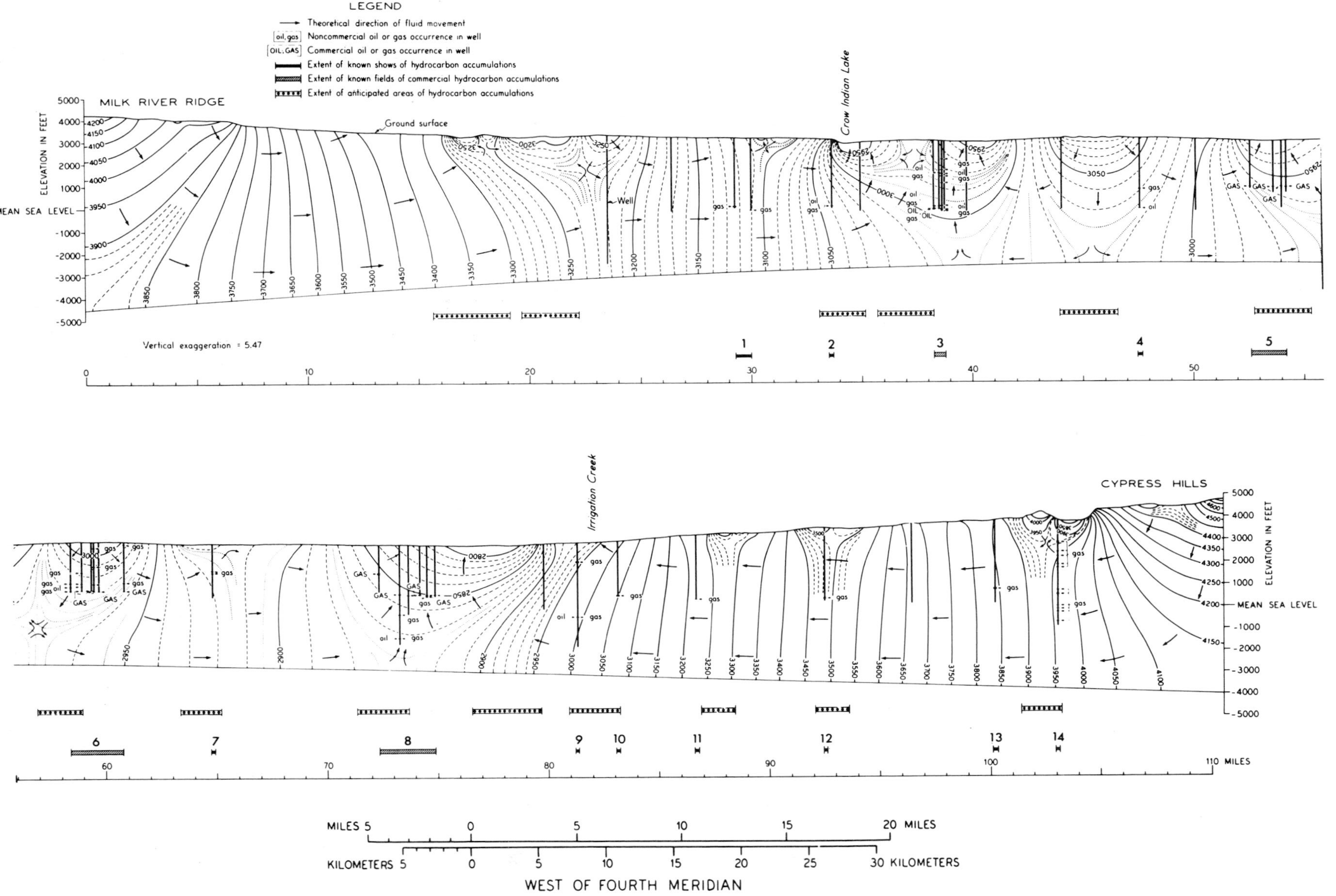

FIG. 39 Electric analogue pattern of hydraulic heads, and known and predicted locations of hydrocarbon deposits, Etzikom Coulee, southern Alberta, Canada (After Tóth, 1970, Fig. 8b).

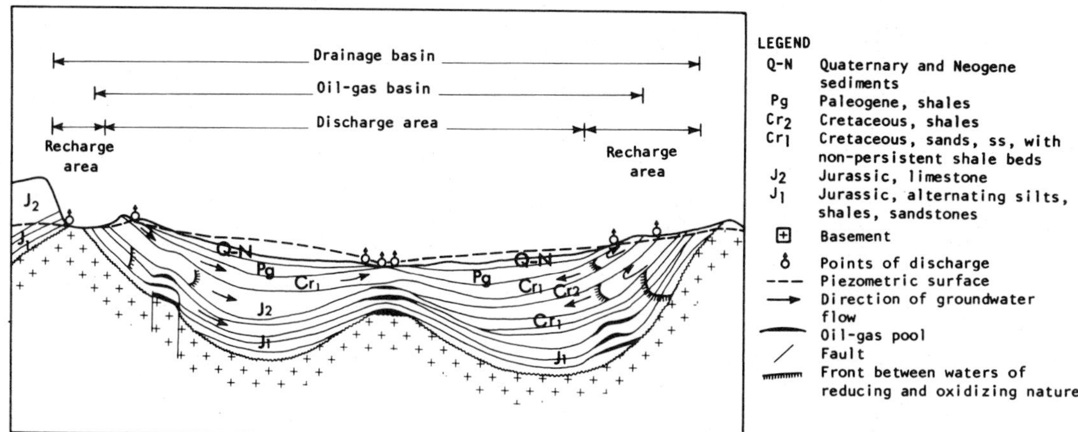

FIG. 40 Diagrammatic summary of groundwater-hydrocarbon relations observed in the Caspian region (from Bars et al, 1961, Fig. 2, p. 582).

240) . . . "They are paleohydrogeologic indicators of oil and gas in that they are areas of artesian systems into which, over certain periods of geologic time, groundwater carrying free and dissolved hydrocarbons tends to flow" (p. 241).

And finally, Meinhold gave a remarkably clear-sighted summary of his observations concerning the relations of hydrocarbon occurrences and groundwater flow, illustrated by examples from the Caucasus including the Baku region, Rocky Mountains, Carpathians, Russian Platform, Hungary, and North American Platform, in the abstract of his paper (Meinhold, 1971): "Numerous examples over the world show that large accumulations exist in regions of groundwater discharge, accompanied by geothermal and hydrodynamic anomalies which can be explained only as the results of groundwater moving in open hydrodynamic systems" (writer's translation).

Summary of Observed Relations Between Petroleum Accumulations and Regional Groundwater Flow

It is apparent from the previous section that certain systematic relations exist between the position of petroleum deposits, on the one hand, and regional groundwater flow patterns, energy conditions, and hydrogeologic phenomena, on the other. Thus, hydrocarbons are associated with low or zero lateral hydraulic gradients combined frequently with saddles or closed relative minima of the fluid potential; positive dynamic pressure increments (hydraulic heads or pressures increasing downward at a rate higher than hydrostatic and possibly reaching flowing well conditions); stepwise change in the potentiometric surface; relatively high contents of dissolved salts, primarily of sulfate, chlorine, and sodium; and positive temperature anomalies. On the other hand, relatively few petroleum deposits are found in regions characterized by: relative maxima of fluid-potential; decrease in hydraulic head with depth (negative Δp); high lateral gradients of hydraulic head; low salinity of formation fluids, combined normally with carbonic acid and calcium-magnesium chemical facies; and negative temperature anomalies.

The first set of conditions was shown in the section on regional groundwater flow to be indicative of: discharge zones (focal points of converging and ascending flow systems forming hydraulic traps); relatively slow or stagnant flow conditions; an abrupt decrease in flow velocity either due to a decrease in permeability or to an increase in the vertical flow component; and reducing chemical environment. However, in areas where the second set of conditions prevails groundwater is likely to be descending from and divergent under upland recharge regions, relatively fast moving and oxidizing in nature.

Generalized Hydraulic Theory of Petroleum Migration

Formulation of the Theory

In his "hydraulic theory of oil and gas accumulation" Munn (1909) envisaged water descending from the land surface across beds of sandstone and shale and pushing ahead parts of the oil and gas contained in these beds. Permeability differences would cause different parts of the fluid front to advance at different rates ". . . which would finally result in zones of conflicting currents of water between which the bodies of oil and gas would be trapped and held" (Munn, 1909, Figs. 77, 79, p. 525; see also Figs. 42a, b).

Rich (1921, 1923, 1931, 1934) expanded the hydraulic theory by introducing long distance lateral migration through highly permeable carrier beds. These beds collected fluids both across and along the bedding and conveyed them from relatively high elevations to low-lying foci of discharge where "Anticlines and other structural disturbances, by causing fissures through which fluids can escape upward from the carrier beds, become centers toward which rock fluids,

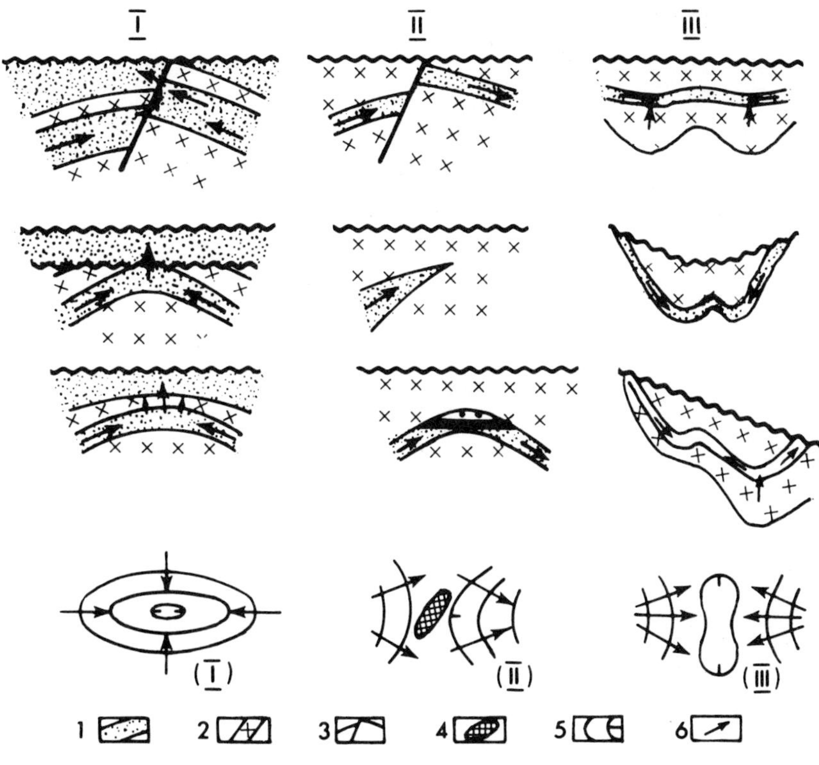

I) overflow type; II) barrier type; III) frontal type. Bottom: configuration of piezometric surfaces for the corresponding minima: 1) aquifers; 2) aquicludes; 3) faults; 4) screening zones; 5) reduced isopiestic lines; 6) direction of ground-water flow.

FIG. 41 Types of piezometric minima and configurations of piezometric surfaces (after Kudryakov, 1974, Fig. 1, p. 241).

including oil, move from all directions and thereby become localizers of oil accumulations from wide areas" (Rich 1931, p. 911).

Hubbert's (1953) "Entrapment of petroleum under hydrodynamic conditions" treats of the required energy conditions under which oil and gas can accumulate and become trapped at a particular locality within the carrier beds. According to this theory, the fundamental criterion of entrapment is a local energy minimum which is expressed by a tilt of the interface between the trapped and carrier fluid that is less than the slope of a sealing bed. Although Hubbert presented rigorously the principles of three-phase fluid movement in the subsurface, recognized flow across the beds, and referred to his work as "the general theory of migration and entrapment of oil and gas under hydrodynamic conditions . . . " (op. cit., p. 1956) he did not consider regional migration patterns of groundwater and examined entrapment only in unidirectional flow restricted to laterally extensive but vertically confined aquifers (Hubbert, 1953, p. 1,973, Fig. 11; see also Fig. 43).

Obviously, the three theories represent milestones in a progressive development of thought which is based on the recognition that hydrocarbons are transported by moving groundwater in the rocks' pores. However, notwithstanding the acceptance in general of hydraulic communication between the land surface and flow region, and of gravity as the main source of the motive force, none of the three versions of the hydraulic theory considered quantitatively the effects of the topography on the flow distribution, i.e. the resulting patterns of subsurface fluid migration. This omission appears to have been the chief factor responsible for not recognizing the regularities in the spatial distribution of hydrocarbon desposits.

Thus, the purpose of the present work is the addition of regional patterns of formation-fluid flow to the hydraulic theory of petroleum migration and accumulation as developed by Munn, Rich, and Hubbert. As it is believed that, thus complemented, the hydraulic theory includes all mechanical aspects of the migration and accumulation process in areas of a stable rock framework, it will be referred to as the generalized hydraulic theory of petroleum migration.

Conceptually the theory is simple. It is based on the observation that in geologically mature drainage basins the rock framework is hydraulically continuous.

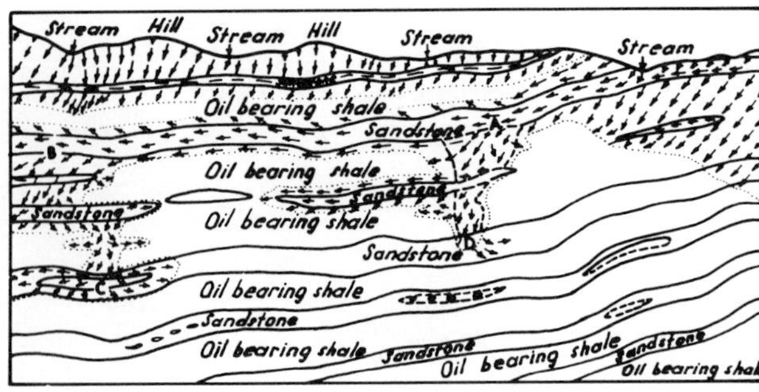

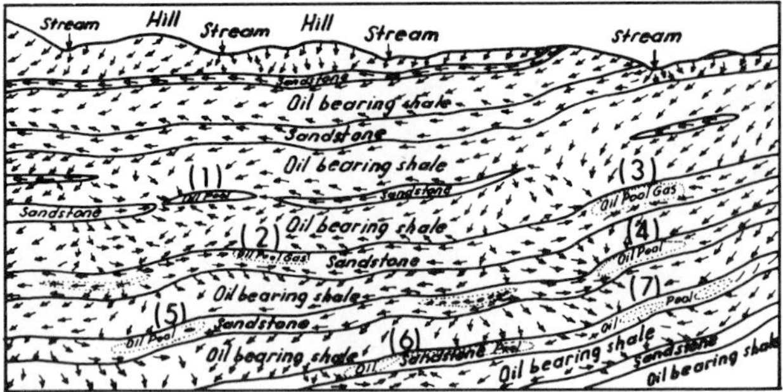

FIG. 42 Conceptual illustration of Munn's hydraulic theory of oil and gas accumulation (after Munn, 1909, Figs. 77, 79, p. 526).

In such an environment the major (universally present and effective) driving force of subsurface fluids is that generated by differences in elevation of the water table which, in most areas, closely conforms to the land surface. Consequently, groundwater flow is distributed into regionally unconfined systems which are generated by and adjusted to the relief of the water table or, approximately, to the land surface and modified by permeability differences in the rock framework. These systems of groundwater flow may function in different capacity along different parts of their lengths and, depending on local energy conditions, may (like conveyor belts) mobilize, transport, or deposit hydrocarbons. Mobilization and transport are most probable in high energy environments, primarily in recharge areas and regions of shallow and/or lateral flow. However, segregation and deposition of hydrocarbons are associated with areas of local or regional energy minima. Local energy minima are associated with the various types of hydrodynamic traps caused characteristically by anticlinal and/or domal structures, structural terrace or anticlinal nose, homoclinally dipping sands containing marked increases in permeability (Hubbert, 1953, Figs. 24, 25, 26, p. 1,997, 2,001), or highly permeable lenses enclosed in slightly permeable matrix (Tóth, 1970, Fig. 9, p. 999). Regional energy minima, on the other hand, represent hydraulic traps in which flow systems converge from opposite directions, turn vertically up to discharge into shallower zones, and leave their hydrocarbon cargo under the retaining screens of shales or other poorly permeable rock.

On the basis of this theory the outlines of a petroleum collecting area coincide with the natural boundaries of the subsurface drainage basin (Fig. 44). The frequency of hydrocarbon accumulations increases toward and is maximum in the centers of discharge and hydraulically stagnant subjacent zones. Similarly, salinity of formation fluids and specific gravity of hydrocarbons will increase in the same direction, hydraulic heads will increase with depth faster than the normal hydrostatic rate, positive temperature anomalies generated by ascending warm fluids and phenomena indicating excess moisture, such as marshy conditions, salt deposits, mud volcanoes, and so on, will characterize the areas most favorable for the principal hydrocarbon accumulations (Figs. 10, 44). Indeed, according to the generalized hydraulic theory of petroleum migration, deposits of hydrocarbons are but one of the great variety of indicators of regional groundwater flow.

Principal Component Processes and Conditions of the Migration and Accumulation of Petroleum by Regional Groundwater Flow

General Comments—The generalized hydraulic theory of petroleum migration is based on the relative positions of petroleum deposits with respect to regional flow patterns. Therefore, it is a statement of the observable integrated effects of the various physical, chemical, and dynamic processes and conditions to which petroleum or its protoforms are subjected from the time of mobilization to the time of accumulation. Thus, the theory's validity is not dependent on the details and on the type and nature of the individual component processes and conditions, such as the timing of hydrocarbon generation or the form in which oil migrates. Nevertheless, a brief review, even though qualitative, of the various processes and conditions that can be accommodated by the theory should serve to show its generality.

Type and Location of Petroleum Source Materials—The only constraint imposed on the type of the source material by the theory is that it be mobile in moving water. As long as this requirement is satisfied, petroleum, its constituents, and its protoforms may be dispersed or pooled virtually anywhere within the supply area, including the land surface, aquitards, carrier

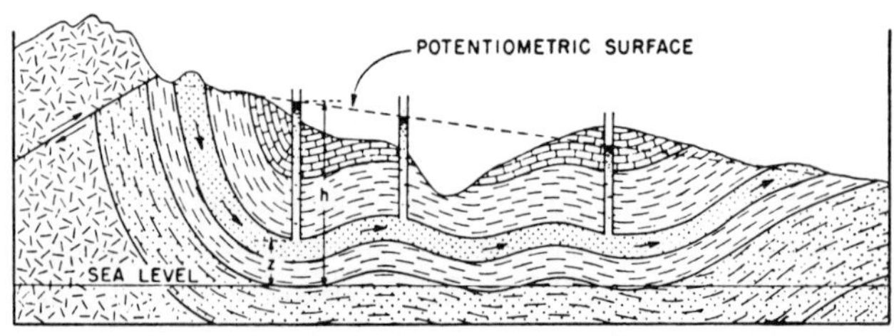

FIG. 43 Groundwater flow in regional carrier beds (after Hubbert, 1953, Fig. 11, p. 1973).

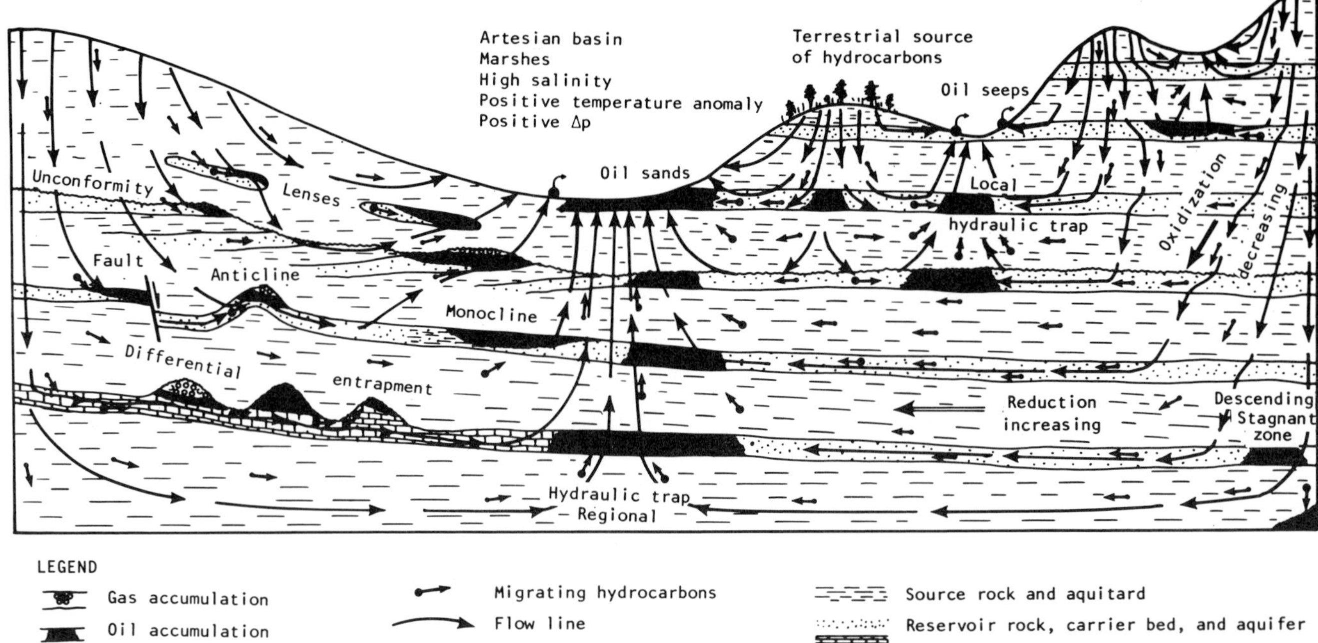

FIG. 44 Conceptual illustration of the generalized hydraulic theory of the migration and accumulation of petroleum.

beds, and existing deposits; gravity-induced cross-formational flow systems will redistribute them according to a topography dominated regular pattern.

Length and Direction of Transport—The migration path of hydrocarbons can be as long as that of the transporting flow systems unless trapping occurs en route. These lengths depend on the configuration of the basin's topography and range from a few hundred yards to tens or, rarely, hundreds of miles. The deeper a system penetrates the longer the route and, at the same time, the larger the gathering area. It is important to realize that the hydrocarbons themselves do not have to transmigrate slightly permeable formations in order to be transported by regional groundwater flow systems; it is sufficient if only water crosses aquitards. Indeed, owing to capillary displacement pressures they probably do not commonly do so except at permeable avenues such as fracture zones, faults, stratigraphic windows, and so on. The main direction of oil migration appears, therefore, to be chiefly lateral and towards regions of low energy.

Types of Traps and Methods of Entrapment—The present theory envisages continuous flow of formation fluids to import hydrocarbons into regions of relative energy minima where petroleum is segregated by buoyancy or capillary filtration or both (Fig. 44). These mechanisms are particularly effective in regions of slow movement or quasistagnant zones, sudden decrease in pore pressure, or sudden change in flow direction, or across permeability boundaries. Gravity separation of migrating fluid components and their differential entrapment (Gussov, 1953) along the flow route (according to increased density) is inherently implied in the generalized hydraulic theory.

All trap types recognized in petroleum geology are, therefore, considered to be effective. These include structural and stratigraphic traps, closed lenses, and hydrodynamic traps. In addition, "hydraulic" traps are postulated which develop in areas where opposing regional systems meet.

Stability of Accumulations—Changes in the pattern of fluid flow may be caused by epeirogenic tilting, orogenic deformations, and erosional changes in a basin's surface. These changes may result in the diversion of feedstock from developing accumulations, destruction of existing deposits by attrition or chemical and biochemical means, and in deposition and redeposition of hydrocarbons at locations that have become new regions of low energy. However, all or parts of previous deposits may remain in place, witnessing conditions of bygone times. It is because of these relict manifestations of earlier flow distributions that an understanding of the historical evolution of groundwater flow patterns is of importance to the explorationist.

Discussion

In retrospect it seems that the hydraulic theory of petroleum migration has enjoyed two periods of relative popularity: first through the years of 1909 into the 1930s (Munn, 1909; Shaw, 1917; Rich, 1921, 1923, 1931, 1934), and then some years following the publication of Hubbert's theory on hydrodynamic entrapment (1953). Now, when what could be considered a new attempt is made to revive interest in this theory the lack of its general acceptance before may be rightfully questioned. Three primary reasons appear to have contributed to this: (1) lack of predictive strength; (2) popularity of the theory of migration by compaction currents; and (3) proliferation of other concepts and theories of migration. Discussion of the third point is beyond the scope of this paper.

The lack of predictive strength seems to be due to the failure of the previous versions of the theory to take into account the actual traffic patterns of formation fluids. Plausible theories have been developed for most component processes of the migration phenomenon, including mechanisms of expulsion, transport, and entrapment, yet the true geometry of the migration paths was never before considered. A complete understanding of the migration and accumulation process requires the correct spatial distribution of the appropriate component mechanisms along the flow lines of the migrating fluids. The need for this has been felt intuitively by every student of the problem but the lack of an appropriate conceptual framework inhibited the solution: Munn (1909, Figs. 77-79, p. 526) developed the concept of hydraulic entrapment but his flow field (Figs. 42a, b) is entirely fictitious; Rich recognized the importance of discharge areas but did not realize that local relief also can generate them; Hubbert calculated the hydrodynamic criteria of entrapment but failed to place them in the context of the regional flow field (Fig. 43); Coustau et al allowed fluids to move through confining aquitards yet paid no attention to whatever effect local topography may have had (Fig. 33); even the Russian researchers who appeared to have developed the clearest understanding of the effect of the topographic relief on subsurface fluid migration omitted references to quantitatively established flow patterns (e.g. Fig. 40).

As to the second point, contrary to common belief there is no conflict between the theories of migration by compaction currents and by gravity-flow of formation fluids. Instead of these theories being mutually exclusive they represent different stages of hydrodynamic evolution as is clearly recognized, for instance, by Coustau et al (1969), Juhász (1976), Tóth (1978), and the Russian petroleum hydrogeologists (Kartsev and Vagin, 1964). Compactive forces prevail in oceanic basins during subsidence and sedimentation, inducing landward and predominantly ascending cross-formational flow (Magara, 1969; Jacquin et Poulet, 1973). One form of primary migration is probable at this stage. However, upon emergence and completion of compaction, gravity-induced flow takes over. Existing and new accumulations of petroleum now will be adjusted to the topography-controlled flow systems. Along the borders of continents, landward-directed compaction flow meets oppositely oriented continental systems, rendering these areas particularly favorable for hydraulic entrapment of hydrocarbons derived from both sides (the oceanic basin and the continent).

It probably is not coincidental that major improvements in the understanding of the migration of petroleum by moving groundwaters follow major advances in the understanding of regional groundwater flow itself. The birth of the hydraulic theory (Munn, 1909) was preceded 10 to 15 years by the pioneering papers of King (1892) and Slichter (1897/1898); Hubbert developed his theory of hydrodynamic entrapment (1953) from his own theory of groundwater motion published 13 years previous (Hubbert, 1940); the present paper and its precursor (Tóth, 1970) are direct outgrowths of analyses of regional groundwater flow patterns (Tóth, 1962, 1963, 1968; Freeze and Witherspoon, 1966, 1967, 1968); a rapidly increasing appreciation of the role of gravity-flow systems on petroleum migration has become apparent in the relevant French literature (Chiarelli, 1973; Coustau et al, 1975) after a general acceptance in France of the concept of leakage ("drainance") through aquitards (BRGM, 1969); and apparently the acceptance of topography-generated groundwater hydraulics as a working principle of hydrocarbon exploration by the Russian petroleum geologists postdates their hydrogeologist colleagues' recognition of cross-formational gravity flow by about one decade. However, in North America a general recognition of the role of regional groundwater flow in the transport and accumulation of petroleum appears this time to take longer than elsewhere.

SUMMARY AND CONCLUSIONS

The intended message of this paper is epitomized in Figure 44 and may be stated in one sentence: topography-induced cross-formational flow of groundwater has a significant genetic effect on the distribution of hydrocarbon deposits in geologically mature drainage basins. The conceptual framework of this assertion has been called the generalized hydraulic theory of the migration of petroleum and is based on the theory of regional groundwater flow.

The development of the hydraulic theory of petroleum migration includes three successive conceptual steps: (1) on a regional scale the rock framework of geologically mature (tectonically stable and noncompacting) areas is hydraulically continuous; (2) in a hydraulically continuous continental area, groundwater flow is generated by the relief of the water table and modified by permeability differences, flow systems of a different order develop and the natural unit of subsurface fluid-hydraulics is the drainage basin; and (3) groundwater mobilizes, transports, and deposits hydrocarbons as it moves along its flow paths from regions of high energy to regions of low energy.

The cornerstone of these theories is, therefore, regional hydraulic continuity. As it is intrinsically impossible to prove the existence of regional hydraulic continuity by inductive reasoning or local observations, the method of field verification of a working hypothesis was used. Distribution patterns of four dynamic parameters, namely: hydraulic head, flow, vertical pressure gradient, and dynamic pressure increment, were calculated for hydraulically continuous hypothetical subsurface flow regions and compared with regional observations from several parts of the world. As well, physical and chemical phenomena including surface and subsurface accumulations of salts, temperature anomalies, springs, and seeps of water and oil, mud volcanoes, wetland vegetation and so on (theoretically expected to result from regional groundwater flow), were examined. The unequivocal conclusion from the comparison of theoretical expectations with actual situations was that large drainage basins do function as hydraulically continuous regions in which groundwater flow-distribution is determined by the configuration of the water table. Furthermore, it was found that slightly permeable extensive formations modify the flow distribution and also may delay the transmission of changes in fluid potential from the basin's surface to lower zones. Lenses of relatively high permeability encased completely in slightly permeable matrix act as local collectors of groundwater flow and cause subnormal pressures to develop at their upstream extremity and supernormal pressures at their downstream extremity.

Groundwater flow in a basin is distributed in systems which can be classified as local, intermediate, and regional. Water in each system descends at an upland recharge area, moves laterally, and then ascends towards the surface under topographically low discharge areas. The patterns of pressures, pressure gradients, and hydraulic heads are diagnostically different in different parts of a flow system. Systems may be oriented in any direction with respect to one another. Where they meet or part or change direction significantly, zones of relatively slow water movement or stagnancy develop. The flow and energy distributions may be conveniently portrayed by means of two-dimensional groundwater flow patterns.

The reasoning and procedure used to examine the possiblity of genetic relation of petroleum accumulations with groundwater flow patterns were similar to those of verifying regional hydraulic continuity: positions of known hydrocarbon accumulations were determined with respect to established groundwater flow patterns. Generally, accumulations were found to be preferentially associated with ascending limbs and relatively stagnant zones of present or previously existing groundwater flow fields and their associated natural phenomena. These relations are interpreted to reflect the cumulative effect of a large number of component processes, including the various mechanisms of mobilization, transport, and emplacement of hydrocarbons. Each of those component processes already has been considered by previous theories. The major contributions of the present work are therefore thought to be the integration of the concept and methods of analysis of regional groundwater flow patterns with known conditions of petroleum accumulations, and the recognition of the hydraulic trap: i.e. the quasistagnant regions between oppositely directed flow systems. Because only approximately one half of a basin's surface is underlain by regions of quasistagnant or ascending flow conditions, the application of the generalized hydraulic theory may mean the doubling of efficiency for the explorationist.

As a hydrogeologist, the present writer has not yet had a chance to apply the theory in exploration. However, the evaluation of the hypothetical prospects along the Etzikom Coulee in southern Alberta (Tóth, 1970; Fig. 36) would have located, out of fourteen known occurrences, all four accumulations of commercial size, five or six shows, would have missed four shows, and would have indicated four areas as favorable that could not be verified for want of information. The fact that no subsurface data of any kind were used in the predictive phase of this test study only underscores the validity of the generalized hydraulic theory and its potential as a philosophy of exploration.

REFERENCES CITED

Albinet, M., and S. Cottex, 1967, Utilisation et interpretation des cartes de differences de pression entre nappes superposees: Chronique d'hydrogeologie, no. 12, p. 43-48.

Al-Shahristani, H., and M. J. Al-Atyia, 1972, Vertical migration of oil in Iraqi oil fields: Evidence based on vanadium and nickel concentrations: Geochim. et Cosmochim. Acta, v. 36, p. 922-938.

Astié, H., R. Bellegarde, and M. Bourgeois, 1969, Contribution a l'etude des differences piezometriques entre plusieurs aquiferes superposes — Application aux nappes du tertiaire de la Gironde: Chronique d'Hydrogeologie, no. 12, p. 49-59.

Back, W., 1960, Origin of hydrochemical facies of groundwater in the Atlantic Coastal Plain: 21st Session, Int. Geol. Cong., Part 1, p. 87-95.

——— Back, W., 1966, Hydrochemical facies and groundwater flow patterns in northern part of Atlantic Coastal Plain (Hydrology of aquifer systems): U.S. Geol. Survey Prof. Paper 498-A, 42 pages.

Bars, Ye. A., et al 1961, Genetic relationship of oil-gas basins to basins of subsurface water surrounding them: Petroleum Geology, v. 5, no. 11, p. 579-586.

Besbes, M., G. de Marsily, and M. Plaud, 1978, Bilan des eaux souterraines dans le bassin Aquitain, *in* Hydrogeology of great sedimentary basins, Conference of Budapest, 1976: Memoires 11, Int. Assoc. Hydrogeologists; Budapest, Hungarian Geol. Inst., p. 294-303.

Bredehoeft, J. R., et al, 1963, Possible mechanism for concentration of brines in subsurface formations: AAPG Bull., v. 47, no 2, p. 275-269.

BRGM, 1969, Chronique d'Hydrogeologie, no. 12, Paris, 229 pages.

——— 1970, Atlas des eaux souterraines de la France, Chapter 11: Aquitaine.

Chebotarev, I. I., 1955, Metamorphism of natural waters in the crust of weathering: Geochim. et Cosmochim. Acta, v. 8, p. 22-48; 137-170; 198-212.
Chiarelli, A., 1973, Etude des nappes aquiferes profondes, Contribution de l'hydrogeologie a la connaissance d'un bassin sedimentaire et a l'exploration petroliere: D.Sc. thesis, no 401, l'Universite de Bordeaux I, Bordeaux, 187 pages plus references and appendices.
Clayton, R. N., et al, 1966, The origin of saline formation waters, 1, Isotopic composition: Jour. Geophys. Research, v. 71, no. 16, p. 3,869-3,882.
Coustau, H., et al, 1969, Essai sur les aquiferes du tertiaire et du cretace superieur en Aquitaine: Cronique d'Hydrogeologie, no. 12, p. 127-144.
——— et al, 1975, Classification hydrodynamique des bassins sedimentaires, Utilization combinee avec d'autres methodes pour rationaliser l'exploration dans des bassins non-productifs: World Petroleum Cong. Proc., v.2, no. 9, p. 105-119.
Davis, S. N., and R. J. M. DeWiest, 1966, Hydrogeology: New York, John Wiley and Sons, Inc., 463 p.
Defence Mapping Agency Aerospace Center, 1972a, 1:1,00 0,000 Topographic map Series ONC, Sheet H-7, Edition 6.
——— 1972b, 1:1,000,000 Topographic Map Series ONC, Sheet H-6, Edition 4.
——— 1973, 1:1,000,000 Topographic Map Series ONC, Sheet J-7, Edition 6.
——— 1974, 1:1,000,000 Topographic Map Series ONC, Sheet F-7, Edition 3.
Dodd, H. V., 1922, Some preliminary experiments on the migration of oil up low-angle dips: Econ. Geology, v 17, no. 4, p. 274-291.
Erdélyi, M., 1972, Hydrology of deep groundwaters, chapter 4, *in* Hydrological methods for developing water resources management: Res. Inst. for Water Resources Development (Budapest), p. 90-158.
——— 1976, Outlines of the hydrodynamics and hydrochemistry of the Pannonian basin: Acta Geologica Academiae Scientiarum Hungaricae, v. 20, nos. 3-4, p. 287-309.
Freeze, R. A., 1972, Subsurface hydrology at waste disposal sites: IBM Jour. Res. Devel., v. 16, no 2, p. 117-129.
——— and P. A. Witherspoon, 1966, Theoretical analysis of regional groundwater flow, 1. Analytical and numerical solutions to the mathematical model: Water Resources Research, v. 2, no. 4, p. 641-656.
——— ——— 1967, Theoretical analysis of regional groundwater flow, 2. Effect of water-table configuration and subsurface permeability variation: Water Resources Research, v. 3, no. 2, p. 623-634.
——— ——— 1968, Theoretical analysis of regional groundwater flow. Quantitative interpretations: Water Resources Research, v. 4, no 3, p. 581-590.
Grossman, I. G., 1968, Origin of the sodium sulfate deposits of the northern great plains of Canada and United States: U.S. Geol. Survey Res. Paper 600-B, Chapter B, p. 104-109.
Guiraud, R., 1973, Reflexions concernant la dynamique des eaux souterraines et la notion de chott, *in* Evolution post-triassique de l'avant-pays de la chaine alpine en Algerie d'apres l'etude du Bassin du Hodna et des regions voisines: Thesis l'Universite de Nice, p. 235-241.
Gussov, W. C., 1953, Differential trapping of hydrocarbons: Alberta Soc. Petroleum Geologists Bull., v. 1, p. 4-5.
Hanshaw, B. B., and G. A. Hill, 1969, Geochemistry and hydrodynamics of the Paradox basin region, Utah, Colorado and New Mexico: Chem. Geol., v. 4, p. 263-294.
Hantush, M. S., 1956, Analysis of data from pumping tests in leaky aquifers: Am. Geophysical Union Trans., v. 37, p. 702-714.
——— and C. E. Jacob, 1954, Plane potential flow of groundwater with linear leakage: Am. Geophysical Union Trans., v. 35, p. 917-926.
Hitchon, B., 1969, Fluid flow in the western Canada sedimentary basin, 1. Effect of topography: Water Resources Research, v. 5, no. 1, p. 186-195.
——— and J. Hays, 1971, Hydrodynamics and hydrocarbon occurrences, Surat basin, Queensland, Australia: Water Resources Research, v. 7, no. 3, p. 658-676.
Holmquest, H. J., 1965, Deep pays in Delaware and Val Verde basins, *in* Fluids in subsurface environments: AAPG Memoir 4, p. 257-279.
Hubbert, M. K., 1940, The theory of ground-water motion: Jour. Geology, v. 48, no. 8, p. 785-944.
——— 1953, Entrapment of petroleum under hydrodynamic conditions: AAPG Bull., v. 37, no. 8, p. 1954-2026.
Illing, V. C., 1933, The migration of oil and natural gas: Inst. Petroleum Tech. Jour., v. 19, no. 114, p. 229-274.
Institut de Geographie et d'Etudes Regionales de l'Universite de Bordeaux III, 1973, Atlas d'Aquitaine:Paris Editions Technip.
Issar, A., and E. Rosenthal, 1968, The artesian basins of the Central Plateau of Iran, *in* Proc., Int. Accoc. Hydrogeologists, 23rd Session, Int. Geol. Cong. (Prague), p. 1-5.
Jacquin, C., and M. Poulet, 1973, Essai de restitution des conditions hydrodynamiques regnant dans un bassin sedimentaire au cours de son evolution: Rev. Inst. Francais du Petrole, v. 27, no. 3, p. 269-297.
Juhász, J., 1976, Hidrogeologia: Budapest, Akademiai Kiado, 767 p.
Kartsev, A. A., and S. B. Vagin, 1964, Paleohydrogeological studies of the origin and dissipation of oil and gas accumulations in the instance of cis-Caucasian Mesozoic deposits: Int. Geology Review, v. 6, no. 4, p. 644-655.
King, F. H., 1892, Fluctuations in the level and rate of movement of ground water: U.S. Dept. Agriculture Weather Bureau Bull., no. 5, 75 p.
Kolesov, G. D., 1965, On the question of artesian feeding of rivers: Soviet Hydrology, Selected Papers, no. 3, p. 195-203.
Kudryakov, V. A., 1974, Piezometric minima and their role in the formation and distribution of hydrocarbon accumulations: Doklady, Acad. Sci. USSR, v. 207, p. 240-242.
Levorsen, A. I., 1958, Geology of petroleum: W. H. Freeman and Co., 703 p.
MÁFI, 1958, Hydrogeological atlas of Hungary: Budapest, 102 p.
Magara, K., 1969, Upward and downward migrations of fluids in the subsurface: Bull. Canadian Petroleum Geology, v. 17, no. 1, p. 20-46.
——— 1974, Aquathermal fluid migration: AAPG Bull., v. 58, no. 12, p. 2513-2521.
Margat, J., 1969, Remarques sur la signification des surfaces piezometriques des nappes captives: Chronique d'Hydrogeologie, no. 12, p. 13-17.
Martin, B., 1969, Variations des caracteres chimiques des eaux dans les niveaux poreux du Bassin de Parentis: Chronique d'Hydrogeologie, no. 12, p. 77-89.
Meinhold, R., 1971, Hydrodynamic control of oil and gas accumulation as indicated by geothermal, geochemical and hydrological distribution patterns: Proc., 8th World Petroleum Cong., v. 2, p. 55-66.
Mifflin, M. D., 1968, Delineation of ground-water flow systems in Nevada: Univ. Nevada Desert Research Inst. Tech. Rept. (Ser. H-W), no. 4, 111 p.
Munn, M. J., 1909, The anticlinal and hydraulic theories of oil and gas accumulation: Econ. Geology, v. 4, no. 6, p.

509-529.

Muskat, M., 1932, Problems of underground water-flow in the oil industry: Am. Geophysical Union Trans., p. 399-401.

——— 1946, The flow of homogeneous fluids through porous media: J. W. Edwards, Inc., 763 p.

Neuman, S. P., and P. A. Witherspoon, 1969, Transient flow of groundwater to wells in multiple-aquifer systems; Univ. California-Berkeley, Geotechnical Engineering Publication no. 69-1, 182 p.

Rich, J. L., 1921, Moving underground water as a primary cause of the migration and accumulation of oil and gas: Econ. Geology, v. 16, no. 6, p. 347-371.

——— 1923, Further notes on the hydraulic theory of oil migration and accumulation: AAPG Bull., v. 7, no. 3, p. 213-225.

——— 1931, Function of carrier beds in long-distance migration of oil: AAPG Bull., v. 15, p. 911-924.

——— 1934, Problems of the origin, migration, and accumulation of oil; *in* W. E. Wrather and F. H. Lahee, eds., Problems of petroleum geology: Tulsa, AAPG, p. 337-345.

Róna, A., 1961, Az Alfold talajvizterkepe (The groundwater map of the Great Lowland, Hungary): Budapest, MÁFI, map and report, 102 p.

Ryzhkov, O. A., et al, 1968, Paleozoic oil in Uzbekistan and adjacent territories: Int. Geol. Review, v. 10, no. 6, p. 717-725.

Schmidt, E. R., and E. Almássy, 1958, Cross sections of pressure conditions, Great Lowland, Hungary, *in* Hydrogeological atlas of Hungary: Budapest, MÁFI.

Shaw, E. W., 1917, Discussion — The absence of water in certain sandstones of the Appalachian oil fields: Econ. Geol., v. 12, p. 610-628.

Skrotskiy, S. S., 1974, Evidence of vertical hydrocarbon migration through salt in the western part of the Caspian basin: Doklady, Acad. Sci. USSR, Earth Sci. Sect., v. 217, nos. 1-6, p. 201-202.

Slichter, C. S., 1897-1898, Theoretical investigations of the motion of ground waters: US Geol. Survey 19th Annual Report, p. 294-384.

Szebényi, L., 1965, Az artezi viz forgalmanak mennyisegi meghatarozasa (Quantitative determination of the artesian water budget): Hidrol. Közl., v. 45, no. 3, p. 125-130.

Tóth, J., 1962, A theory of groundwater motion in small basins in central Alberta, Canada: Jour. Geophys. Research, v. 67, no. 11, p. 4375-4387.

——— 1963, A theoretical analysis of groundwater flow in small drainage basins: Jour. Geophys. Research, v. 68, no. 16, p. 4795-4812.

——— 1968, A hydrogeological study of the Three Hills area, Alberta: Research Council Alberta Bull. 24, 117 p.

——— 1970, Relations between electric analogue patterns of groundwater flow and accumulation of hydrocarbons: Canadian Jour. Earth Sci., v. 7, no. 3, p. 988-1007.

——— 1972, Properties and manifestations of regional groundwater movement: Proc., 24th Int. Geol. Cong., Section 2, p. 153-163.

——— 1978, Gravity-induced cross-formational flow of formation fluids, Red Earth region, Alberta, Canada: Analysis, patterns, evolution: Water Resources Research, v. 14, no. 5, p. 805-843.

——— 1979, Patterns of dynamic pressure increment of formation-fluid flow in large drainage basins, exemplified by the Red Earth region, Alberta, Canada: Bulletin of Canadian Petroleum Geology, v. 27, no. 1, p. 63-83.

Tuan Yung-hou, and Chao Hsueh-tun, 1968, Hydrogeological features of Dzungarian artesian basin: Int. Geol. Review, v. 10, no. 8, p. 918-933.

Urbancsek, J., 1963, A foldtani felepites es retegviznyomas kozotti osszefugges az Alfoldon (Relation between geology and pore pressure in the Hungarian Plain): Hidrol. Kozl., v. 43, no. 3, p. 205-218.

USSR, Chief Administration of Geodesy and Cartography, 1967, The world atlas, 2nd ed:, Moscow, 250 maps.

Walton, W. C., 1960, Leaky artesian aquifer conditions in Illinois: Illinois Water Survey Rept. Inv. 39, 27 p.

Williams, R. E., 1970, Groundwater flow systems and accumulation of evaporate minerals: AAPG Bull., v. 54, no. 7, p. 1290-1295.

Witherspoon, P. A., and S. P. Neuman, 1967, Evaluating a slightly permeable caprock in aquifer gas storage: 1. Caprock of infinite thickness: Jour. Petroleum Technology (July), p. 959-955.

Yerofeyev, V. F., 1969, Nature of thermal anomalies of Volga-Ural oil-gas basin: Int. Geol. Review, v. 11, no. 12, p. 1382-1389.

Zatenatskaya, N. P., 1969, The relation between the chemical composition of subsurface waters and the composition of the interstitial waters of "impervious" clay rocks: Doklady, Acad. Sci. USSR, Earth Sciences Section, v. 138, p. 642-645.

Origin of Petroleum: In-Transit Conversion of Organic Compounds in Water[1]

By G. W. Hodgson[2]

Abstract Generation of hydrocarbons from dissolved organic matter in migrating water appears to be the explanation of the problem of how hydrocarbons of source sediments actually enter the water for transport to evolving reservoirs. In the hypothesis evaluated here, two of the processes required for the formation of petroleum, generation and migration, are collapsed into a single process to overcome fundamental problems of mobilizing hydrocarbons from source sediments. In the in-transit hypothesis, hydrocarbons are pictured as being simply generated from organic matter dissolved in the water, and as such appear in a molecularly dispersed form. Such dispersions are very stable and the hydrocarbons are freely transported with the water toward the point of accumulation. Laboratory simulations indicate the mechanisms by which hydrocarbons and associated compounds of petroleum may be generated in transit. Thus, dissolved organic matter including humic and fulvic acids readily produce alkanes (normal, branched, and cyclic) and aromatics at temperatures well below 200°C. Similarly, organic sulfur compounds are generated from organic matter and elemental sulfur; and petroleum porphyrins become available in homologous and decarboxylated forms under mild thermal conditions from biogenic sources. Quantitatively, the rates and quantities appear to be in the necessary order of magnitude for the transport for the evolution of substantial volumes of crude oil.

INTRODUCTION

One of the most difficult questions in the origin of petroleum is the mechanism of transport of the hydrocarbons and associated compounds from source sediments to reservoir. A great body of information exists regarding the generation and existence of hydrocarbons in recent marine sediments. In current theories on the origin of petroleum, these hydrocarbons are somehow mobilized and transported to the point of accumulation. Two problems arise. First, if the hydrocarbons are inadequately soluble in the migrating water, how are they to be incorporated into the water? Secondly, the distribution of hydrocarbons in the source sediments is clearly different from that in the reservoir. This difference in hydrocarbon distribution lies in the distribution of the n-alkanes. In the source the alkanes primarily display odd-carbon preferences; in the reservoir, there is no odd-carbon preference. Attempts to explain the decay of the off-carbon preference lead to explanations as follows: either the source-sediment hydrocarbons were swamped with a much larger flow of petroleum-like hydrocarbons, or the source-sediment hydrocarbons were seriously modified on the way to the reservoir. In either case something more than a simple step of mobilization is involved (Hodgson, 1972).

An associated concern is the mechanism by which hydrocarbons are dispersed in the water to result in a stable suspension of the hydrocarbons for migration. Baker (1960) and others drew attention to the role that surfactants might play in dispersing the hydrocarbons without calling for an improbable mechanism such as mechanical dispersion—a geochemical blender! Accordingly, Baker (1960) developed hypotheses of micellar dispersion and was able to demonstrate the nature of the dispersions. The results appeared to be important in offering explanations for particular patterns of distribution of hydrocarbons within petroleum. However, the micellar theory failed to gain wide support because the ranges of concentrations of hydrocarbons in the water appeared to be too high to be relevant in the context of source sediments.

Very dilute dispersions were generated by Peake and Hodgson (1966, 1967) without surfactants but with substantial input of mechanical energy. The principal point was that dilute dispersions, 0.1 to 100 mg/l, of hydrocarbons in the C_{20}-C_{33} range of n-alkanes were found to be remarkably stable. The stabilities were great enough in laboratory simulations

[1]Manuscript received October 5, 1978; accepted for publication, October 1, 1979.

[2]Kananaskis Centre for Environmental Research, The University of Calgary, Calgary, Alberta, T2N 1N4, Canada.

Grateful acknowledgement is extended to colleagues who have contributed to the development of the concept over the years, notably B. L. Baker, Eric Peake, M. S. Strosher, and S. A. Tilang.

Article Identification Number:
0149-1377/79/SG10-0009/$03.00/0

to indicate stable homogeneous flow through sediments, with failure only when salinities were sharply increased, and with greatest effects from multivalent ions.

While these and other related studies were being conducted, other investigators, notably Tóth (1970, 1978), Hitchon (1969a, 1969b), Hitchon and Gowlak (1972), Hitchon et al (1971) and Roberts (1972, 1978) were developing fundamental concepts of basin-wide migration of subsurface waters. In general, new principles emerged based on charge and discharge involving flow into and through all subsurface formations in updip directions to the basin edge, as in the case of the western Canada sedimentary basin from southwest to northeast. In detail, the migratory flows are responsive to topographic and stratigraphic features, as well as to permeability parameters.

The purpose of this paper is to link concepts of hydrocarbon generation with those of basin flow for a solution to the question of hydrocarbon mobilization. The objective is to test the hypothesis that migrating hydrocarbons are generated in the migrating water. In this concept organic matter in the migrating water is the source of the hydrocarbons. On generation from the dissolved organic matter the hydrocarbons enter the water in a molecularly dispersed form. In this in-transit generation, the hydrocarbons may reach levels well above their solubility limits and yet be present as very stable dispersions for long-distance migration. One of the corollaries of the hypothesis, if it has merit, is that the concept of source beds or source rocks is no longer required, and those organically enriched strata can simply be recorded as repositories in the greater petroliferous continuum. The in-transit hypothesis is evaluated on the basis of data already in hand. The evaluation in the present communication is conducted not only for hydrocarbons but also for other constituents of petroleum including sulfur compounds and petroleum porphyrins. The general findings are that the hypothesis has merit, but cannot be strongly developed because of paucity of field data.

SOLUBLE ORGANIC MATTER

Data for organic constituents of subsurface waters are limited and the most productive approach initially is to examine the organic compounds of surface waters.

River waters commonly have 1 to 24 mg/l of total organic carbon. Being constituted of waters that have drained from the soils of the basin, both above and below the surface, the river waters contain organic acids (commonly recognized as humic and fulvic acids) as their principal components. The acids in general are circumscribed by functional definitions of Ogner and Schnitzer (1970) in their studies on soil-derived humics. The majority of the organic

Table 1, Organic Compounds in River Waters.

Total Organic Carbon	1 to 24 mg/l
Humic & Fulvic Acids	1 to 10 mg/l
Tannins & Lignins	0.1 to 1.7 mg/l
Fatty Acids	0.006 to 0.03 mg/l
Amino Acids	0.003 to 0.17 mg/l
Phenolics	0.002 to 0.01 mg/l
Hydrocarbons	0.0001 to 0.003 mg/l

material dissolved in river waters (Table 1) comprises humic and fulvic acids. Tannins and lignins account for about 10% of the total organic matter; nitrogenous compounds are about the same. Specific organic compounds of phenols, amino acids, and fatty acids are each commonly about 0.1% of the total organic matter. Hydrocarbons are an order of magnitude less abundant, roughly 0.0001 mg/l. If these data for small streams in west central Alberta are typical for water entering groundwater and subsurface reservoirs, it is clear that the bulk of the dissolved organic matter is complex in molecular structure. Further, it is equally clear that the entrained hydrocarbons, whether in solution or dispersed, are trivial in abundance.

Limited data for groundwaters relative to stream waters were developed by Wallis in this laboratory for a mountain stream basin. Total organic contents were higher than the corresponding stream values in the mountain basin. So also, a special case of groundwaters was reported by Baker for waters associated with sanitary landfills. In four such samples, total organic values were about 100 mg/l for near-neutral leachates, and were very much higher in leaching groundwaters that had become strongly alkaline.

Turning from freshwater to seawater, Collins (1975) noted that dissolved organic matter ranges from 2 mg/l in the open sea to 15 mg/l in near-shore locations. The relevance of this observation in the present context is very limited because seawater is not likely to be in a position to charge underlying aquifers connected with oil fields. However, it is important to consider the dissolved organic material in pore water of recent marine sediments because it may reveal the effect of contact of saline waters with organic-rich unconsolidated sediments. The effect is clearly to increase the level of dissolved organic matter as shown by the data of Krom and Sholkovitz (1977) for the pore water in sediments from a fjord-type estuary on the northwest coast of Scotland. In the upper oxic sediments the dissolved organic matter was 8 to 16 mg/l, roughly double that in the overlying sea water. At depth (in the anoxic waters) the content was much greater, reaching 70 mg/l. The dissolved organic matter was determined to be largely fulvic acids with only 1% humic acids. The remaining organic matter was presumed to be melanoidins—complexes of amino acids and sugars (Hedges, 1978).

Similar data for dissolved organic matter in recent marine sediments were obtained by Nissenbaum and Kaplan (1972). In their studies, attention was directed to the formation of humic/fulvic substances in the maturation of the recent sediments during accumulation of the sedimenting materials.

It is important to note that the foregoing discussion has dealt with dissolved organic matter of which hydrocarbons account for only a trivial amount. The waters also contain (presumably) traces of sulfur compounds and biogenic porphyrins. Sulfur compounds exist in soils and surface waters (Hsü and Hodgson, 1977). So also, chlorophyll derivatives exist in soils and surface waters. Concentrations generally are very low—of the order of micrograms per litre in each case, several orders of magnitude lower than the total organic carbon.

Subsurface waters have not been examined rigorously for dissolved organic matter. However, oil field waters are known to contain naphthenic acids (Davis, 1969) of widely varied concentrations. In addition it was reported by Shabarova et al (1961) that organic acids are sometimes very abundant in oil field waters. For example, in one well the acids ranged from about 700 to over 2000 mg/l. Corrosion engineers are well aware of naturally occuring propionic acid in coproduced waters.

GENERATION OF HYDROCARBONS

As noted above, although hydrocarbons already exist in depositing sediments, the odd-carbon configuration of these hydrocarbons indicates that they do not represent an important source of reservoir hydrocarbons. On the other hand, there is reason to believe that hydrocarbons may be generated from the soluble organic matter in pore waters. Simplistically one would believe that they might be generated from the organic matter present in the greatest abundance—the humics.

Much work has been done on determining the composition and structure of humic substances, for example by Falk and Smith (1963), Cheshire et al (1967), Ogner and Schnitzer (1970), and Rashid and King (1970). The basic structure seems to be that of condensed lignin fractions, with ether linkages and carboxylic acid functional groups. Degradative treatment leads to identifiable fragments. Gross degradation by thermal or chemical means results in the formation of hydrocarbons—alkanes, normal, branched and cyclic, and aromatics with condensed ring structures.

A program was designed in Kananaskis laboratories in which data were generated to explain the formation of hydrocarbons in testing the present hypothesis. The basic framework recognized that subsurface temperatures commonly reach 100°C and occasionally 200°C, and that residence times at these temperatures could be taken to be at least thousands of years. In studying the evolution of hydrocarbons from humic/fulvic acids, Baker (1973) extracted humics from recent marine sediments of the Beaufort Sea. At 300°C alkanes in the C16 to C30 range were generated in a matter of hours from a solution containing 110 mg/l of humic substances. At the same time a variety of aromatic hydrocarbons including anthracene and benzperylene were formed. The reaction solutions for these studies were 0.5 M sodium hydroxide, therefore limiting the application of the results. Although the generality of the results was limited by the high pH at which the experiment was carried a number of significant observations were made. The addition of suspended sediments to the aqueous solution of the humics inhibited the reaction to some degree; the addition of behenic acid (as used by Jurg and Eisma [1964] to demonstrate hydrocarbon generation) increased the yields of both alkane and aromatic hydrocarbons. The carbon preference index of the n-alkanes produced showed that the hydrocarbons were being generated, not just released, from the humic material; and the generation, as opposed to release, of the hydrocarbons was confirmed by parallel experiments in which the humics were methylated to alkanes and aromatics.

The foregoing work was extended to lower temperatures and to in-place organic materials. Further experiments were conducted at 200 to 300°C and at 100 to 150°C. In all cases hydrocarbons were generated in a matter of days. C1 to C6 alkanes were formed with yields increasing directly with temperature and duration of the experiment. For example, organic matter in the dark shales from a Hoodoo Dome core sample yielded C1 to C6 alkanes at 80 μg/g of shale. At only 200°C for 24 hours the C9 to C36 yield was one fifth that obtained at 300°C. Isoprenoids (pristane and phytane) were present in the product in very low amounts, independent of the severity of the treatment. They appeared to be released by the treatment rather than being generated. Aromatic compounds increased in yield with increased thermal exposure. At 300°C for 84 hours the yield was about 13 μg/g of shale, roughly double the yield of C9 to C36 n-alkanes.

Of particular importance in the present evaluation of in-transit generation, the hydrocarbons generated in a typical experiment above were 54% in the aqueous phase. This corresponded to a dispersion of hydrocarbons in the water at the level of 2.1 mg/l.

At the lower temperatures, 100 to 150°C, C9 to C36 alkanes were generated to the extent of 20 μg/g of shale in 15 to 30 days. Aromatics were generated in much smaller amounts, roughly 2 μg/g of shale. Dispersion of hydrocarbons in the water phase for an experiment conducted at 150°C for 15 days showed that the bulk of the hydrocarbons remained in the solid matrix. The solubilization in the water phase appeared to be 0.2 mg/l.

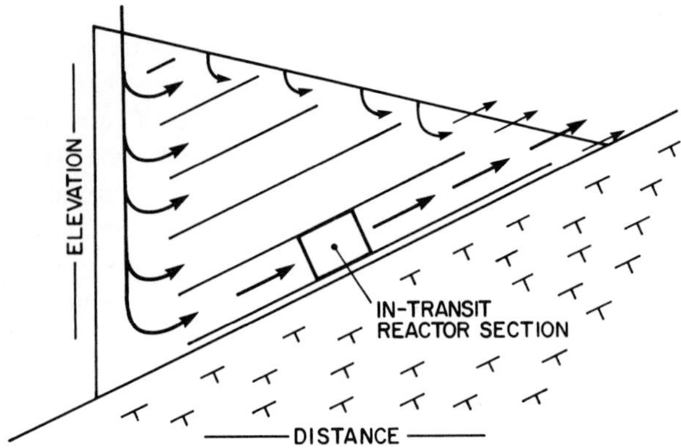

FIG. 1—Idealized vertical section of a regional sedimentary basin.

Little attempt was made in the foregoing work to relate the nature of the hydrocarbon product to that of the starting material. However, it is clear that humic and fulvic acids have a wide range of properties depending on the source material and mechanism of evolution of the acids. It would seem likely that acids with one set of characteristics would generate hydrocarbon mixtures of one pattern and those of another would generate a different mixture. It would be feasible therefore for a single basin to support generation, transport, and accumulation of several families of crude oils, consistent for example with those described by de Roo et al (1977) for the western Canada sedimentary basin.

In summary, Ishiwatari et al (1977) notwithstanding, hydrocarbons are generated by thermal treatment of soluble humic/fulvic materials; generation of alkanes and aromatics occurs in a matter of days at temperatures as low as 150°C; soluble organics in addition to humics also produce hydrocarbons; and dispersions of the hydrocarbons produced in this way are effected in the range of 0.2 to 2 mg/l.

It is important to note that the laboratory simulations described above were conducted over short periods of time, and as a result, the conversions of soluble organics to hydrocarbons were very limited (yields were low). In the geologic record, the yields would be expected to be much greater as a result of prolonged exposure to effective reaction conditions. Accordingly, the indicated concentration of dispersed hydrocarbons should be taken as very conservative figures.

ESTIMATION OF IN-TRANSIT YIELDS

To evaluate the in-transit hypothesis in terms of the foregoing data, an attempt at quantification of these hypothetical concepts must be made. That is, if hydrocarbons are generated from water-soluble organic materials, are they generated and transported in sufficient quantities to be relevant to oil field accumulations? The following section deals with this question on the basis of reasonable estimates for the various parameters involved. The basic calculation treats the question in terms of volumetric flow of migrating water and concentration of hydrocarbons in the migrating water.

Consider a section of blanket sandstone in a basin as an aquifer (Fig. 1) receiving surface waters and discharging artesian water undip at the edge of the basin. The volume of flow through the aquifer can be estimated by assuming 20% porosity and a linear flow rate of, say, 20 m/year (Roof and Rutherford, 1958). If the section of immediate interest is 20 m thick and 10 km wide (Fig. 2) the flow of water during 50 m. y., for example, would be 2×10^{13}cu m. If the hydrocarbon content of the water was 1 mg/l due to in-transit generation, the volume flow of hydrocarbons during the period would be 2×10^{7}cu m. This, in more familiar terms, is 100 million bbl, roughly equivalent to the oil in a small oil field. Accordingly, on an order-of-magnitude basis there is merit in pursuing the hypothesis further.

The parameter of greatest doubt in the foregoing calculation is the concentration of hydrocarbons in the migrating water. For credibility in the hypothesis the concentration must be near the 1 mg/l value chosen for the evaluation, because the figures given by McAuliffe (1963, 1966) for solubilities of hydrocarbons in water are five and six orders of magnitude lower. In more general terms, the in-transit system may be evaluated by considering an input/output model for a unit length of the flow system (Fig. 3). The increase in hydrocarbon transport over the length of the unit is given by $\Delta Q = Q_{out} - Q_{in} + Q_{generated} + Q_{added} - Q_{lost}$.

Q out and *Q in* account for the movement of hydrocarbons through the unit section, perhaps a kilometer in length for scale, due to the steady undip movement of migrating waters in the basin. In the general case they are nearly equal and are very much larger than any of other Q's.

Q generated refers to the formation of dispersed hydrocarbons from dissolved organic matter in transit. It

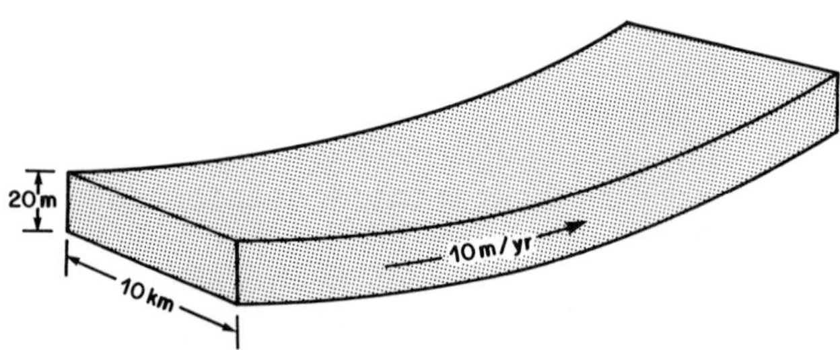

FIG. 2—Hypothetical aquifer illustrating the transport of in-transit generated hydrocarbons, assuming 20% porosity, and 50 m.y. burial. A hydrocarbon concentration of 1 mg/l would amount to about 100 million bbl of hydrocarbons.

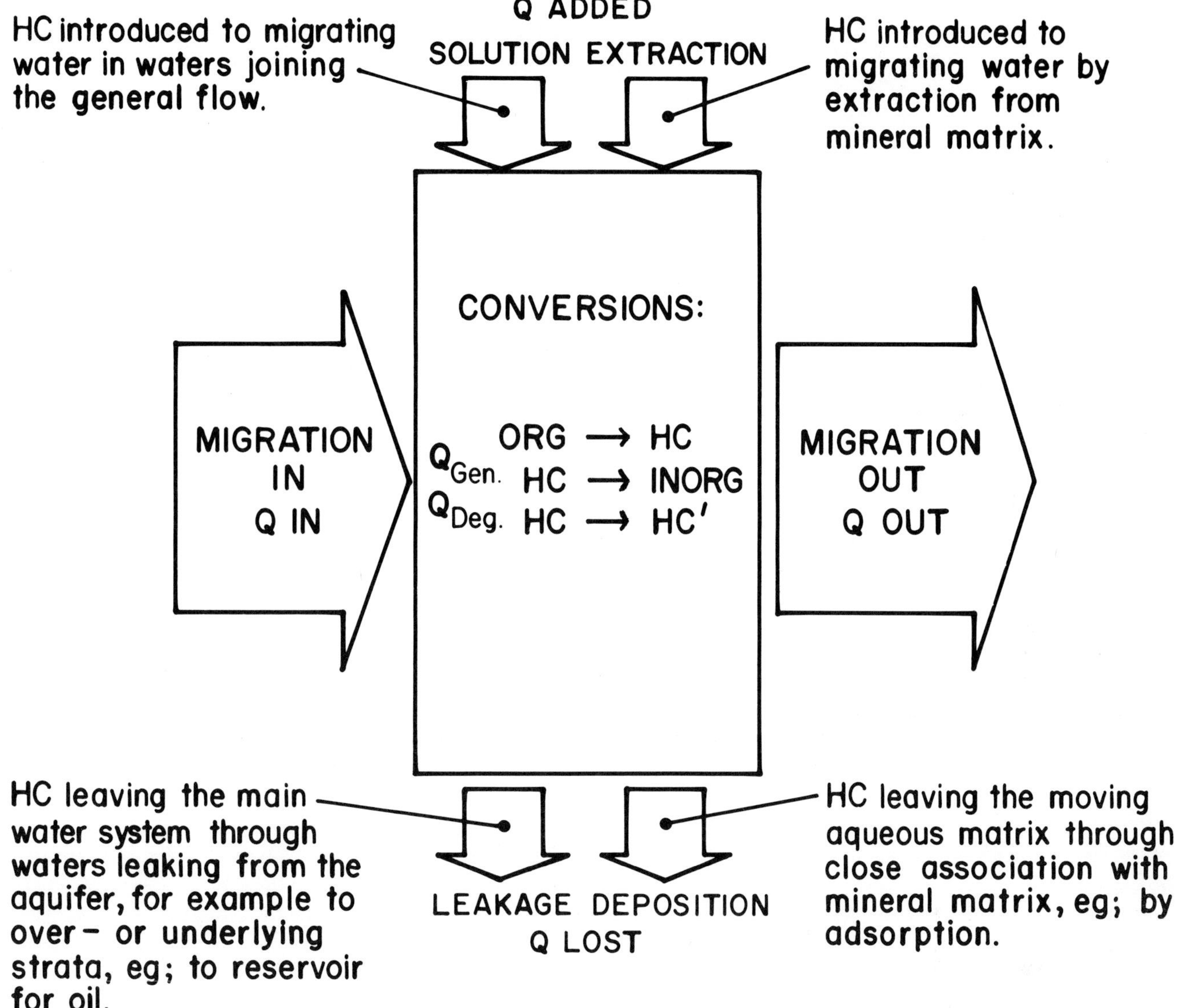
HC introduced to migrating water in waters joining the general flow.
Q ADDED
SOLUTION EXTRACTION
HC introduced to migrating water by extraction from mineral matrix.
CONVERSIONS:
MIGRATION IN Q IN
$Q_{Gen.}$ ORG → HC
HC → INORG
$Q_{Deg.}$ HC → HC′
MIGRATION OUT Q OUT
HC leaving the main water system through waters leaking from the aquifer, for example to over- or underlying strata, eg; to reservoir for oil.
LEAKAGE DEPOSITION
Q LOST
HC leaving the moving aqueous matrix through close association with mineral matrix, eg; by adsorption.

FIG. 3—Model for generation, transport, conversion, and loss of hydrocarbons in in-transit reactor section of sedimentary basin.

depends directly on (1) the availability of organic matter in solution; and (2) the extent of conversion under the prevailing conditions. The limiting factor is difficult to discern from the results of the laboratory data for thermal generation of hydrocarbons reported above but temperature and residence time probably are not limiting. Conversely, dissolved organic matter may be limited, either by supply or by solubility considerations. Current data indicate most likely values are in the range of 1 to 20 mg/l unless the pH is significantly higher than 7, under which circumstances higher concentrations could be expected.

Q degraded refers to the destruction of hydrocarbons in migrating subsurface waters. The major mechanism of destruction is by oxidation, either by simple chemical conversion or by metabolism in microorganisms in the subsurface environment. Metabolic degradation is due in many cases to indigenous aerobic microorganisms using dissolved organic matter, including hydrocarbons, as their carbon source. In other cases, intruding oxygen-containing groundwaters introduce both oxygen and aerobes from surface waters with the same effect. In aquifer systems lying below the oxidized zone (i.e. in reducing environments), the degradation by oxygen or aerobes is minimal, and *Q degraded* becomes a very small term.

The entry of co-migrating water through lateral or vertical movement would account for adding hydrocarbons to the system: the *Q added* term. In most instances this item would be of little concern, but in some spectacular cases when a number of migrating flows come together the effect could be dramatic. This

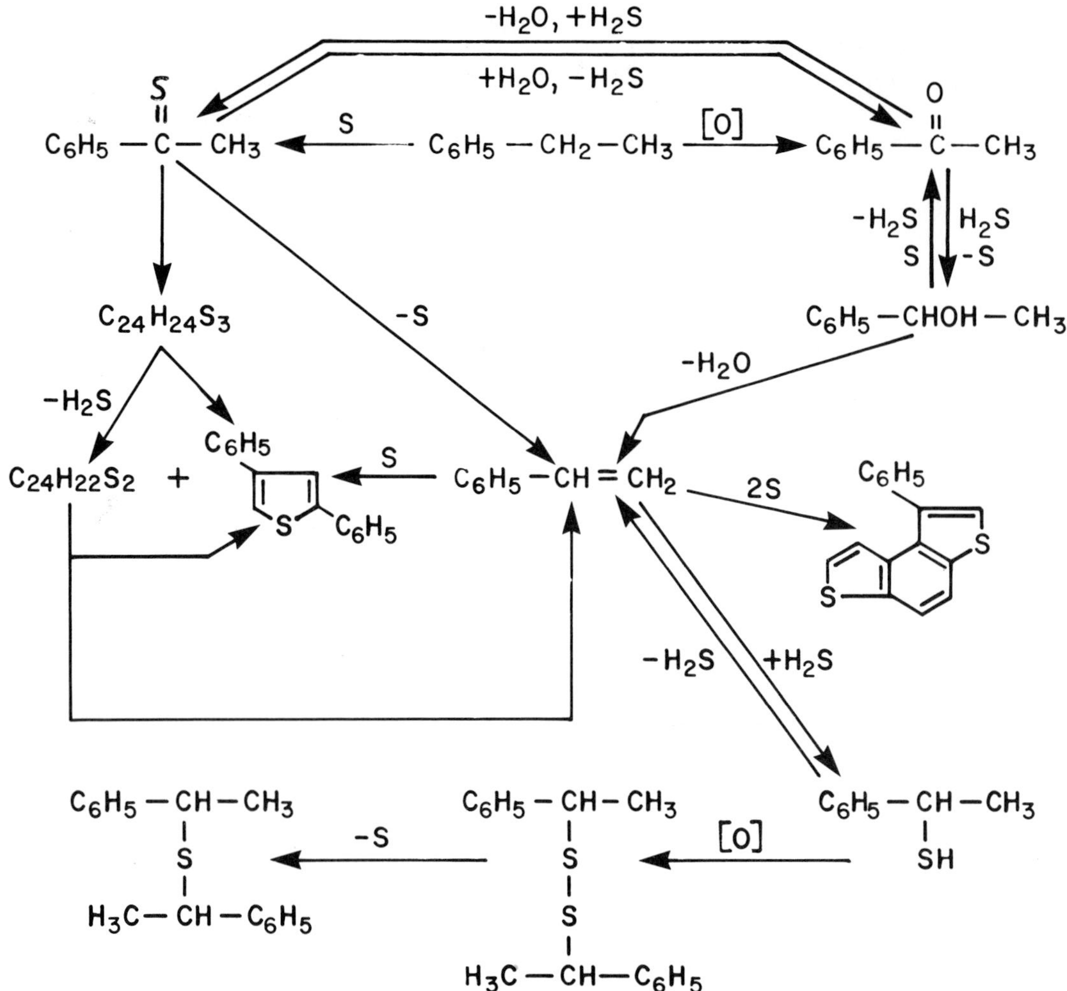

FIG 4—Generation of organic compounds of sulfur illustrated by reaction between ethyl benzene and elemental sulfur at 130 °C for 100 days (from de Roo and Hodgson, 1978).

is believed by a number of workers, notably Hitchon (1969a, 1969b) and Roberts (1972, 1978), to account for the occurrence of heavy oil belts at the edges of sedimentary basins.

The *Q lost* term in the general case refers to leakage from the aquifer, a very minor item. However, in the special case it refers to the accumulation of hydrocarbons in a local reservoir — an oil field. As developed by Roberts (1978) and Tóth (1978), the accumulation process is basically one in which the stability of the migrating system is sharply reduced. This may be a result of physical change — temperature or pressure — or a chemical change in pH or salinity, as the migrating water moves from the aquifer to a different stratum. As noted earlier, increases in salinity have dramatic effects of reducing the accommodation of dispersed hydrocarbons in water.

Summation of the gains and the losses provides a framework for estimating the mass generation and transport of migrating hydrocarbons. The principal limiting factor for generation appears to be the paucity of dissolved organic matter in the migrating water. Even so, the level of dispersed hydrocarbons could reach 10 mg/l and possibly higher, well beyond the 1 mg/l used to calculate the transport of 100 million bbl of petroleum in the hypothetical section of a basin described above.

Thus the in-transit concept of hydrocarbons appears to have merit of simplicity and some justification in quantification. To have continuing validity it must be appropriate for other components of petroleum, notably organic sulfur compounds and petroleum porphyrins.

GENERATION OF SULFUR COMPOUNDS

In-transit generation of organic compounds of sulfur appears possible because of the ease with which elemental sulfur reacts with organic substances. Douglas and Mair (1965) reported the facile reaction of sulfur with terpenes, and Toland (1961) drew attention to the reactions between hydrocarbons and elemental sulfur. Turning to soluble organic matter, Martin and Hodgson (1973) described the formation of a wide spectrum of organic sulfur compounds in reactions of elemental sulfur and phenylalanine. Subsequent studies showed similar reactions between sulfur and benzylamine, with the formation of heterocyclic compounds (Martin and Hodgson, 1976). Indicated reaction mechanisms included condensation of unstable intermediates to cyclic compounds. This was confirmed later by reactions between ethylbenzene and sulfur in which styrene was generated as a reactive intermediate which condensed with a second styrene molecule and sulfur to produce a diphenyl thiophene product (deRoo and Hodgson, 1978; Fig. 4). The most significant aspects of these studies involving elemental sulfur were (a) the ease of the reactions: a few days at 30 to 130°C in media containing water, and (b) the wide spectrum of sulfur-containing products that emerged.

With the advent of sulfur gas chromatography, sulfur is reported more widely in the environment than before. For example, elemental sulfur is common in surface waters. Sometimes it is clearly a result of pollution outfalls (Hsü and Hodgson, 1977), but more commonly it appears to be related to ordinary geochemical processes. Thus, sulfate is reduced in anoxic environments to sulfide. In turn, if conditions permit, the sulfide is oxidized to elemental sulfur. This appears to account for its presence in surface waters. The solubility of sulfur in water is very low, but sulfur does exist in stable colloidal dispersions in water. In a manner analagous to the generation of hydrocarbons from soluble organic matter, elemental sulfur may thus be generated in a dispersed form from soluble (inorganic) sulfides in water. The opportunity for in-transit evolution of organic sulfur compounds from sulfur and organic matter apparently parallels the evolution of hydrocarbons in the same aqueous system.

When sulfur was added to the alkaline reaction system (Baker, 1973) the dissolved humic material gave rise to detectable amounts of organic sulfur in simple organic compounds.

Sulfur compounds in petroleum rarely exceed 5% and it would appear that some limiting factor(s) must be operating in the genesis of the sulfur compounds. Again as in the case of hydrocarbons, the rates of reaction seem unlikely to provide such control because the reactions take place very readily at low temperatures and in relatively short times ($\cong$ 100°C and days). The more likely limiting factor is the availability of sulfur in subsurface waters.

Because the oxidation state of sulfur appears to be important in the introduction of sulfur into organic compounds, it may be significant to note that the transition metals could have a role to play in the formation of sulfur compounds. This is indicated by the relationship between heavy oils, sulfur, and vanadium, and by the observation that vanadate appears to accelerate the generation of organic sulfur compounds (deRoo and Hodgson, 1978).

Trace metals obviously exist in subsurface waters at levels that reflect the trace element composition of the sediments. Further, trace metals are sequestered by soluble humic substances as noted for example by Kribek et al (1977). Where this is important in the petroleum geochemistry of sulfur, it has even more direct importance in dealing with the porphyrin pigments of petroleum.

GENERATION OF PETROLEUM PORPHYRINS

In-transit generation of hydrocarbons and sulfur compounds are supported in validating the hypothesis of this in-transit generation and mobilization by a study of the mechanisms by which petroleum porphyrins may also be generated, because porphyrins are universal components of accumulated petroleum (Hodgson, 1973).

The porphyrins of petroleum are nonpolar heterocyclic compounds comprising four pyrrole rings bridged around a central metal atom. They exist in petroleum in two major homologous series: one having molecular masses of 476 $\pm$ 14n (phyllo series), and the other 478 $\pm$ 14n (etio series). The central cations are VO^{+2}, and Ni^{+2}. Simplistically the members of the phyllo series appear to be derived from chlorophyll and the etio series from heme.

As in hydrocarbons, porphyrins of petroleum are only slightly soluble in water, while their precursor compounds have much higher solubilities. In the first instance the precursor compounds are still porphyrins and have one or more carboxylic acid functional groups. Thus the precursor compounds for the phyllo series are almost certainly pheoporphyrin, pheophorbide, pheophytin, and ultimately chlorophyll. Of these, pheoporphyrin and pheophorbide both have (single) carboxylic acid groups. In pheophytin and chlorophyll the acid group is esterified with a long-chain alcohol, phytol. Thus the latter two compounds are less soluble in water than pheophorbide and pheoporphyrin. Accordingly, if pheophorbide and pheoporphyrin were available, they would tend to enter the in-transit migrating water system more readily.

So also, the precursors of the etio series are carboxylated. In fact, the precursors compounds are recognized as having two carboxyl groups rendering them more soluble than the chlorophyll analogs. In addition, their solubility may be enhanced even further by relict peptide moieties on the tetrapyrrole ring (Hodgson et al, 1969).

As a result of the real, but limited, solubilities of the phyllo and etio precursors, they would be expected to be present in the migrating water system. To convert

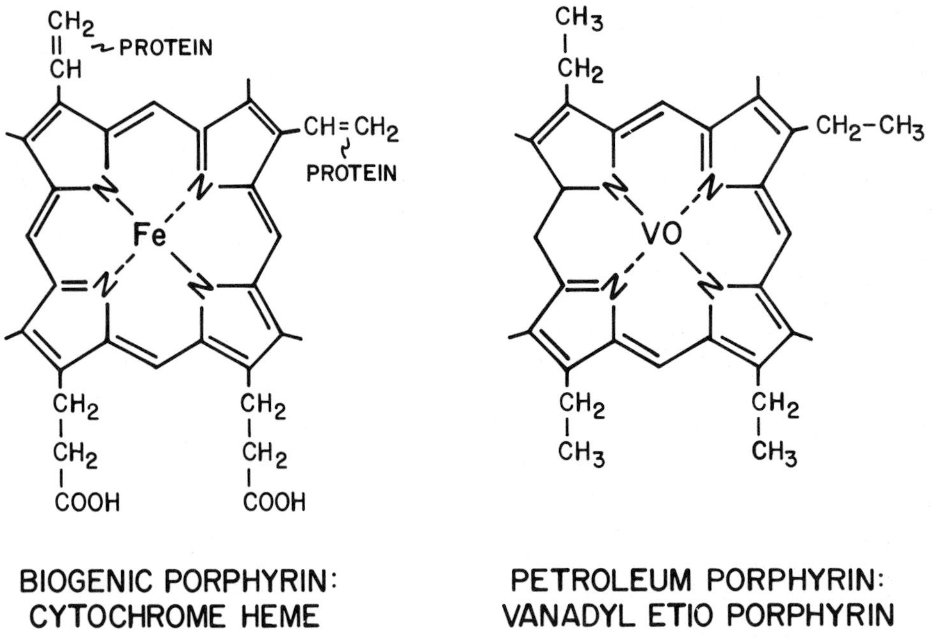

BIOGENIC PORPHYRIN: CYTOCHROME HEME

PETROLEUM PORPHYRIN: VANADYL ETIO PORPHYRIN

FIG. 5—Basic structures of etio porphyrins of petroleum related to biogenic heme pigments.

the in-transit precursors to the petroleum pigments, several steps are required (as illustrated in Fig. 5): (1) decarboxylate the soluble pigments; (2) introduce a complexing cation into the phyllo series; (3) replace (to a greater or lesser degree) the iron in the etio series with either a vanadyl or nickel cation; (4) reduce relict hydroxyl groups; and (5) oxidize (dehydrogenate) the single bond in ring IV in the phyllo series to strengthen conjugation in the tetrapyrrole structure.

Whereas the foregoing steps appear to be required to convert the soluble pigments to petroleum porphyrins, it is important to note that the following step is not required: "Establish one or more homologous series from a single mass precursor, e.g. from chlorophyll a or heme."

Where it has been shown that such a homologous series can be generated by thermal treatment as in the case of mesoporphyrin (Casagrande and Hodgson, 1971), the temperatures required may be considered a barrier. More importantly Purdie and Holt (1965) and others (Hodgson and Whiteley, 1980) demonstrated that the biogenic porphyrins in chlorophylls already exist in homologous series. Similarly Hodgson and Whiteley (1980) found evidence for homologous porphyrins in the etio precursors (i.e. hemes of cytochromes). In addition it has been shown that homologous porphyrins exist in recent sediments (Casagrande and Hodgson, 1976).

The steps listed above appear to be consistent with the reaction products demonstrated in the conversion of soluble humic material to hydrocarbons. Thus, decarboxylation and reduction present no problems. The oxidation step is an intramolecular oxidation-reduction step which is easily achieved. The metal complexing step is facilitated by a polar solvent and elevated temperatures, and could readily be expected to occur in the in-transit system, particularly because soluble humics tend to absorb di- and trivalent cations (Hitchon et al, 1975; Kribek et al, 1977).

In summarizing the possible in-transit generation of the porphyrins of petroleum, it is important to note that the general conversion is from soluble organic substances to nonpolar substances that are highly insoluble, in close analogy with the case for hydrocarbons and sulfur compounds. The hypothesis is summarized schematically in Figure 6.

CONCLUSIONS

Evaluation of the hypothesis that petroleum hydrocarbons are generated in-transit in the migrating water from soluble organic matter has shown that the hypothesis has merit in terms not only of the hydrocarbons but also of sulfur compounds and petroleum porphyrins. In this manner it is possible to avoid the perplexing questions concerning source beds in which the mechanism is fundamentally obscure for introducing basically insoluble hydrocarbons, sulfur compounds, and porphyrins into the migrating water. Although insufficient data are available to test the hypothesis rigorously, existing data support the in-transit principle, both qualitatively and quantitatively, at least for geological basins in which artesian aquifers are clearly operative on the regional scale discussed in

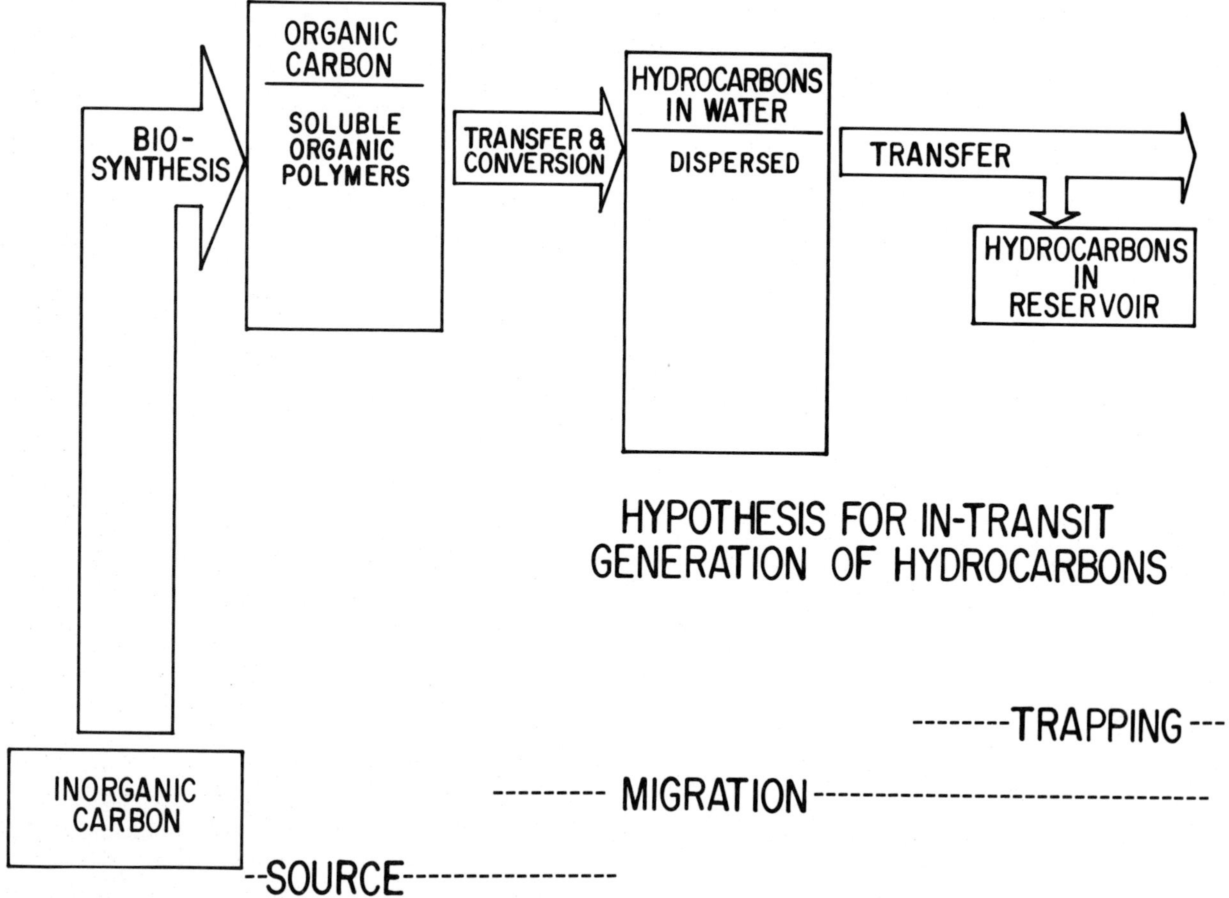

FIG 6—Generalized scheme for illustrating the origin of petroleum in terms of source conversion and transport; vertical dimension refers to potential energy of carbon-containing substances.

this analysis. However, care must be taken not to claim in-transit generation as the sole explanation for generation and transport of hydrocarbons but it seems to offer a plausible mechanism, particularly for mobilization and transport of hydrocarbons.

REFERENCES CITED

Baker, B. L., 1973, Generation of alkane and aromatic hydrocarbons from humic materials in arctic marine sediments: Advances in Organic Geochemistry, p. 137-152.

Baker, E. G., 1960, A hypothesis concerning the accumulation of sediment hydrocarbons to form crude oil: Geochim. et Cosmochim. Acta, v. 19, p. 309-317.

Casagrande, D. J., and G. W. Hodgson, 1971, Geochemical simulation evidence for the generation of homologous decarboxylated porphyrins: Nature Physical Science, v. 233, p. 123-124.

——— ——— 1976, Geochemistry of porphyrins: The observation of homologous tetraphrroles in a sediment sample from the Black Sea: Geochim. et Cosmochim. Acta, v. 40, p. 479-482.

Cheshire, M. V., et al, 1967, Humic acids, II. Structure of humic acids: Tetrahedron, v. 23, p. 1669-1682.

Collins, A. G., 1975, Geochemistry of oil field waters: New York, Elsevier, 358 p.

Davis, J. B., 1969, Distribution of naphthenic acids in an oil-bearing aquifier: Chem. Geol., v. 5, p. 89-95.

de Roo, J., and G. W. Hodgson, 1978, Geochemical origin of organic sulfur compounds: Thiophene derivatives from ethylbenzene and sulfur: Chem. Geol., v. 22, p. 71-78,

de Roo, G., et al, 1977, The origin and migration of petroleum in the western Canadian sedimentary basin, Alberta: Canadian Geol. Survey Bull. 262, p. 1-136.

Douglas, A. G., and B. J. Mair, 1965, Sulfur: Role in genesis of petroleum: Science, v. 147, 499-501.

Falk, M., and D. G. Smith, 1963, Structure of carboxyl groups in humic acids: Nature, v. 200, p. 569.

Hedges, J. I., 1978, The formation and clay mineral reactions of melanoidins: Geochim. et Cosmochim. Acta, v. 42, p. 69-76.

Hitchon, B., 1969a, Fluid flow in the western Canadian sedimentary basin, 1. Effect of topography: Water Resources Research, v. 5, p. 186-195.

——— 1969b, Fluid flow in the western Canadian sedimentary basin, 2. Effect of geology: Water Resources Research, v. 5, p. 460-469.

——— and M. Gawlak, 1972, Low molecular weight hydrocarbons in gas condensates from Alberta, Canada: Geochim. et Cosmochim. Acta, v. 36, p. 1043-1059.

——— G. K. Billings, and J. E. Klovan, 1971, Geochemistry and origin of formation waters in the western Canada sedimentary basin, III. Factors controlling chemical composition: Geochim. et Cosmochim. Acta, v. 35, p. 567-598.

——— R. H. Filby, and K. R. Shah, 1975, Geochemistry of trace elements in crude oils, Alberta, Canada, *in* T. F. Yen, ed., The role of trace metals in petroleum: Ann Arbor (Mich.) Science, p. 111-120.

Hodgson, G. W., 1971, Origin of petroleum: Chemical constraints: Advances in Chemistry Series, No. 103, p. 1-29.

——— 1972, Hydrocarbons, *in* R. W. Fairbridge, ed., Encyclopedia of geochemistry and environmental sciences: New York, Van Nostrand, v. 4A, p. 495-503.

——— 1973, Geochemistry of porphyrins—reactions during diagenesis: New York Acad. Sci. Annals, v. 206, p. 670-684.

——— and C. G. Whiteley, 1980, The universe of porphyrins; 4th Int. Symp. Environmental Biogeochemistry, Proc. (Canberra).

——— J. Flores, and B. L. Baker, 1969, The origin of petroleum porphyrins: Geochim. et Cosmochim. Acta, v. 33, p. 532-535.

——— B. Hitchon, and K. Taguchu, 1964, The water and hydrocarbon cycles in the formation of oil accumulation, *in* Y. Miyake and T. Koyama, eds., Recent researches in the fields of hydrosphere, atmosphere and nuclear chemistry: Tokyo, Maruzen Company, p. 217-242.

Hsü, H., and G. W. Hodgson, 1977, Organic compounds of sulfur: Initial data for soils and streams in Alberta, *in* H. S. Sandhu and M. Nyborg, eds., Proceedings of Alberta sulphur gas research workshop, III: Edmonton, Alberta Environment, p. 233-251.

Ishiwatari, et al, 1977, Thermal alteration experiments on organic matter from recent sediments in relation to petroleum genesis: Geochim. et Cosmochim. Acta, v. 41, p. 815-828.

Jurg, V. W., and E. Eisma, 1964, Petroleum hydrocarbons: generation from fatty acid: Science, v. 146, p. 1451-1452.

Kribek, B., J. Kaigl, and V. Oruzinsky, 1977, Characteristics of di- and trivalent metal-humic acids complexes on the basis of their molecular-weight distribution: Chem. Geology, v. 19, p. 73-81.

Krom, M. D., and E. R. Sholkovitz, 1977, Nature and reactions of dissolved organic matter in the interstitial waters of marine sediments: Geochem. et Cosmochim. Acta, v. 41, p. 1565-1573.

Mair, B. J., 1964, Terpenoids, fatty acids and alcohols as source materials for petroleum hydrocarbons: Geochim. et Cosmochim. Acta, v. 28, p. 1303-1321.

Martin, T. H., and G. W. Hodgson, 1973, Geochemical origin of organic sulfur compounds: Reaction of phenylalanine with elemental sulfur: Chem. Geology, v. 12, p. 189-208.

——— ——— 1976, Geochemical origin of organic sulfur compounds: Precursor products in the reactions of phenylalanine and benzylamine with elemental sulfur: Chem. Geology, v. 20, p. 9-25.

McAuliffe, C., 1963, Solubility in water of C1-C9 hydrocarbons: Nature, v. 200, p. 1092-1093.

——— 1966, Solubility in water of paraffin, cycloparaffin, olefin, acetylene, cycloolefin, and aromatic hydrocarbons: Jour. Phys. Chem., v. 70, p. 1267-1275.

Nissenbaum, A., and I. R. Kaplan, 1972, Chemical and isotopic evidence for the in situ origin of marine humic substance: Limnology and Oceanography., v. 17, p. 570-582.

Ogner, G., and M. Schnitzer, 1970, The occurrence of alkanes in fulvic acid, a soil humic fraction: Geochim. et Cosmochim. Acta, v. 39, p. 921-928.

Peake, E., and G. W. Hodgson, 1966, Alkanes in aqueous systems, I. Exploratory investigations on the accommodation of C20-C33 n-alkanes in distilled water and occurrence in natural water systems: Jour. Amer. Oil Chem. Soc., v. 43, p. 215-222.

——— ——— 1967, Alkanes in aqueous systems, II. The accommodation of C12-C36 n-alkanes in distilled water: Jour. Amer. Oil Chem. Soc., v. 44, p. 969-702.

——— B. L. Baker, and G. W. Hodgson, 1972, Hydrogeochemistry of the surface waters of the Mackenzie River drainage basin, Canada, II. The contribution of amino acids, hydrocarbons and chlorins to the Beaufort Sea by the Mackenzie River system: Geochim. et Cosmochim. Acta, v. 26, p. 867-883.

——— et al, 1972, The potential of arctic sediments: 24th Int. Geol. Congr. (Montreal), Proc. Sect. 5, p. 28-37.

Purdie, J. W., and A. S. Holt, 1965, Structures of Chlorobium chlorophylls (650): Canadian Jour. Chemistry, v. 43, p. 3347-3353.

Rashid, M. A., and L. H. King, 1970, Major oxygen containing functional groups present in humic and fulvic acid functions isolated from contrasting marine environments: Geochim. et Cosmochim. Acta, v. 34, p. 193-201.

Roberts, W. H., III, 1972, Hydrodynamic and analysis in petroleum exploration: Encyclopedia del Petroleio e die Gas Naturali.

——— 1978, Design and function of oil and gas traps as clues to migration: this volume (presented 1978).

Roof, J. G., and W. M. Rutherford, 1958, Rate of migration of petroleum by proposed mechanisms: AAPG Bull., v. 42, p. 963-980.

Shaborova, N. T., A.P. Tunyak, and M. B. Nektarova, 1961, Study of organic acids in underground waters: Geol. Nefti Gaza, v. 5, p. 50-51 (in Russian); reported by Collins (1975).

Toland, W. G., 1961, Oxidation of alkylbenzenes with sulfur and water: Jour. Organic Chemistry, v. 26, p. 2929-2932.

Tóth, J., 1970, Relation between electric analogue patterns of groundwater flow and accumulation of hydrocarbons: Canadian Jour. Earth Sciences, v. 7, p. 988-1007.

——— 1978, Cross-formational gravity flow of groundwater: a mechanism of the transport and accumulation of petroleum: this volume (presented 1978).

Methane Generation and Petroleum Migration[1]

By Hollis D. Hedberg[2]

Abstract The outstanding contribution of methane generation to petroleum migration is as a major source of internal energy to move fluids within the petroleum source system.

In the early stages of sediment consolidation, methane generation is largely the result of bacterial action, and most of the gas produced escapes readily along with large volumes of compaction water, either to the surface or to associated reservoirs. However, with increasing depth of burial, bacterial activity diminishes and is overlapped and replaced by thermochemical activity which increases in vigor with depth because of increasing temperature, and becomes responsible for generation of both gas and oil.

If the volume of methane generated exceeds the capacity of interstitial water to take it into solution, free gas bubbles develop in the pore spaces and an internal fluid pressure begins to build up *within* the already dense, relatively impermeable sediment, in addition to the external pressure due to overburden. At the same time, a further inhibiting effect on fluid escape is caused by the presence now of two phases—liquid and gaseous—in the sediment which may effectively impede intergranular flow (Jamin effect).

Eventually something will have to "give." As pressures rise to equal or exceed the weight of rock overburden, microfractures will begin to develop in the rock, allowing relief of pressure and permitting fluids to migrate to reservoir strata. The movement of hydrocarbons may be in part as a water solution and even as a continuous oil phase, but an important part may be as a solution of higher hydrocarbons in methane gas.

Methane generation also may (under certain circumstances) play an inhibitive role in petroleum migration through the formation of impermeable methane-hydrate barriers and through development of impermeable, overpressured shale zones and diapirs.

Illustrations of shale diapir and mud volcano activity are given to demonstrate the magnitude of the power that may be built up by methane generation.

INTRODUCTION

Most abundant and best known of the fluids filling rock pore space is *water*. But next in abundance is *petroleum*—liquid petroleum (*oil*) and gaseous petroleum (*natural gas*). And the major component of natural gas is *methane*—probably the most abundant and widespread of all gaseous fluids in the sedimentary rocks of the Earth.

PROPERTIES OF METHANE

Methane (CH_4) is a colorless, odorless, hydrocarbon gas. It is the first and simplest member of the straight-chain homologous paraffin series of hydrocarbons. It is much lighter than air: only about half as dense. Its solubility at 20°C (at atmospheric pressure) is only about 34.7 ml per liter in pure water and 28 ml per liter in sea water (Yamamoto et al, 1976). This is about 0.194 cu ft/bbl in pure water. Solubility increases rapidly with pressure, but decreases with increasing temperature to a minimum at about 82°C, above which, strangely enough, solubility increases (Culberson and McKetta, 1951). Methane is highly flammable, and when burned breaks down into carbon dioxide and water vapor.

The element carbon has three isotopes: ^{12}C, ^{13}C, and ^{14}C. Different conditions of origin and mobility of methane result in differences in the proportions of these isotopes, and thus carbon isotope ratios, particularly the $^{12}C/^{13}C$ ratio, are helpful in identifying the source of a specific methane occurrence.

OCCURENCE OF METHANE

Methane is widespread throughout the world—in the atmosphere, hydrosphere, and lithosphere. To fully appreciate the frequency and abundance of its occurrence, and hence the ubiquity of the role it may play, some of its principal sources in nature are outlined below:

Marsh gas. Methane has been identified since ancient times (under the name of "marsh gas") as a product of bacterial decay of vegetation under anaero-

[1]Manuscript read before the Association, April 11, 1978; submitted for publication, August 31, 1978; accepted for publication, February 6, 1979.

[2]Consulting geologist, 118 Library Place, Princeton, New Jersey 08540.

Article Identification Number:
0149-1377/79/SG10-0010/$03.00/0.

bic conditions in swamps and marshes, lake and river beds, beaches, lagoons, etc. Methane generations of this sort, when associated with reservoir rocks and buried by sealing sediments, have formed commercial gas deposits in many parts of the world, commonly, in Recent, Quaternary, and Tertiary strata. The city of Frederikshavn, Denmark, for many years depended for lighting and heating on biogenic methane from shallow deposits in Quaternary strata.

Petroleum. Methane is probably most widely known in its association with petroleum where it is the main constituent of petroleum natural gas accumulations, and is almost always a dissolved constituent of petroleum crude oil accumulations (on the average of several hundred to a thousand cu ft/bbl). In many petroleum fields, producing gas/oil ratios may be as high as several thousand cu ft/bbl, grading upward all the way to so-called *condensate* fields where production is largely methane and higher gaseous hydrocarbons with only minor proportions of liquid-range hydrocarbons. In addition to the methane in subsurface petroleum reservoirs, surface seeps of methane from such accumulations are widespread both on land and in the oceans.

Coal. Some methane is invariably associated with lignite and coal. The common-but-dangerous, flammable, and explosive "fire-damp" of coal mines is methane. So common is this occurrence that projects are now under way to commercially extract methane from coal seams in advance of (or during) mining, thus adding to the mine safety. Other projects are under way to artificially convert coal seams into methane in situ, thus anticipating a natural conversion.

Carbonaceous and Bituminous Shales. Many thick carbonaceous and/or bituminous muds, shales and siltstones throughout the world have their pore space and fracture space charged with methane. A well-known example is the dark Devonian shale so widespread in North America under the local names of Ohio Shale, Tennessee Shale, Chattanooga Shale, Woodford Shale, etc. The Cretaceous shales of northwestern Siberia and of northern Spain are also well known for their methane content. It may be that all carbonaceous and bituminous shales are at least to some degree methane-bearing. Because of the low permeability of these shales, economic extraction of methane is difficult. However, at certain localities commercial production has been carried on for years (e.g., Big Sandy Field in eastern Kentucky with some 4,000 wells producing from Devonian shale; Castillo Field in northern Spain producing from several thousand feet of Cretaceous shale). Many shale gas projects are currently in prospect.

Surficial Waters. Methane is present in most ocean waters, but in very low concentration. However, in gulfs and fiords and in restricted waters such as the Black Sea, it may be considerably higher. Some lake waters, such as Lake Kivu in Africa, are highly charged with methane.

Subsurface Waters. Subsurface waters of petroliferous sedimentary basins are commonly charged with methane. The methane capacity of these deep waters is high because of the high pressures and temperatures, but precise information on degree of saturation is commonly inadequate. A well in Louisiana is reported to have produced 100 barrels of water an hour at a surface temperature of 420°F with a yield of 107 cu ft of methane per barrel (Cochran, 1977). There is high interest in the prospect of commercially recovering methane from the large supply which undoubtedly exists in solution in the subsurface waters of the Gulf of Mexico basin, but practical difficulties are substantial. Several parts of Japan have for many years used solution gas from shallow subsurface waters for lighting and heating.

Sediments Beneath the Oceans. Cores of ocean floor sediments along continental margins and in small restricted ocean basins commonly yield substantial quantities of methane, showing its widespread generation and occurrence in young marine sediments.

Solid Gas Hydrates. As will be discussed later (under Solid Methane-Hydrate Barriers) much methane is believed to be locked up in sediments in the deep oceans and in arctic regions in solid gas-water molecular structures that form under appropriate temperature and pressure conditions.

Digestive Tracts of Animals. An abundance of methane is generated in the digestive tracts of ruminants and other animals, even man, and is released to the atmosphere as a result of their flatulence.

Sewage Sludge. Large amounts of methane are generated from bacterial decomposition of the organic matter in sewage. Projects are under way to use this source of methane commercially.

Combustion of Wood, Coal, etc. It seems probable that methane is generated (but also largely consumed in the reducing flame) whenever wood or coal is burned under restricted-oxygen conditions. Methane is produced industrially by burning coal in the manufacture of coal-gas and coke-oven-gas.

Volcanic Activity. Methane has sometimes been recorded as a constituent of volcanic gases, although in minor quantities. There still is doubt to what extent it may be derived from the magma and to what extent it is generated thermochemically by the effect of volcanic heat on associated sedimentary rocks.

GENESIS OF GEOLOGICAL METHANE

We are concerned here only with "geological methane"—that which has originated in the rocks of the Earth's crust. Although it may be agreed that some minor part of the geological methane has had a high-temperature igneous-related origin (Weisman, 1971), it still can be said with assurance that most of it was derived from the relatively low-temperature decomposition of organic matter in sedimentary rocks, either through biochemical or thermochemical processes. If geological methane is largely derived from organic matter, it is then important to consider what rocks carry important quantities of organic matter.

Organic Matter in Sedimentary Rocks

Organic matter—the remains of once-living plants and animals—forms some part of almost all sedimentary rocks. The organic content may vary from almost 100 percent of the rock (in the case of some coals) to a mere trace in some chemically deposited sediments, but almost all processes and environments of sedimentation result in the inclusion of some organic matter.

Degens calculated (1965, p. 203) that the total organic matter in sedimentary rocks of the world runs to nearly 4 quadrillion metric tons. Surprisingly, less than one percent of this is accounted for by the coal deposits of the world. Instead, the great bulk lies in muds, shales, and silty shales, finely disseminated and associated with fine-grained inorganic sediments of both marine and nonmarine origin. Hunt (1972) put the organic carbon of clays and shales as 8,900 times that of reservoired oil.

This organic matter deposited in sediments has been temporarily preserved, but is inevitably on a path toward its own ultimate destruction. And this ultimate destruction involves the generation in the sediments (by biochemical and thermochemical processes) of natural gases, among which methane is preeminent. Organic matter is perishable, but as it perishes it produces gas. Given the immense amount of organic matter in the sediments of the Earth's crust, it is evident that the geologic consequences of this immense subterranean generation of gas must also be immense (Hedberg, 1974).

Biochemically-Generated Methane

The first step in the decomposition of organic matter in marine sediments occurs immediately after deposition, just below the sediment-water contact, and is biochemical (microbial). This is the attack on the organic matter by many kinds of bacteria under the aerobic conditions that commonly prevail at or just below the depositional surface. The principal product here is carbon dioxide, which readily escapes to the atmosphere, or (being readily soluble) to the hydrosphere. In many situations the organic matter is almost completely used up by this kind of bacterial action under oxidizing conditions.

However, if the organic matter is covered rapidly by succeeding deposition, aerobic conditions are rapidly replaced by anaerobic at a depth which may be as little as a few centimeters, and may even prevail right at the surface of deposition in the case of stagnant restricted waters and marsh environments. Under anaerobic conditions the aerobic bacteria can no longer survive but are replaced by anaerobic types that use sulfate, carbon dioxide, or other oxygen-bearing compounds rather than oxygen as an energy source.

In the upper part of the anaerobic domain, a principal bacterial activity is the reduction of sulfate with generation of carbon dioxide and hydrogen sulfide (sulfate-reducing zone of Fig. 1). Nitrates are also reduced, with the evolution of nitrogen. Deeper down, however, where the sulfate has already been removed, bacterial methane production becomes dominant, largely by consuming the carbon dioxide and hydrogen already present in the waters and being continually renewed by the action of other anaerobic bacteria on the organic matter itself (methane-producing zone of Fig. 1).

Thus, as pictured by Claypool and Kaplan (1974) and shown on Figure 1, there is a succession of biochemical zones in the open-marine, organic-rich sedimentary environment, from top downwards: (1) an aerobic oxidizing zone at and just below the surface of deposition, which produces carbon dioxide; (2) an

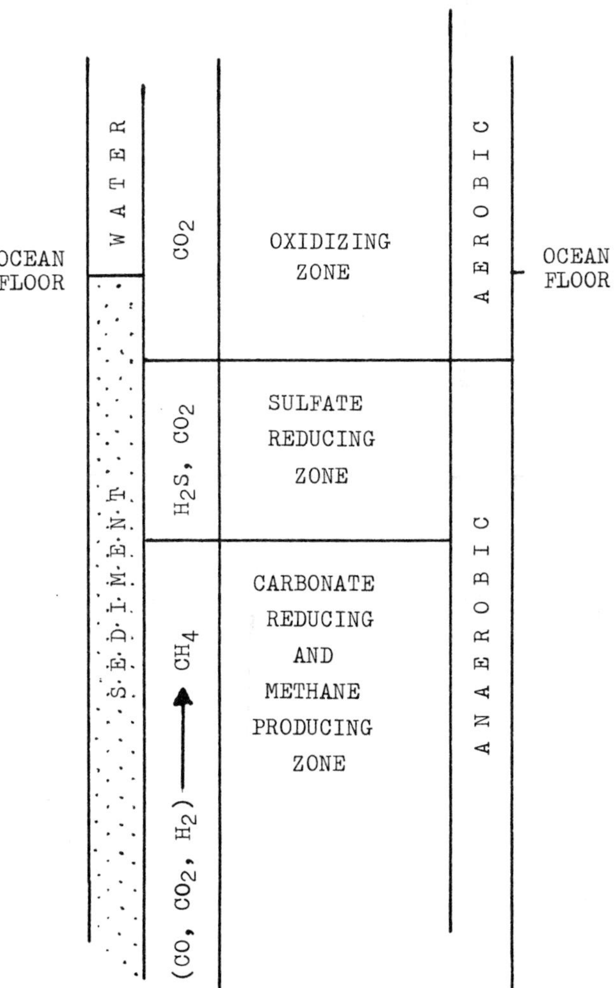

FIG. 1—Biogeochemical zones just below the ocean floor. (Slightly modified from Claypool and Kaplan, 1974).

anaerobic sulfate-reducing zone, commonly generating carbon dioxide and hydrogen sulfide; and (3) an anaerobic methane-producing zone, extending on down to the lower limits of bacterial action in the sediments.

Eventually, with depth, bacterial action of any sort ceases, due either to rising temperature, increased compactness of the rock, build-up of toxic products, or exhaustion of the supply of organic matter. What this limiting depth is, remains controversial; and it is probably quite variable. In any case, we can expect biogenic methane to be found to depths of thousands of meters, due to subsequent deep burial of originally near-surface generations in younger sediments.

That the quantity of biogenic methane thus generated and stored in the rocks is great, is demonstrated by the large and growing number of commercial gas fields now recognized as producing biogenic methane, because of a relatively high ratio of ^{12}C to ^{13}C (for examples, see Rice, 1975).

Thermochemically Generated Methane

A second mode of "geological methane" generation is through thermochemical or thermocatalytic action—the spontaneous chemical decomposition of organic matter under the influence of rising temperature. This process, probably already in operation at the sediment surface although at a scarcely detectable rate, is strongly temperature dependent so that it becomes increasingly prominent with depth of burial. Thus, thermochemical methane production overlaps bacterial methane production and eventually supersedes it: biochemical action diminishing with depth and thermochemical action increasing with depth. Conditions and reactions involved in thermochemical production of methane from kerogen are discussed by Erdman (1967), McIver (1967), and others.

A rough diagrammatic relationship of the generation of liquid and gaseous hydrocarbons to temperature and depth of burial is shown on Figure 2 (from Hedberg, 1974). This is a modification and amplification of a well-known diagram by Sokolov (1967). The original Sokolov diagram (reproduced from Sokolov, 1968) is shown for comparison as an inset in the right corner of Figure 2.

Several other published modifications of the Sokolov diagram have been reproduced and compared by Vassoevich (1975, Fig. 1). Most of these have indicated a much lower relative production of methane in the biochemical stage, and in the early thermochemical oil-generating stage, than is shown by my curve. However, increasing records of extensive accumulations of biogenic methane seem to me to fully justify the emphasis given to it in Figure 2. Moreover, the importance shown for thermochemically generated methane during the early oil-generating stage is amply supported by (1) the almost invariable presence of substantial quantities of methane dissolved in oil, and

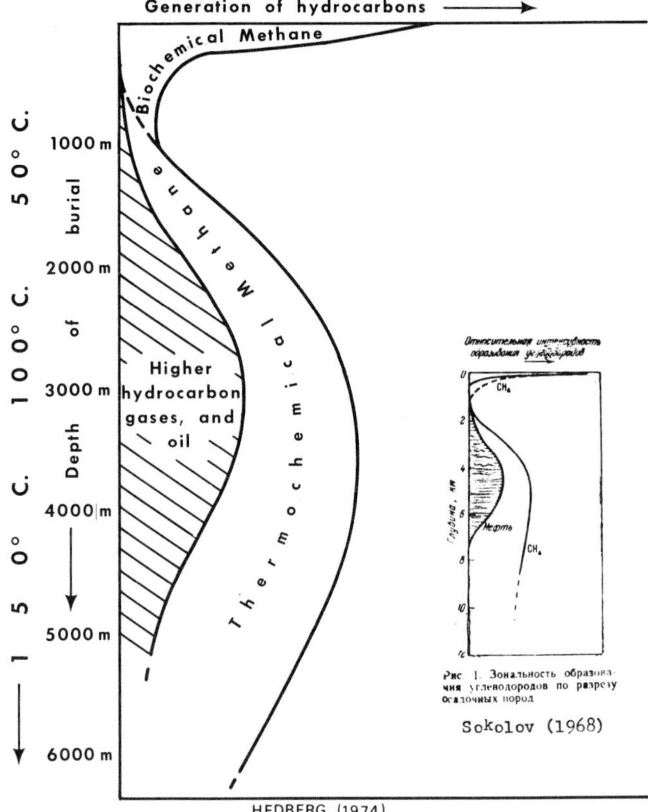

FIG. 2—Relation of generation of liquid and gaseous hydrocarbons to depth of burial and temperature (Hedberg, 1974; modified from Sokolov, 1968). The original Sokolov diagram is shown as an inset at the right of the figure.

(2) the frequency of gas caps, gas-only accumulations, and high GOR reservoirs, interspersed among oil accumulations in the upper parts of oil-bearing sequences. (Note also the emphasis given by Stroganov [1974] on relatively early generation of methane.)

In my opinion, geologic evidence supports abundant generation of methane throughout the whole thermal range of oil production. For rocks carrying certain types of organic matter, hydrocarbon generation through this temperature span may even be almost entirely methane. As indicated on Figure 2, at temperatures above about 150°C regardless of the type of organic matter, hydrocarbon generation is completely dominated by methane. It seems to the writer that laboratory pyrolysis experiments which have attempted to closely simulate nature by gradual low-temperature heating (e.g., Hoering and Abelson, 1963), have tended to show a more substantial early thermochemical genesis of methane from organic matter than have higher temperature, more rapid, and accordingly less realistic, laboratory experiments.

Methane is, by no means, the only gaseous product of thermochemical decomposition of organic matter. Carbon dioxide (and monoxide), nitrogen, hydrogen, hydrogen sulfide, and higher hydrocarbon gases may be formed in substantial volumes at various points along the way, depending on local conditions.

ROLE OF METHANE GENERATION IN PROMOTING PETROLEUM MIGRATION

The general conditions of genesis of petroleum (both liquid and gaseous) appear fairly well understood, and the dominating role of thermochemical decomposition of organic matter has been accepted for more than a century.

However, the manner of *migration* of petroleum—how it has moved out of a dense nearly impermeable organic-rich shale or carbonate mother-rock to get to reservoir strata and eventually to form commercial petroleum accumulations—has been a much more difficult problem and is still highly controversial.

In order to attain a temperature sufficient to stimulate substantial thermochemical conversion of organic matter to petroleum, an organic-rich mud must be buried under such a thickness of overburden as to compact it to a rock so dense as to have little or no effective permeability. It thus becomes a problem to reconcile conditions favorable to petroleum genesis with conditions which will allow the generated petroleum (or its precursors) to migrate from the source rock to the reservoir rock to form commercial petroleum accumulations.

The outstanding contribution of methane generation to petroleum migration results from the fact that methane generation is a major source of internal energy and pore pressure within sedimentary rocks.

Migration of fluids within the rock column requires energy. Some of this energy may come through weight of overburden, some through density differential of fluids, some through water expansion with increasing temperature, some through osmosis, and some from other causes. But, whatever else, the generation of methane gas is an inevitable and potent source of internal energy in any organic-rich sediment, as it attains progressively higher temperatures with increasing depth of burial (Hedberg, 1974, p. 666-668).

Figure 3 is an attempt to depict roughly by numbered stages the probable development of various factors affecting the origin and migration of petroleum in a common type of source sediment—for example, an interval of organic-rich laminated silty clay-mud—deposited in a subsiding basin of active marine sedimentation.

Stages 1 and 2 involve largely bacterial action on organic matter, generating primarily carbon dioxide under surface and near-surface oxygenated conditions, but later, *methane*, under the anaerobic conditions attained in Stage 2. At the relatively low pressures prevailing near the depositional surface only a very limited amount of this early methane can go into solution. However, both in solution and as free gas bubbles, it escapes readily to the surface or to interbedded aquifers, through the porous and relatively permeable, abundantly water-bearing, unconsolidated sediment of these stages. Faas and Wartel (1977) concluded, "The effect of methanogenesis on previously deposited sediment is to disrupt the developed fabric through gas bubble generation and swelling and to cause weakening of the sediment."

Stage 3 (Fig. 3) sees the gradual cessation of microbial activity and its gradual replacement by thermochemical generation, principally of methane, but also carbon dioxide and small amounts of higher hydrocarbons from suitable constituents of the organic matter. Increasing compaction under the weight of increasing overburden has now formed a silty shale from the originally silty clay-mud, with a great reduction of porosity (water content) and permeability. However, both methane and other gases, now more soluble under increasing pressure, are still largely taken into solution and expelled with the outward fluid flow resulting from compaction.

Stage 4 (Fig. 3) marks the onset of substantial generation of hydrocarbons of the liquid petroleum range, at temperatures now of 50°C, or more. However, this generation of oil, or its precursors, is still commonly accompanied by unabated generation of methane. With a diminishing volume of pore water, due to compaction, even with increasing solubility, the amount of both liquid and gaseous hydrocarbons that can go into water solution begins to be exceeded and tiny bubbles and globules begin to appear.

In Stage 5 (Fig. 3) a temperature has been attained at which the formation of liquid petroleum (oil), or the precursors of liquid petroleum, reaches a peak. Methane generation now has reached the point where solubility in available water (or oil) is far exceeded, even at the now increased pore-water pressure; and continued generation tends to form an abundance of minute coalescing gas bubbles in the pore spaces.

These bubbles cause greatly increased *internal* fluid pressure which is added to the external pressure from the weight of overburden, and in spite of the increasing solubility of hydrocarbons, tends strongly toward the expulsion of fluids from the rock. However, at the same time, acting *against* the strong tendency of the fluids to escape, is the ever-diminishing matrix permeability of the shale, as compaction squeezes the platy clay particles and the organic matter more closely together and reduces the size of the pore openings.

Morever, as illustrated in Figure 4, a further very effective inhibiting factor to fluid movement now comes into play. With the formation of gas bubbles, there are now *two* phases, liquid and gaseous, in the reduced pore spaces. The resulting so-called Jamin effect tends to destroy the little remaining permeability of the rock. These gas bubbles cannot move through the increasingly narrow pore constrictions except by distortion of their spherical form (Fig. 4)—a distortion

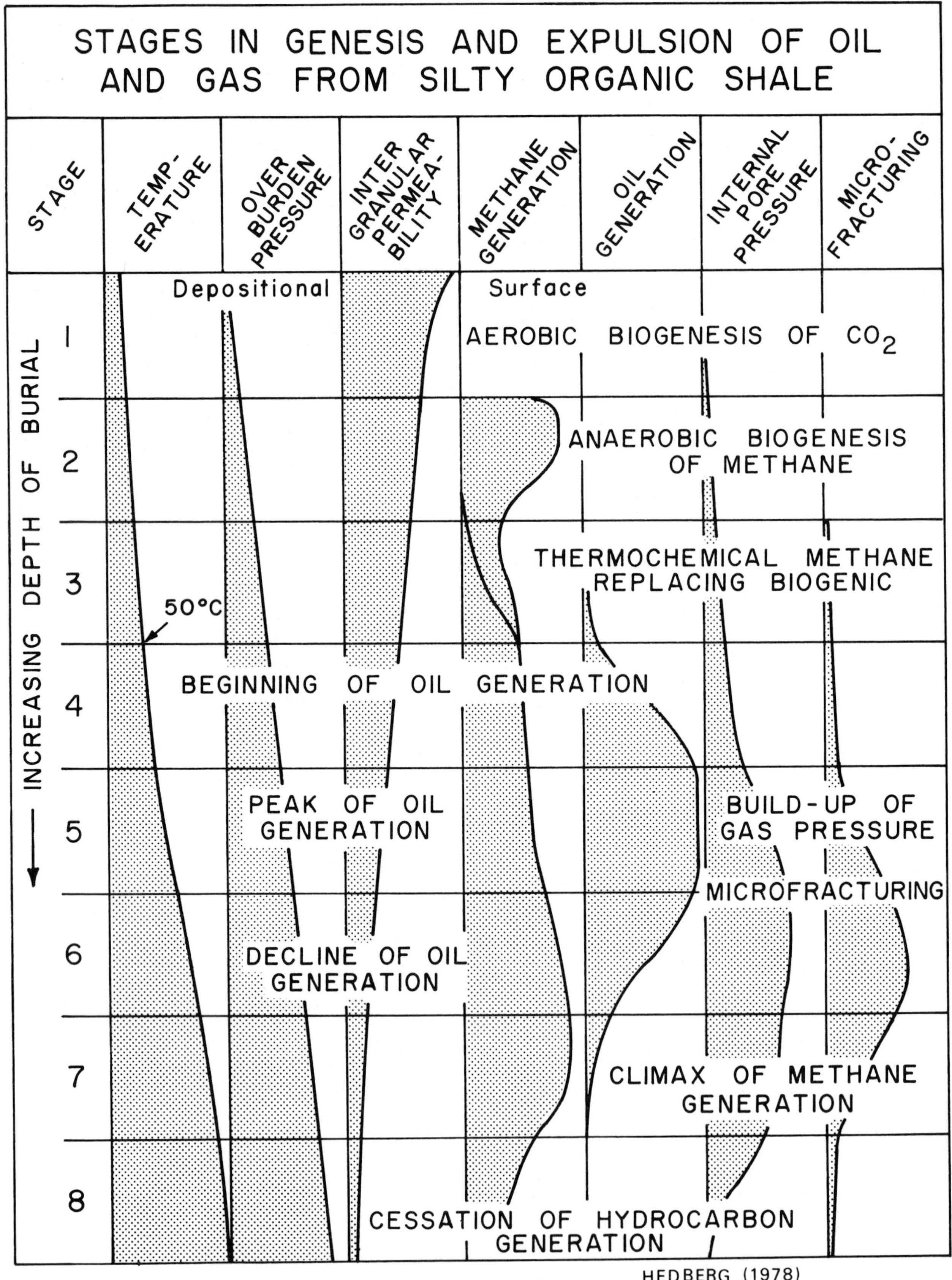

FIG. 3—Stages in genesis and expulsion of oil and gas from a silty organic-rich mud or shale.

that is strongly resisted by interfacial forces and thus results in a blocking of gas escape and in a build-up of internal pressure[3].

Eventually something has to give somewhere; and, as the fluid pressure rises to equal or exceed the weight of rock overburden, microfractures begin to develop in the now well-consolidated rock. This allows relief of pressure by escape of fluids to strata of higher permeability such as silt laminae or interbedded sand layers.

Dickey (1972, p. 14) recognized the possibility that interstitial water in deeply buried sediments may be under sufficient pressure "to burst the rock, and the water can therefore move out through fissures which it forms itself." It seems probable that such disruptive fluid pressure may be due, to a large extent, to gas generation within the sediment.

Movement of hydrocarbons at this stage, through a combination of intergranular pore connections and developing microfractures, may be still to a considerable extent in water solution. Coincidentally, this stage would probably span the main periods of clay-mineral dewatering and thermal aquapressuring, for whatever significance these may have.

The montmorillonite (smectite)-illite transformation, with consequent release of bound water, has frequently been invoked to explain the overpressuring of shales and also as an important factor in the movement of petroleum through source rocks (Powers, 1967; Burst, 1969). However, it appears improbable that this change will cause any significant difference in fluid pressure. Certainly the clay-mineral change cannot provide any general answer to petroleum migration, because carbonates and many source-rock shales lack montmorillonite.

Likewise, aquathermal pressuring (Barker, 1972) is said to be a cause of overpressured shales. The effect of this process may be significant at certain stages but the increase in water volume must be only so gradually translated into increased pressure, in any but a completely closed system (which a compacting shale is not), and distributed through so long a vertical sequence of strata, that it is difficult to see how it can be more than a modest contributing factor to shale overpressuring and petroleum migration.

Where the source rock is sufficiently rich in kerogen of a suitable type, oil or oil precursors might now begin to move out of the rock by some such mechanisms of continuous flow as proposed by Dickey (1975). Moreover, and perhaps particularly important, the opening of the microfractures would begin to allow the escape of previously trapped and now coalesced bubbles of gas, rich in dissolved higher hydrocarbons.

One of the first published papers to propose a mechanism for escape of fluids from source rocks through natural rupture appears to have been that of A. N. Snarskiy (1961), entitled "Relationship Between Primary Migration and Compaction of Rocks." According to the English translation of this paper (published in 1964), Snarskiy said (p. 364): "As a result of the compaction of oil-source rocks, the elastic deformation of rocks, and the elasticity of oil and water, and also due to the increase in temperature and tectonic forces, the intra-pore pressure increases and becomes considerably higher than the rock presure. This pressure leads to rupture of the rock and extension of the fractures. Oil and gas move along the paths thus formed into porous and permeable strata having a pressure equal to the hydrostatic pressure."

In this same paper, Snarskiy referred to two previous papers of his, published in 1959 in "Problems of the Migration of Oil and Gas Accumulations" (Gostoptekhizdat) which may deal with the same subject but which were not available to me at this writing. A subsequent paper by Snarskiy (1962; English translation, 1967) elaborated somewhat further on the effects of "intra-pore pressure due to elastic deformation of the rock and elasticity of the oil, gas, and water, and also due to an increase in temperature," in "rupturing the rock and opening existing fractures," and the relation of this mechanism to clay-mineral alteration and changes in the adsorption properties of the rock. Snarskiy's papers on microfracturing have received very little citation; the only American references which I have seen were in Cordell (1972) and Hedberg (1974)[4].

Sokolov et al (1964) discussed the migration of liquid-range petroleum hydrocarbons dissolved in compressed gas, citing publications of experimental results by Zhuse and Ushkevich (1959), Sokolova et al (1954), and others. They also emphasize the importance of microfractures in the "filtration" of gas and oil through shales.

Tissot and Pelet (1971, p. 41) invoked both (1) the development of internal pore pressure through gas generation, and (2) consequent microfissurization of shale source rock, as necessary to petroleum migration. They referred to Snarskiy's (1961) conclusions on the importance of microfractures in migration and also to experiments of their own showing that internal gas generation in a shale can cause microfissurization. They stated: "The pressure within the fluids formed in the pores of the source rock increases constantly in proportion as the products of the evolution of the kerogen are formed. If this pressure comes to exceed the mechanical resistance of the rock, microfissures will be produced which are many orders of size greater than the natural channels of the rock, and will permit the escape of an oil or gas phase, until the pressure has

[3]Muskat (1949, 296-302) discussed the Jamin effect and its impeding action on fluid flow, but suggested that, "the multiplicity of lateral pore interconnections in actual rocks will permit continued flow of the liquid phase even though the free-gas phase may be locked in place." However, he was speaking of reservoir sands, quite a different situation from that imposed by the already tight pore-structure of a compacted shale.

[4]J. W. Momper (1978) in an AAPG Course Note distributed at the same Oklahoma City symposium at which my paper was presented orally, has given recognition to Snarskiy's contribution. However, Momper listed only one literature reference: Snarskiy, A. N., 1970, The nature of primary oil migration: Neft Gag 13, no. 8, p. 11-15 (translated by Assoc. Technical Service, Inc.), which I have been unable to identify.

fallen below the threshold which allows the fissures to be filled and a new cycle to commence."

Mirchink et al (1971) stressed the role of compressed gases in oil migration, referring to Sokolov (1948). They stated (p. 13) that recent experiments (Zhuse and Safronova, 1967) "have shown that the compressed gases, when passing through stiff clay and other rock, extract those liquids which are soluble in gas. This extraction of the liquid oil-type (oily) hydrocarbons is one of the primary migration factors." They also cited the intensity of oil and gas generation as a factor in migration which causes "microbreaks" in clays and also predetermines the dissolution of liquid hydrocarbons in compressed gas.

Cordell (1972) reviewed both microfracturing and migration in gas solution. Hedberg (1974) proposed that methane gas generation could be an internal source of pore pressure responsible for the phenomenon of overpressured shales and for eventual microfracturing of these rocks.

The writer recently has seen unpublished notes of Grover Schrayer (Research Laboratory, Gulf Oil Corp.) dated July 6, 1965, which show that Schrayer envisioned much the same processes saying: "The gaseous products formed by coalification-like maturation of kerogen in petroleum-source rocks may produce sufficient pressure both to escape from the source rocks and to simultaneously expel liquid petroleum components . . .In addition to providing a driving pressure, gas formation during maturation would facilitate the loss of oil from the source rock by dissolving to an appreciable extent in the liquid petroleum components, thereby lowering their viscosity and increasing their volume . . .If light petroleum components are present in the rock, these could dissolve in the gas phase to a certain extent and leave the source rock with the gas, thus yielding accumulations of wet gas, and perhaps, condensate."

Fine-scale fracturing has frequently been observed in petroleum reservoir rocks where it is generally attributed to tectonic forces, shrinkage, or weathering. It has less commonly been identified in source rocks, although this may be only due to their greater healing capacity. In neither case has much incontrovertible evidence been found of such fracturing being due to internal fluid pressure. However, at the same time it is well known that such fractures can readily be produced by artificial hydraulic fracturing with pressures approaching or only slightly in excess of hydrostatic. Bredehoeft et al (1976) showed that in a tectonically relaxed area, vertical fractures can be induced at hydraulic pressures as low as 0.6 of the overburden stress. Secor (1965) showed that in theory even hydrostatic fluid pressure will allow fracturing to several thousand feet and that "as the ratio of fluid pressure to overburden weight approaches one, the maximum depth to which fracturing can occur becomes very great, even for exceedingly weak rocks." Currie (1977) recently commented on work by a number of investigators on the relation of changes in pore pressure to fracturing.

Pending more conclusive direct evidence of the existence of effective gas-generated microfractures in source rocks, one can only for the present cite as support (in addition to such experimental evidence as has been given): (1) the circumstantial evidence of the ubiquity in sediments of methane-producing organic matter; (2) the huge internal pressure potential in methane generation; (3) the actual demonstration of this force in overpressured shales, shale diapirs, and mud volcanoes; (4) the known large-scale disruptive effects of these phenomena on sedimentary strata; and (5) the consequent probability that substantial fine-scale microfracturing must also have occurred.

Stage 6 (Fig. 3) shows continued strong build-up of gas pressure. The rate of gas generation has greatly increased in response to rising temperatures while intergranular pore space continues to become more and more restricted. Oil generation relative to gas is already declining. During this stage there might be repeated cycles of pressure build-up, microfracturing, and fluid migration, through a progressively developing network of microfractures leading toward carrier horizons of lower pressure potential. Microfractures might periodically open and heal depending on internal pore pressure. Gas, condensate, and high GOR oil accumulations would commonly be formed as a result of migration during this stage.

Stage 7 (Fig. 3) represents relatively high temperature conditions at which generation of oil-forming hydrocarbons has been largely completed, and the high-hydrogen part of the organic matter has been exhausted. In this stage, methane may be generated not only from such remaining organic matter as still possesses adequate hydrogen, but also from previously generated oil itself. The rock has now attained a degree of induration where remaining water is almost entirely *bound* water and intergranular pores are almost totally closed. Hence the escape of fluid from the interval, other than very slowly by molecular diffusion, is almost wholly as free gas through microfractures, and the resulting reservoir accumulations are largely dry methane gas.

An ultimate stage—Stage 8 (Fig. 3)—after exhaustion of hydrogen sources even sufficient only to make methane, theoretically terminates in the development of a carbon- or graphite-bearing slate or argillite.

This, then, is a rough outline of what, it seems to me, may be the role of methane in promoting petroleum migration—principally through the building up of internal pressure in the sediment, which, in spite of increasing solution capacity, acts in the direction of expelling fluids, and eventually creating microfracture passageways for fluids. Although I have used shale source rocks here as an example, essentially the same important role in promoting migration could be played by methane generation in the case of carbonate rocks, modified somewhat by their different pore structure and their different response to overburden compaction.

However, methane generation is only one of many factors which may be involved in petroleum migration. Others include compaction pressure, clay-mineral changes, aquathermal pressuring, osmosis, molecular diffusion, etc.—each of which may be more or less im-

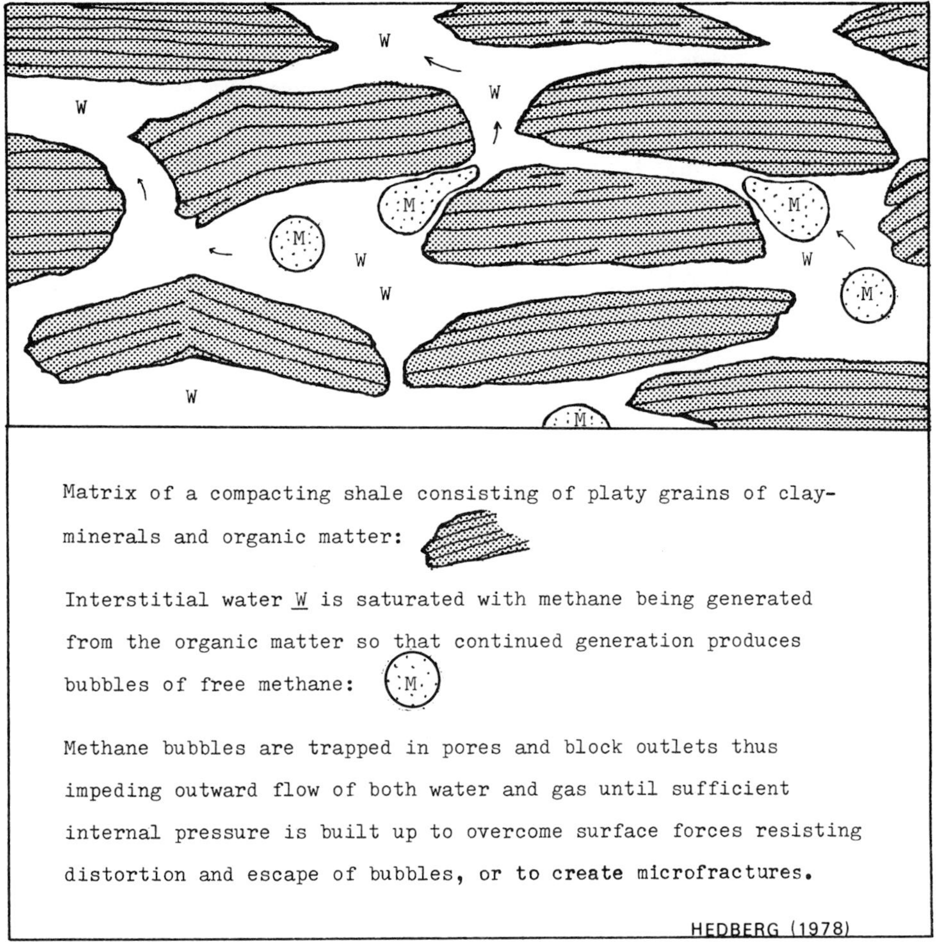

FIG. 4—Methane bubbles impeding expulsion of fluid from pores of shale.

portant depending on the character of the source rock and the stage of petroleum evolution represented. As with so many other natural phenomena, there is no reason to require that petroleum migration take place only in one certain way under all the variety of natural conditions that may obtain. Consider, for example, the many ways by which water is known to migrate at or near the Earth's surface: as a liquid in surface streams and in ground water percolation, as a vapor in solution in air, as a condensate in falling rain, as a solid in falling snow and floating icebergs, etc.

Particularly intriguing is the thought that the internal pore pressure resulting from methane generation may be responsible for creating microfracture passageways which have facilitated escape from source rocks, not only of methane itself, but also of petroleum in gaseous solution, in water solution, and even as a liquid oil phase. Although there is already considerable theoretical and experimental evidence that such microfracturing can have taken place and although the difficulties of moving oil or gas through the water-soaked, dense matrix of shales in any other way make the microfracture route very attractive, more concrete observational evidence of the existence of such fracture pathways in rocks under natural conditions should be sought. Perhaps the intricate, fine-scale patterns into which shales break up at the weathered outcrop surface are an inheritance from lines of weakness, *microfractures*, along which fluids have moved for relief during periods of internal pressure. These could then subsequently have been closed, as internal pressures relaxed, although still leaving their traces in the rock as lines of weakness.

ROLE OF METHANE GENERATION IN IMPEDING PETROLEUM MIGRATION

Methane generation may play not only a promotional role, but also, under certain circumstances, a distinctly *inhibitive* role with respect to petroleum migration, through (1) the formation of impermeable, solid methane-hydrate barriers, and (2) through the development of likewise impermeable over-pressured shale bodies.

Solid Methane-Hydrate Barriers

It long has been known that certain gases, such as methane, under appropriate low temperatures and high fluid pressures, after reaching solubility saturation in water, will enter into a crystal structure with

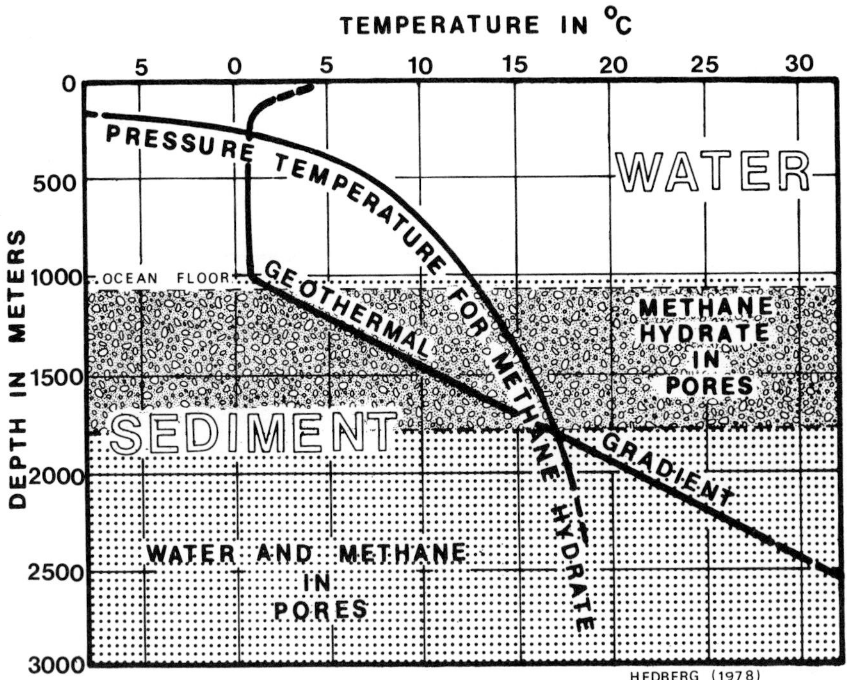

CONDITIONS FOR METHANE–HYDRATE IN SEDIMENTS

FIG. 5—Diagrammatic geologic section in the oceans showing theoretical relation of zone of methane-hydrate formation to pressure (depth), temperature, and geothermal gradient.

water molecules to form an ice-like solid—a solid gas-hydrate—which can exist at temperatures well above the freezing point of water. The fact was discovered by Sir Humphrey Davy in 1810.

The practical application of knowledge of the phenomenon as regards methane was realized only as late as the 1930s when it was discovered that solid methane hydrate was formed in natural gas pipelines under certain conditions and could be an obstacle to the transmission of natural gas. The geologic significance of the discovery came about with the development of Russian gas fields in the Arctic in the 1960s when it was found that methane gas hydrates existed in nature in the subsurface of these fields; and in 1971 when a paper by Stoll, Ewing, and Bryan identified anomalous seismic wave velocities in sediments in the deep oceans as probably due to a zone of solid methane hydrate. Now, with the realization of the wide distribution of methane in organic-rich sediments, and the many established occurrences of methane-hydrates on land in the Arctic, it is only reasonable to presume that there could be substantial volumes of sedimentary rock filled with methane-hydrate beneath deep-water areas of continental margins and in small ocean basins. Many occurrences of gas-hydrates beneath the oceans have already been indicated by seismic work and many have been suspected during DSDP drilling. Some cores of methane hydrate-bearing sediments have been taken successfully on land in the Arctic with a pressure core barrel, but none has yet been obtained beneath the oceans, although efforts are currently being made to develop satisfactory tools and procedures. Attention is called to the English translation of a very comprehensive work on *Hydrates of Natural Gas*, by Yu F. Makogon (1978).

Figure 5 shows a hypothetical geologic section in the oceans with 1,000 m of water overlying 2,000 m of methane-rich sediment. A grid showing *depth* (= pressure), *temperature*, a curve representing the local *geothermal gradient*, and a second curve showing theoretical *P-T relations for methane hydrate*, have been superimposed on the geologic section (The temperature scale applies only to superimposed curves and not to the picture of the sediments.). The pebbled interval on the figure, extending from the lower intersection of the theoretical P-T curve for methane hydrate with the curve for the local geothermal gradient, upwards nearly to the ocean floor, is roughly the zone in which methane-hydrate might be expected, wherever conditions have favored the saturation of the interstitial water with methane.

It is evident that, for the conditions shown on Figure 5, the methane-hydrate zone extends from just below the ocean floor to about 1,800 m below the ocean surface, giving a sediment zone some 700 m thick in

Author's Note: As of January 1980, methane hydrate now has been recovered on DSDP Legs 64 and 66 cruises.

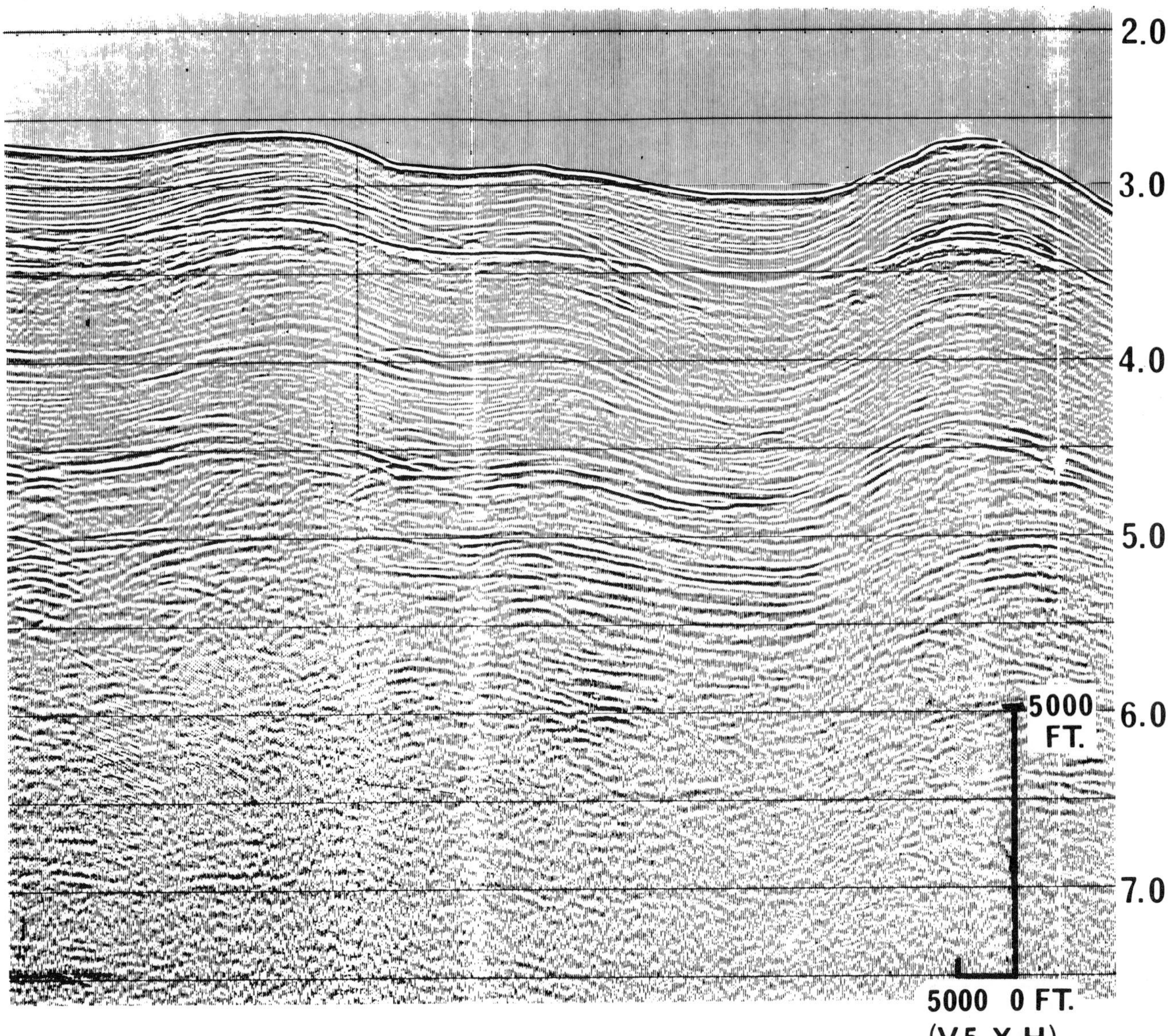

FIG. 6—Seismic reflection profile in Gulf of Mexico showing a clathrate (methane-hydrate) horizon, about half-second below the seafloor, cutting across structural attitude of strata but paralleling the topography of the seafloor (Courtesy of Ed Driver, Gulf Oil Corp.).

which methane-hydrate might be expected.

The importance of methane-hydrate with respect to petroleum migration is that it probably forms an impermeable barrier to vertical fluid cross-flow, situated parallel with the present depositional surface and thus often cutting boldly across the normal layering of lithologic strata. The base of the methane-hydrate zone theoretically could form the ceiling of accumulation traps.

Figure 6 (courtesy of Ed Driver, Gulf Oil Corp.) is a seismic profile in the Gulf of Mexico, with a clathrate (methane-hydrate) horizon about a half-second below the sea floor, cutting the structural attitude of strata but paralleling the topography of the sea floor.

Figure 7 is a sketch of a seismic profile across the Blake Outer Ridge off the U.S. Atlantic coast (slightly modified after Tucholke, Bryan, and Ewing, 1977). It shows a probable methane-hydrate zone (dotted pattern) paralleling the profile of the ocean floor but cutting across the gentle dips of the strata. The possibilities of free gas accumulations at the base of the methane-hydrate zone, sealed by its own impermeable character, are evident.

Overpressured Shale Barriers

Overpressured, undercompacted shales now are commonly recognized worldwide and from many parts of the stratigraphic column (Fig. 8). They occur as stratiform bodies several hundred to thousands of meters thick and many kilometers in extent, and also in the form of diapirs. Overpressures that have been recorded are commonly somewhere in the range between normal hydrostatic pressure for seawater and that for rock overburden pressure, but a number of cases have been reported where fluid pressures were even greater than the weight of overburden.

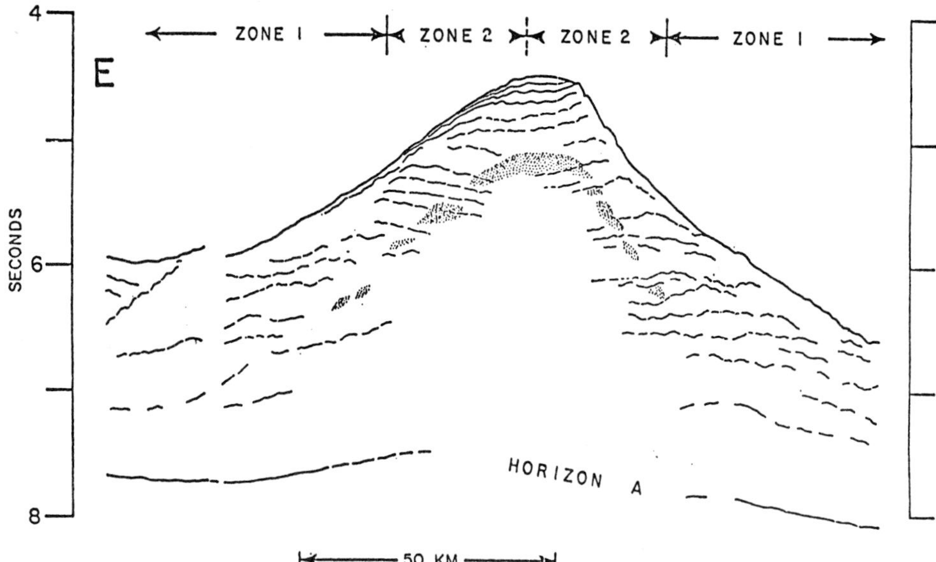

FIG. 7—Sketch of a seismic reflection profile across the Blake Outer Ridge off the U.S. Atlantic Coast (slightly modified from Tucholke, Bryan, and Ewing, 1977). A probable methane-hydrate zone (dotted pattern) parallels the topography of the ocean floor but has an angular relation to the gently dipping strata.

Characteristic features and probable causes of overpressuring of muds and shales have been discussed at length (Hedberg, 1974). Among many factors which may contribute to overpressuring and undercompaction, the importance of methane generation was stressed. It is a source of internal energy responsible not only for overpressured muds and shales but also for the related phenomena of mud and shale diapirs and mud volcanoes

The mechanism involved in the pressuring of muds and shales by methane generation has already been detailed earlier (see also Stage 5, Fig. 3) and also in the writer's 1974 publication. Salient points are: (1) the huge amount of organic matter in muds and shales, (2) the great potential for methane gas generation as this organic matter is decomposed by increasing earth temperature, (3) the difficulty of fluid escape from the fine pores of a compacting mud or shale, (4) the further blocking action of methane gas bubbles in a dense, water-saturated mud or shale matrix, and (5) the inevitable internal pressure buildup that must take place.

The end result of the methane gas pressure, thus developed, may often be the creation of microfractures in the rock, relieving overpressure and at the same time constituting an important factor in promoting petroleum migration. However, before such relief is obtained, methane generation could be an impediment to migration (rather than an aid) through its effects in reducing permeability by addition of a gas phase to the liquid content of the pore space (Jamin effect). This impeding effect might not be very important in the case of a sand sediment, but in a fine-grained, highly plastic shale it would be a major factor in retarding fluid migration until pressure was gradually bled off by diffusion or very gradual leakage or the sediment had reached a consistency that would allow microfracturing.

There is generally, of course, some escape of methane around the periphery of a methane-overpressured shale mass even while its interior stays relatively sealed. This may be due to pressure fracturing or faulting, melting of associated clathrate, diffusion, or simply pressure buildup to the point of intergranular leakage. Figure 9 (courtesy of J. L. Worzel of the University of Texas Marine Science Institute) is an echogram showing the acoustic detection of methane bubbles or plumes escaping to the water surface of the Gulf of Mexico above a section containing overpressured shales. It is noteworthy that the well-known Mississippi mud lumps in the Gulf of Mexico persistently leak methane gas (Shaw, 1917). The association of high-pressure, dominantly methane gas with melanges and mud diapirs on the Washington (state) coast has been described by Rau and Crocock (1974).

Not only methane gas but also some shows of liquid petroleum are generally found in the proximity of overpressured shale bodies. However, large oil accumulations in this association seem not as frequent as might at first have been anticipated, although there are important exceptions. The thought comes to mind that one possible reason for this deficiency, if indeed it does exist, is that it may be only the natural result of the "strainer mechanism" postulated by Roberts (1978) as a requisite to oil accumulation. Whereas overpressured shales (and likewise sediments whose pore

PROPERTIES OF UNDERCOMPACTED OVERPRESSURED SHALE

1. Fluid pressure higher than normal calculated hydrostatic pressure.
2. Porosity higher than normally expected from weight of overburden.
3. Density lower than normal density-overburden relation.
4. Velocity of seismic waves lower than normally to be expected.
5. Occasionally strong seismic reflections from boundary of zone but lack of continuous reflections from within.
6. Increased rate of drilling penetration.
7. Increased temperature above normal.
8. Reversal on sonic logs in the normal trend of decreasing transit time with depth.
9. Reversal on electric logs in the normal trend of decreasing conductivity and increasing resistivity with depth.
10. "Trip gas" and small pockets of methane frequently encountered during drilling.

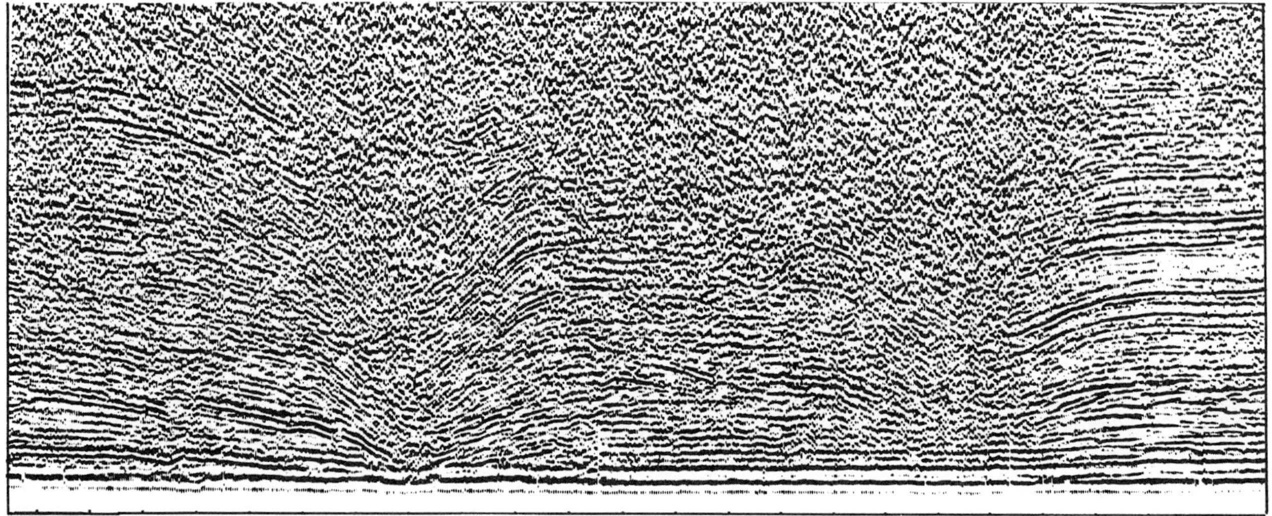

FIG. 8—Some distinctive properties of overpressured shales. Seismic reflection profile at base of figure shows overpressured diapiric shale at right changing laterally to normally pressured strata at left.

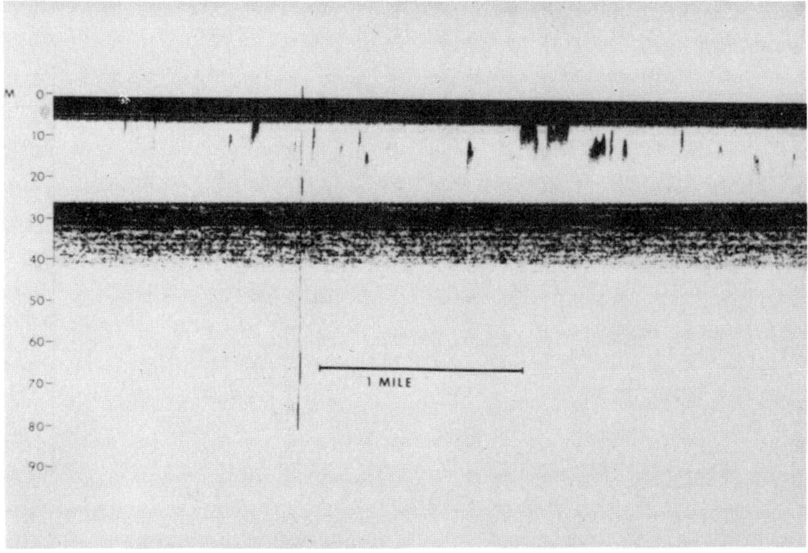

FIG. 9—Echogram in Gulf of Mexico showing methane bubble-plumes escaping to water surface (Courtesy of J. L. Worzel, University of Texas Marine Science Institute). Overpressured shales are common in underlying section.

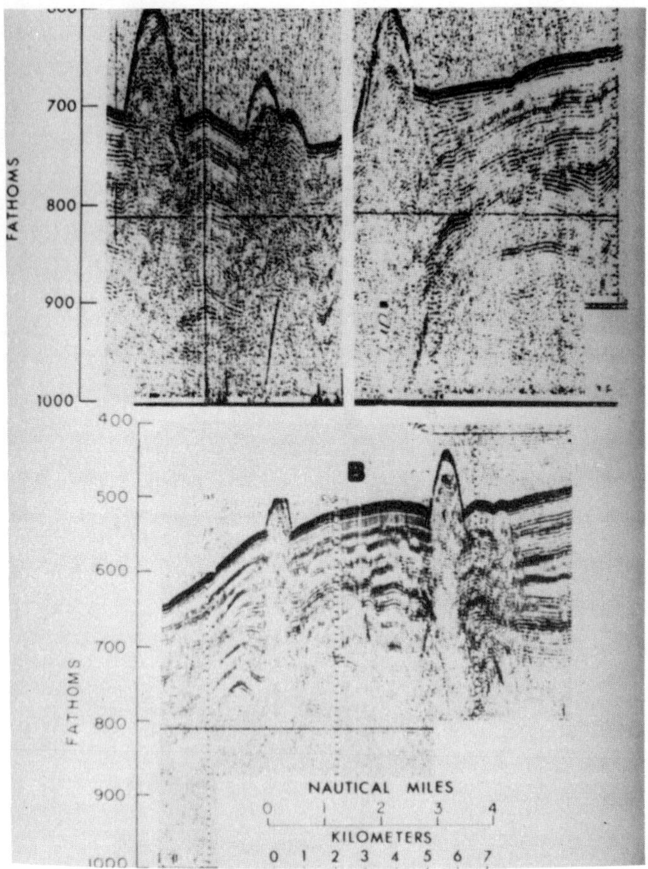

FIG. 10—Echogram showing mud diapirs off the coast of northern Colombia (after F. P. Shepard, 1973). Methane occurrences are common along this coast.

space is filled with methane-hydrate) constitute highly effective barriers to fluid movement and traps for gas accumulations, at the same time under some circumstances, they may have formed too effective a fluid block to permit an adequate "strainer mechanism" to function for the segregation and concentration of liquid oil or its precursors.

OBSERVATIONS OF SHALE DIAPIRS AND MUD VOLCANOES SHOWING MAGNITUDE OF PRESSURES RESULTING FROM METHANE GENERATION

The phenomena of mud and shale diapirs and mud volcanoes, associated with overpressured shales, often provides dramatic evidence of the magnitude of forces resulting from the internal generation of methane gas in muds and shales, which may help to make it convincing to ascribe microfracturing and fluid migration to this cause.

Figure 10 is an echogram (Shepard, 1973) showing mud diapirs off the northern coast of Colombia, pushing up through the sedimentary section to form topography on the sea floor. This coastal area is notorious for methane shows.

Figure 11 is a reflection seismic profile (courtesy of R. V. Brodine of Gulf Oil Corp.) showing numerous mud diapirs in deep water off the petroliferous coast of equatorial West Africa. The thick sedimentary section of the region has been extensively disrupted by these features, and the intricately broken and confused structure within each individual diapir is evident.

Figure 12 is an example of mud or shale diapirs from the Gulf of Mexico (courtesy of Ed Driver of Gulf Oil Corp.). Again, the diapir not only cuts through a thick section of sediments but also is expressed by topography on the seafloor. The internal structure of the diapir is highly broken and confused. The area is again highly petroliferous, with numerous methane shows (Fig. 9), and there is even good evidence in Figure 12 of a methane-hydrate reflection horizon at about 2.75 seconds, cutting across the diapir.

Mud volcanoes represent a final stage in the sequence from undercompacted, overpressured muds and shales, to mud and shale diapirs, to an actual outbreak of the gas and mud at the surface (Hedberg, 1974). Commonly the activity of a mud volcano is simply a mild surface upwelling of mud, water, and methane gas bubbles. However, many instances are known of highly explosive eruptions where large masses of rock have been violently blown out hundreds of feet into the air and scattered widely over the countryside—not exactly what might be expected alone from such phenomena as thermal expansion of water, osmosis, or montmorillonite-illite transformation. The violence of these intermittent eruptions and the huge quantities of flammable gas emitted, also suggest that the motive force is not merely the weight of gradually increasing overburden, or buoyancy adjustment, but is due to the periodic release of internal pressure built up within a shale mass or diapir by the generation of methane gas.

Mud volcanoes are not an isolated phenomenon but are well known from all parts of the world, generally in Cenozoic or late Mesozoic sediments. Hundreds of descriptive papers have been published. Among the more celebrated areas are Colombia, Trinidad, eastern Venezuela, south Texas, eastern Mexico, Peru, Ecuador, the Makran coast of Pakistan and Iran, Indonesia, Timor, Taiwan, Raukumara Peninsula of New Zealand, Arakan coast of Burma, Andaman Islands, Caspian area of Azerbaijan, Kerch-Taman basin near the Black Sea, Rumania, Sakhalin, and Sicily. Not just coincidentally, all of these are areas of deep basins filled with shaly, organic-rich sediments usually of Mesozoic-Cenozoic age, where thermal generation of gaseous hydrocarbons might be expected to continue long after original permeability had been greatly reduced by compaction.

Figure 13 shows two small mud volcano cones surrounded by a large area mud flow in the Volcancitos area south of Cartagena, Colombia. Figure 14

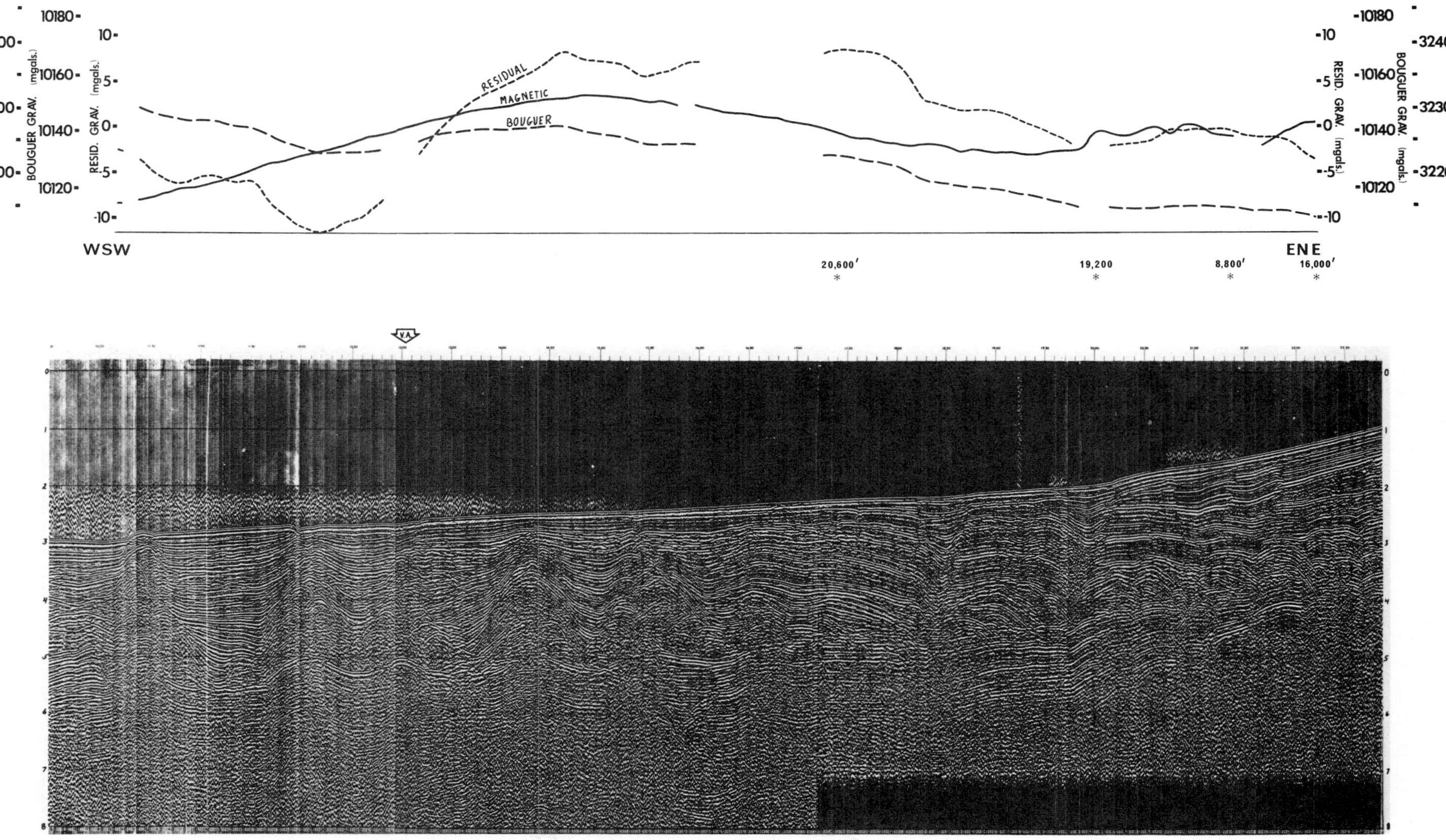

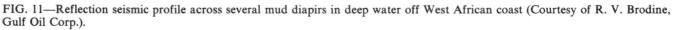

FIG. 11—Reflection seismic profile across several mud diapirs in deep water off West African coast (Courtesy of R. V. Brodine, Gulf Oil Corp.).

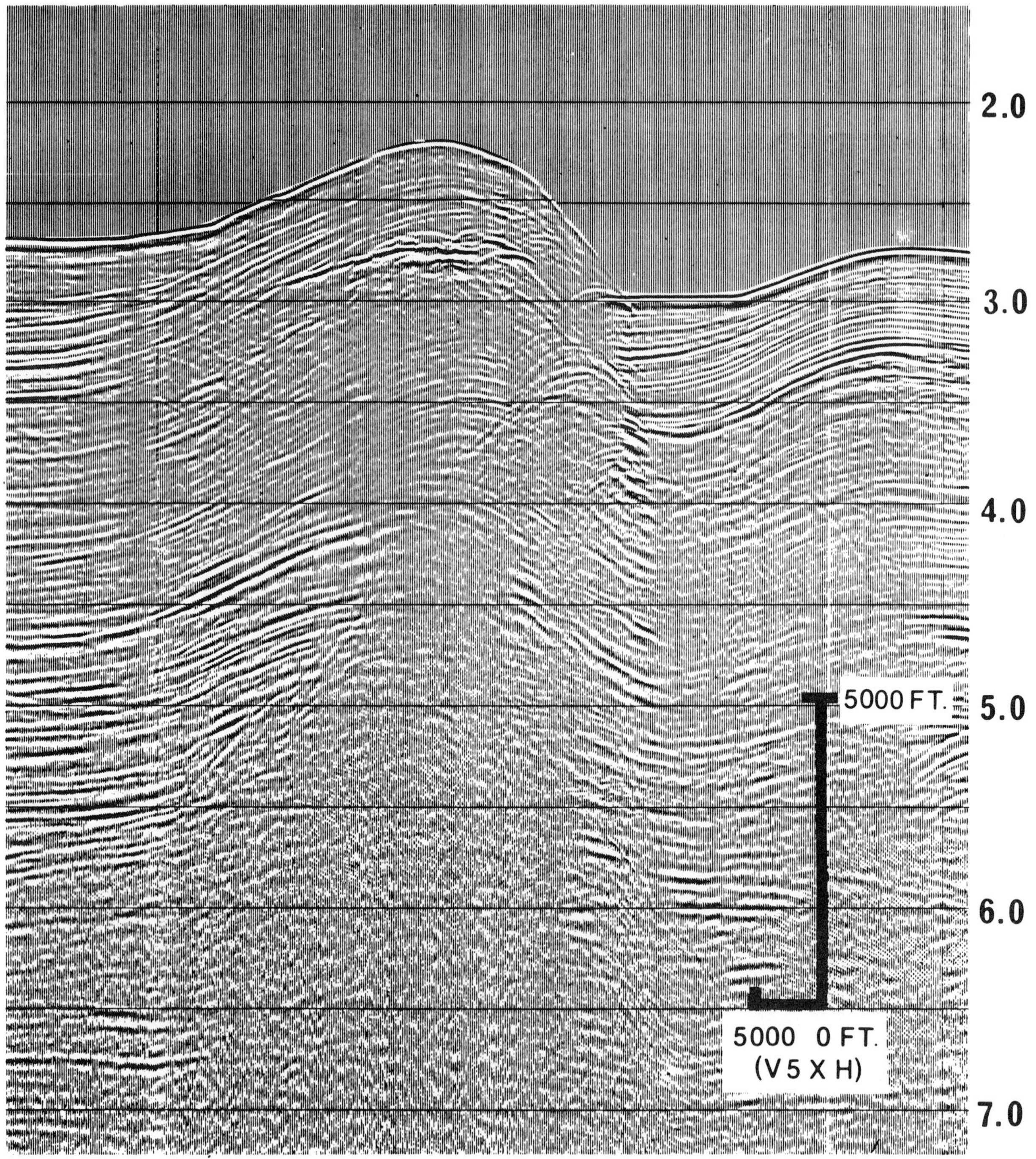

FIG. 12—Seismic reflection profile across mud or shale diapir in Gulf of Mexico. (Courtesy of Ed Driver, Gulf Oil Corp.) Methane-hydrate horizon cuts across diapir at about 2.75 seconds.

FIG. 13—Small mud volcano in Volcancitos area, south of Cartagena, northern Colombia.

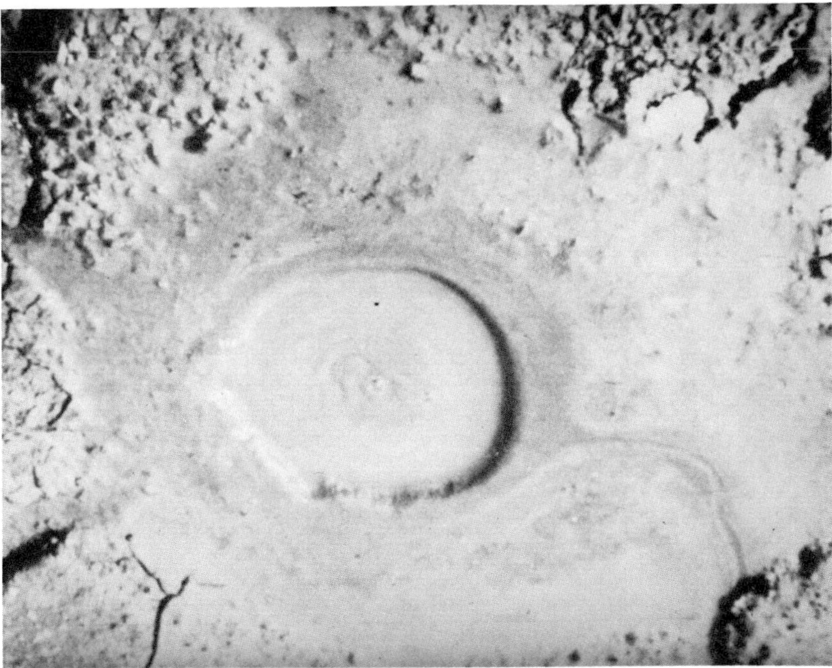

FIG. 14—Gas bubble in crater of small volcano in Turbaco area of south of Cartagena, northern Colombia.

FIG. 15—Totumo mud volcano north of Cartagena, northern coastal Colombia.

FIG. 16—Cluster of mud volcanoes, Moruga Bouff area, southern Trinidad, West Indies (Courtesy of John Saunders).

shows methane gas bubbling up from a crater-like depression in the Turbaco area of this same northern Colombia coastal region.

Figure 15 is a typical mud volcano, known as El Totumo, in the Zamba area of coastal northern Colombia north of Cartagena. This feature is some 20 m high and obviously it took more than hydrostatic pressure and more than weight of overburden to build it. Alexander von Humboldt noted El Totumo when he visited Colombia in 1800. The botanist-geologist, Hermann Karsten, in 1852, described a recent mud volcano eruption along the coast at Galera Zamba, near Totumo, as follows (translated from German):

> This last fire, according to the word of the inhabitants, began on a night in October and burned unceasingly for 11 days, lighting up the whole countryside to a distance of 20 miles and blowing out heated masses of earth which as glowing spheres showered on the neighboring land and far out to sea. After the fire, this part of the peninsula began to sink, and finally two years ago disappeared wholly beneath the surface of the sea, although the position of this former Volcan de Zamba can still be recognized by outpourings of gas bubbles in the sea. (And, incidentally, still can. . . author.)

After reading Karsten's spectacular 1852 account, and visiting the area myself some years ago, it was interesting to see in a recent copy of Geotimes the following news note from this same general region of northern Colombia, under date of October, 1976:

> "Mud eruption and fire from La Lorenza, a vent about 70 km north of Monteria, Colombia; grey mud buried two houses and many farm animals; an hour later methane gas from the vent ignited with a flame 100 m high and burned for several days, destroying houses and trees up to 8 km away."

Perhaps even more interesting is the very recent discovery of the Galerazamba gas field just offshore near the site of the eruption described by Karsten (Franco, 1978, p. 74).

Figure 16 is a cluster of mud volcano cones from near Moruga in southern Trinidad (Higgins and Saunders, 1974), and Figure 17 is a mud volcano in the Moruga Bouff area of Trinidad with the crater rim breached by an extensive mud flow (courtesy of John Saunders). Figure 18 (Higgins and Saunders, 1967) is a closeup and two distant profiles of the spectular Chatham Mud Island off the south coast of Trinidad. This mud island rose out of the sea in 1964 to a height of 25 ft (7.6 m) above sea level and had an area of 10 acres (4 ha.). The uplift was accompanied by blows of methane from several vents. The island disappeared about 8 months later. Several other mud islands of this sort have appeared and disappeared off the south coast of Trinidad in historic times, reflecting a continuing subsurface turmoil due to methane generation, accumulation, and escape. Kugler (1933,p. 13) quoted the following account:

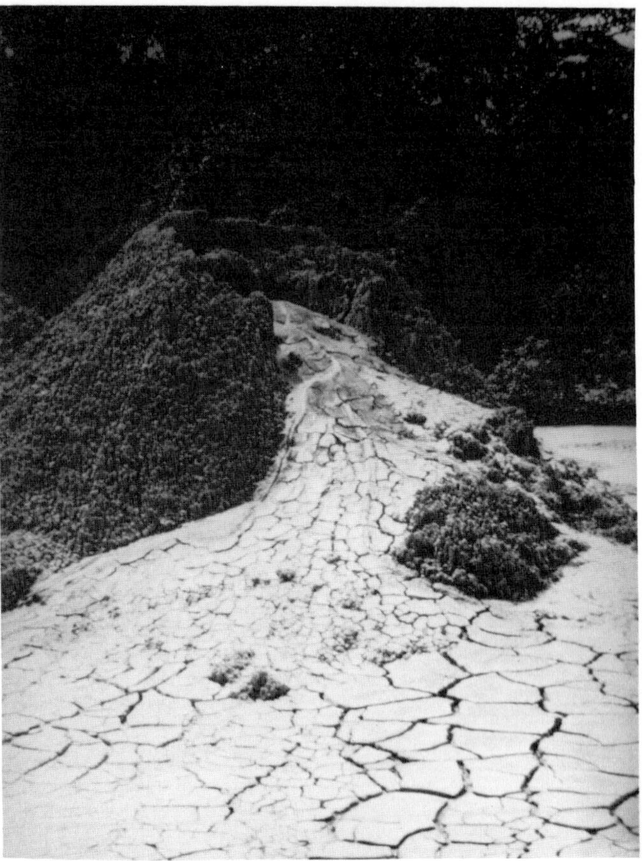

FIG. 17—Mud volcano cone breached by mud flow, Moruga Bouff area, southern Trinidad, West Indies. (Courtesy of John Saunders.)

> "An exceptionally violent eruption occurred off Erin on the southern coast of Trinidad on November 4th, 1911. Masses of material were observed being ejected from the sea, accompanied by enormous volumes of gas which became ignited, and by considerable noise. The flames could be seen many miles away, as they rose to a height of 100 feet or more. An island was noticed to be steadily rising out of the sea as the material increased in quantity, until after several hours of activity it had an area of 2½ acres (1 ha.) and was 14 ft (4.3 m) above the sea level at its highest point . . .Cadman states that the flames have been 300 feet high and burnt for 15½ hours, (and) that the shock of the explosion was felt in Cedros (on the mainland)."

Many phenomena analogous to those of igneous volcanic activity—sills, dikes, mud-flows, etc.—are associated with gas-produced "sedimentary volcanism," a term happily coined by Hans Kugler (1933). Figure 19 is a photograph of a sill and dike of gas-cut mudflow in the La Brea Formation of Trinidad.

Higgins and Saunders (1974), based on well logs, have reproduced a cross section through a fossil gascut mud-flow plug encountered in the drilling of the Grand Ravine oil field in Trinidad. Figure 20 shows the main plug cutting through the upper Tertiary strata, with various sills of gas-cut mud extending from it. The base of the plug as shown on the figure is about 1 km in diameter and the thickness of the section pene-

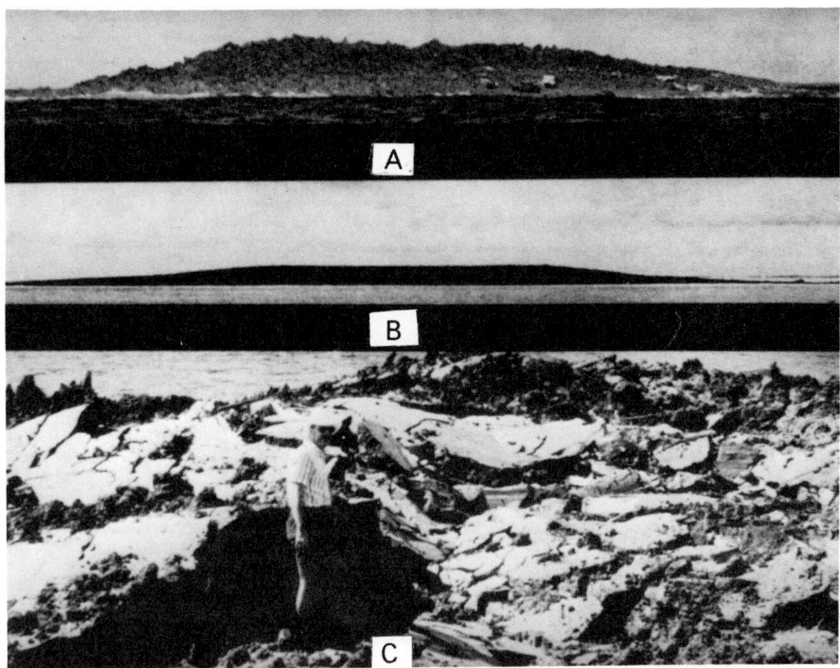

FIG. 18—Chatham Mud Island, off south coast of Trinidad, West Indies. **A.** Moderately close profile of island. **B.** Distant silhouette of island. **C.** Close-up of mudflow on island (After Higgins and Saunders, 1967).

FIG. 19—Sill and dike of gas-cut mudflow injected into La Brea Formation, near Pitch Lake, Trinidad, West Indies (After H. G. Kugler, 1933).

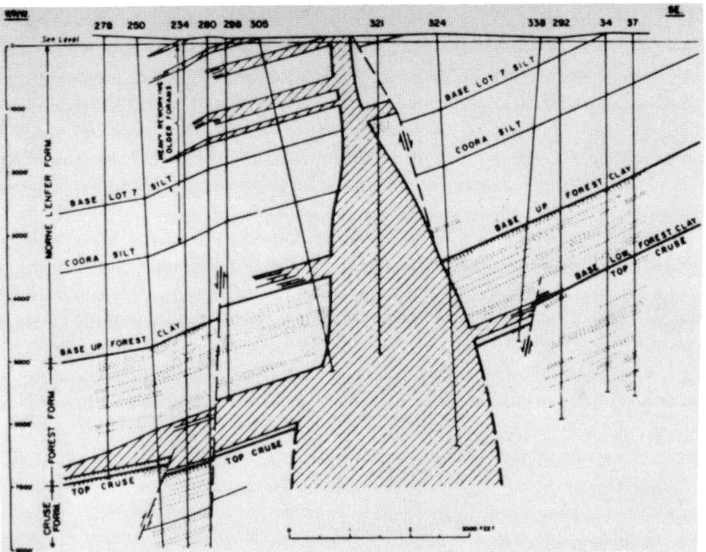

FIG. 20—Cross section from wells in Grand Ravine oil field, Trinidad, West Indies, showing mudflow plug and sills (cross-hatched) intruding middle Tertiary strata (After Higgins and Saunders, 1974).

FIG. 21—Active mud volcano 150 ft (45.7 m) high on Mekran Coast of Pakistan (After Freeman, 1968).

FIG. 22—Napag mud volcano 200 ft (60.9 m) high, Mekran coast of Iran (After Harrison, 1941).

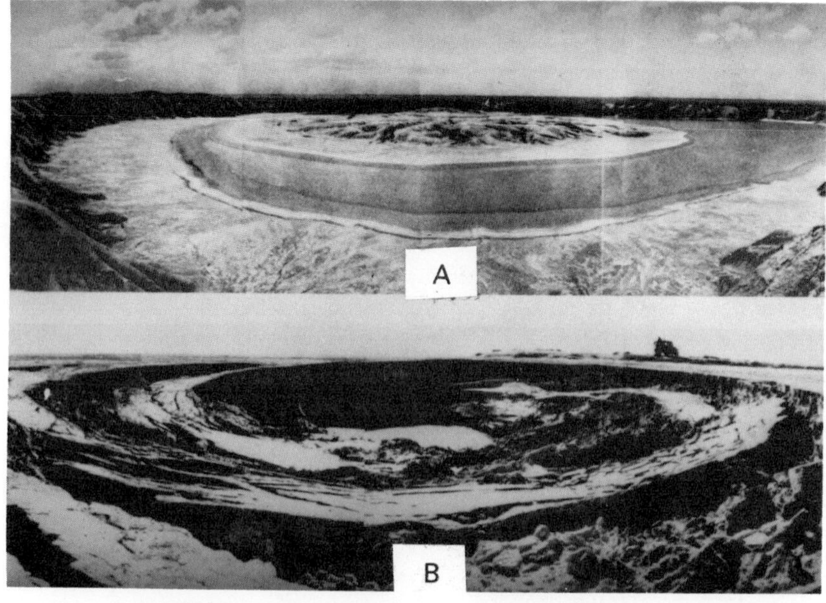

FIG. 23—Mud volcanoes in Gorgan area of northern Iran near the Caspian. **A.** Garniarik Tepe with crater 500 m in diameter. **B.** Naftli Djeh (After Gansser, 1960).

FIG. 24—Waimata mud volcano area, North Island, New Zealand. Large block of Cretaceous limestone in center surrounded by bluish gray diapiric mud. Mud volcano area is surrounded by upper Tertiary sediments and block was presumably brought up from a horizon 5,000 m stratigraphically below.

trated is about 2,500 m. The volume of the central plug is calculated at about 515 million cu m. This is not exactly a microfracture but perhaps is significant of a force sufficient to create microfractures.

Figure 21 (Freeman, 1968) is a photo of a mud volcano along the Mekran Indian Ocean coast of Pakistan, and Figure 22 shows the solitary Napag mud volcano rising 200 ft (60.9 m) above the desert landscape of the Iranian side of the Mekran coast (Harrison, 1941). The Mekran coast has had some drilling exploration for petroleum and is well known for its high pressure shales.

Iran also has mud volcanoes in the north in the Gorgan area of the Caspian. Figure 23 shows (top photo) the Garniarik Tepe mud volcano with a crater 500 m in diameter and a central eruptive high, and (bottom photo) the crater of the Naftli-Djeh mud volcano near the USSR border (Gansser, 1960).

There are several extensive mud volcano areas in the Gisborne region of North Island, New Zealand. In the Waimata area (Fig. 24), a block of dense, Cretaceous limestone some 20 to 30 ft (6.1 to 9.1 m) in dimensions surrounded by diapiric mud appears to have been brought up in the mud diapir from a zone some 5,000 m stratigraphically below that of the upper Tertiary sediments surrounding the present mud volcano orifices. Although the New Zealand mud volcano areas (at the moment) are very quiet and subdued, they have been recurrently very exciting. Thus Ridd (1970) quoted old published reports as follows:

> The eruption of July, 1908, commenced at 9:00 pm and lasted about an hour accompanied by a noise that was audible in the Waimata Valley 2 km away, and was reported by Adams (1908) to have resembled the 'snorting of a huge beast.' The top of the fountain of ejected material was visible from the old Waimata store and, knowing the elevation of an intervening hill, Adams calculated that the debris was hurled to a height of 250-300 feet.

Adams (1908) commented further that many boulders weighing from ½ to 60 pounds (¼ to 27 kg) were found to have been thrown fully 420 ft (128 m) from the vent. (Also see Strong, 1931).

Strong (1933) described the 1933 eruption on the Hangaroa River as "A loud report accompanied by a flash of flame as some of the escaping gas ignited, was immediately followed by the ejection of a large quantity of finely crushed Tertiary mudstone and sandstone. The erupted mass covered an area of approximately 2¼ acres, and has been estimated to be 20,000 tons in weight."

Figure 25 shows impressive mud volcanoes associated with methane in southwestern Taiwan (Lin, 1965).

Probably the most spectular mud volcano region in the world is the Baku region of the Caspian in USSR. Some 220 huge mud volcanoes are known, both on land and in the sea. The Azerbaijan Academy of Sciences published (1971) an outstanding atlas of photographs, a few of which are reproduced here. Figure 26 brings out the huge size of these mud volcano "mountains." Figure 27 is a map taken from the atlas showing the geographic distribution of mud volcanoes in the Baku area and their structural alignment. The distance across the belt of mud volcano trends shown on the map is about 150 km. Figures 28 and 29 are photographs of representative mud volcanoes in the Baku area. The lower picture of Figure 29 is a re-

FIG. 25—Mud volcanoes in southwestern Taiwan (After Lin, 1965).

FIG. 26—Mud volcano mountains of Baku area on the Caspian, USSR (From Atlas of Azerbaijan Academy of Sciences, 1971).

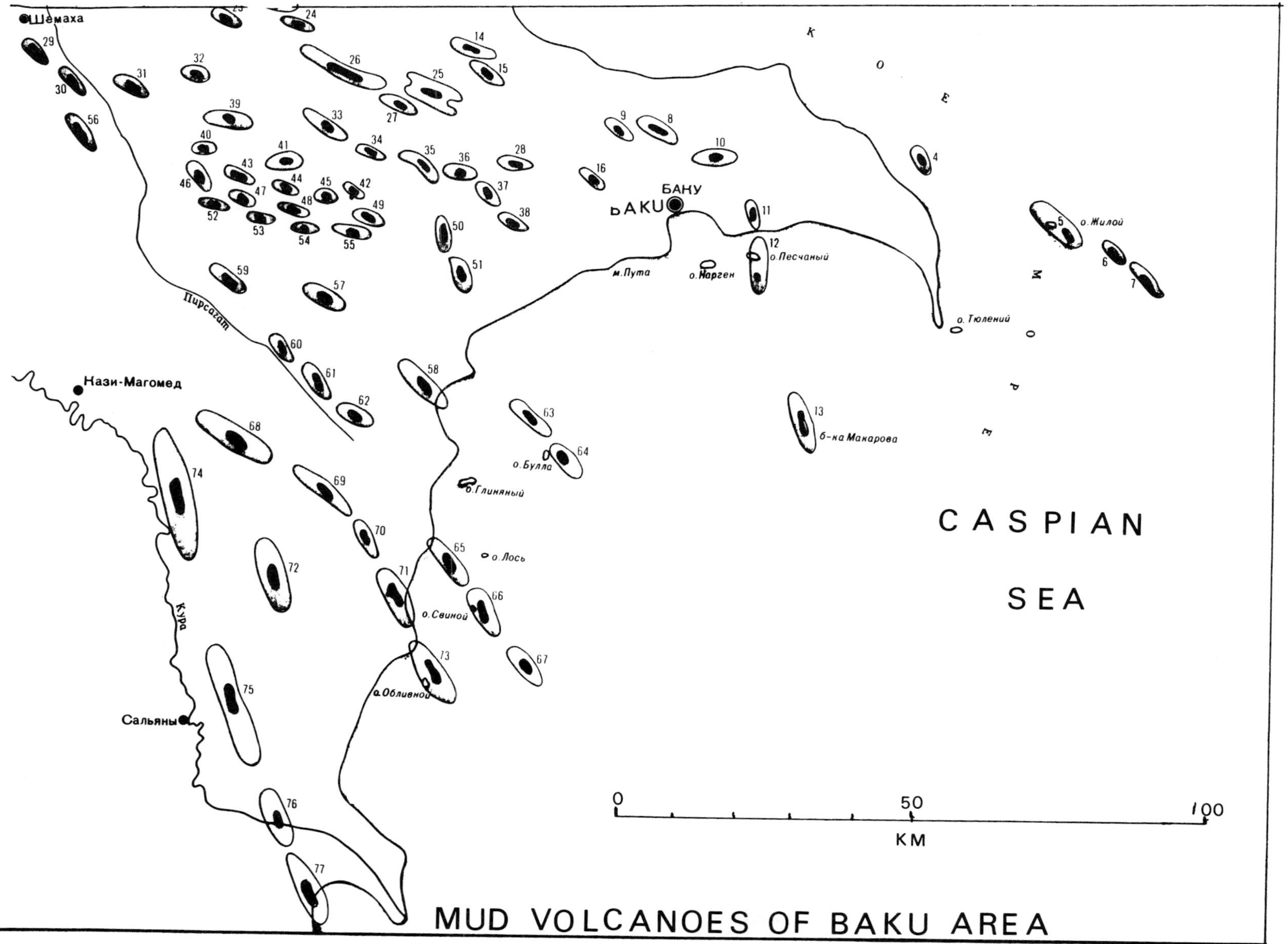

FIG. 27—Sketch map showing distribution of lines of mud volcanoes in Baku area, USSR (From Atlas of Azerbaijan Academy of Sciences, 1971).

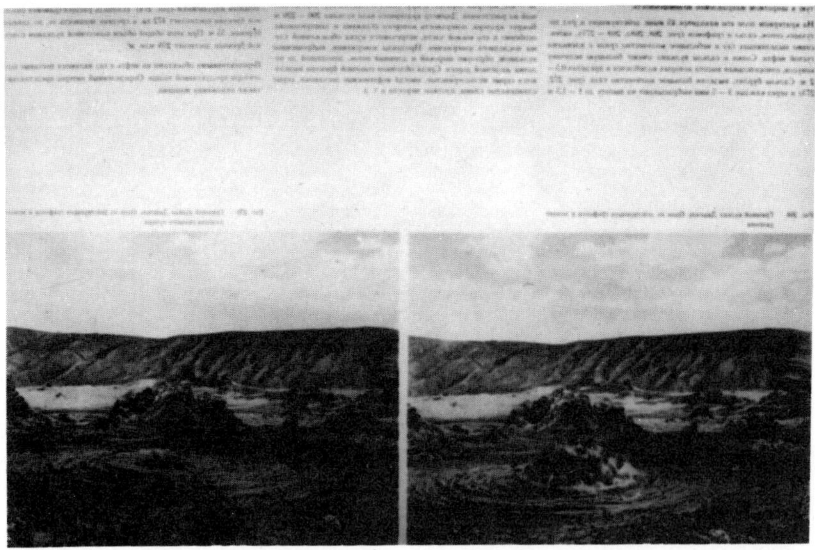

FIG. 28—Active mud volcanoes of Baku area (From Atlas of Azerbaijan Academy of Sciences, 1971).

FIG. 29—Mud volcanoes of Baku area. (From Atlas of Azerbaijan Academy of Sciences, 1971). **A.** Gas-mud bubble. **B.** Violent eruption.

markable shot of an eruption throwing huge rock masses into the air. Sokolov et al (1969) commented on eruptions in the Baku area as follows:

> Eruption of mud volcano on the bank of Makarov, located in the sea at the distance of 20 km from the coast opposite Baku, took place on 15th November, 1958. According to the observers' determinations, the height of the gas column originally thrown out and flaming was estimated at several km. Later the flame had the diameter of 120 m and height of 500 m. The volume of gas expelled by the eruption was estimated. . . as 300 million cubic meters."

And again:

> The eruption of Irantekjan mud volcano took place at night on 6th October, 1965. The observers state that they first heard the underground rumbling, and after that the explosion took place; the column of flame rose to a height of 100 to 150 m over the mountain. The vicinity of the volcano for 10 to 15 km around was brightly lightened. The temperature of the atmosphere rose so much, that it was hard to breathe near the borehole, which was located within 1 km of the center of the eruption. (The authors add an estimate that in the South Caspian region "the total quantity of gases ejected during the Quaternary period from the volcanoes . . . reach approximately 10^{11} tons; sic!)

The foregoing illustrations and word descriptions of only a few of the many mud diapir and mud volcano areas throughout the world may help to give some impression of the magnitude of the power built up in overpressured shales and shale diapirs by methane generation, and its potential rock disruptive force.

REFERENCES CITED

Adams, J. H., 1908, The eruption of the Waimata mud spring: New Zealand Mines Record, v. 12, p. 97-101.

Akademiya Nauk Azerbaijan SSR, 1971, Mud volcanoes of the Azerbaijan, SSR (Atlas): Baku, Akad. Nauk Azerbaijan, SSR, 257 p. (in Russian, with English summary).

Barker, C., 1972, Aquathermal pressuring—role of temperature in development of abnormal-pressure zones: AAPG Bull., v. 56, p. 2068-2071.

Bredehoeft, J. D., et al, 1976, Hydraulic fracturing to determine the regional in situ stress field, Piceance Basin, Colorado: Geol. Soc. America Bull., v. 87, no. 2, p. 250-258.

Burst, J. F., 1969, Diagenesis of Gulf Coast clayey sediments and its possible relation to petroleum migration: AAPG Bull., v. 53, p. 73-93.

Claypool, G. E., and I. R. Kaplan, 1974, The origin and distribution of methane in marine sediments: *in* Natural gases in marine sediments: I. R. Kaplan, ed., Plenum Press, New York, p. 99-139.

Cochran, W., 1977, Quotes sans comment: Geotimes, September 1977, p. 15.

Cordell, R. J., 1972, Depths of oil origin and primary migration: a review and critique: AAPG Bull., v. 56, p. 2029-2067.

Culberson, O. L., and J. J. McKetta, Jr., 1951, The solubility of methane in water at pressures to 10,000 PSIA: Petroleum Trans., AIME (T.P. 3082), v. 192, P. 223-226.

Currie, J. B., 1977, Significant geologic processes in development of fracture porosity: AAPG Bull., v. 61, p. 1086-1089.

Degens, E. T., 1965, Geochemistry of Sediments: Englewood Cliffs, N.J., Prentice-Hall, Inc., 342 p. (see p. 202-203).

Dickey, P. A., 1972, Migration of interstitial water in sediments and the concentration of petroleum and useful minerals: 24th Int. Geol. Cong. (Canada), Section 5, p. 3-16.

——— 1975, Possible primary migration of oil from source rock in oil phase: AAPG Bull., v. 59, p. 337-345.

Erdman, J. G., 1967, Geochemical origins of the low molecular weight hydrocarbon constituents of petroleum and natural gases: 7th World Petroleum Cong. (Mexico), Proc., v. 2, p. 13-24.

Faas, R. W., and S. Wartel, 1977, The effect of gas bubble formation on the physical and engineering properties of recently deposited fine-grained sediments: Geologie en Mijnbouw, v. 56, no. 3, p. 211-218.

Franco, A., 1978, Latin America's petroleum surge gathers momentum: Oil and Gas Jour., June 5, 1978, p. 67-74.

Freeman, P. S., 1968, Exposed middle Tertiary mud diapirs and related features in South Texas: *in* J. Braunstein and G. D. O'Brien, eds., Diapirism and diapirs: AAPG Memoir 8, p. 162-182.

Gansser, A., 1960, Uber Schlammvulkane und Salzdome: Vierteljahrschrift d. Naturf. Ges. Zurich, Jahrg. 105, p. 1-46.

Harrison, J. V., 1941, Coastal Makran: Geog. Jour., v. 97, no. 1, p. 1-17.

Hedberg, H. D., 1974, Relation of methane generation to undercompacted shales, shale diapirs, and mud volcanoes: AAPG Bull., v. 58, p. 661-673.

Higgins, G. E., and J. B. Saunders, 1967, Report on 1964 Chatham Mud Island, Erin Bay, Trinidad, West Indies: AAPG Bull., v. 51, p. 55-64.

——————1974, Mud volcanoes—their nature and origin: Naturforsch. Ges. Basel, Verh., v. 84, no. 1, p. 101-152.

Hoering, T. C., and P. H. Abelson, 1963, Hydrocarbons from kerogen: Ann. Rept. Dir. Geophys. Lab., 1962-1963, Carnegie Inst., Washington, p. 229-234.

Hunt, J. M., 1972, Distribution of carbon in crust of earth: AAPG Bull., v. 56, p. 2273-2277.

Karsten, H., 1852, Geognostische Bemerkungen uber die Nordkuste Neu-Granada's, insbesondere uber die songenannten Vulkane von Turbaco und Zamba: Zeit. Deutsch. Geol. Ges., 4/3 (Mai, Juni, Juli, 1852), p. 579-583.

Kartsev, A. A., et al, 1971, The principal stage in the formation of petroleum: 8th World Petroleum Cong. (Moscow), Proc., v. 2, p. 3-11.

Kugler, H. G., 1933, Contribution to the knowledge of sedimentary volcanism in Trinidad: Inst. Petroleum Technology Jour., v. 19, no. 119, p. 743-760.

Lin, C. C., 1965, The naming of the Akungtien Formation, with discussion of the origin of the fossils in the mud ejected from the Kunshupi'ing volcanoes near Ch'iao-t'ou, Kaohsiunghsien, Taiwan: Petrol. Geol. Taiwan, no. 4, p. 107-145.

Makogon, Yu. F., 1978, Hydrates of natural gas (Transl. from Russian by W. J. Cieslewicz): Denver, Geoexplorers Associates, Inc., 178 p.

McIver, R. D., 1967, Composition of kerogen—clue to its role in the origin of petroleum: 7th World Petroleum Cong. (Mexico) Proc., v. 2, p. 25-36.

Mirchink, M. F., et al, 1971, Main concepts of the theory of oil and gas origin and their accumulation in the light of the most recent investigations: 8th World Petroleum Cong. (Moscow), Proc., v. 2, p. 27-33.

Momper, J. A., 1978, Oil migration limitations suggested by geological and geochemical considerations: AAPG Course Note Series 8, p. B-1 to B-60.

Muskat, M., 1949, Physical principles of oil production: New York, McGraw Hill, 922 p. (See p. 298-299).

Powers, M. C., 1967, Fluid-release mechanisms in compacting marine mudrocks and their importance in oil exploration: AAPG Bull., v. 51, p. 1240-1254.
Rau, W. W., and G. R. Grocock, 1974, Piercement structure outcrops along the Washington coast: Inf., Circular 51, Div. Geol. and Earth Resources, State of Washington, 7 p.
Rice, D. D., 1975, Origin of and conditions for shallow accumulations of natural gas: Wyoming Geological Association Guidebook, 27th Ann. Field Conf. (1975), p. 267-271.
Ridd, M. F., 1970, Mud volcanoes in New Zealand: AAPG Bull., v. 54, p. 601-616.
Roberts, W. H., III, 1978, Design and function of oil and gas traps as clues to migration: AAPG Bull., v. 62, p. 558 (abstract).
Secor, D. T., Jr., 1965, Role of fluid pressure in jointing: Am. Jour. Sci., v. 263, p. 633-646.
Shaw, E. W., 1914, Gas from mud lumps at the mouths of the Mississippi: U. S. Geol. Survey Bull. 541, p. 19-22.
Shepard, F. P., 1973, Sea floor off Magdalena Delta and Santa Marta area, Colombia: Geol. Soc. America Bull., v. 84, n. 6, p. 1955-1972.
Snarskiy, A. N., 1961, Verteilung Von Erdgas, Erdol und Wasser im profil: Zeit. angewandte Geologie, v. 7, n. 1, p. 2-8.
——— 1964, Relationship between primary migration and compaction of rocks; *in* Petroleum geology (An English translation of the Russian "Geologiya Nefti i Gaza", v. 5, n. 7, 1961, p. 362-365): published Oct. 1964 by J. W. Clarke, McLean, Virginia.
——— 1967, Primary migration of oil; *in* Petroleum geology (An English translation of the Russian "Geologiya Nefti i Gaza", v. 6, n. 11, 1962, p. 700-703); published March 1967 by J. W. Clarke, McLean, Virginia.
Sokolov, V. A., 1948, Essays on the genesis of Petroleum: Moscow, Gostoptekhizdat.
——— 1967, Organic and inorganic formation of hydrocarbons in nature: *in* Origin of petroleum and gas: Moscow, Nedra (In Russian). (See Fig. 5, p. 120). (Fide Vassoevich, 1975).
——— 1968, Theoretical basis for the formation and migration of oil and gas: *in* Origin of petroleum and gas: Nauka, Moscow. (In Russian). p. 4-24. (See Fig. 1, p. 6).
———et al, 1964, Migration processes of gas and oil, their intensity and directionality: 6th World Petroleum Cong. (Frankfort, 1963) Proc., Section 1, p. 493-505.
———et al, 1969, The origin of gases of mud volcanoes and the regularities of their powerful eruptions; *in* Advances in Organic Geochemistry (1968): Oxford, Pergamon Press, p. 473-484.
Sokolova, M. N., M. A. Kapelyushnikov, and C. L. Sax, 1954, On the possibiity of the extraction of hydrocarbons of argillaceous rocks by means of solutions in compressed gases: Doklady Akad, Nauk, v. 108, n. 4 (in Russian), (fide Sokolov et al, 1964).
Stoll, R. D., J. Ewing, and G. M. Bryan, 1971. Anomalous wave velocities in sediments containing gas hydrates: Jour. Geophys. Research, v. 76, no. 8, p. 2090-2094.
Stroganov, V. P., 1974, Principal phases in origin of gaseous and liquid hydrocarbons and conditions of formation of zones of oil-gas accumulations: Int. Geol Rev., v. 16, n. 7, p. 769-776. (Trans. from Russian of Sovetskaya Geologiya, n. 9, p. 65-75, 1973).
Strong, S. W. S., 1931, Ejection of fault breccia in the Waimata Survey District, Gisborne: New Zealand Jour. Sci., v. 12, n. 5, p. 257-267.
——— 1933, The Sponge Bay uplift, Gisborne, and the Hangaroa mud blowout: New Zealand Jour. Sci., v. 15, n. 1, p. 76-78.
Tissot, B., 1966, Problemes geochimiques de la genese et de la migration du petrole: Rev. Inst. Francais du Petrole et Annales des Combustibles Liquides, v. 21, n. 11, p. 1621-1671.
——— and R. Pelet, 1971, Nouvelles donnees sur les mecanismes de genese et de migration du petrole simulation mathematique et application a la prospection: 8th World Petroleum Cong. (Moscow) Proc., v. 2, p. 35-46.
Tucholke, B. E., G. M. Bryan, and J. I. Ewing, 1977, Gas-hydrate horizons detected in seismic-profiler data from the western North Atlantic: AAPG Bull., v. 61, p. 698-707.
Vassoevich, N. B., 1975, Origin of petroleum: Vestnick Moskovskogo Universiteta (Geologiya) n. 5, p. 3-23 (in Russian).
Weismann, T. J., 1971, Stable carbon isotope investigation of natural gases from Sacramento and Delaware-Val Verde Basins—possible igneous origin: AAPG Bull., v. 55, no. 2, p. 369 (abstract).
Yamamoto, S., J. B. Alcauskas, and T. E. Crozier, 1976, Solubility of methane in distilled water and seawater: Jour. Chem. and Eng. Data, v. 21, n. 1, p. 78-80.
Zhuse, T. P., and G. H. Ushkevich, 1959, Solubility of oil and its heavy fractions in compressed gases: Trudy Instituta Neft Akad. Nauk SSR, v. 13 (in Russian). (fide Sokolov et al, 1964).
——— and T. P. Safronova, 1967, Experimental studies of regularities governing transportation of carbons (of bitumens) through sedimentary rocks by compressed gases; *in* Genesis of oil and gas: Moscow, Nedra, (In Russian). (Fide Mirchink et al, 1971).

Role of Geopressure in the Hydrocarbon and Water System [1]

by Paul H. Jones [2]

Abstract Hydrodynamic regimes in deep sedimentary basins can be described in terms of the depth-pressure gradient, as (1) the hydropressure zone, (2) the geopressure zone, and (3) the zone of transition between (1) and (2). In the hydropressure zone the aquifer systems are open to the atmosphere, and the fluid pressure at any depth reflects roughly the weight of the superincumbent fluid column. In the geopressure zone the aquifer systems are cut off from the atmosphere, and fluid pressure at any depth reflects a part of, or all of the weight of the superincumbent rock column. In the transition zone, field data are seldom adequate to define the conditions.

In noncarbonate sedimentary basins these three pressure zones are associated with characteristic geothermal and formation-water salinity regimes, both of which are keyed to water flux. Confinement of fluids (mainly water) in the geopressure zone causes thermal build-up, partly because the rate of upward movement of water is not great enough to carry away geothermal heat added to the system, and partly because water has a relatively high specific heat and a low thermal conductivity. Interstitial fluid pressure in the geopressure zone is increased by thermal expansion of fluids and rock matrix, by organic matter transformations, and by the loss of load-bearing strength of shale beds as a consequence of the thermal diagenesis of expandable clay minerals. Progressive loading further increases interstitial fluid pressure by compacting the clay-shales and causing shale-water influx into adjacent sand-bed aquifers. Low-salinity water from shales progressively dilutes and displaces the more saline "connate" water in the sand beds, which moves towards growth faults—the principal avenues of escape from the geopressure zone. Accordingly, marked decrease of water salinity in sand-bed aquifers with depth in the geopressure zone is common.

Fluid loss from the geopressure zone, and associated fluid movements in all three pressure zones, is episodic, being triggered by fluid pressure increase in the geopressure zone sufficient to exceed the weight of the rock overburden load. This increase may be the result of molecular forces generated within the geopressure zone, or of progressive loading of the geopressured deposits. Fluid loss results in shale compaction in the geopressure zone and provokes fault movement. The shear zone of resulting growth faults continues to propagate downward into the massive underlying marine shales until the dehydration process is completed.

Associated with these interacting regimes are phase changes of fluids, changes in mineral and hydrocarbon solubility, and chemical reactions triggered by temperature, pressure, water chemistry, and clay mineral catalysis. Mass transfer of soluble rock constituents from depth and their upward redistribution as precipitates (cements), as liquids (saline solutions and oil), and as gases (mainly methane) thus occur.

INTRODUCTION

Large commercial accumulations of petroleum and natural gas in many of the world's deep Mesozoic and Cenozoic sedimentary basins can be explained in terms of a fluid pressure cycle which occurs early in their structural evolution. This fluid pressure cycle is transient lasting 10 to 20 million years or more, during which rapid deposition and/or crustal tectonics compartmentalize sand-bed sequences and cause progressive rise of pore-fluid pressure. As the rate of interstitial water loss is reduced with deepening burial, fluid pressure reflects some part of the overburden rock load and the sediments are said to be geopressured (Stuart, 1970).

Sediments in which pore water is immobile are excellent thermal insulators, and it has been observed that the 100°C isothermal surface follows closely the rising upper boundary of the geopressure zone through the sedimentary pile (Jones, 1975). The rate at which petroleum and natural gas are formed from insoluble organic debris in the shales (kerogen) accelerates rapidly at temperatures above 80°C (LaPlante, 1974). Fluid hydrocarbons formed in geopressured source beds are dissolved in the high-pressure, high-temperature pore waters of the sediments (Schmidt, 1973). As solutes, they move with water out of the source beds and into adjacent sandstone beds or open

[1]Presented before the Association, April 1978, at the Oklahoma City Annual Meeting. Received for publication, September 21, 1978; accepted for publication, August 20, 1979.

[2]P. H. Jones Hydrogeology, Inc., Baton Rouge, Louisiana 70809.

Article Identification Number
0149-1377/79/SG10-0011/$03.00/0.

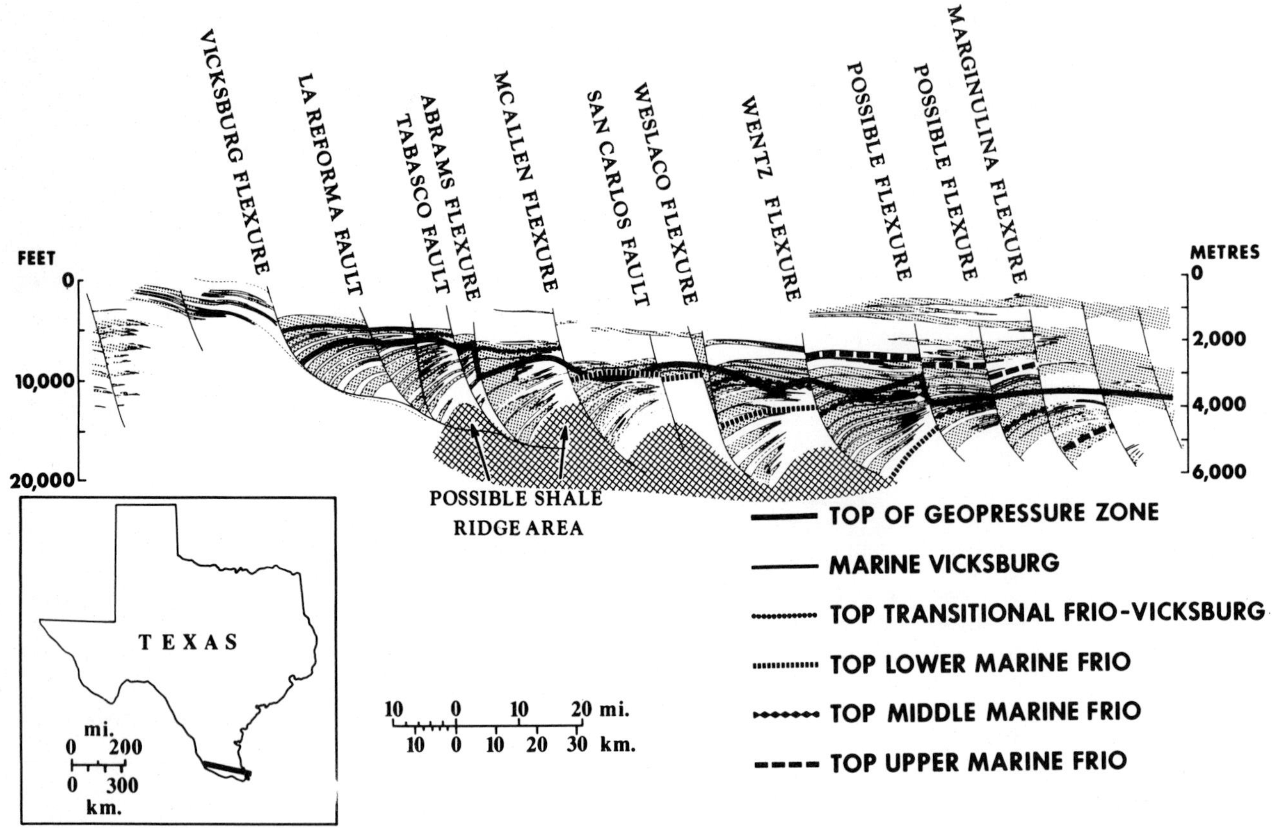

FIG. 1—Generalized geologic section along the axis of the Rio Grande embayment showing prograding deltaic deposits, effects of associated growth faulting, and the top of the geopressure zone (from Jones, 1975).

fractures in shale, and ultimately escape from the geopressured zone. Fluid hydrocarbons are exsolved from host water in the transition zone from geopressure to hydropressure; moving as oil and gas, they accumulate in the first structural trap. Pressure differential in the geopressure transition zone has been recognized as a trapping mechanism for some time (Myers, 1968). According to Fertl and Timko (1973), 90% of the oil in the Gulf of Mexico basin is found in reservoirs having fluid pressure reflecting pressure gradients less than 0.7 psi/ft.

ORIGIN AND PRESERVATION OF GEOPRESSURE

Sedimentary basins rapidly filled by prograding deltas are geopressured below some minimum depth, mainly as a consequence of faulting due to differential compaction keyed to sediment facies distribution (Fig. 1). Progressive subsidence of the downthrown block with continuing deposition is known as "growth faulting" (O'Camb,1961); faulting of this type is characterized by marked increase in thickness of correlative beds across the fault, and increasing displacement with depth. Reversal of dip and blockage of escape routes for water in sandstone beds offset against shale is accompanied by increasing fluid pressure beneath some regionally-extensive clay or shale bed. Beneath this "seal" pore-fluid pressure reflects part or all of the weight of superincumbent rocks.

Restriction of upward water movement retards geothermal heat flow, increasing the geothermal gradient (Fig. 2). The maturation rate of kerogen, which averages about 1.2% by weight of shale source beds and ranges upwards of 5% in the North Slope of Alaska (L. L. Lundell, Atlantic Richfield Co., Dallas, personal commun., 1977) accelerates with depth below the top of the geopressured zone (Fig. 3). As kerogen is converted to petroleum and natural gas, a marked increase in volume occurs.

In the temperature range from 80° to 120°C, at which petroleum maturation accelerates, certain clay minerals in the shale source beds undergo chemical changes and dehydration. Bound and intracrystalline water become free pore water, in an amount equal to about half the volume of the clay mineral altered (Burst, 1969). This water is fresh, and it is dispersed within the shale bed in close proximity to the newly-formed fluid hydrocarbons. Released water and hydrocarbon fluids occupy space previously filled by solids which supported a part of the overburden rock load; as a consequence of this loss of load-bearing strength, and of the pressure gradients resulting from compaction stress, fluids begin to move out of the shale source beds and into the more competent adjacent sandstone beds. However, fluid pressures in adjacent shale and sandstone beds soon reach equilibrium because escape routes from sandstone beds have been sealed by faulting. The deposits are now geopressured, and the pressure gradient increases with depth as shown in Figure 4.

Kerogen and clay-mineral conversion continues with increasing depth of burial in the geopressured zone at a rate indicated by the geothermal gradient. Both reactions are endothermic, and both convert heat-conductive solids to less conductive fluids. Consequently, geothermal gradients in the geopressure zone are commonly 3 to 5 times greater than in the hydropressure zone above. Geothermal gradients and fluid pressure gradients increase hand in hand (Lewis and Rose, 1969), but no direct numerical correlations are apparent (Jones, 1969). Fluid losses by leakage from the geopressured zone are offset for millions of years by the continuing generation of new fluids from organic and clay-mineral solids, by the progressive thermal expansion of fluids and rock matrix, and by the upward redistribution of fluids from the metamorphic zone deep in the basin.

The upper boundary of the geopressured zone is by far the most important physical interface in deep sedimentary basins. Downward across this interface, in a depth range of a few tens of meters (hundreds of feet), fluid pressure commonly increases an order of magnitude or more; temperature gradients triple or quadruple; porosity of both sandstone and shale increases by 10 to 25% (Fig. 5); and formation-water salinity commonly decreases by 50,000 mg/l or more (Fig. 6). Most of the world's commercial oil and much of its natural gas is believed by the writer to originate by exsolution as formation water escapes across this interface.

The depth of occurrence of the geopressured zone is indicated during drilling by an abrupt increase in the temperature of the drilling-mud returns; by decrease and then abrupt increase in bit-penetration rate; by "kicks" in the circulating mud fluid pressure—thin sandstones discharge water and gas into the hole as they are penetrated by the drill bit; and by decrease in the density of shale drill-cuttings on the shaker. It can also be identified on the electric log of the well using the amplified short normal resistivity curve for "clean" shale sections, by noting reversal of the depth-resistivity trend line. The depth of occurrence of the geopressure zone can be mapped regionally by seismic methods, by noting the depth of reversal of the depth-velocity trend line. The map, Figure 7, shows the depth of occurrence of the geopressure zone in an area of some 190,000 sq km (about 75,000 sq mi) in southern Louisiana and adjacent parts of the Gulf continental shelf; it is based on seismic data, supported by electric-log and acoustic-log data.

GEOTHERMAL REGIME OF THE GEOPRESSURE ZONE

The three modes of heat transport through fluid-saturated sediments are (1) conduction through mineral grains and interstitial fluids, (2) convective and advective flow of interstitial fluids, and (3) radiation. Conductive, convective, and advective heat flow are most important in the relatively cool deposits of the hydropressure zone. Ray transport of heat becomes increasingly important with depth in the geopressure zone—partly because of the marked increase in thermal radiation that occurs at elevated temperature, and partly because both porosity and fluid movement approach zero with increasing depth. The critical temperature of water (705.38°F or 374.1°C) is exceeded at depths no greater than 30,000 ft (9.3 km) in parts of some deep sedimentary basins. At those depths, radiant heat flow is the dominant mode.

In the range of temperatures common at depths less than 25,000 ft (7.5 km) in most sedimentary basins, even where the deposits are geopressured below 10,000 ft (3.1 km), heat flow can be analyzed effectively on the assumption that radiant heat transfer is negligible. Generally, the thermal conductivity of the mineral

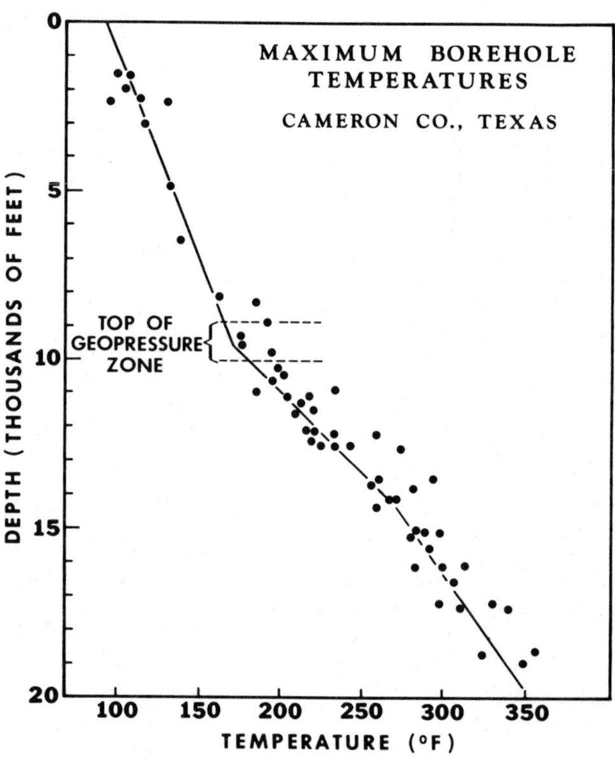

FIG. 2—Maximum temperatures recorded in boreholes in Cameron County, Texas, showing effect of geopressure on geothermal gradient (from Jones, 1975).

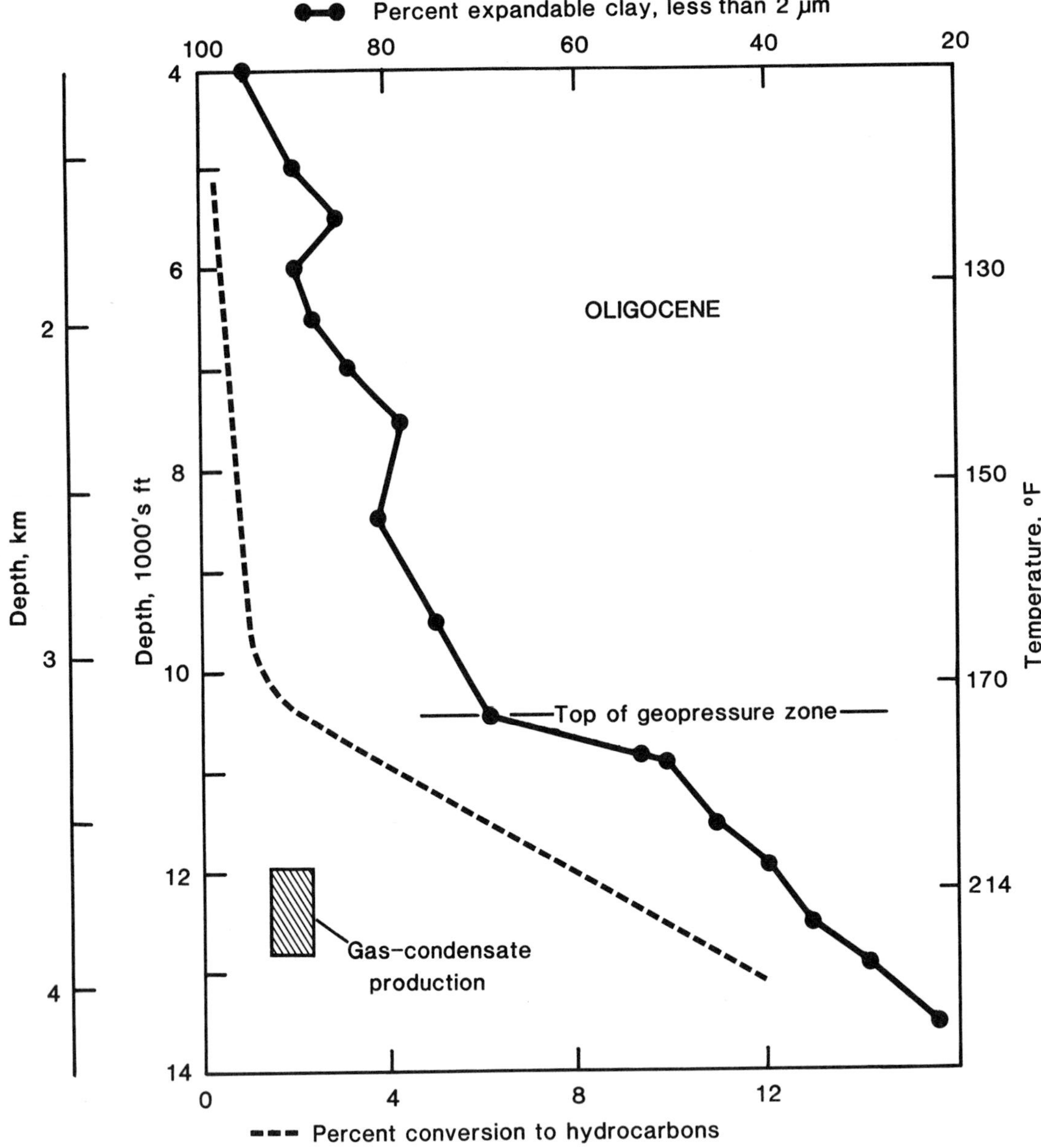

FIG. 3—Rate of kerogen conversion to hydrocarbon, and the rate of clay mineral conversion as a function of increasing depth and temperature, Manchester Field, Calcasieu Parish, Louisiana (from Schmidt, 1973, and LaPlante, 1974).

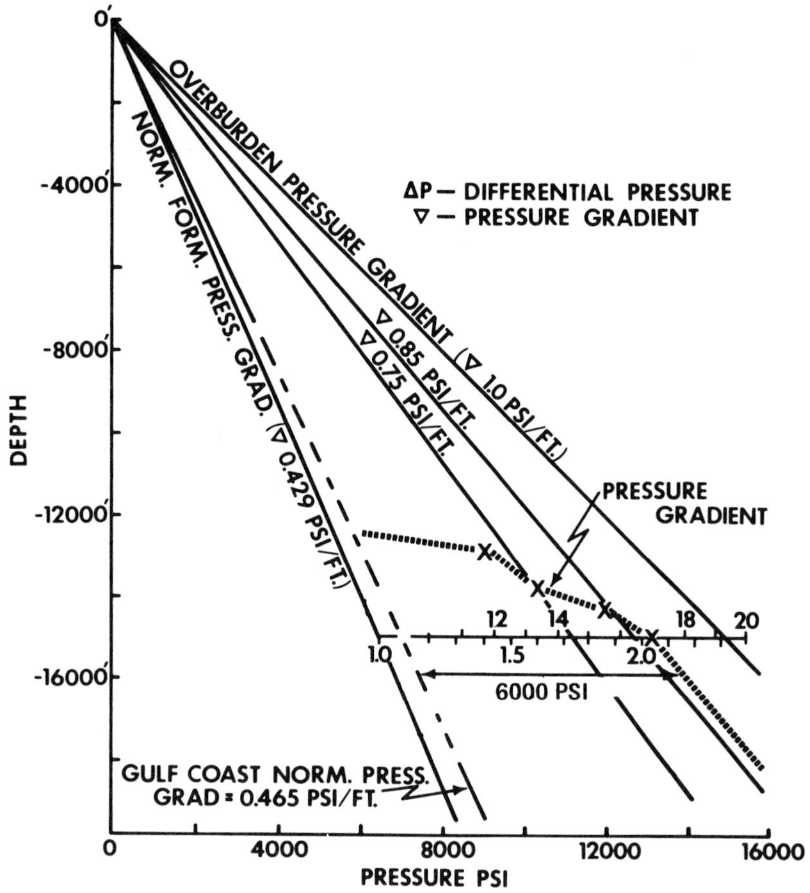

FIG. 4—Formation fluid pressure gradient observed in an offshore Louisiana well (from Myers, 1968).

grains of sediments is four to five times greater than that of the pore water, and their specific heat is only about one-fifth as great. Accordingly, the thermal conductivity of shale or clay varies inversely with porosity, because pore-water movement is very slow. The thermal conductivity of sandstone varies directly with its porosity, owing to the occurrence of convective heat transport in the wider pores (Zierfuss and van der Vliet, 1956). Large heat-flow rates observed in geothermal resource areas are not possible without mass transfer by thermal convection, according to Elder (1965).

The principal source of the heat in sedimentary basins is the earth's upper mantle, and the principal mechanism of its migration is mass transfer. Igneous intrusion of geosynclinal deposits commonly occurs along the axis of maximum subsidence (Bernhard Kummel, Harvard Univ., personal commun. 1976). Very great thicknesses of rock salt commonly occur near the base of the sediment pile in the world's petroliferous basins. By virtue of its thermophysical properties, rock salt is an ideal heat transfer agent in such settings; it is an excellent conductor of heat by comparison with other rock-forming minerals; its density is low, and its load-bearing strength does not exceed 50 kg/sq cm. Salt becomes plastic in one direction at temperatures above 200°C, in two directions at 300°C, and is completely plastic above 350°C (Gussow, 1970). Salt melts at 800.8°C, with an increase in volume of 18.6%. Diapiric salt, forming hundreds of domes scattered widely throughout a sedimentary basin, and flowing laterally in response to depositional loading (Fig. 8), can be a highly effective heating agent where it is mobilized by igneous activity at depth in the basin (Lehner, 1969).

Salt diapirs may cause the thermal diagenesis of enormous volumes of montmorillonite-rich clay or shale, releasing a flood of hot, high-pressure water into overlying deposits. Structural downwarp of salt withdrawal areas and continued sedimentary loading prolongs salt mobilization, and may cause detachment of salt diapirs from the mother salt bed and subsequent freezing, by cutting off the principal source of heat. Cycles of salt mobilization and associated structural deformation, sedimentary filling, dehydration of shales, and flush of overlying deposits by hot, high-pressure water move basinward in a series of waves, each keyed to a major progradational cycle of deltaic

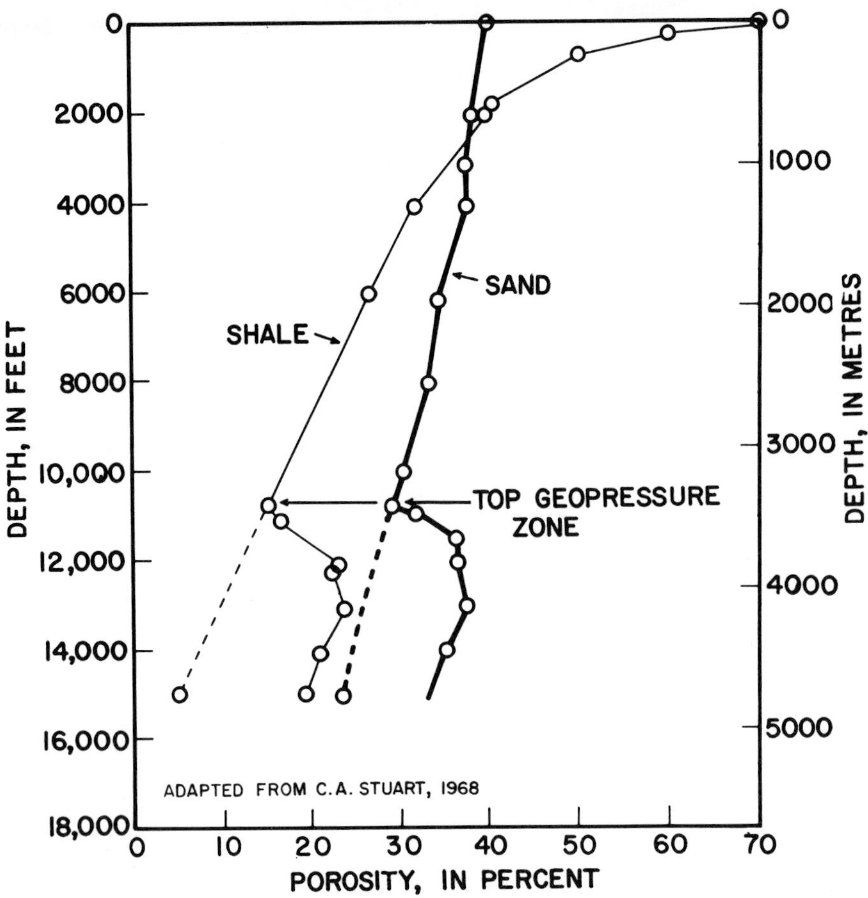

FIG. 5—Relation of porosity to depth of burial and fluid pressure, in Cenozoic deposits of South Louisiana (from Stuart, 1970).

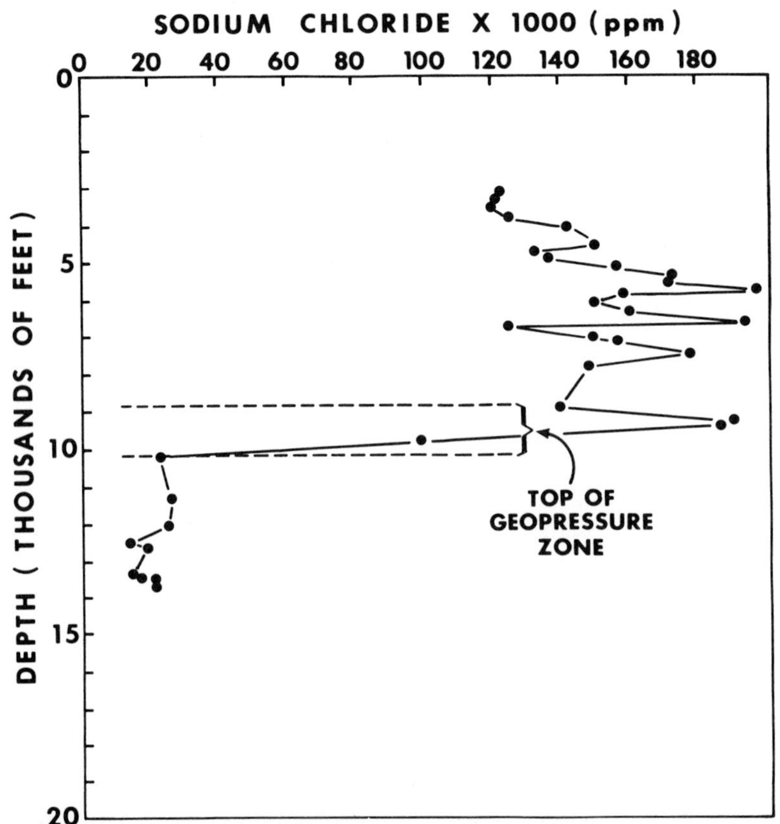

FIG. 6—Change in formation water salinity with depth, in relation to the geopressure zone, Manchester Field, Calcasieu Parish, Louisiana (from Schmidt, 1973).

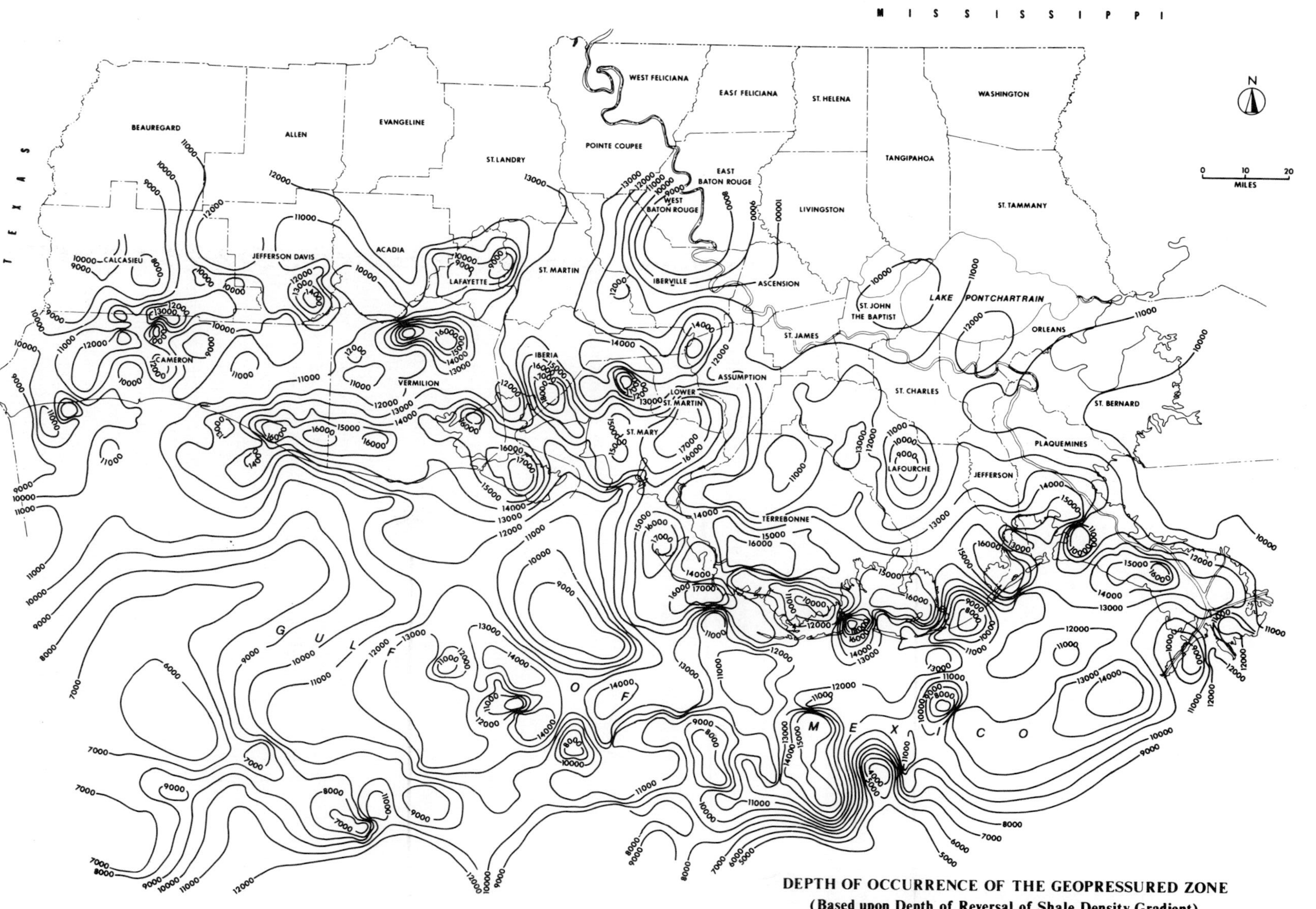

FIG. 7—Depth of occurrence of the geopressured zone based upon depth of reversal of shale density gradient.

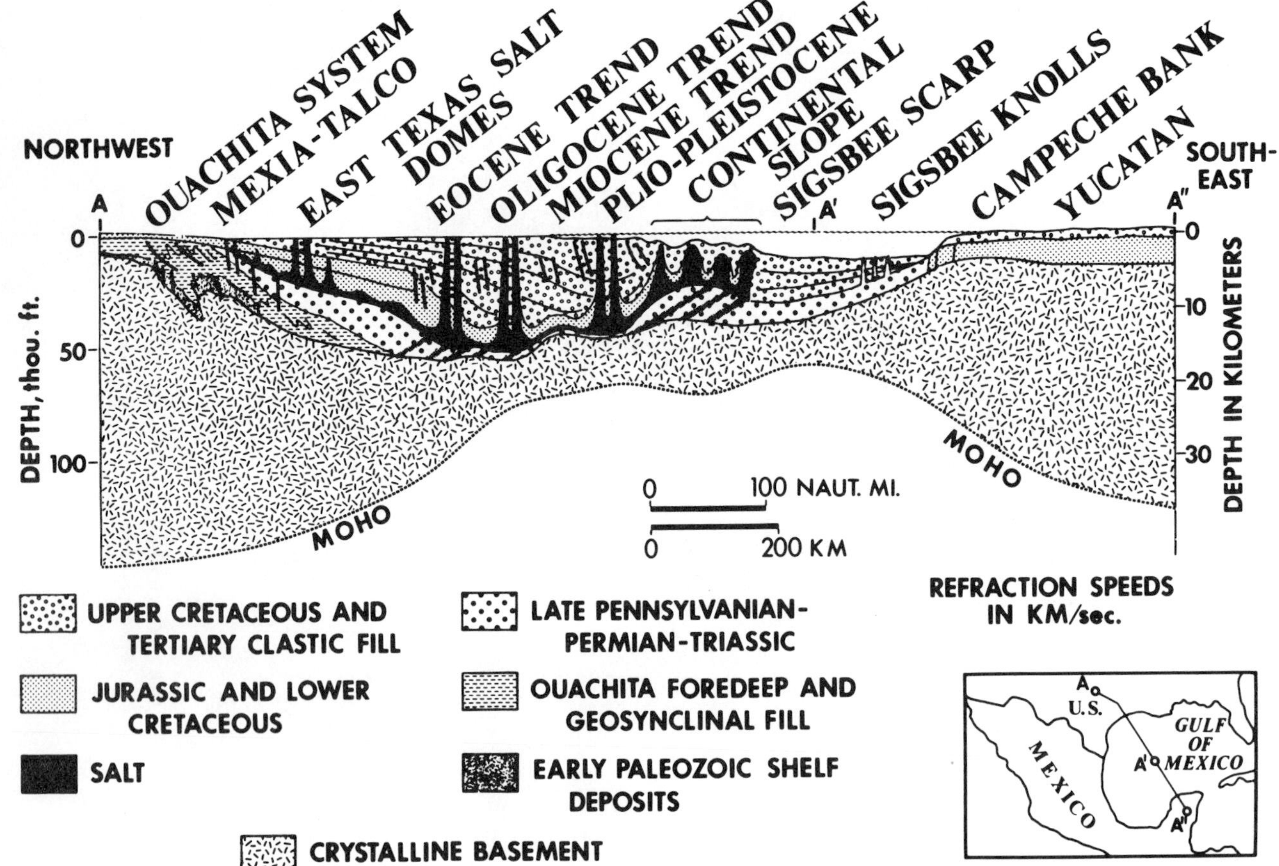

FIG. 8—Profile section through the earth's crust and upper mantle from East Texas to Yucatan showing salt instrusions and diapirs penetrating overlying sediments, and the salt roller belt beneath the continental slope (after Lehner, 1969).

sedimentation. Eight such cycles have occurred in the Gulf of Mexico basin during the past 40 million years (Jones and Wallace, 1974).

HYDROCARBON REGIME OF THE GEOPRESSURE ZONE

Progressive conversion of kerogen to fluid hydrocarbons with increasing depth and temperature in the geopressure zone is evidenced by progressive decrease in abundance of unconverted kerogen. However, very little information is available on the composition, occurrence, and abundance of dissolved petroleum hydrocarbons in formation waters of the geopressure zone. In the northern Gulf of Mexico basin, almost no commercial oil is found in reservoirs having initial fluid pressures reflecting gradients greater than 0.7 psi/ft; and almost none is found in reservoirs at temperatures greater than 150°C (302°F). More than 80% of the known natural gas reserves in the basin is found in the geopressure zone not associated with oil. Recent laboratory studies by Price (1976, 1977) indicated that the absence of liquid petroleum in the geopressure zone is the consequence of its rather high solubility in methane-saturated water at elevated pressure and temperature—1 to 10 wt.% at 300 to 350°C (572 to 662°F).

Observed rates of kerogen conversion in the geopressure zone suggest that liquid hydrocarbons formed in shale source beds are immediately dissolved in high-temperature pore water. In water solution they move in episodic flow pulses through the geopressure zone, escaping from it up or across faults. Petroleum as liquid oil makes its first appearance where the temperature and pressure of the host water are reduced and exsolution occurs. As the solubility of whole oil is much enhanced by methane saturation, simultaneous loss of methane by exsolution in zones of pressure drop accelerates the exsolution of oil in those zones.

The importance of pressure differential as a key to the occurrence of commercial oil is now commonly recognized (Myers, 1968). But the significance of the high solubility of methane in water at elevated pressure and temperature (Sultanov, et al, 1972), is not. Methane solubility in fresh water is described by the curves of Figure 9, and is summarized in Table 1.

Temperature and pressure gradients in the geopressure zone result in progressive conversion of kerogen to fluid hydrocarbons with depth, the cracking of

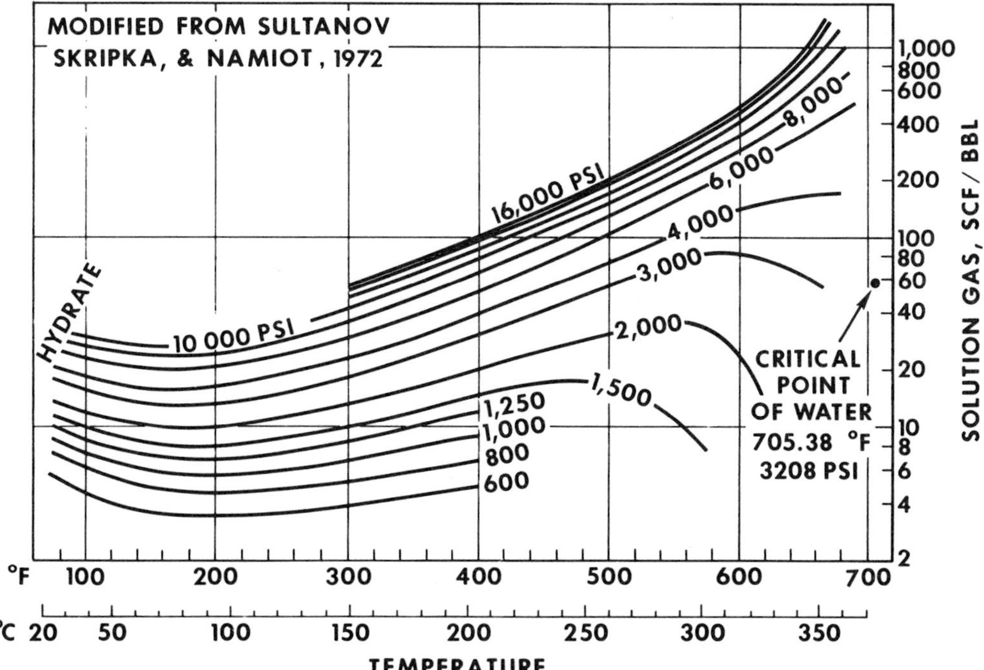

FIG. 9—Solubility of methane in fresh water in the temperature range 30° to 360°C (86° to 680°F) at pressures of 600 to 16,000 psi. Solution gas expressed in standard cubic ft per barrel (scf/bbl) (after Sultanov, Skripka, and Namiot, 1972).

heavier hydrocarbons to lighter hydrocarbons and gases, and the dissolution and transport of these fluid hydrocarbons out of the geopressure zone. However, at great depth in geopressured basins, where fluid movement may be essentially nil for long periods,

> . . . both the reactions which lead to the generation of hydrocarbons from kerogen and to the thermal conversion of hydrocarbons to methane and graphite will come to a steady state kinetic equilibrium. This equilibrium apparently will persist over geologic time unless: (1) products from the reaction are removed from the system, (2) temperature is raised, or (3) pressure is lowered (Price, 1977).

Price examined deep basin fault systems in the largest oil-producing basins of the world. According to his findings, most of the major fields are on or near regional faults that dip toward the axes of basin depocenters. Although his work made plausible a hot, deep origin for much of the world's oil and gas, it is well known that fluid hydrocarbons can be generated and mobilized at any depth in a sedimentary basin where (1) source rock is present, (2) temperature exceeds 80°C, and (3) fluid pressure gradients in source rock sequences is sufficient to drive hydrocarbon fluids into reservoir rocks.

The phenomena of geopressure play critical roles in the fluid hydrocarbon regime. Reduced rates of water loss increase geothermal gradients by reducing mass transfer rates. Temperatures required for kerogen conversion also result in clay mineral conversion; these processes generate fluids from solids, increasing porosity, fluid pressure, and the geothermal gradient. As fluid pressure approaches lithostatic pressure, natural hydrofracturing releases fluids into the hydropressure zone. Shearing and sediment compaction, accompanying fluid releases, close avenues of escape (faults), and fluid pressure again rises. Each release of high-pressure, high-temperature water carries with it dissolved hydrocarbons that come out of solution in the pressure transition zone and migrate updip as petroleum and natural gas, or accumulate in traps. But by far the largest part of the fluid hydrocarbon resource

Table 1 Solubility of methane in water at selected temperatures and pressures, in scf/bbl (Sultanov et al, 1972)*

PRESSURE PSI	TEMPERATURE °F 200	300	400	500	600	656
2,000	10	12	20	30	17	
3,000	13	17	30	52	80	
4,000	15	23	40	76	135	
6,000	20	29	52	105	230	380
8,000	24	35	64	130	285	440
10,000	28	41	77	149	340	620
12,000		47	86	168	400	800
14,000		53	95	186	440	900
16,000		58	104	200	480	1,000

* values approximate

in young sedimentary basins may be found dissolved in formation waters (Jones, 1976).

The cycle of geopressuring, heating, clay-mineral conversion, maturation of fluid hydrocarbons, and escape of fluids to the hydropressure zone continues until clay-mineral conversion is complete, released waters have been expelled from compacting shales, and the equilibrium compaction gradient has been restored. As this occurs, heat flow by mass transfer diminishes and isotherms fall with the isopressure surfaces.

REFERENCES CITED

Burst, J. F., 1969, Diagenesis of Gulf Coast clayey sediments and its possible relation to petroleum migration: AAPG Bull., v. 53, p. 73-93.

Elder, J. W., 1965, Physical processes in geothermal areas, *in* Terrestrial heat flow: American Geophys. Union Mon. Series 8, p. 211-239.

Fertl, W. H., and D. J. Timko, 1973, How downhole temperatures, pressures affect drilling: World Oil, (June 1972-March 1973).

Gussow, W. C., 1970, Heat, the factor in salt rheology: Louisiana State Univ. Symposium, Geology and Technology of Gulf Coast Salt (Baton Rouge), Proc., p. 125-148.

Jones, P. H., 1969, Hydrodynamics of geopressure in the northern Gulf of Mexico basin: Jour. Petroleum Tech., v. 21, p. 803-810.

——— 1975, Geothermal and hydrocarbon regimes, northern Gulf of Mexico basin: 1st Conf., Geopressured Geothermal Resources of the Gulf Basin (Univ. Texas, Austin), Proc., p. 15-89.

——— 1976, Natural gas resources of the geopressured zones in the northern Gulf of Mexico basin, *in* Natural gas from unconventional geologic sources: Nat. Aca. Sciences Board of Mineral Resources (committee on natural resources), p. 17-33.

——— and R. H. Wallace, 1974, Hydrogeologic aspects of structural deformation in the northern Gulf of Mexico basin: U. S. Geol. Survey Jour. Research, II, no. 5, p. 511-517.

LaPlante, R. E., 1974, Hydrocarbon generation in Gulf Coast Tertiary sediments: AAPG Bull., v. 58, p. 1281-1289.

Lehner, P., 1969, Salt tectonics and Pleistocene stratigraphy on continental slope of northern Gulf of Mexico: AAPG Bull., v. 53, p. 2431-2479.

Lewis, C. R., and S. C. Rose, 1969, A theory relating high temperatures and overpressures: Soc. Petroleum Engineers, SPE 2564.

Myers, J. D., 1968, Differential pressures, a trapping mechanism in Gulf Coast oil and gas fields: Gulf Coast Assoc. Geol. Socs. Trans., v. 18, p. 56-80.

O'Camb, R. D., 1961, Growth faults of south Louisiana: Gulf Coast Assoc. Geol. Socs. Trans., v. 11, p. 139-175.

Price, L. C., 1976, Aqueous solubility of petroleum as applied to its origin and primary migration: AAPG Bull., v. 60, p. 213-244.

——— 1977, Crude oil and natural gas dissolved in deep, hot geothermal waters of petroleum basins—a possible significant new energy source: U. S. Geol. Survey (draft manuscript), Denver, Colo.

Schmidt, G. W., 1973, Interstitial water composition and geochemistry of deep Gulf Coast shales and sandstones: AAPG Bull., v. 57, p. 321-337.

Stuart, C. A., 1970, Geopressures: Louisiana State Univ., 2nd Symposium on Abnormal Subsurface Pressure (Baton Rouge), 121 p.

Sultanov, R. G., et al, 1972, Solubility of methane in water at high temperatures and pressures: Gazovaia Promyshlennost, v. 17, May, p. 6-7.

Zierfuss, H., and G. van der Vliet, 1956, Laboratory measurements of heat conductivity of sedimentary rocks: AAPG Bull., v. 40, p. 2475-2488.

Design and Function of Oil and Gas Traps[1]

By W. H. Roberts[2]

Abstract It is only in *traps* that oil and gas are found as distinct, coherent fluids; thus traps should yield the most positive information about these fluids. If we can understand what is happening in the traps, we should be able to look back along the migration trail with special insight to what has occurred. That insight may even extend all the way back to the "source."

This study concludes that traps are the most logical places for hydrocarbon components to be put together as the distinct fluid mixtures which we call oil and gas. It follows that traps are not just passive receivers or containers of hydrocarbon mixtures put together elsewhere. Effective oil and gas traps of different well-known types have a very important feature in common: structurally and stratigraphically, they are designed to discharge waters from depth. Thus they function as active focal mechanisms to gather and process feedstock waters carrying hydrocarbons and other organic derivatives. It is a forced-draft system. The concept reinforces the anticlinal theory, but without total dependence on fluid buoyancy. Moreover, it honors all factual observations made around oil and gas deposits.

Very simply, the most important function of a trap is to leak water while retaining hydrocarbons. The water can leak through the enclosing membranes and covering strata because they are water-soaked, like a wick. The hydrocarbons and other organics are separated from the waters as they pass through the trap. The separation is caused by abrupt changes in pressure, temperature, and probably salinity—which in turn are jointly related to the basic change in direction of feedstock (water) movement from lateral to upward. Coalescence of hydrocarbons makes tiny bubbles or globules which grow in size and move less easily than water. The ultimate composition of a trapped hydrocarbon mixture depends on the respective residence times of the various components of that mixture which in turn depend on: (1) what the water carried to the trap, (2) what the trap retained, and (3) what was the pore-volume exchange rate.

INTRODUCTION

For those with traditional ideas about origin, migration, and accumulation of oil and gas, the arguments that follow may sound like heresy. Actually, few of the ideas are original. Many are borrowed from some of the earliest oilseekers (e.g. Evans, 1866; Munn, 1909; Rich, 1921; Heald et al, 1930; and Pratt, 1942). In a few cases, the present writer may be guilty of tying certain ideas together in certain ways to produce concepts which seem radical. The departure from orthodoxy is not without risk. To help the reader find credibility in new ways of thinking about petroleum origin, a special effort is made to list some of the more relevant, though not so well known, literature.

All of us would like to know how oil and gas happen to be where they are. From our research and some of our mistaken exploration efforts, we have to conclude that petroleum origin is a complex thing. The clues we have tell us that it probably involves countless interrelated processes, using widely variable materials, and going on at an infinitely variable rate. We don't know definitely how it starts or how it ends, but there is ample reason to believe there are varieties of beginnings and varieties of endings. We think we've found some "windows" for observation. We speak of *time* windows and *temperature* windows, but we can't agree on what we see in those windows. Most of our reasoning is based on models rather than direct observations. Sometimes we seem to be drowning in conjecture, or bogged down in conceptual inertia.

What we need is a conceptual break-through; something like what plate tectonics is thought to have done for structural geology. We'd like to have a realistic, unifying theory that explains *all* oil and gas deposits. Unfortunately, such a goal seems to be unrealistic. It appears wise to beware of any unique explanation of anything about the origin of petroleum. This paper merely considers some simple observations about oil and gas traps about which many people are unaware. Some readers will be disturbed by apparent contradictions of long-held concepts. They are invited to judge

[1]Read before the Association, April 11, 1978 at the Oklahoma City Annual Meeting; received for publication, August 29, 1978; accepted for publication, August 20, 1979.

[2]Houston Technical Services Center, Gulf Research and Development Co., 77036. Permission to publish is gratefully acknowledged.

Article Identification Number:
0149-1377/79/SG10-0012/$03.00/0.

the evidence. Some may decide to rethink those long-held concepts.

The "trap" is the best-known part of the oil and gas environment. It is in traps, prospective or proven, that we have the best information from samples, tests, and production histories. Most importantly, it's where the best exploration tools are used to define the trap geology. Few people realize how important that information may be to solving the mystery of petroleum origin. If we understand what happens in a trap—how it works or doesn't work—it should help us to understand how the materials of the oil and gas deposits got there (how they migrated) which has been the most vexing part of the mystery. The trap may indeed play an unexpectedly important role in the origin of oil and gas.

The general conclusion of the present discourse is that, by design and function, the most logical place for oil and gas to be formed as distinct fluids is in the "trap" rather than the "source." Such a conclusion disposes of a lot of mechanical problems related to the travel of oil and gas between source and trap as separate, distinct fluids. However, on the face of it, it appears that mechanical problems are traded for chemical problems related to the waterborne transport of organic components from sources to traps. If we're dealing strictly with hydrocarbons (HCs), of the full spectrum found in most crude oils, we have a quantitative impass because of limited molecular solubilities in water.

The problem of accommodating hydrocarbons and other organic compounds in water has had limited attention by explorationists in this country although there is considerable literature available among chemists and engineers (e.g. see reference group "A" in the Appendix to this paper). How we look at solubility and migration problems depends on what materials we are talking about. Little thought has been given to the water-soluble organic acids and alcohols, or carbohydrates, so easily accommodated (and found) in oil-field waters, although some literature is available (see reference group "B" in the Appendix to this paper). Many such compounds are known (see reference group "C" in the Appendix to this paper) to be transformable into hydrocarbons by decarboxylation. It has been shown (see reference group "D" in the Appendix to this paper) that significant quantities of liquid hydrocarbons can be accommodated in water by solution in gases (especially in the presence of carbon dioxide), which are amply soluble in water. Regardless, it is logical that quantitative constraints on the waterborne transport of the organic raw materials for petroleum should be held in abeyance while consideration is given to the apparent functions of entrapment.

The arguments in the present paper are chiefly based on evidence observed in and around the traps. That evidence may include clues about migration, and even clues about source and origin. The ensuing discussion will treat six basic topics of practical importance: (1) what a trap is, (2) what a trap does, (3) what traps have done—field evidence, (4) hydrothermal effects, (5) other trap functions, and (6) what to look for.

WHAT A TRAP IS

The definition of a trap should include its functional characteristics. The following statements have functional relevance and will be supported by evidence:

1. A trap is a paradox, because it must leak to function as a trap.
2. A trap is a forced-draft, flow-through system.
3. A trap is a center of deep-water-discharge.
4. A trap is an active focal mechanism, not a passive, sealed, dead-end container.
5. A trap is a hydrocarbon separator, in a sense a filter.
6. A trap is an imperfect and hydrochemically noisy system.

It is useful to consider what traps look like. The easiest way is to sketch the different types schematically in profile, realizing of course that they often occur in combination. Figure 1 shows twelve basic types. The first eight are essentially structural, and the bottom four are essentially stratigraphic. The stratigraphic traps depict from left to right a prism of sand lenses or other porous bodies, a reef, a facies change or pinch-out, and an aquifer truncated by unconformity.

It may require conceptual reorientation to think of a trap as an apparatus for *creating* an oil or gas deposit—not just for holding it. In the twelve trap types of Figure 1, each case is a reservoir space leading upward. Considering the transmissability of the porous reservoirs in relation to ambient rock, the reservoirs are thought of as *shunting* that part of the rock section, making it easier for water (and other fluids) under pressure to move upward. For the so-called stratigraphic cases, each one might be regarded as equivalent to one side of an anticline.

WHAT A TRAP DOES

Guide Planes

Consider what is important about the *structural* and the *stratigraphic* traps in general. In Figure 2 the essence of the two basic styles from Figure 1 is represented by a simple anticline and a simple pinchout. In both there is upward convergence of the boundaries of porous media, as emphasized by the dashed lines. One convergence is upright and the other is inclined. If the porous media are assumed to contain water under pressure, we can think of this as the convergence of fluid guide planes. Guide planes, of course, imply motion, as though the waters are going somewhere. Actually, they are. They are working their way up through the sedimentary cover; the traps focus and promote that movement. To put it another way, the traps are

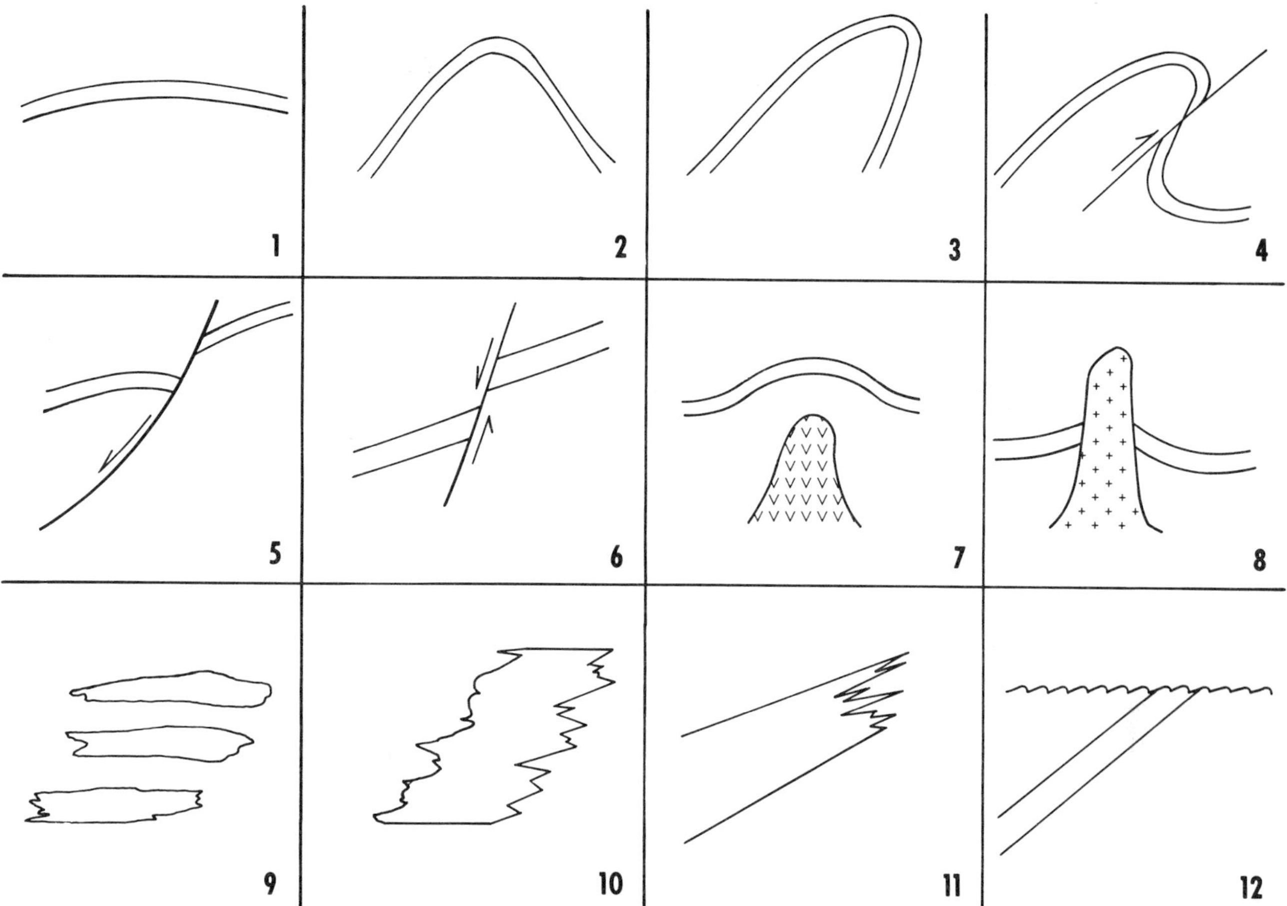

FIG. 1—Common trap styles in schematic profile. Numbers 1 through 8 are normally classed as structural; numbers 9 through 12 are stratigraphic.

functioning as valves, to gather and discharge waters from depth.

Membranes

It appears that a good working trap has more vertical transmissibility than the surrounding rock framework. Evidence will show that a working trap is a dynamic forced-draft system, not a passive cul-de-sac. It has been commonly assumed that the convergent boundaries and the caprocks commonly found over oil and gas pools are essential for retaining the trapped fluids. Fortunately, if those enclosures have the proper membraneous qualities, they can retain the oil and gas as separate phases, while at the same time passing water as if through a wick. If they couldn't, the trap wouldn't work. So the boundaries function not only as guide planes, but also as membranes. It so happens that most of the sedimentary media have been water-soaked from the beginning, so that even the very fine-grained facies will pass water at *some appreciable rate.* Of course, microfractures and joints may help—if they don't let too much oil and gas escape.

Local Weakness

For the confluent waters to move upward through a trap there must be a continued path of least resistance through the covering strata, at least until an aquifer with lesser pressure is reached. The full evaluation of a functioning trap, therefore, should include an analysis of the covering strata. This discussion does not treat that detail, or the mechanics of transmissibility; but as a significant general rule, it is observed that *good traps are commonly related to local weaknesses* in one or more of the otherwise competent, covering strata (aquitards, like shales or evaporites). The most critical stratum may be directly above the pay zone, in the "caprock" position, or it may be somewhere higher up in the covering section. Faulting in the cover, as well as below the trap, may be an important factor.

Basically, a trap is a deep-water-discharge mechanism. If a working trap situation allows (in fact depends on) upward water movement, some traps will work better than others. Figure 2 shows that the upward convergence is much broader for the structural trap than for the stratigraphic trap. This probably means that in most cases of comparable dimensions the anticline has more focal power and more gathering area than the pinchout. Exceptions occur for some stratigraphic traps of regional dimensions (e.g., East Texas, Athabasca).

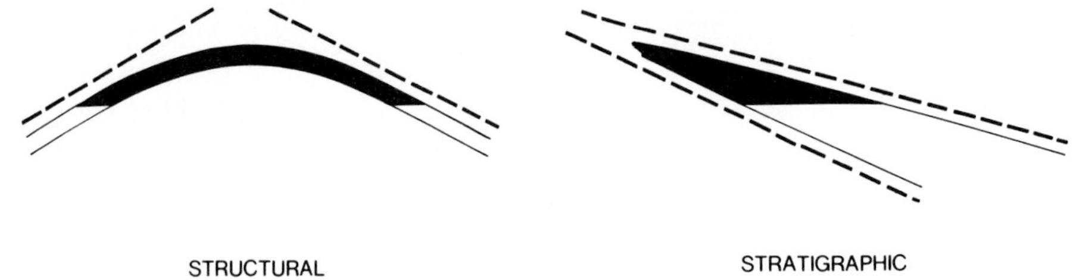

FIG. 2—The general case: basic design shows upward convergence (dashed lines) of reservoir boundaries which function both as fluid guide planes and as membranes. As a general rule, convergence is broad for structural traps and narrow for stratigraphic traps.

Anticlinal Supremacy

Figure 3 (from a paper by Moody, 1973) shows the relative amounts of oil held by different styles of traps among the world's 198 giants, each having more than ½ billion bbls (75 million metric tons) of oil recoverable. It seems to be a clear case of anticlinal supremacy. There is a valid objection, of course. Fewer stratigraphic traps get drilled because they are harder to find and usually thought to be riskier. Even so, on a prospect-for-prospect basis the "structural" traps probably can be expected to function more efficiently. One of the chief reasons is depicted in Figure 2. Not only does the anticlinal trap have a wider entrance, but the effects of the structural condition may persist in deeper and(or) shallower strata, offering multiple objectives as well as reinforcing the focal power and the vertical transmissibility which make the trap work.

Deep-Water-Discharge

The most obvious clues of trap function are related to deep-water-discharge. Evidence in the form of salt solution, hydrochemistry, and hydrothermal anomalies will be shown by field examples in this paper.

The upward movement and discharge of waters from depth has to be supplied, of course, by lateral water movements (gathering) toward the trap. It is that basic change in the sense of movement, from lateral to vertical, which causes separation of the organic components (or precursors of oil and gas) from the feedstock water. The unloading of the organic materials in the trap probably depends on three internal trap functions, acting in concert and caused by that same change in direction: (1) a sudden pressure drop, (2) a sudden temperature drop, and (3) a probable water salinity increase. Together these make a convincing argument that the trap *is* a logical place for organic components to separate from water, and for oil and gas to form as coherent mixtures.

Pressure Change

Figure 4 depicts the pressure environment which would be experienced by waterborne organic compounds if the direction of movement turned from lateral to vertical. The lateral pressure gradient in the aquifer is 0.002 psi/ft (0.0005 Atm/m), equivalent to about 25 ft of "head" per mile (4.75 m/km) which is reasonable for the average sedimentary basin. The picture intends to show by arrows a progressively increasing vertical component of water movement toward the center of the structure. We should bear in mind, too, that in *three* dimensions the lateral movement is semicentripetal toward the center of the trap. Depending somewhat on directional permeabilities, the lateral flow rate for each aquifer will tend to increase semigeometrically toward the center of the structure. If there is a vertical sequence of aquifers, each one feeding the center of the stack, the rates of upward water movement will be additive. Obviously, for a functioning trap with deep-water-discharge, the vertical rate of flow is very likely to exceed the lateral rate. For this reason the pore-volume-exchange rate in certain prospective reservoirs may prove to be excessive for the retention (residence time) of hydrocarbons.

In Figure 4, the dashed numberless isobars show a two-dimensional pressure sink. The shading represents pressure intensity conforming to the isobars. The pressure change experienced by the water in *vertical* movement is essentially the hydrostatic gradient, or 0.465 psi/ft (0.116 Atm/m) for an average brine. That's about 233 times the lateral gradient, so an abrupt pressure drop is imposed on the feedstock water as it

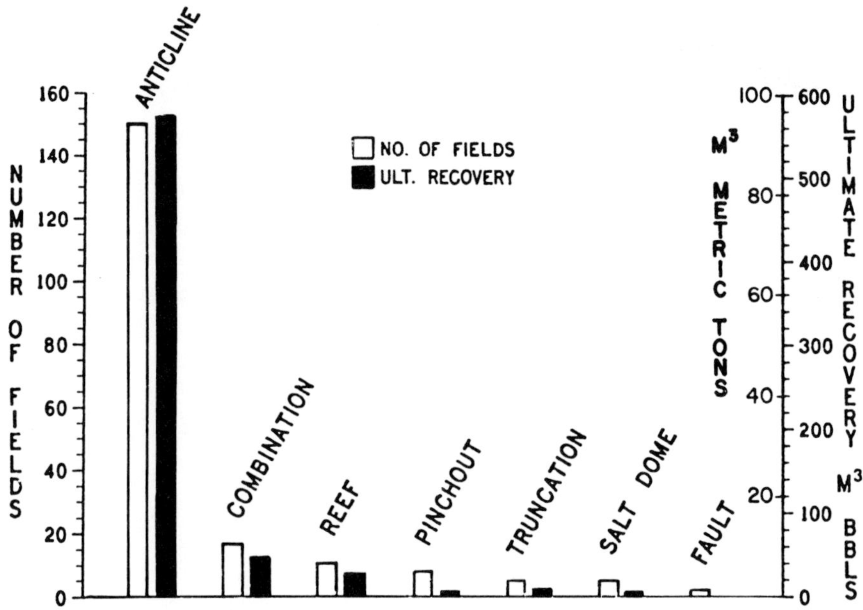

FIG. 3—Compared oil volumes held by different types of known giant traps (from Moody, 1975).

passes upward through the trap.

Temperature Change

The temperature diagram in Figure 5 is similar to the pressure diagram (Fig. 4). Once again the shift from lateral to vertical movement causes a sudden change in gradient. Except where vertical water movements occur, lateral temperature gradients at a given depth are commonly gentle. Vertical gradients are appreciable, however, as recorded from down-hole temperature measurements. Figure 5 shows a typical vertical gradient of about 1.4°F/100 ft (24°C/km) or 0.014°F/ft (0.024°C/m) as compared to a lateral gradient of 0.00014°F/ft (0.00024°C/m). Thus a reasonable change in rate of temperature drop from lateral to vertical movement might be at least 100-fold. Incidently, whereas in Figure 4 the pressure isobars are depressed in the anticlinal discharge area, the temperature isotherms in Figure 5 are raised in the central zone, in a typical hydrothermal attitude. The shading for thermal intensity conforms to the isotherms.

Salinity

The salinity diagram in Figure 6 is more conceptual than quantitative. It is assumed to be the same aquifer as in the pressure and temperature diagrams, but the arrows showing sense of water movement are left off for simplicity. The several strata above the aquifer represent aquitards which possibly contain clays and serve as membranes or filters with respect to the cross-flow of salty waters. Laboratory and field observations (see reference group "E" in the Appendix of this paper) support the assumption that some salt is apt to be held back on the high-pressure side of the membranes. Therefore, the tendency in a trap is to build up salinity in the area of greatest vertical flow. The effect is likely to spread laterally near the base of the system, as suggested by the shading here. The upward-converging dashed lines represent outwardly diminishing isosalinity values.

The above interpretation of a trap as an active, salt-concentrating apparatus may help to explain the fact that waters closely associated with oil and gas are commonly more saline than those elsewhere. Inasmuch as increased salinity may cause organic materials to be exsolved ("salted out") from the water, a trap which is effective in concentrating salt probably will be effective also in concentrating hydrocarbons if they are available.

WHAT TRAPS HAVE DONE—FIELD EVIDENCE

Field evidence of deep-water-discharge related to important known oil deposits is needed. The most visible and available evidence is related to hydrology, salt solution, water chemistry, pressure, and temperature.

Yates Field, West Texas

Figure 7 is a simplified cross-section of a giant oil deposit, the Yates Field in West Texas, which produces from the Lower Permian "Big Lime." The removal and thinning of salts in the cover by solution, and the proximity of important surface drainage (Pecos River), fed in part by the discharging waters, are clearly indicative of deep-water-discharge which prob-

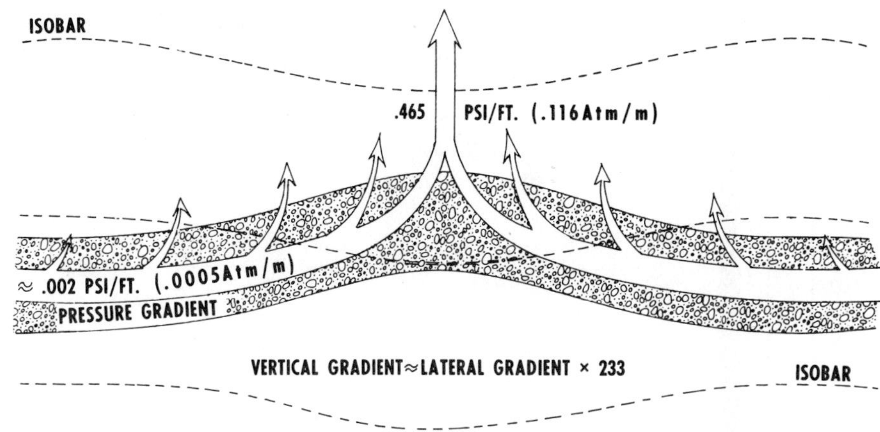

FIG. 4—Pressure change imposed on convergent, upward-moving waters and their contents.

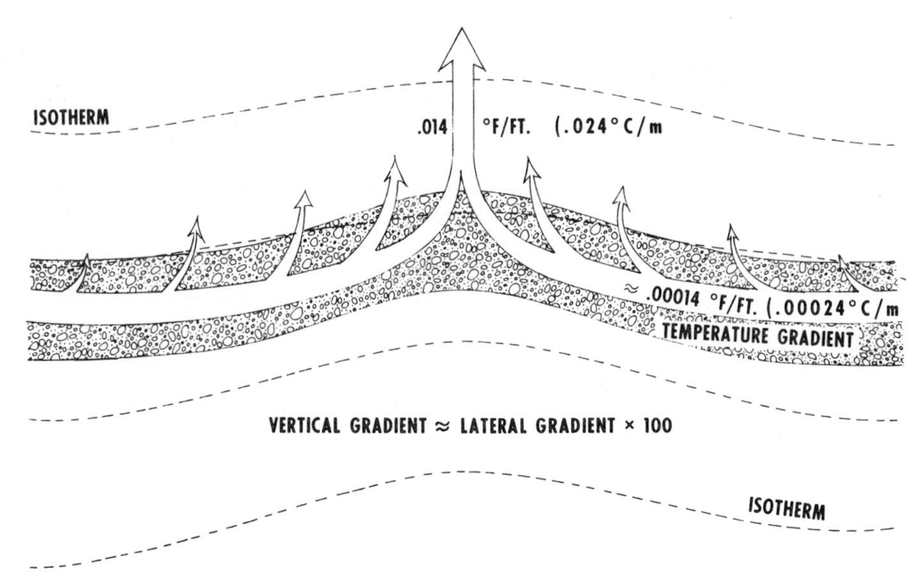

FIG. 5—Temperature change imposed on convergent, upward-moving waters and their contents.

ably continues in the present. It would be of interest to know if modern stream flow measurements reflect a local contribution from depth; also if brine springs have been noted in the area.

Ochoa Sag, Permian Basin

Figure 8 (after Hiss et al, 1969) also shows removal of salt by deep-water-discharge. In this case it is the Ochoa Series on the west side of the Central basin platform in the Permian basin of West Texas. Here the anomalous vertical transmissibility is related to an extensive, deep, fault zone plus the grouping of permeable reservoir facies at the west edge of the platform. Due to solution and collapse, there is a pronounced sag in the strata (Triassic, Quaternary) above the salt. Figure 9 (after Hiss et al, 1969) is a map showing that salt removal trend which closely parallels a highly productive trend of oil and gas deposits in those permeable reservoir facies referred to above.

Athabasca Area

A profile of the second largest oil deposit in the world, the Athabasca heavy oil deposit, is shown in Figure 10 (Carrigy, 1974). This is a very large trap, functionally related, it seems, to the retreating eastern edge of the Devonian Elk Point evaporite series on the west flank of the Canadian shield. The waters from pre-Elk Point aquifers appear to have broken through and dissolved the Elk Point salts, allowing the Upper Devonian carbonates and evaporites and younger strata to sag and reverse the regional dip of the post-Elk Point strata near the Athabasca River. It's entirely possible that waters from the Alberta basin have been discharging in the Athbasca River area through much

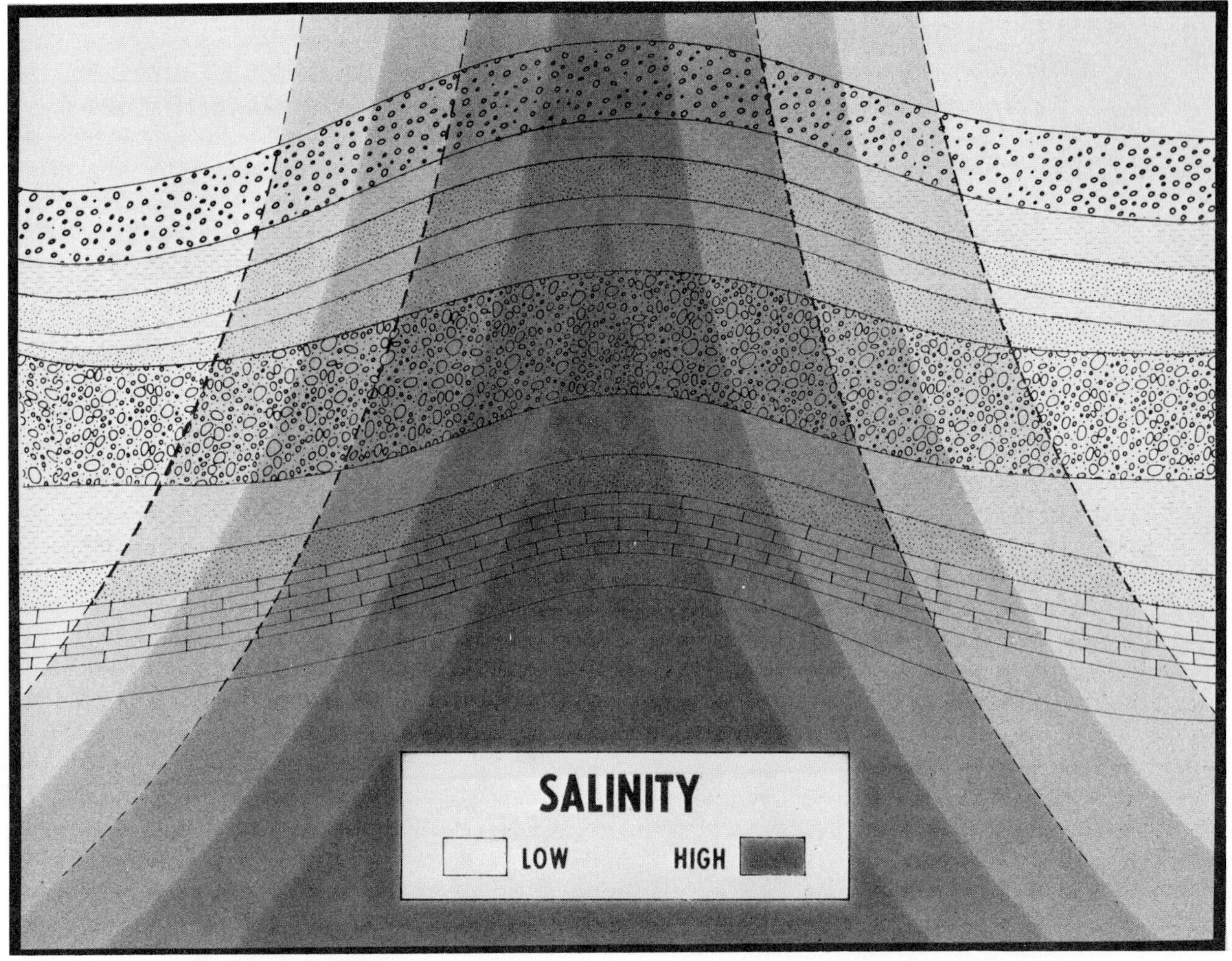

FIG. 6—Water salinity concentration as a trap function. In general, maximum salinity results from maximum vertical water movement through covering argillaceous membranes.

of the basin's history (pre-Elk Point and post-Elk Point deposition), as they continue to do in the present. It is, in fact, a tremendous swamp in which brine springs are common today. It is also a center of confluent surface drainage. It seems to be a classical, enduring, hydrocarbon seepage, functionally related to a major, basinal, water discharge system. Actually, the same conditions rimming the shield extend many miles southeastward through the Cold Lake and Lloydminster heavy oil areas as shown in Figure 11 (Orr et al, 1977). Note again the collapse and dip-reversal in post-evaporite strata. The oil-soaked Manville sandstones fill the depression caused by salt removal, just as at Athabasca.

East Texas Field

Another example of a major oil deposit related to increased vertical transmissibility is the East Texas field, schematically drawn in Figure 12. The Woodbine aquifer of the East Texas basin rising on the west flank of the Sabine Uplift is regionally covered to the west by the low-permeability Eagle Ford Shale, an effective aquitard. The shale pinches out by unconformity near the west side of the East Texas field, bringing the Woodbine sand aquifer in direct contact with the overlying chalks which are assumed to transmit water more easily than the shale. This weakness of cover above the truncated Woodbine aquifer is presumed to have allowed the discharge of large volumes of water ever since pre-Austin Chalk deposition. Plummer and Sargent (1931), prior to the development of the East Texas field, reported low pressures "along the east edge of the Woodbine sand on the flank of the Sabine uplift where the sand is not effectively sealed but thins out into an unconformity marked by a more or less porous erosion surface." Reports of paraffin dirt in the soils overlying the East Texas field suggest the leakage of some hydrocarbons with the water.

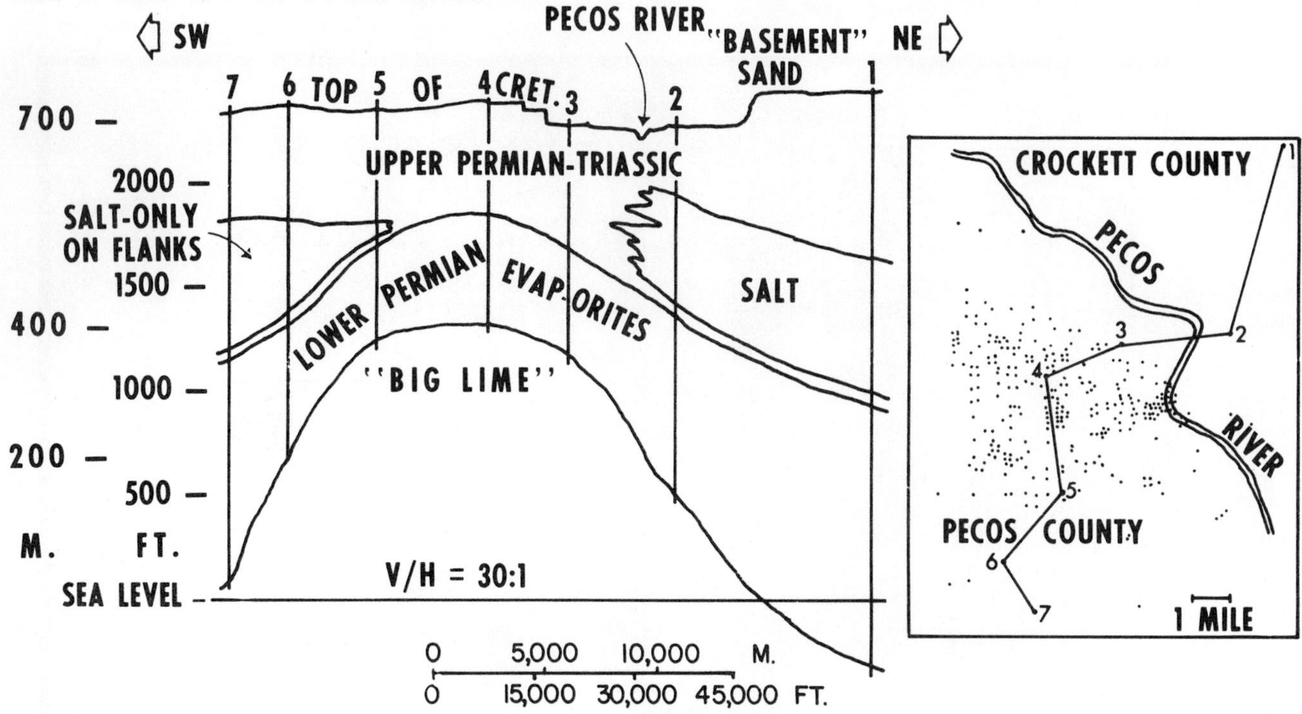

FIG. 7—Profile of Yates field, West Texas, showing removal of Permo-Triassic salts by deep-water discharge into Pecos River (from McCoy, 1934).

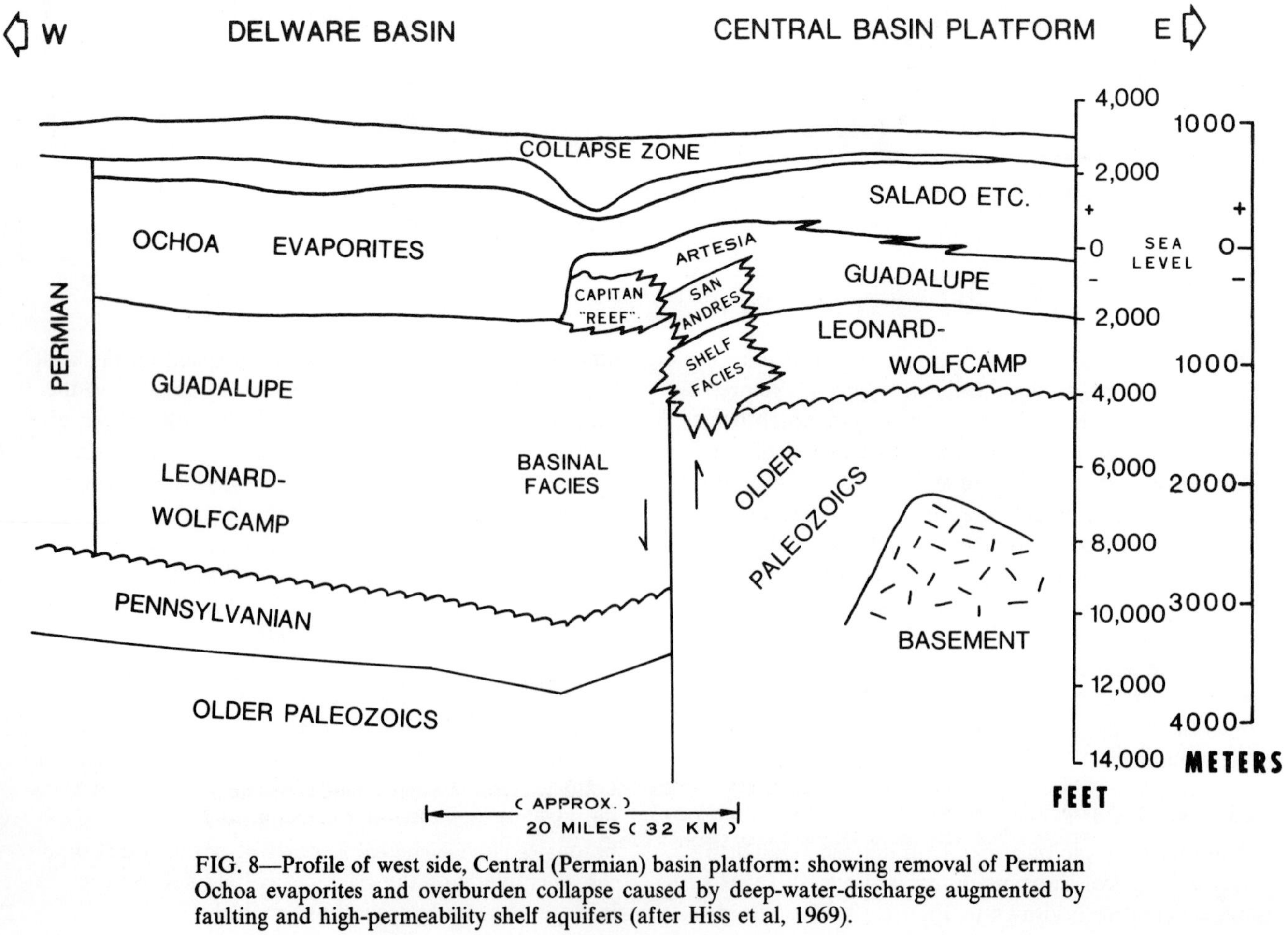

FIG. 8—Profile of west side, Central (Permian) basin platform: showing removal of Permian Ochoa evaporites and overburden collapse caused by deep-water-discharge augmented by faulting and high-permeability shelf aquifers (after Hiss et al, 1969).

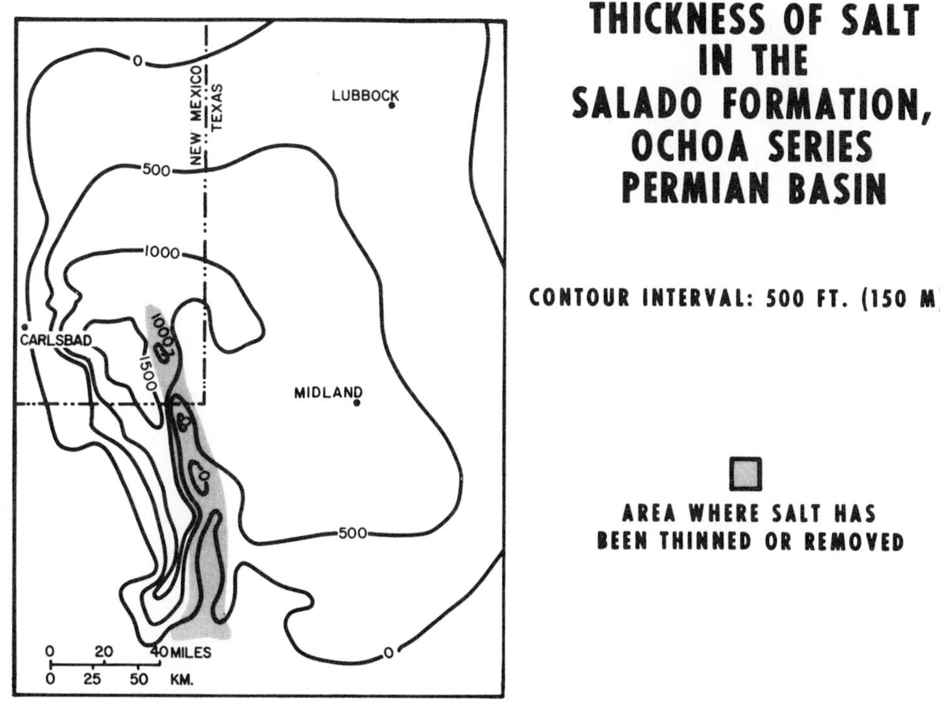

FIG. 9—Thickness of Salado (Ochoa) salt series in Permian basin. Contour interval 500 ft (153 m). Trend of thin or absent salt coincides closely with strong oil and gas production (after Hiss et al, 1969).

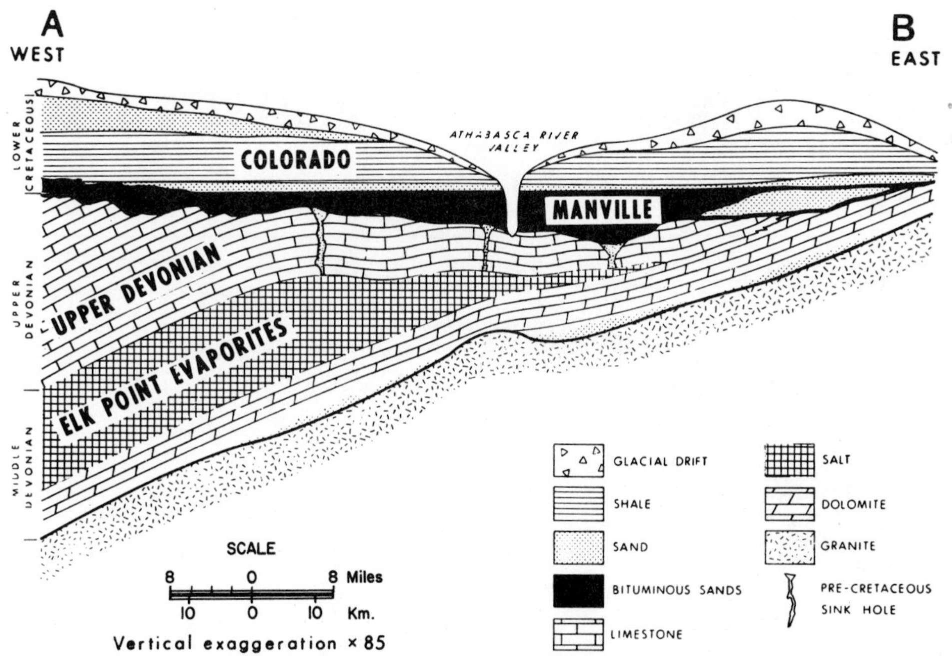

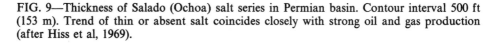

FIG. 10—Profile of Athabasca heavy oil deposit. Oil-saturated Manville sandstones lie in depression formed by solution and collapse of Elk Point salts. Deep-water-discharge and hydrocarbon seepage continue today (from Carrigy, 1974).

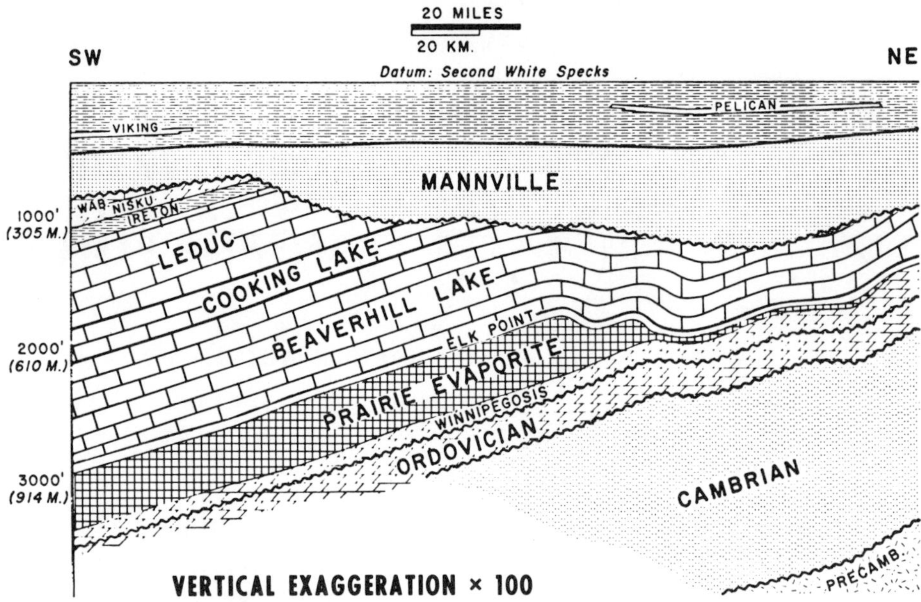

FIG. 11—Profile of Lloydminster heavy oil deposit, closely analogous to Athabasca (Fig. 10), except more pre-evaporite aquifer section is present and the Mannville sandstones are more effectively covered (from Orr et al, 1977).

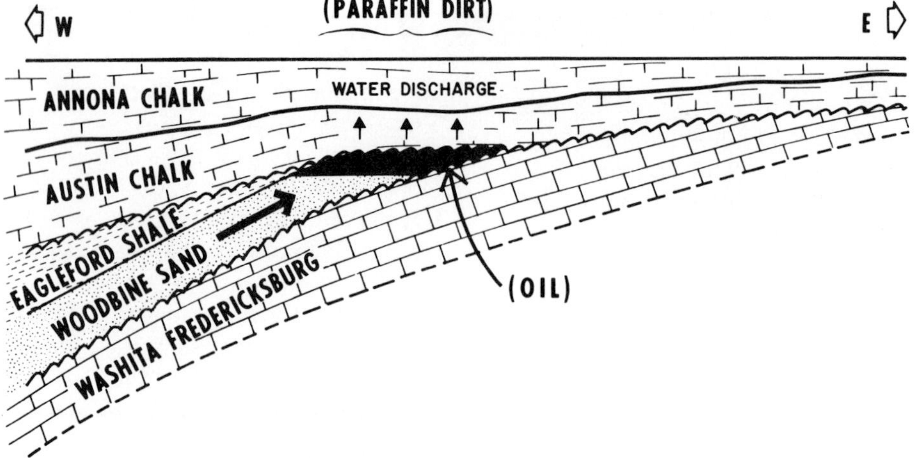

FIG. 12—Schematic profile of the East Texas field: showing local contact of the Woodbine aquifer with overlying chalks to allow escape of waters and accumulation of hydrocarbons east of the Eagle Ford Shale pinchout. (Gulf Research & Development Co., unpublished internal memorandum, 1968).

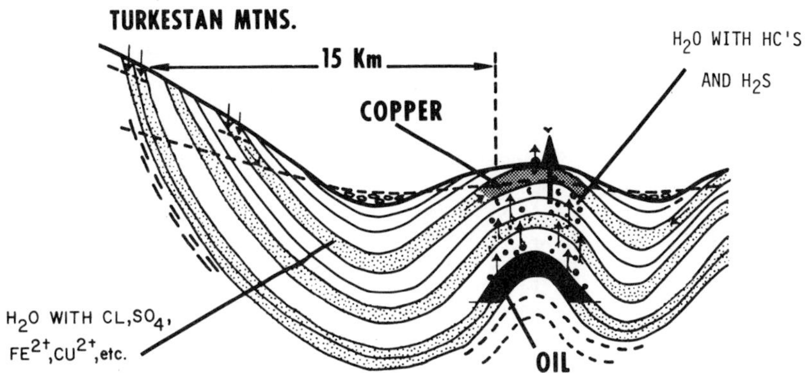

FIG. 13—Oil and copper ore trapped due to deep-water-discharge in the same anticline at Ferghana, USSR. Note oxidated waters outside trap, and reduced waters inside trap (from Germanov, 1963).

Ferghana

An example of a combined mineral and petroleum trap at Ferghana from the Russian literature (Germanov, 1963) is shown in Figure 13. Here, the deepest aquifer contains the oil and gas. The shallowest contains a sedimentary copper deposit. Water springs occur on the crest of the producing structure, and recharge is in the basinal borderlands. Note the waters off structure carry sulfates whereas those on structure are reduced. A reason for the reduction will be suggested later, in a discussion of the Don Medvedica area.

HYDROTHERMAL EFFECTS

In the preceeding examples, most of the evidence of deep-water discharge associated with oil and gas traps depends on cumulative process over some length of geologic time. However, there are two types of evidence which can have real time relevance to deep-water-discharge, namely *pressure* and *temperature*. The use of pressure, or potentiometric, data in monitoring basinal water movements is well known and needs only brief mention here. For practical reasons concerning water supply, most of the pressure work has been two-dimensional and applied chiefly to lateral flow. Three-dimensional studies (see reference group "F" in the Appendix to this paper) are of special interest because in a typical basin vertical transmissibility greatly exceeds lateral transmissibility.

Temperature related to low-speed water movements in oil- and gas-bearing sedimentary media seems to be a much-neglected area of study. Although temperature anomalies have been observed over a number of producing oil and gas fields, the anomalies often are erroneously attributed to factors other than water movement. Hydrothermal activity has long been genetically associated with mineral deposits, but this writer is not aware of any precise, quantitative field work on the heat exchange capacity of those mineralizing waters. It is obvious from temperature work (see reference group "G" in the Appendix to this paper) in oil-producing areas that ascending water velocities are sufficient to cause geothermal anomalies. Bredehoeft and Papadopoulos (1965) determined that the minimum rate of water movement thermally detectable was 0.1 cm per day or 36.5 cm (about 1 ft) per year. Updated techniques may have reduced the detectable minimum.

Some experimental data are available (Adivarahan et al, 1962) on the heat exchange capacity of brine moving through natural sedimentary and comparable synthetic porous media. In Figure 14, heat exchange in BTU/hr/sq ft (kg cal.Hr/234 sq cm) is plotted against brine flow in lbs/hr/sq ft (kg/Hr/420 sq cm) for five different porous media. As the brine flow is reduced toward zero, the heat exchange is also reduced toward nearly zero for all samples. This suggests that the heat exchange capacity of brine moving through porous media can dominate the conductivity effect of the matrix in governing the thermal regime. Translating this deduction into the subsurface, we can expect that wherever there is appreciable upward movement of warmer waters from depth there should be a thermal signal, like a very mild hot spring. It is reasonable to assume that the residual effects of such flow would be to some extent cumulative as long as the flow continues. However, it should be noted that unsupported thermal anomalies would be expected to have a rapid decay rate. Therefore, any anomaly observed today must represent an active system.

Rangely Field

Figure 15 (from Cupps et al, 1951) is a profile of temperature data in the Rangely field of northwestern Colorado. The data came from a rather tight Pennsylvanian quartzitic sandstone which depends on fracture permeability for production. The steep side of the anticline, of course, is more fractured; in fact a thrust

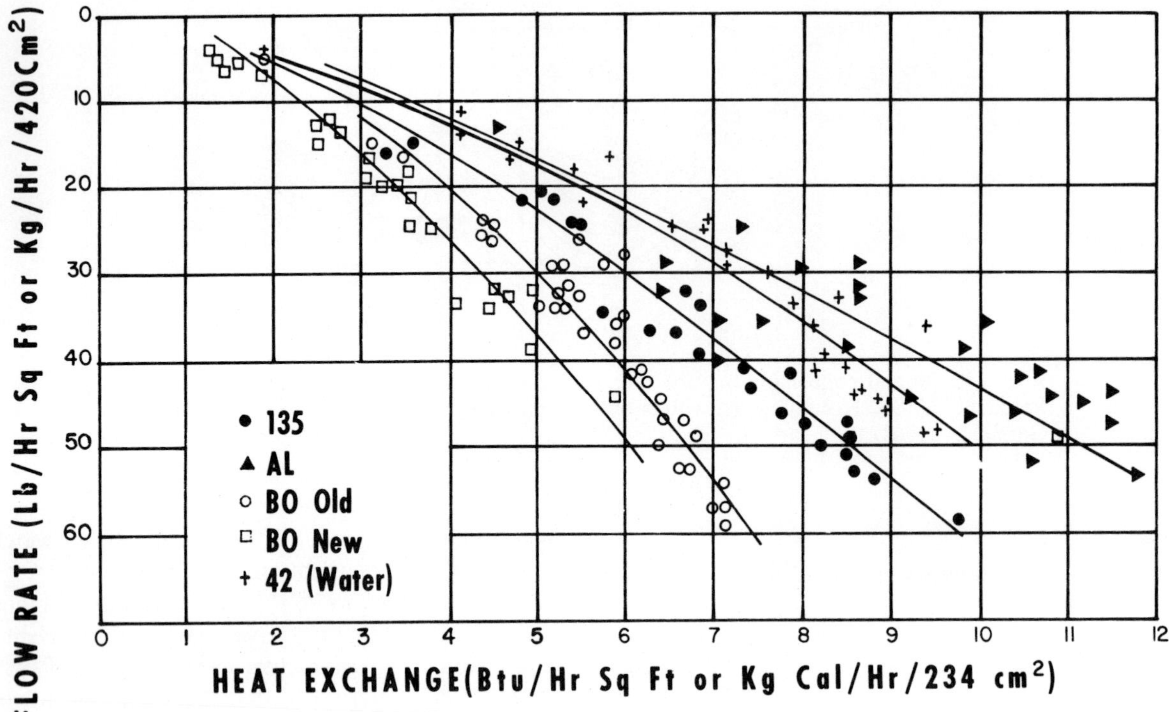

FIG. 14—Hydrothermal effects of brine flow. Graph shows rate of heat exchange compared to rate of flow by 10% NaCl solution in five different porous media. Heat exchange is almost directly proportional to brine flow (from Adivarahan et al, 1962).

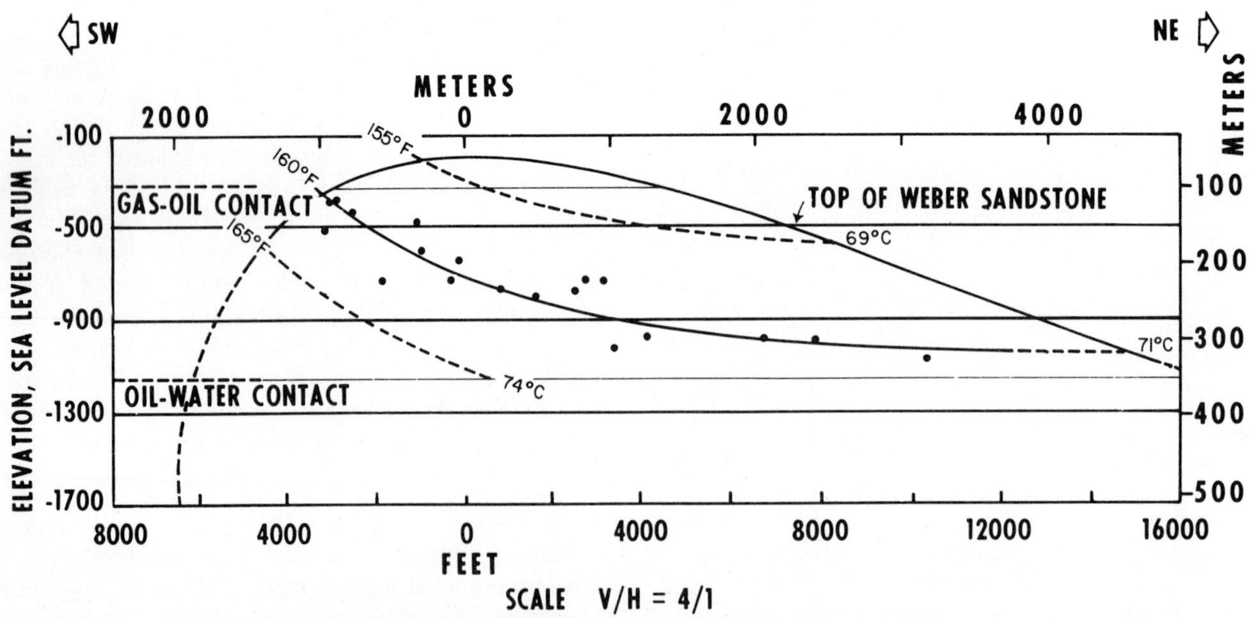

FIG. 15—Geothermal profile of the Rangely field, Northwest Colorado. Data are stable temperature measurements in Weber fractured quartzitic reservoir (from Cupps et al, 1951).

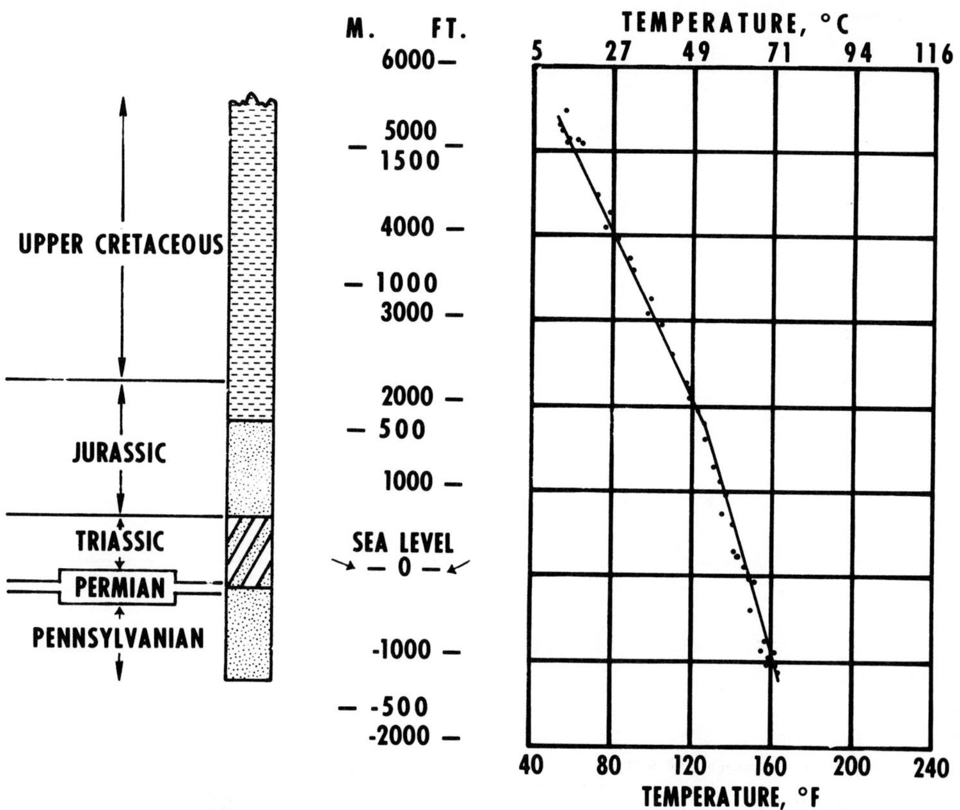

FIG. 16—Temperature-depth graph compared to generalized stratigraphic section in the Rangely field, Northwest Colorado. Note change of pressure-depth gradient opposite change of lithology at 1,800 ft (554 m) (after Cupps et al, 1951).

fault exists at greater depth. Apparently there is enough extra vertical water movement to lift the isotherms on that side. The writer is not aware of any hydrodynamic tilting of the oil/water contact at Rangely, though it is suspected.

In Figure 16 (from Cupps et al, 1951) there is a relation between the stratigraphic column and a temperature-depth plot at Rangely. At about 1,800 ft (550 m) above sea level, the lithology changes from mostly sandstone below to mostly shale above. At the same level, there is a distinct change in the slope of the temperature data. The simplest interpretation is that temperatures tend to be equalized in the more transmissible sandy media, where fluids are relatively free to move, whereas there is greater temperature difference across the less transmissible shale. With respect to the lower sandstone section, temperatures probably are above average because of a heat-bank effect, due to confinement below the shale aquitard. The case illustrates our need for *interval* temperature data, rather than merely average geothermal gradients based on maximum temperatures from total depths.

Ob Riber Area

Temperature maps (Koshlyak, 1963) of a producing area on the Ob River in West Siberia are shown at two different depths in Figure 17. The strength of the temperature anomaly seems to be about 10°C as mapped. The structure here overlies an igneous basement ridge, shown in Figure 18. It might be said that conductivity from the basement ridge is causing the high temperature. However, the fact that the system is discharging water into the Ob River suggests rising waters are dominating the temperature regime.

Trans Caucasus and Baku

Some impressive regional temperature work (Sukharev, 1964) in the Trans Caucasus region between the Black Sea and the Caspian Sea is shown in Figure 19. Up to 70°C of thermal relief between high and low structural positions is shown at a common depth of 2,000 m. At the southeastern corner of the map is the Aspheron Peninsula and the First Baku area on the west side of the Caspian Sea where additional temperature detail and other data are of special interest. Shown by dashed lines on Figure 20 (from Ovnatanov and Tamrazyan, 1962) are a number of temperature anomalies related to structure and showing as much as about 15°C of relief at a 500 m datum in the Baku area. This is near the south shore of the Aspheron peninsula. The temperature control is based on over 2,300 stabilized, down-hole measurements. The anomalies on the major, south-plunging anticline are closely associated with local oil producing closures and ter-

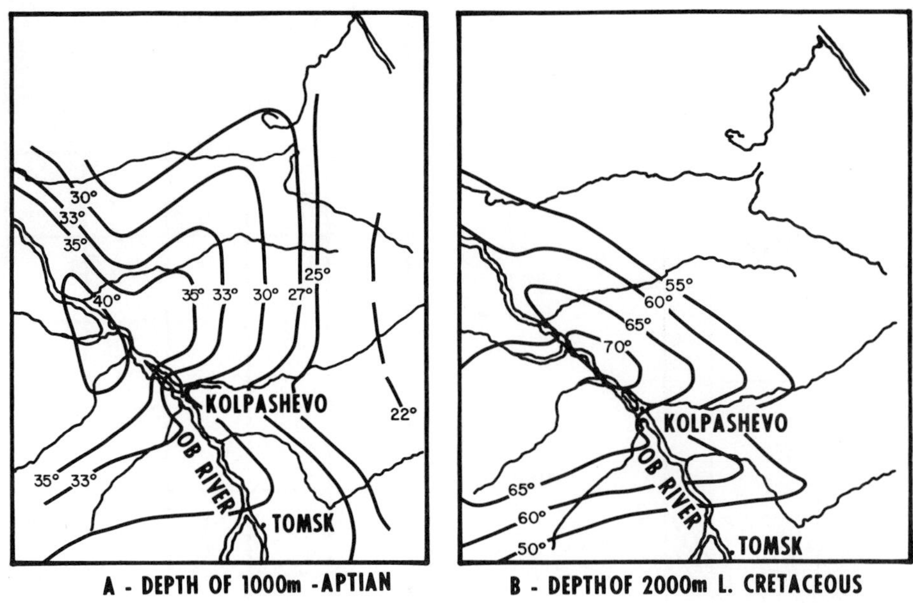

FIG. 17—Geothermal maps near Kolpashevo on the Ob River in West Siberia (from Koshlyak, 1963).

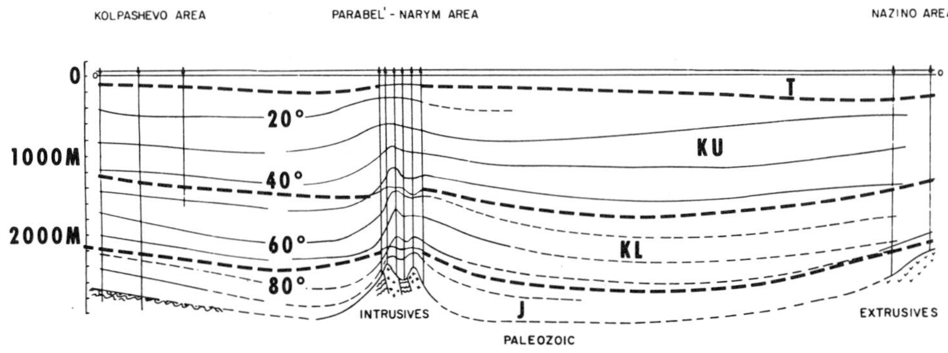

FIG. 18—Geothermal profile of area near Kolpashevo shown in Figure 17 (from Koshlyak, 1963).

races on that feature. A more deeply buried anticlinal trend running east-west near the shoreline also supports several temperature anomalies which seem to be associated with seismically defined structures below an unconformity.

Zykh Field, Baku Area

The 40°C anomaly (Oil Field No. 3) near the shoreline in Figure 20 is detailed in Figure 21. This map, of the Zykh field, instead of showing temperature variation at a common depth, shows depth variation at a common temperature. The contours show about 300 m of relief on the 40°C isothermal surface. The trap here is also expressed by surface hydrology. Lake Zykh, which is believed to be in part spring-fed from depth, lies almost directly over the intersection of the older anticline and the younger anticline. Both structural trends, north-south and east-west, are reflected by lineations in the shorelines of the lake. Near the center of the geothermal closure there is a large, active mud volcano expelling gas, brine, and microfossils identified as coming from considerable depth.

The upward movement of water at depth in the Zykh field also seems to be proven by the water chemistry shown in Figure 22. The graphs are based on more than 5,000 water analyses (Zargaryan, 1962) from producing zones. The two producing zones, upper and lower, are separated by about 200 m of dense carbonate rock, a barren zone which has been regarded as a fluid barrier based on real-time interference

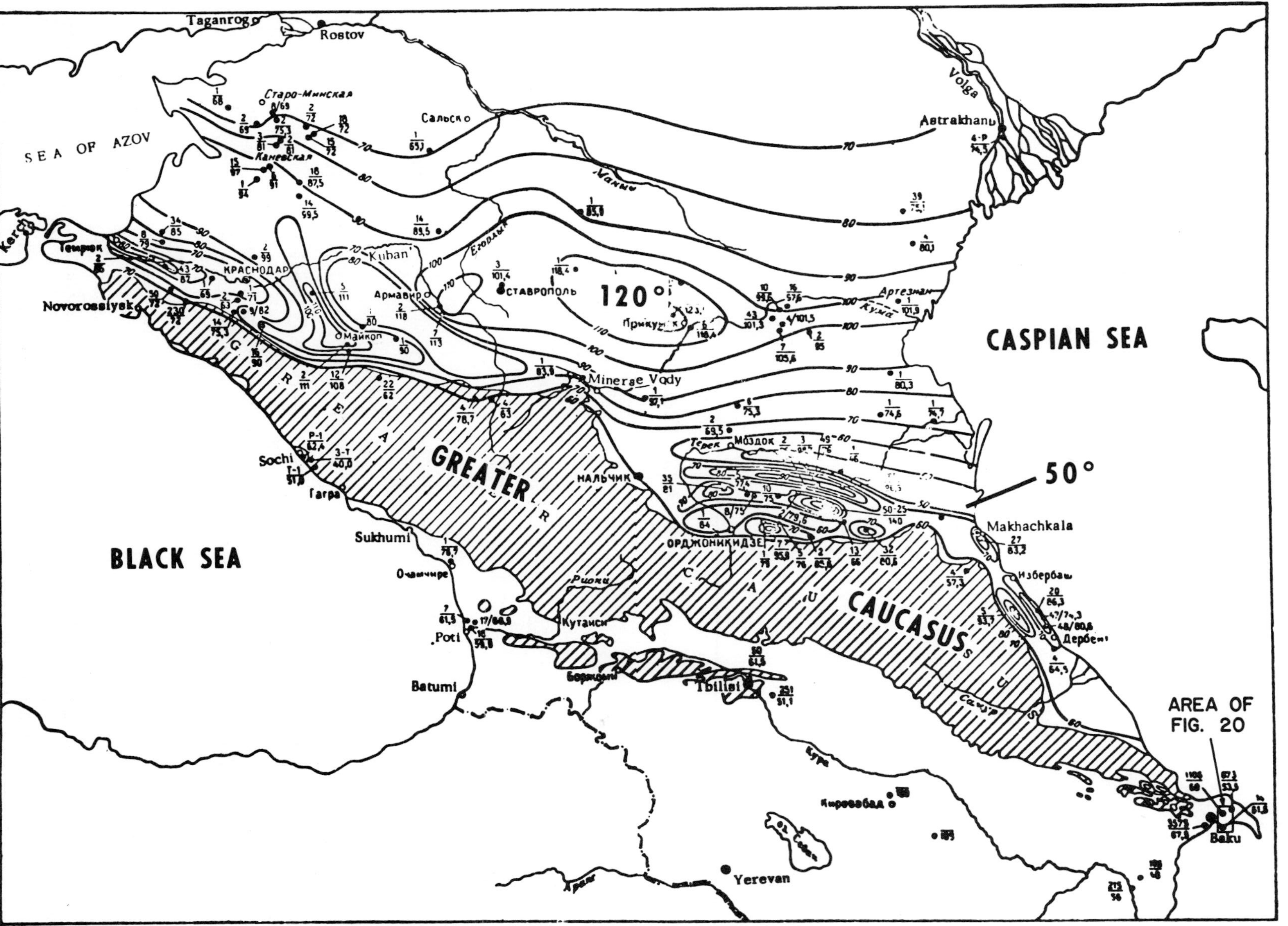

FIG. 19—Regional geothermal map of Trans Caucasus at 2,000 m depth. Note difference between maximum and minimum is 70°C (from Sukharev et al, 1964).

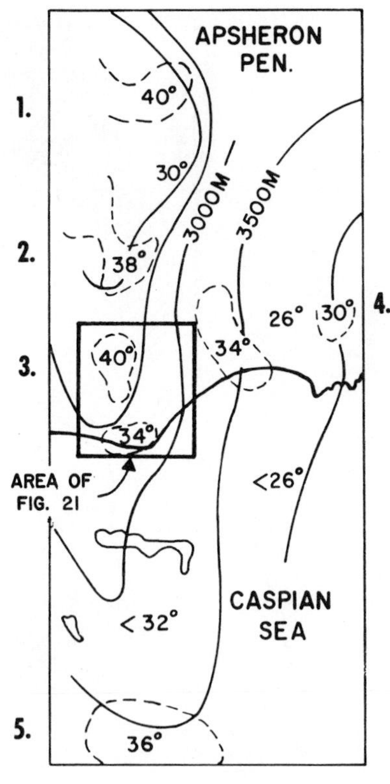

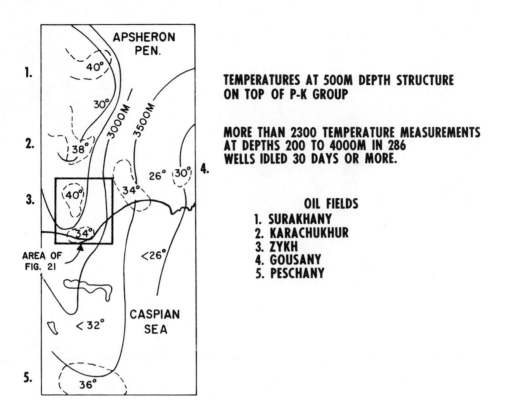

FIG. 20—Baku area temperature, structure, and oil fields. Location shown on Figure 19 (after Ovnatanov and Tamrazyan, 1962).

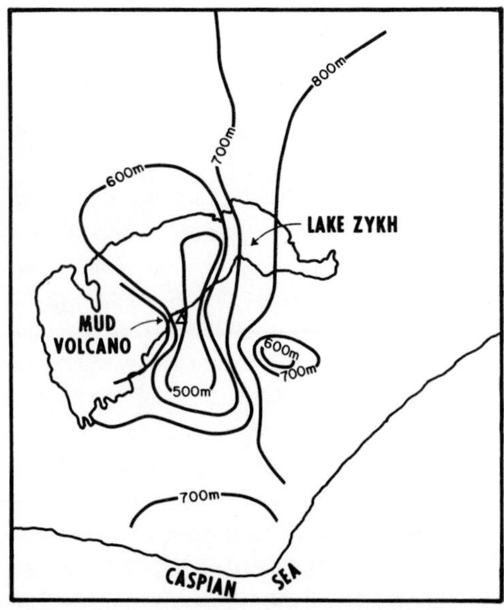

FIG. 21—Isothermal depth map at 40°C, Zykh field, Baku area. Location shown on Figure 20 (from Ovnatanov and Tamrazyan, 1962).

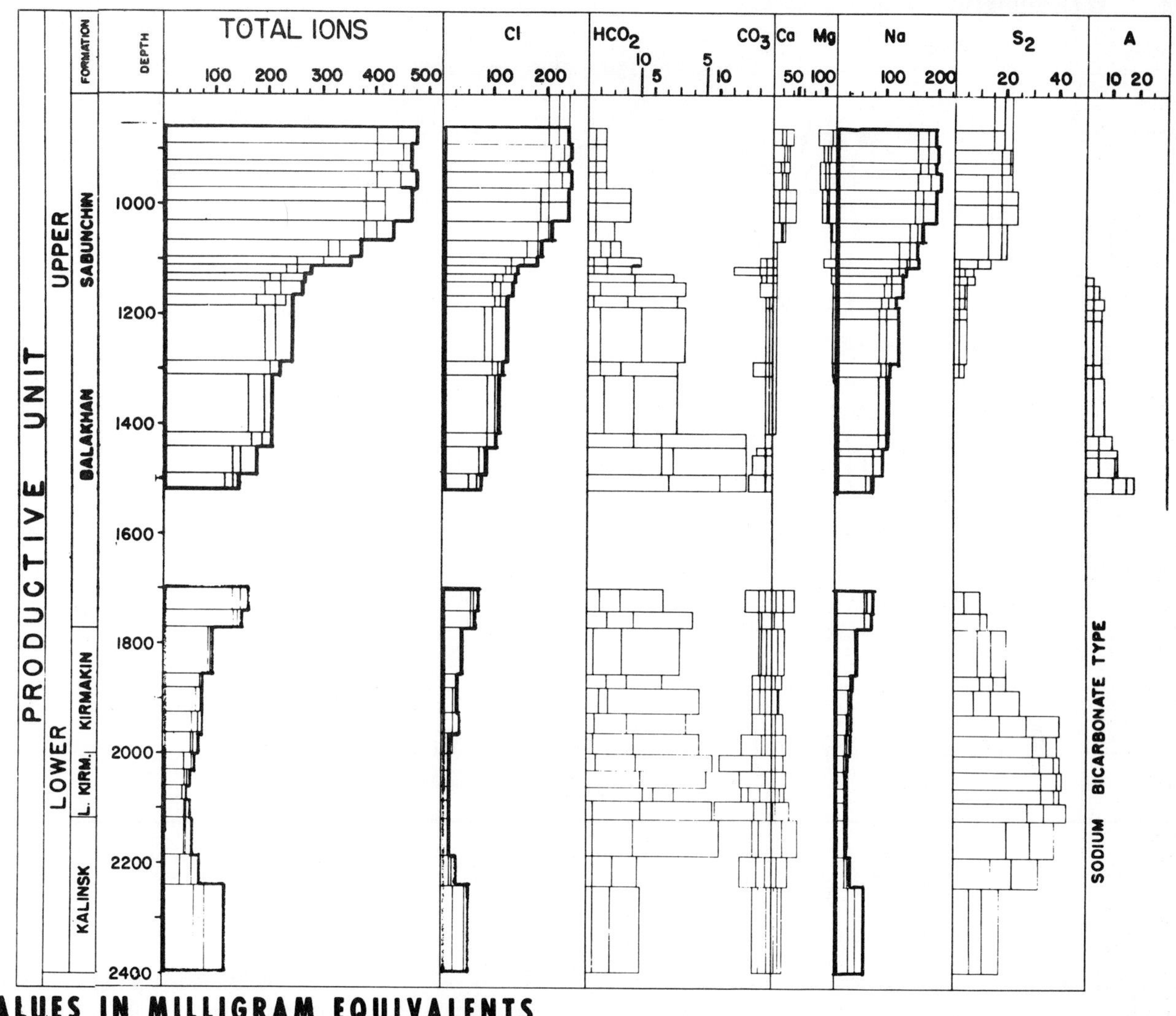

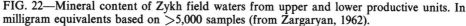

FIG. 22—Mineral content of Zykh field waters from upper and lower productive units. In milligram equivalents based on >5,000 samples (from Zargaryan, 1962).

tests. The progressive, upward increase in water mineralization (total ions) however, ignores the so-called barrier, suggesting that in geologic time the ascending waters do not "see" it. The heaviest most mineralized waters are uppermost in the reservoir system. This contradicts gravity and shows the trap to be a forced-draft, salt-concentrating mechanism. It is quite uncommon to find a producing oil field with such a convincing combination of structural, hydrological, hydrothermal, and hydrochemical evidence of deep-water-discharge. It provides incentive to seek similar data where they may be available elsewhere.

Don Medvedica Area, on the Volga

A producing area with several signs of deep-water-discharge is shown in Figure 23 (from Meinhold, 1971). It is an uplift paralleling the Volga River and, like the river, it probably is associated with deep faulting. The structural outline is shown by the dip symbols. Temperature relief is shown to be at least 15°C at 2,000 m depth on the structure. About 18 oil deposits occur in the area.

Of special interest in Figure 23 is the discharge from depth of chemically reduced waters, shown by crosses on the map. It is proposed that a reduced condition is characteristic of oil and gas deposits and is a reliable criterion of deep-water-discharge, because the ascending waters bring the reduced condition with them and thus impose it on the traps as they pass through. It does not seem necessary that the enriched brines or the reducing environments found in traps should be regarded as proof of fossil, sealed-in (original) conditions. Moreover, if the geologic structuring of the cover remains coherent and convergent, the reducing effect should persist upward, perhaps reaching the surface. It seems reasonable to assume that compared to ambient waters, the ascending waters are reduced because they have been hotter, and pressured longer, at greater depths where they have given up or exchanged all available oxygen. In relation to the am-

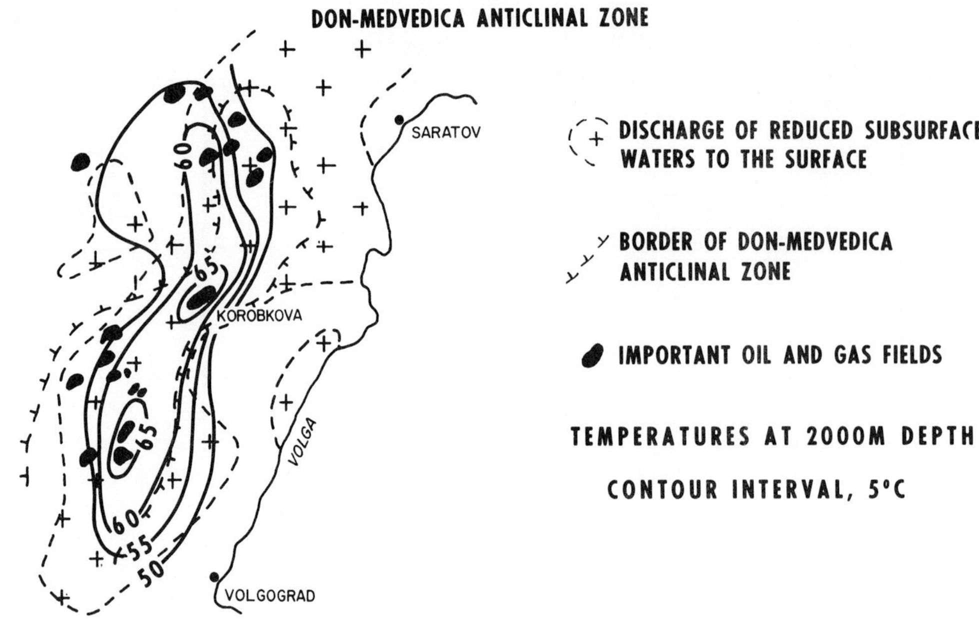

HYDROLOGIC CONTROL OF OIL AND GAS ACCUMULATION SHOWN BY GEOTHERMAL, AND HYDROCHEMICAL PATTERNS

FIG. 23—Temperature, structure, and reduced waters in Don Medvedica oil and gas area (from Meinhold, 1971).

bient shallower environment, these waters are oxygen hungry, or reduced. As criteria of reducing conditions in areas where evaporites occur, and sulfates thus are easy to come by, we typically find reduced compounds like pyrite and hydrogen sulfide associated with areas of entrapment. It may be of interest to consider the possible magneto-electric effects of a reduced plume of water ascending through sedimentary cover.

Other Trap Functions

Accumulation

The foregoing evidence shows clearly that oil and gas traps are appreciably and continuously active systems. The conditions for unloading organic material from a water feedstock passing upward through traps do seem capable of gradually, perhaps molecule by molecule, building the oil and gas deposits. As the organic materials coalesce to form bubbles or globules, gradually growing in size, they cannot have the same freedom of movement as the water. If eventually pore-to-pore continuity and appreciable saturation of the pore space by hydrocarbons is achieved, then movement of the separate phase could occur through the more permeable media *if* enough pressure differential is applied.

Trap Content

The discussion of trap function in the formation of oil and gas deposits can be extended to explain wide variations of trap efficiency, degrees of saturation, oil and gas composition, maturation, and modification of the primary organic source imprint on the trapped hydrocarbon mixtures. None of this is treated in any detail here, but based on what has been discussed, a rather astounding deduction can be made. The amount and the composition of a trapped hydrocarbon mixture at any geologic moment will depend largely on the residence times of the various components of that mixture, which in turn depend on three things: (1) what the water brings , (2) what the trap retains, and (3) the pore-volume-exchange rate.

Maturation and Residence Time

Experience shows that a tighter (commonly more deeply buried) trap associated with an average organic source system is likely to retain gas. A looser, more open (commonly shallower) trap associated with a similar source system is likely to release gas and retain oil. The tight trap has a very slow rate of pore-volume-exchange for its feedstock water. The loose trap has a more active water exchange. On the one hand, increased water exchange means the more water-sensitive organic compounds (i.e., lighter hydrocarbons) are more likely to be washed on through, thus having a shorter residence time in the trap. On the other hand, the increased flow has a better chance to bring into the trap the heavier, more complex organic compounds which are less easily accommodated in water and therefore less available in the migrating system. However, once in the trap and separated from the water, such heavier organic compounds are apt to have a longer residence time than the others. The resident mixture, or aggregate, thus tends to "residualize" and evolve toward a heavier product.

Differential Entrapment

Given a typical spectrum of organic source materials available in the feedstock waters, the resident mixture in the trap will gradually evolve or "mature" from lighter to heavier. Such maturation can be a function of either time or pore-volume-exchange rate, or both. Obviously, this type of maturation *in the trap* should have at least as much control over the content of the trap as would the conditions of an oil or gas source system outside the trap.

From the above, it follows that the density, weight, or richness of a trapped hydrocarbon mixture may be proportional to the cumulative pore-volume-exchange by feedstock waters. A trap that is too loose may contain only heavy residual oil staining. That doesn't prove that a complete, pre-existent oil deposit has been washed out, or flushed. It could mean that because of excessive water exchange, the trap was never able to hold anything except the heaviest, least mobile hydrocarbons (probably in the least accessible crevices or pore spaces), the light ends having been continuously washed right on through. This is differential entrapment in a very logical sense, based on the discriminating power of the trap.

Because traps which are either tighter or younger will have experienced less feedstock exchange, the trapped mixtures are apt to be dominated by the organic components most abundant in the available waters, namely the volatiles or light ends. The tendency for traps to contain gases or condensates, therefore, may be related to either deep burial or a short period of trap function (accumulation).

It is considered of utmost significance that oil and gas deposits of all depths and apparent ages, and regardless of the degrees of enrichment (from "shows" to giant fields), appear to be closely adjusted to their present physical and chemical environment. At the same time, it should be pointed out that the composition of resident hydrocarbon mixtures shows evidence of continual adjustment to changing conditions rather than static, thermodynamic equilibrium.

WHAT TO LOOK FOR

Table 1 is a list of surface criteria signaling active entrapment. Most of the surface phenomena are easily observed. Sometimes they are associated with oil and gas deposits, but perhaps the reasons for the association have never been very clear. No one or two of these criteria will promise an oil field; however, if we see several of them in one place it does reinforce the probability of deep-water-discharge, which is favorable. That, combined with other favorable information, might give us the courage to drill for a discovery.

Table 1. Deep-Water Discharge: Surface Observations

Creekology	Hydrocarbon Seepage
Topography	Mineral Haloes
Brine Springs	Magneto-Electric Anomalies
Fresh Springs in Arid Areas	Soil Changes (Non-Facies)
Water Seepage	Fault Intersections
Reduction	Fracture and Joint Swarms
Vegetative Closures	Landsat Features

This paper has presented examples of some of the surface criteria associated with oil and gas production; namely creekology, brine-springs, reduction, and hydrocarbon seepage. The reader should have no trouble identifying and interpreting the others.

Table 2 is a list of subsurface criteria signaling active entrapment. The subsurface criteria are less available because of drilling expense. But if wells are to be drilled, and the subsurface geology is going to be worked up anyway, these phenomena should be included in the observations.

Among the subsurface criteria, this paper has shown examples of cover weakness, thinning aquitards, solution and collapse, temperature anomalies, strong brines, and reduction. These and other signs of ascending water should be observed whenever possible.

CONCLUSIONS

Because of the ability of the rock apparatus to regulate and control water movements, and because of the signs of active water movement in old as well as young basins, it seems very possible that the waters

Table 2. Deep-Water Discharge: Subsurface Observations

Cover Weakness	Water Chemistry
Thinning Aquitards	Inorganic
Shales	Strong Brine
Evaporites	Reduction
Solution and Collapse	Organic
Permeable Carbonates	Acids, Carbohydrates
Increased SD/SH Ratio	Hydrocarbons
Low Pressure	Caprock Mineralization
High Temperature	Anomalous Velocity

could carry a lot of organic raw material from source to trap. From the preceding evidence, it is quite clear that the most common types of traps are designed to collect and retain that material. The trap is indeed the most logical place for oil and gas to take form as such.

It seems quite possible that trapped organic compounds, in fine-grained as well as coarse-grained rocks, may act as secondary sources. Maybe the enriched fine-grained rocks commonly called "source rocks" should be regarded as *organic banks* from which withdrawals can be made by exposure to heat and percolating waters. Perhaps we should be thinking and talking about source *systems* instead of about source *rocks*.

Information on the effects of changes in pressure, temperature, and water chemistry on the accommodation of various organic materials in water is still very limited. Additional field sampling and measurement are clearly needed. Many oil-field waters and test waters are simply disposed of as a nuisance, or else they're merely typed for identification by mineral content. Few investigators seem to care about the organic content, except the ecologists and the corrosion engineers.

The forced-draft water-model seems to solve many of the problems of petroleum migration. It may even cast a different light on source problems—including what a truly primary source might be expected to deliver to the system. It has been argued here that a trap is a collective system. Such a concept does not require whole oil to be issued from the source end of the system. One can think of the primary organic derivatives, whatever they are, as being scattered and entrained in the restless waters. A lot of them will find their way into the traps, if that's where the waters are going.

The importance of trap function has been stressed. There are compelling reasons to re-think some common sense geology, hydrology, and geochemistry. The discharge of chemically imprinted waters from depth shows up on air photos and satellite imagery in the form of anomalous soil and vegetation patterns. Basic "creekology" can show where to go for ascending waters and how to conserve with the more expensive exploration tools. Understanding how traps work can also help us to interpret the inevitably noisy geochemical data which are coming in such a rush today.

The rock framework should be regarded as an apparatus, in which things are happening because it is a restless system. Pressures and thermal energies are continuously applied to it. All we have to do is read the signs. In a few words, the advice seems to be—look for the leaks. Look not only at the traps, but at the cover. That's where the vital leakage occurs.

APPENDIX: REFERENCE GROUPS CITED IN THIS PAPER

The writer has grouped together convenient reference sections for the interested reader. These seven groupings are recommended reading, according to topic, as cited in this manuscript at various points. The papers cited here are all included in the *References Cited* section immediately following.

Reference Group "A"

The reader is referred to the following: Amirijafari and Campbell (1970), Baker (1973), Blumer (1965), Blumer and Cooper (1967), Bray and Foster (1979), Brod (1960a, 1960b, 1963a, 1963b), Brooks (1969), Carothers and Kharaka (1978), Caudle (1977), Collins (1975), Colombo (1967), Colombo et al (1965), Cooper and Bray (1963), Degens et al (1964a, 1964b), Duke and Seppi (1977), Eganhouse and Calder (1976), Fokeev (1956), Germanov (1963), Gullickson et al (1961), Hedberg (1968,1978), Hodgson (1978), Hunt (1973a, 1973b), Illing (1959), Johannes and Webb (1965), P. H. Jones (1976, 1978), R. W. Jones (1978), Kartsev et al (1959), Kidwell and Hunt (1958), Kuznetsova (1963), Matusevich and Shvets (1974), McAuliffe (1969a, 1969b, 1978), Meinhold (1971, 1977), Meinschein (1959), Neglia (1979), Ogner and Schnitzer (1970), Parker (1967), Peake and Hodgson (1965, 1968, 1973), Peake et al (1972), Price (1973, 1975, 1976, 1978), Rich (1921), Roberts (1960, 1966, 1968), Seifert and Howells (1969), Seifert et al (1969, 1970), Sellers (1965), Sever and Parker (1969), Sokolov et al (1963), Stewart and Nielsen (1954), Sukharev et al (1964), Swanson et al (1968), Van Orstrand (1934), Van Tuyl et al (1945), Wershaw and Pinckney (1977), Wershaw et al (1977), Witherspoon and Saraf (1965), Yushkevich and Zhuze (1958), Zarrella et al (1963), Zhuze (1960, 1967), and Zhuze et al (1971, 1977).

Reference Group "B"

The reader is referred to the following: Abelson (1963), Abelson et al (1962), Amirijafari and Campbell (1970), Blumer (1965), Blumer and Cooper (1967), Bray and Foster (1979), Brod (1960a, 1960b, 1963a, 1963b), Brooks (1969), Carothers and Kharaka (1978), Collins (1975), Colombo (1967), Colombo et al (1965), Cooper and Bray (1963), Degens et al (1964a, 1964b), Duke and Seppi (1977), Eglinton (1972), Eglinton et al (1966), Gullickson et al (1961), Hedberg (1968), Hodgson (1978), Johannes and Webb (1965), Kartsev et al (1959), Matusevich and Shvets (1974), Meinhold (1971, 1977), Meinschein (1959), Ogner and Schnitzer (1970), Parker (1967), Peake and Hodgson (1965, 1968, 1973), Peake et al (1972), Seifert and Howells (1969), Seifert et al (1969, 1970), Sellers (1965), Sever and Parker (1969), Sokolov et al (1963), Swanson et al (1968), Wershaw et al (1977), Wershaw and Pinckney (1977), and Zhuze et al (1977).

Reference Group "C"

The reader is referred to the following: Blumer (1965), Carothers and Kharaka (1978), Collins (1975), Colombo (1967), Colombo et al (1965), Gullickson et al (1961), Hodgson (1978), Kartsev et al (1959), Matusevich and Shvets (1974), Seifert and Howells (1969), Seifert et al (1969, 1970), Sellers (1965), Sever and Parker (1969), and Sokolov et al (1963).

Reference Group "D"

The reader is referred to the following: Amirijafari and Campbell (1970), Bray and Foster (1979), Brod (1960a, 1960b, 1963a, 1963b), Brooks (1969), Carothers and Kharaka (1978), Colombo (1967), Colombo et al (1965), Eganhouse and Calder (1976), Fokeev (1956), Hedberg (1968, 1978), Kartsev et al (1959), Kuznetsova (1963), Munn (1909), Price (1976, 1978), Stewart and Nielsen (1954), Yushkevich and Zhuze (1958), Zhuze (1960, 1967), and Zhuze et al (1971, 1977).

Reference Group "E"

The reader is referred to the following: Berry (1969), Bredehoeft et al (1963), Brod (1960b), Collins (1975), Eganhouse and Calder (1976), Fokeev (1956), Gullickson et al (1961), Hedberg (1968), Kartsev et al (1959), McKelvey and Milne (1962), Meinhold (1971), Milne et al (1964), Neglia (1979), Peake and Hodgson (1968), Price (1973), Sokolov et al (1963), and Zargaryan (1962).

Reference Group "F"

The reader is referred to the following: Bredehoeft (1971), Bredehoeft et al (1963), Collins (1975), P. H. Jones (1968, 1974, 1976, 1978), Meinhold (1971), Plummer and Sargent (1931), Roberts (1966, 1968), and Toth (1962, 1963, 1970, 1978a, 1978b).

Reference Group "G"

The reader is referred to the following: Bredehoeft and Papadopulos (1965), Heald et al (1930), P. H. Jones (1974, 1978), Kartsev et al (1959), Koshlyak (1963), Neglia (1979), Ovnatanov and Tamrazyan (1962), Plummer and Sargent (1931), Pratt (1942), Sokolov et al (1963), Stallman (1965), Van Orstrand (1934), and Yushkevich and Zhuze, (1958).

REFERENCES CITED

Abelson, P. H., 1963, Organic geochemistry and the formation of petroleum: Proc. 6th World Pet. Cong. (Frankfort), Sec. 1, paper 41, PD 1, p. 397-407.

——— T. C. Hoering, and P. L. Parker, 1962, Fatty acids in sedimentary rocks: Advances in Organic Geochemistry (1964), p. 285-295.

Adivarahan, P., D. Kunii, and J. M. Smith, 1962, Heat transfer in porous rocks through which single-phase fluids are flowing: Soc. Petroleum Engineers Jour. (Sept.), p. 290-296.

Amirijafari B., and J. M. Cambell, 1970, Solubility of gaseous hydrocarbon mixtures in water: SPE/AIME An. Mtg. (Houston), Paper No. SPE 3106.

Baker, B. L., 1973, Generation of alkane and aromatic hydrocarbons from humic materials in arctic marine sediments: Advances in Organic Geochemistry, p. 137-152.

Baker, E. G., 1967, A geochemical evaluation of petroleum migration and accumulation, *in* B. Nagy and U. Colombo, eds., Fundamentals of Petroleum Geochemistry: New York, Elsevier Publ. Co., p. 299-329.

Berry, F. A. F., 1969, Relative factors influencing membrane filtration effects in geologic environments: Chem. Geology, v. 4, no. 1/2, p. 295-301.

Billings, G. K., B. Hitchon, and D. R. Shaw, 1969, Geochemistry and origin of formation waters in the Western Canada sedimentary basin, 2. Alkali metals: Chem. Geology, v. 4, no. 1/2, p. 211-223.

Blumer, M., 1965, Organic pigments; their long-term fate: Science (August 13), p. 722-726.

——— and W. J. Cooper, 1967, Isoprenoid acids in recent sediments: Science, (Dec. 15), v. 158, p. 1463-1464.

Bray, E. E., and W. R. Foster, 1979, Process for primary migration of petroleum in sedimentary basins (Abs.): AAPG Bull., v. 63, p. 697-698.

Bredehoeft, J. D., 1971, Analysis of flow in fracture systems—a porous medium model: Geol. Soc. America Abs. with Programs, v. 3, no. 7, p. 512.

——— and I. S. Papadopulos, 1965, Rates of vertical groundwater movement estimated from the Earth's thermal profile: Water Resources Research Jour., v. 1, no. 2, p. 325-328.

——— et al, 1963, Possible mechanism for concentration of brines in subsurface formations: AAPG Bull., v. 47, no. 2, p. 257-269.

Breger, I. A., 1960, Diagenesis of metabolites and a discussion of the origin of petroleum hydrocarbons: Geochim. et Cosmochim. Acta, v. 19, p. 297-308.

Brod, I. O., 1960a, Migration and accumulation of oil and gas according to the source-rock theory: Int. Geol. Rev., v. 2, no. 4, p. 330-345.

——— 1960b, On principal rules in the occurrence of oil and gas accumulations in the world: Int. Geol. Rev., v. 2, no. 11, p. 992-1005.

——— 1963a, Diagnostic indications of the processes of formation of bitumens, and petroleum and gases: Compass (May, 1963), p. 246-254.

——— 1963b, Modern view on formation and regularities in distribution of oil and gas accumulations: Petroleum Geology (English Transl. of Russian), v. 4, no. 11-A, p. 607-614.

Brooks, J. D., 1969, The diagenesis of plant lipids during the formation of coal, petroleum and natural gas: Geochim. et Cosmochim. Acta, v. 33, no. 10, p. 1183-1194.

Carothers, W. W., and Y. K. Kharaka, 1978, Aliphatic acid anions in oil-field waters and the origin of natural gas: Proc. SPE/AIME An. Mtg. (Houston), in press.

Carrigy, M. A., 1974, Guide to the Athabasca oil sands area—introduction and general geology: Alberta Res. Inf. Series 65, contr. 628, p. 3-13.

Caudle, D. D., 1977, Analysis of oil content in production discharges: Am. Petroleum Inst. Prod. Dept. Mtg. (Houston), p. C1-C18.

Collins, A. G., 1975, Some effects of water upon the generation, migration, accumulation and alteration of petroleum; *in* Geochemistry of oilfield waters: New York, Elsevier, Publ. Co., p. 293-306.

Colombo, U., 1967, Origin and evolution of petroleum; *in* Fundamental aspects of petroleum geochemistry: Elsevier Pub. Co., p. 331-369.

——— F. Gazzarrini, and G. Sironi, 1965, Carbon isotope composition of individual hydrocarbons from Italian natural gases: Nature, v. 205 (March 27), p. 1303-1304.

Cooper, J. E., and E. E. Bray, 1963, A postulated role of fatty acids in petroleum formation (preprint): Am. Chem. Soc. Div. Petroleum Chem., v. 8, no. 2, p. A17-A28.

Cox, B. B., 1946, Transformation of organic material into petroleum under geological conditions—the geological fence: AAPG Bull., v. 30, p. 645-659.

Cupps, C. Q., P. H. Lipstate, and J. Fry, 1951, Variance in characteristics of the oil in the Weber sandstone reservoir, Rangely field, Colorado: U. S. Bur. Mines Rept. Inv. 4761.

Degens, E. T., G. W. Chilingar, and W. D. Pierce, 1964a, On the origin of petroleum inside freshwater carbonate concretions of Miocene age: Advances in Organic Geochemistry (1962), p. 149-164.

——— et al, 1964b, Data on the distribution of amino acids and oxygen isotopes in petroleum brine waters of various geologic ages: Sedimentology, v. 3, no. 3, p. 199-225.

Duke, R. B., and N. F. Seppi, 1977, Characterization of Wyoming black trona brine: Amer. Chem. Soc. Petroleum Chem. Div. Preprints, v. 22, no. 2, p. 777-784.

Eganhouse, R. P., and J. A. Calder, 1976, The solubility of medium molecular weight aromatic hydrocarbons and the effects of hydrocarbon co-solutes and salinity: Geochim. et Cosmochim. Acta, v. 40, no. 5, p. 555-561.

Eglinton, G., 1972, Laboratory simulation of organic geochemical processes: Advances in Organic Geochemistry, p. 29-48.

——— et al, 1966, Occurrence of isoprenoid fatty acids in the Green River shale: Science v. 153, (Sept. 2), p. 1133-1136.

Evans, E. W., 1866, On the oil producing uplift of West Virginia: Am. Jour. Science, v. 42, p. 334-343.

Fokeev, V. M., 1956, The solubility of petroleum and of gases in water: Trudy Moskov. Geol. Razvedoch. Inst. im. S. Ordzhonikidze, 29, p. 203-213. (English transl. by Assoc. Tech. Services, Inc., RJ-1832).

Germanov, A. L., 1963, Role of organic substances in the formation of hydrothermal sulfide deposits: Internat. Geology, Rev., v. 5, p. 379-394.

Gullickson, D. M., W. H. Caraway, and G. L. Gates, 1961, Chemical analysis and electrical resistivity of selected California oilfield waters: U. S. Bur. Mines Rept. Inv. 5736.

Hanshaw, B. B., W. Black, and M. Rubin, 1965, Radiocarbon determinations for estimating ground water flow velocities in central Florida: Science, v. 148 (April 23), p. 494-495.

Heald, K. C., et al, 1930, Earth temperatures in oil fields: Am. Petroleum Inst. Prod. Bull. 205, Research Proj. 25.

Hedberg, H. D., 1968, Geological controls on petroleum genesis: Proc. 7th World Petroleum Cong. (Mexico), 11 p.

——— 1978, Methane generation and petroleum migration: this volume.

Hiss, W. L., et al, 1969, Saline water in southeast New Mexico: Chem. Geology, v. 44, p. 341-360.

Hitchon, B., A. A. Levinson, and S. W. Reeder, 1969, Regional variations of river water composition resulting from halite solution, Mackenzie River drainage basin, Canada: Water Resources Research Jour., v. 5, p. 1395-1403.

Hodgson, G. W., 1978, Origin of petroleum: in-transit conversion of organic compounds: this volume.

Horvitz, L., 1978, Near-surface evidence of hydrocarbon movement from depth: this volume.

Hunt, J. M., 1973a, Unsolved problems concerning origin and migration of petroleum (abs.): AAPG Bull., v. 57, no. 4, p. 785.

——— 1973b, An examination of petroleum migration processes: Woods Hole Ocean. Inst. Collect. Reprints, Pt. 2, 13 p., (Paper No. 3084).

Illing, L. V., 1959, Deposition and diagenesis of some upper Paleozoic carbonate sediments in Western Canada: Proc. 5th World Petroleum Cong. (New York), sec. 1, no. 2, p. 23-52.

Johannes, R. E., and K. L. Webb, 1965, Release of dissolved amino acids by marine zooplankton: Science, v. 150, 10/1/65, p. 74-77.

Jones, P. H., 1968, Hydrodynamics of geopressure in the northern Gulf of Mexico basin: SPE/AIME Mtg. (Houston), Paper SPE 2207, 10 p.

——— 1974, Energy resources and geothermal regime, northern Gulf of Mexico basin: AAPG-SEPM An. Mtg. Abs., v. 1, p. 50-51.

——— 1976, Natural gas resources of the geopressured zones in the northern Gulf of Mexico basin: U. S. Dept. Commerce NTIS Publication FE-2271-1, p. 17-33.

——— 1978, Interacting dynamics of pressure, temperatures, and water salinity: this volume.

Jones, R. W., 1978, Some mass balance and geologic constraints on migration mechanisms: this volume.

Kartsev, A. A., et al, 1959, The geochemistry of petroleum: properties of petroleum, *in* Geochemical methods of prospecting and exploration for petroleum and natural gas: Univ. California Press, (Russian original, 1954).

Kidwell, A. L., and J. M. Hunt, 1958, Migration of oil in recent sediments of Pedernales, Venezuela *in* L. G. Weeks, ed., Habitat of oil: Tulsa, AAPG, p. 790-817.

Koshlyak, V. A., 1963, Thermal properties of Mesozoic and Cenezoic sediments in the eastern part of the west Siberian Lowland: Int. Geol. Rev., v. 5, p. 1011-1018.

Kuznetsova, N. P., 1963, Geochemistry of pyrrole compounds in the organic matter of sedimentary rocks, petroleum, and solid bitumens: Geochem., no. 10, p. 945-954.

Matusevich, V. M., and V. M. Shvets, 1974, Significance of organic acids of subsurface waters for oil and gas exploration in West Siberia: Petroleum Geology (English Transl. of Russian), v. 11, no. 10, p. 459-464.

McAuliffe, C., 1969a, Solubility in water of normal C_9 and C_{10} alkane hydrocarbons: Science, v. 163, (January), p. 478-479.

——— 1969b, Determination of dissolved hydrocarbons in subsurface brines: Chem. Geology, v. 4, p. 225-233.

——— 1978, Role of solubility in migration of petroleum from source: this volume.

McCoy, A. W., 1934, An interpretation of local, structural development in mid-continent areas associated with deposits of petroleum; *in* Problems of petroleum geology; Tulsa, AAPG, p. 581-627.

McKelvey, J. G., and I. H. Milne, 1962, The flow of salt through compacted clays: Proc. Natl. Conf. Clays & Clay Minerals, v. 9, p. 248-259.

Meinhold, R., 1971, Hydrodynamic control of oil and gas accumulation as indicated by geothermal, geochemical, and hydrological distribution patterns: Proc. 8th World Petroleum Cong. (Moscow), v. 2, p. 55-66.

——— 1977, New knowledge of the diagenesis of disperse organic rock substances: Zeit Agnew. Geol., v. 23, no. 1, p. 9-16, (in German).

Meinschein, W. G., 1959, Origin of petroleum: AAPG Bull., v. 43, p. 925-943.

Milne, I. H., J. G. McKelvey, and R. P. Trump, 1964, Semipermeability of bentonite membranes to brines: AAPG Bull., v. 48, p. 103-105.

Moody, J. D., 1975, Distribution and geological characteristics of giant oil fields; *in* Petroleum and global tectonics: Princeton Univ. Press, p. 307-320.

Munn, M. J., 1909, The anticlinal and hydraulic theories of

oil and gas accumulation: Econ. Geology, v. 4, p. 509-529.

Neglia, S., 1979, Migration of fluids in sedimentary basins: AAPG Bull., v. 63, p. 573-597.

Ogner, G., and M. Schnitzer, 1970, The occurrence of alkanes in fulvic acid, a soil humic fraction: Geochim. et Cosmochim. Acta, v. 34, p. 921-928.

Orr, R. D., J. R. Johnston, and E. M. Manko, 1977, Lower Cretaceous geology and heavy-oil potential of the Lloydminster area: Bull. Canadian Petroleum Geology, v. 25, p. 1187-1221.

Ovnatanov, S. T., and G. P. Tamrazyan, 1962, Thermal conditions in the anticlinal zone of Surakhany-Karachukhur-Zykh-Peschanyy (Apsheron Penninsula): Int. Geol. Rev., v. 4, May, p. 79-88.

Parker, P. L., 1967, Fatty acids in recent sediment: Contrib. Mar. Sci., v. 12, July, p. 135-142.

Peake, E., and G. W., Hodgson, 1965, Alkanes in natural aqueous systems: accommodation of C_{20}-C_{33} n-alkanes in distilled water and occurrence in natural water systems: Preprint, Am. Oil Chem. Soc. 56th An. Mtg. (Houston), 45 p.

——— ——— 1968, Laboratory studies of the disaccommodation of n-alkanes from simulated formation waters: Proc. Am. Chem. Soc., Pet. Chem. Div. (Atlantic City), p. B38-B39.

——— ——— 1973, Origin of petroleum—steranes as products of early diagenesis in recent marine and freshwater sediments (abs.): AAPG Bull., v. 57, p. 799.

——— B. L. Baker and G. W. Hodgson, 1972, Hydrogeochemistry of MacKenzie River basin, Canada, 2: Geochim. et Cosmochim. Acta, v. 36, p. 867-883.

Plummer, F. B., and E. C. Sargent, 1931, Underground waters and subsurface temperatures of the Woodbine sand in northeast Texas: Univ. of Texas Bull. 3138, 178 p.

Pratt, W. E., 1942, Oil in the Earth—4 Lectures: Univ. Kansas Press (1944), 110 p.

Price, L. C., 1973, Solubility of petroleum in water as function of temperature and salinity and its significance in primary petroleum migration: AAPG Bull., v. 57, p. 801.

——— 1975, Evidence for and use of the model of a hot deep origin of petroleum in exploration: AAPG-SEPM An. Mtg., Abstr., v. 2, p. 60-61.

———1976, Aqueous solubility of petroleum as applied to its origin and primary migration: AAPG Bull., v. 60, p. 213-244.

——— 1978, Crude oil and natural gas dissolved in deep, hot geothermal waters of petroleum basins—a possible significant new energy source: Proc. 3rd Geopressured Geothermal Energy Conf., v. 1, p. 167-249.

Rich, J. L., 1921, Moving underground water as a primary cause of the migration and accumulation of oil and gas: Econ. Geology, v. 16, no. 6, p. 347-371.

Roberts, W. H., III, 1960, The fluid environment of petroleum: some elementary reasoning (abs.): Proc. Hydrogeology Gr., Geol. Soc. America An. Mtg.

——— 1966, Hydrodynamic analysis in petroleum exploration; *in* Enciclopedia del petrolio e del gas naturale: Ente Nazionale Idrocarburi, Rome.

——— 1968, Gulf Res. and Dev. Co., internal memo (unpublished).

Seifert, W. K., and W. G. Howells, 1969, Interfacially active acids in a California crude oil. Isolation of carboxylic acids and phenols: 157th Nat. Mtg. Am. Chem. Soc. (Minneapolis), Pap. Petr. 33.

——— et al, 1969, Analysis of crude oil carboxylic acids after conversion to their corresponding hydrocarbons: Anal. Chem., v. 41, no. 12, p. 1638-1647.

——— et al, 1970, Identification of polycyclic naphthenic mono- and diaromatic crude oil carboxylic acids: Anal. Chem., v. 42, no. 2, p. 180-189.

Sellers, G. A., 1965, Thermal degradation studies of humic acid materials: Proc. Geol. Soc. America., Cordilleran Sect. (Seattle, 1964).

Sever, J., and P. L. Parker, 1969, Fatty alcohols (normal and isoprenoid) in sediments: Science, v. 164, 5/30/69, p. 1052-1054.

Sokolov, V. A., et al, 1963, Migration processes of gas and oil, their intensity and directionality: Proc. 6th World Petroleum Cong. (Frankfort), Sec. 1, Paper 47, PD 2, p. 493-505.

Stallman, R. W., 1965, Steady one-dimensional flow in a semi-infinite porous medium with sinusoidal surface temperature: Jour. Geophys. Res., v. 70, p. 2821-2827.

Stewart, W. C., and R. F. Nielsen, 1954, Phase equilibria for mixtures of CO_2 and several normal saturated hydrocarbons: Producers Monthly, v. 18, no. 3.

Sukharev, G. M., et al, 1964, Geothermal features of Caucasian oil and gas deposits: Int. Geol. Rev., v. 6, p. 1541-1546.

Swanson, V. E., et al, 1968, Hydrocarbons and other organic fractions in recent tidal flat and estuarine sediments, northeastern Gulf of Mexico: Am. Chem. Soc., Petroleum Chem. Div., Preprints, v. 13, no. 4, Sept., p. B35-B37.

Toth, J., 1962, A theory of groundwater motion in small drainage basins in central Alberta, Canada: Jour. Geophys. Res., v. 67, no. 11, p. 4375-4387.

——— 1963, A theoretical analysis of groundwater flow in small drainage basins: Jour. Geopys. Res., v. 68, no. 16, p. 4795-4812.

——— 1970, Relation between electric analog patterns of groundwater flow and accumulation of hydrocarbons: Canadian Jour. Earth Sci., v. 7, no. 3, p. 988-1007.

——— 1978a, Cross-formational gravity-flow of groundwater: a mechanism of the transport and accumulation of petroleum: or, the generalized hydraulic theory of petroleum migration: this volume.

——— 1978b, Gravity-induced cross-formational flow of formational fluids, Red Earth region, Alberta, Canada: analysis, patterns, evolution: Water Resources Research Jour., v. 14, no. 5, p. 805-843.

Van der Waarden, M., A. L. A. M. Bridie, and W. M. Groenewoud, 1971, Transport of mineral oil components to groundwater, Part 1. Model experiments on the transfer of hydrocarbons from a residual oil zone to trickling water: Water Res., v. 5, no. 5, p. 213-266.

Van Orstrand, C. E., 1934, Temperature Gradients; *in* Problems of petroleum geology: Tulsa, AAPG, p. 989-1021.

Van Tuyl, F. M., et al, 1945, The migration and accumulation of petroleum and natural gas: Colorado School of Mines Quart., v. 40, January.

Wershaw, R. L., and D. J. Pinckney, 1977, Chemical structure of humic acids, part 2: The molecular aggregation of some humic acid fractions in N, N-Dimethylformamide: Jour. Res. U. S. Geol. Survey, v. 5, no. 5, p. 571-577.

——— ——— and S. E. Booker, 1977, Chemical structure of humic acids, part 1. A generalized structural model: Jour. Res. U. S. Geol. Survey, v. 5, no. 5, p. 565-569.

Witherspoon, P. A., and D. N. Saraf, 1965, Diffusion of methane, ethane, propane, and n-butane in water from 25° to 43°C: Jour. Phys. Chemistry, v. 69, no. 1, p. 3752-3755.

——— et al, 1968, Geological factors affecting regional ground-water flow (abs.): Geol. Soc. America Mtg. (Mexico, D. F.), p. 326.

Yushkevich, G. N., and T. P. Zhuze, 1958, Solubility of petroleum in compressed gases: Khim. i. Tekhnol. Topliv. i.

Masel 3(7) p. 45-53.

Zargaryan, E. L., 1962, Formation waters of the Karachukhur-Zykh Field: Petroleum Geology (English Transl. of Russian), v. 3, p. 717-722.

Zarrella, W. M., et al, 1963, Analysis and interpretation of hydrocarbons in subsurface brines: Am. Chem. Soc. Mtg. (Los Angeles), Preprint, p. A7-A16.

Zhuze, T. P., 1960, Solubility of materials in compressed gases and the significance of this phenomena for petroleum geology: Trisly Instituta, Geologii i Razrabotki Goryuchikh Iskopayemykh, Akad. Nauk SSSR, I, p. 3-12.

——— 1967, Experimental investigation of transfer relations of hydrocarbons (bitumens) through sedimentary rocks by compressed gases: USSR All-Union Oil & Gas Genesis Symp. (Moscow), p. 404-412.

——— et al, 1971, On solubility of hydrocarbons in water in reservoir conditions: Dokl. Akad. Nauk SSSR, v. 198, no. 1, p. 206-209. Engl. Transl. by Assoc. Tech. Services, Inc., RJ-5905.

——— et al, 1977, Distribution of organic matter and bitumens in modern sediments of the Caspian Sea: Petroleum Geology (English Transl. of Russian), v. 14, no. 6, p. 244-247.

Near-Surface Evidence of Hydrocarbon Movement from Depth[1]

By Leo Horvitz[2]

Abstract Tens of thousands of near-surface sediment samples, taken from both onshore and offshore areas, have been analyzed for the light saturated hydrocarbons, methane through pentane. Many samples were collected over gas and oil fields, but most were gathered in unproven areas. Recognizable hydrocarbon distribution patterns were observed over known fields, and similar patterns also were found in the unproven areas. Many of the anomalies that developed in the latter areas have subsequently been found to be associated with petroleum deposits.

One of the land surveys includes the Flomaton-Jay-Blackjack Creek area and was conducted shortly after the discovery of the Jay field but before the Blackjack Creek field was known. Hydrocarbon distribution patterns developed which reflected the Jay field as well as the older gas field at Flomaton and indicated the Blackjack Creek area as prospective. A hydrocarbon survey, conducted offshore Louisiana prior to the March, 1974, Gulf of Mexico sale, produced a hydrocarbon anomaly which now contains the discovery well of the Cognac field.

The mechanism by which the lighter hydrocarbons move from a deposit to the surface is not yet clear but the phenomenon has been validated by evidence beyond that provided by empirical data of near-surface surveys. Carbon isotope data are part of this evidence. Methane, desorbed from a soil sample taken at 12 ft (3.7 m) from an anomaly over the Francitas field in Texas, yielded a δC^{13} per mil value of -44.0 relative to the PDB standard. Interstitial methane, extracted from a sample taken at another location within the same hydrocarbon anomaly, yielded a value of -40.8. Reservoir methane from the Francitas field showed δC^{13} per mil values ranging from -41.0 to -43.8, almost identical to those of the near-surface methane.

Additional evidence of upward movement of hydrocarbons from petroleum accumulations is provided by analyses of well cuttings. Hydrocarbon buildups, observed on logs prepared from cuttings data, have anticipated oil and gas accumulations.

INTRODUCTION

Surface indications of hydrocarbons in the form of visible oil and gas seeps have been noted in nearly every petroleum province of the world. Without these seepages, drilling for oil and gas might never have occurred and the oil industry, as it is known today, would probably be nonexistent.

The title of this paper would be satisfied by a discussion of macroseepages which wind their way to the surface along faults, fractures, or fissures, but it has reference to another, although related subject. It is concerned with invisible microseepages which also are believed to move from oil and gas reservoirs to the near-surface in amounts that are sufficient to detect and measure accurately. Unlike the visible seep which often is found at great distances from its probable subsurface source, microseepages produce recognizable patterns or anomalies in the near-surface that tend to overlie all (or large parts) of the deposits in which they originate.

Early studies of near-surface hydrocarbons in soil air (Laubmeyer, 1933; Sokolov, 1935) and of absorbed hydrocarbons in the soil itself (Rosaire, 1938; Horvitz, 1939) suggested a direct relationship between these hydrocarbons and subsurface petroleum deposits. As more data accumulated (Horvitz, 1945, 1954, 1959, 1969; Fedynsky et al, 1975; and others), this relationship became more apparent. The writer's experience now includes analyses of more than 50,000 soil and sediment samples from many parts of the world, offshore as well as onshore, and the resulting data continue to support the observations made in earlier reports.

The main purpose of this paper is to present additional sedimentary hydrocarbon data in support of the concept that hydrocarbons move from subsurface petroleum deposits to the near-surface. Data obtained in

[1]Read before the Association at the 1978 Oklahoma City Annual Meeting; received for publication, September 6, 1978; accepted for publication, March 12, 1979.

[2]Horvitz Research Laboratories, Inc., Houston, Texas 77063.

The writer is indebted to Amoco Production Company for determining carbon isotope ratios of methane in near-surface soil samples collected over the Francitas oil field and of methane extracted from produced gas of this field. The writer also is grateful to Tenneco Geological Research for carbon isotope measurements on sedimentary methane from the Mobile South No. 2 area.

Article Identification Number:
0149-1377/79/SG10-0013/$03.00/0.

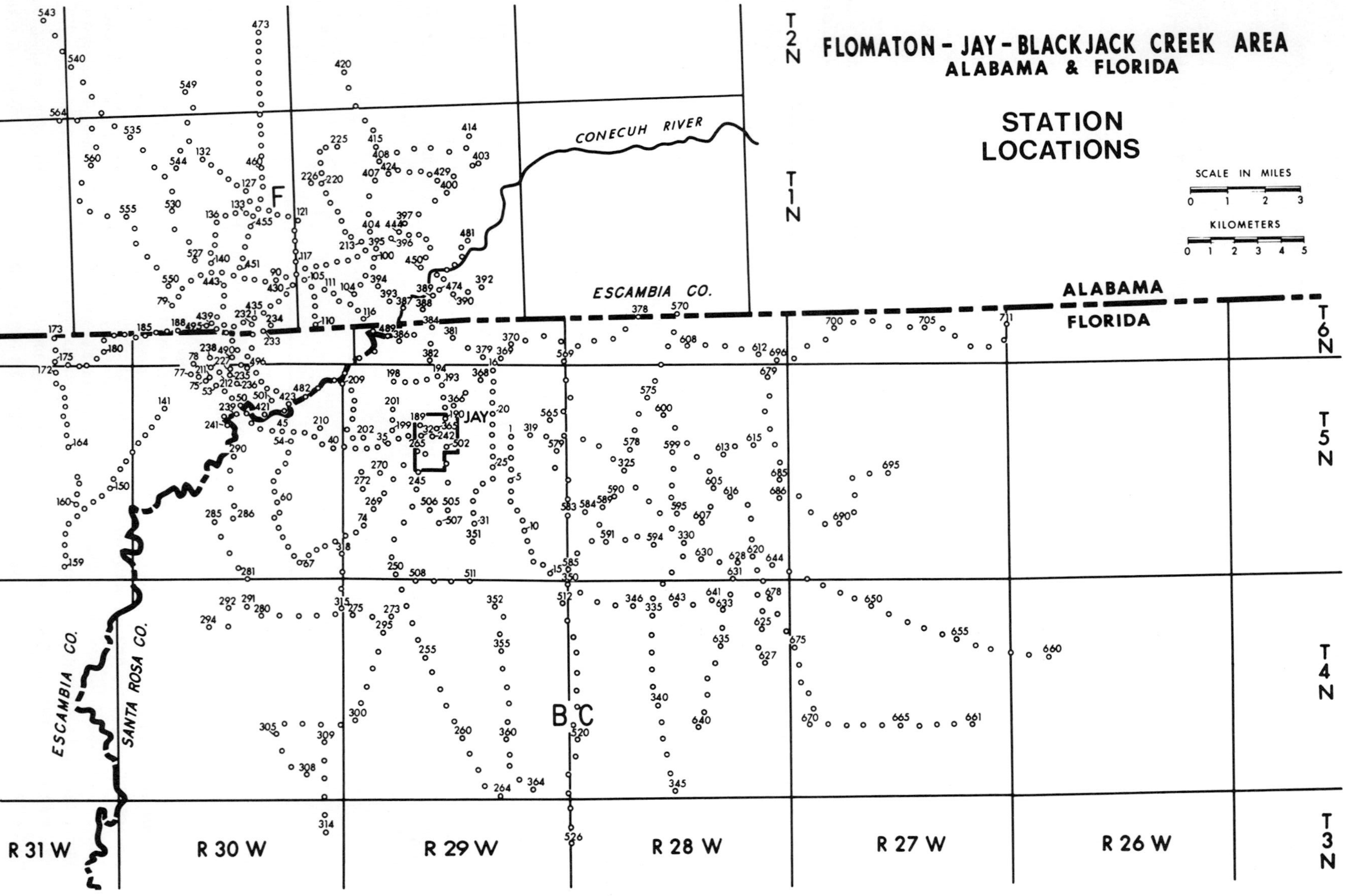

FIG. 1—Locations of stations sampled at 12 ft (3.7 m) in Flomaton (F), Jay, and Blackjack Creek (BC) areas. Station numbers appear above location symbols (o) unless otherwise indicated. See Appendix for analytical data.

the Flomaton, Jay, Blackjack Creek, and adjacent areas in Alabama and Florida, as well as data obtained offshore Louisiana prior to the discovery of the Cognac field are included. Carbon isotope results are cited which link near-surface soil methane with reservoir gas and, finally, data obtained from well cuttings are submitted which also suggest upward movement of hydrocarbons.

ONSHORE HYDROCARBON DATA

Flomaton-Jay-Blackjack Creek and Adjacent Areas

The Flomaton-Jay-Blackjack Creek area is included here for three reasons: (1) the accumulations within this area are below 15,000 ft (4,570 m) and, as far as is known, they occur in the deepest known reservoirs above which detailed near-surface soil hydrocarbon data are available; (2) the most important field within the area, the Jay field, is of a type that is very difficult to find because it is only partially related to structure; and (3) the subsurface geology of both the Jay and Blackjack Creek fields are available for comparison with the near-surface hydrocarbon data.

Adsorbed, saturated hydrocarbons were determined in 711 samples of soil collected in this area from a depth of 12 ft (3.7 m) below the surface. Methane, as well as ethane and heavier hydrocarbons, was measured and the percentage of sand size ($>63\mu$) material in each sample also was determined. The samples were taken from holes dug manually with bucket type augers and locations (Fig. 1) were spaced ¼ mi to ⅓ mi (0.4 km to 0.5 km) apart along the many roads and trails that exist in the area. Care was taken to avoid sampling in ditches, road cuts, or close to producing wells. The overall area that was sampled includes about 270 sq mi (700 sq km) yielding a density of one sample to approximately 250 acres (100 ha.). Flomaton (F) is located in the northwestern part of the area sampled, Jay in the central part, and Blackjack Creek (BC) in the southern part.

Most of the samples (564) in the Flomaton, Jay, Blackjack Creek areas were collected during the Summer and Fall of 1971, after the discovery of the Flomaton and Jay fields but before the discovery of the Blackjack Creek field. Even the Jay field was only partially developed at the time. The samples to the east of Jay (147), in the southeastern part of the area, were collected in May, 1972.

The techniques used to extract and analyze the hydrocarbons have been previously reported (Horvitz, 1969, 1972) but, briefly, the gases were removed from the soil by heating 15 to 50 grams of sample at 100°C in a partial vacuum with a 50% phosphoric acid solution. The extracted hydrocarbons were analyzed by a hydrogen flame gas chromatograph and the values expressed in parts per billion by weight on the dry sample basis.

Ethane and Heavier Hydrocarbons (Ethane+)

Figure 2 shows a contoured map based on the ethane+ fraction which includes ethane through pentane, the hydrocarbons considered most significant because petroleum is believed to be their only known source. Because of its small scale, the hydrocarbon data were not plotted on the map. However, they appear separately in the Appendix. To show the variations in hydrocarbon concentrations over the area sampled, a shading system was used in which values ranging from 25 to 49 parts per billion by weight are in the areas of lightest shading, values ranging from 50 to 99 are in the areas of intermediate shading, and values of 100 parts per billion and above are in the areas of darkest shading. The ethane+ values range from 1 to 354 parts per billion and their average value is 45. The ratio of the average of the anomalous values (shaded areas) to the average of the background values (unshaded) is 14.2 indicating anomalies of high contrast.

Although the ranges of values considered anomalous were selected by visual inspection and experience, histograms, or frequency distribution curves, can also be useful in distinguishing anomalous from background values. In Figure 3, such a curve is shown in which ethane and heavier hydrocarbon (C_2 to C_5) values are plotted along the horizontal axis and the number of samples in specific concentration ranges are plotted along the vertical axis. The shaded areas correspond to the concentration ranges used in contouring the ethane and heavier hydrocarbon map (Fig. 2). Approximately 48% of the samples yielded values less than 15 parts per billion. This group is considered the normal, or background, population. The histogram indicates that a value, between 15 and 20, could have been selected as the lowest anomalous ethane+ value instead of 25. Using lower values, however, would have produced only minor changes in the interpretation.

Effect of Sand Size Material

A map showing contours based on percentages of sand size material in the samples taken in the Flomaton-Jay-Blackjack Creek area is shown in Figure 4. For many years, it was realized that hydrocarbon concentrations in near-surface soils are dependent on lithology. For example, experiments indicated that sands have little attraction for hydrocarbons while clays are highly adsorptive. Although sand acts as a diluent and reduces the magnitude of anomalies, it does not eliminate them entirely unless the sand content is extremely high, on the order of 90% or higher. This condition is encountered frequently in offshore areas. After finding that clays do not lose their adsorbed hydrocarbons even after being subjected to wet sieving and filtration procedures, a sand removal process was finally adopted. Although it is time consuming and, therefore, costly, the resulting improvement

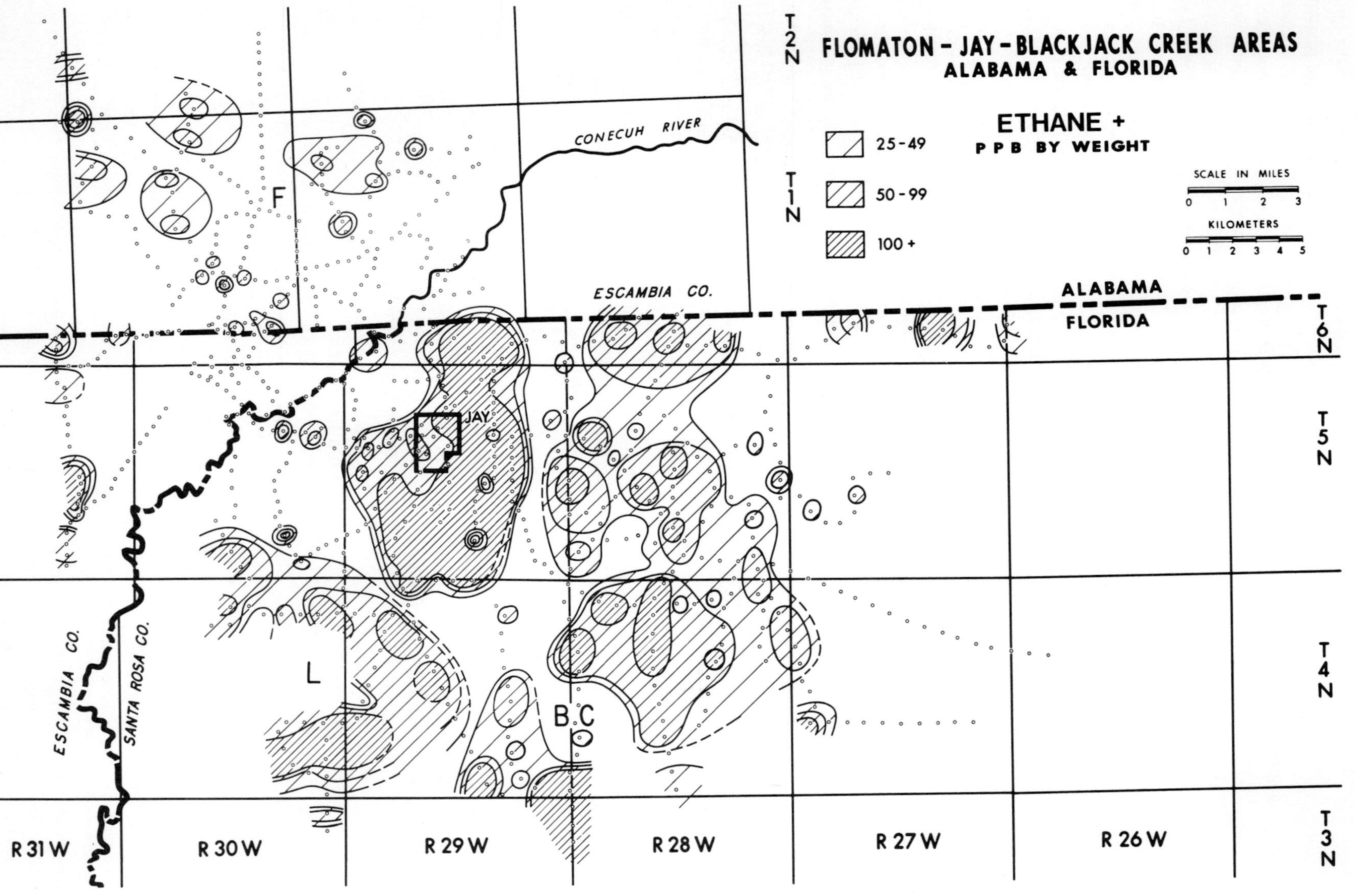

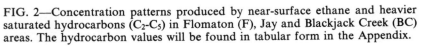

FIG. 2—Concentration patterns produced by near-surface ethane and heavier saturated hydrocarbons (C_2-C_5) in Flomaton (F), Jay and Blackjack Creek (BC) areas. The hydrocarbon values will be found in tabular form in the Appendix.

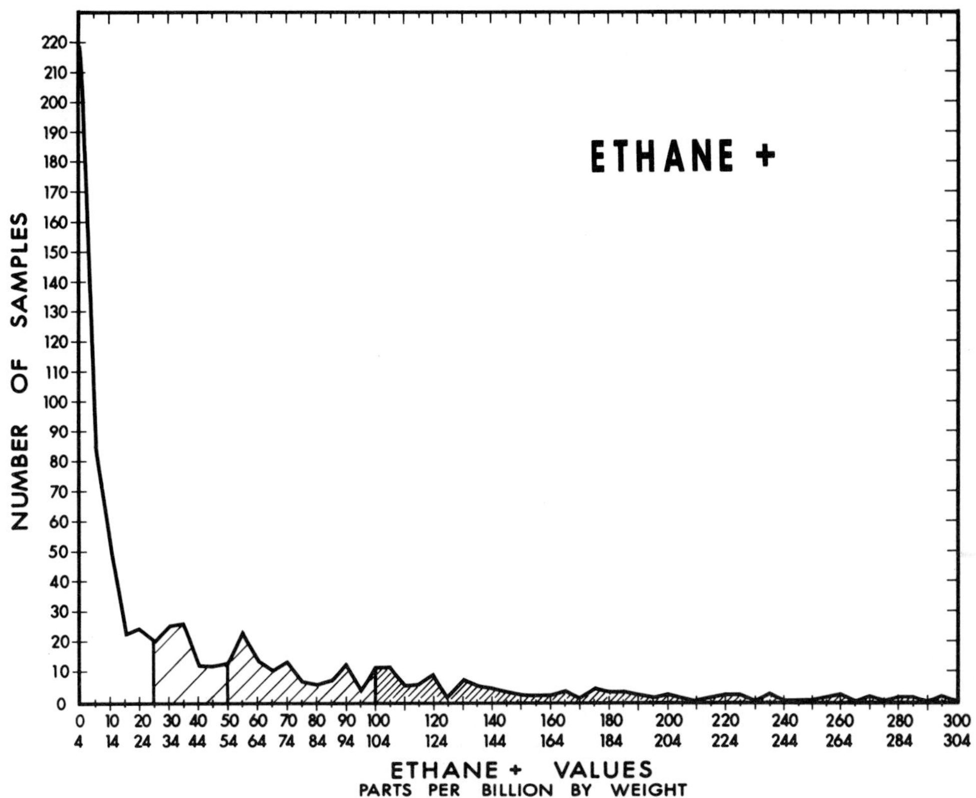

FIG. 3—Histogram prepared from ethane and heavier hydrocarbon values of Flomaton-Jay-Blackjack Creek area. Hatched areas contain values corresponding to those of Figure 2.

in the data justified its use. At first, it was applied only to samples containing 25% or more of sand size material but it is now used routinely on all near-surface samples after unexpected results were obtained where used on sediment of low sand content from a Louisiana swamp. The exceedingly large amounts of free marsh gas which interfered with chromatographic analysis were eliminated by the process while the adsorbed methane remained on the samples. Surprisingly, the adsorbed methane correlated very well with the ethane and heavier hydrocarbon fraction.

The process for removing sand size material involves stirring 300 g to 1,000 g of the bulk sample in a blender with distilled water for about five minutes. By wet sieving, the coarse material is then removed and the fraction that passes through a 63 μ sieve is used for analysis. The distilled water which collects with the fine-grained fraction is removed by filtration. In the Jay area, the sand size ($>63\mu$) material is mainly sand and the fine-grained ($<63\mu$) material is predominantly clay.

Comparison of ethane+ data from bulk samples with data from fine-grained fraction—The solid curve on the bottom of Figure 5 was obtained by analyzing the undried bulk, or whole, sample before sieving; the dashed curve, immediately above, was produced from data obtained from the fine-grained ($<63\mu$) part of the same sample. The patterns are very similar but the anomalous ethane+ values found in the fine-grained material are much higher, up to four times those of the bulk samples. The amounts of sand size material in the samples used in this experiment have been plotted in the upper part of Figure 5 and range from 15% to 99%. For the whole area, the sand size material ranges from 6% to 99% and averages about 69%. The ethane+ data shown in Figure 2 were obtained from the fine-grained fractions of the samples.

Interpretation of Ethane and Heavier Hydrocarbon Data

The most significant hydrocarbon distribution pattern, observed over important petroleum accumulations, is one in which the anomalous hydrocarbons align themselves in the form of a halo (Horvitz, 1939; Rosaire, 1939, p. 304). In this type of pattern, relatively low hydrocarbon concentrations occur in the soil over the main part of the deposit with relatively high values in the sediment over the edges. Low values also are obtained in the background areas. This pattern was observed over important fields including Hastings, Friendswood, and Heidelberg. Another important pattern is of the crescent type, commonly found where an accumulation is trapped against a fault. In this case, high concentrations are found in

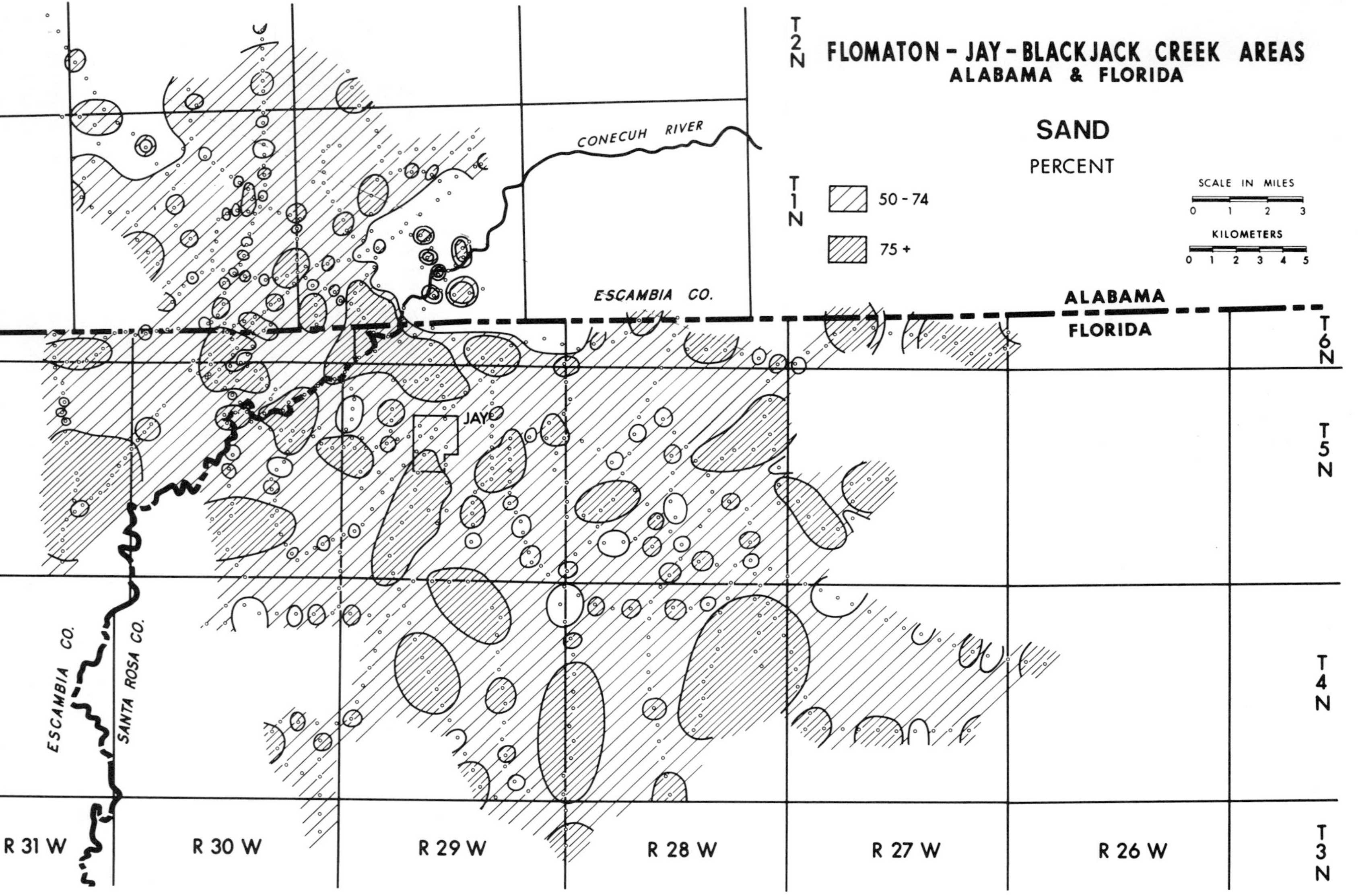

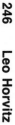

FIG. 4—Concentration patterns produced by sand ($>63\mu$) in 12-ft (3.7-m) samples collected in Flomaton-Jay-Blackjack Creek area. Sand values, in percent by weight, appear in the Appendix.

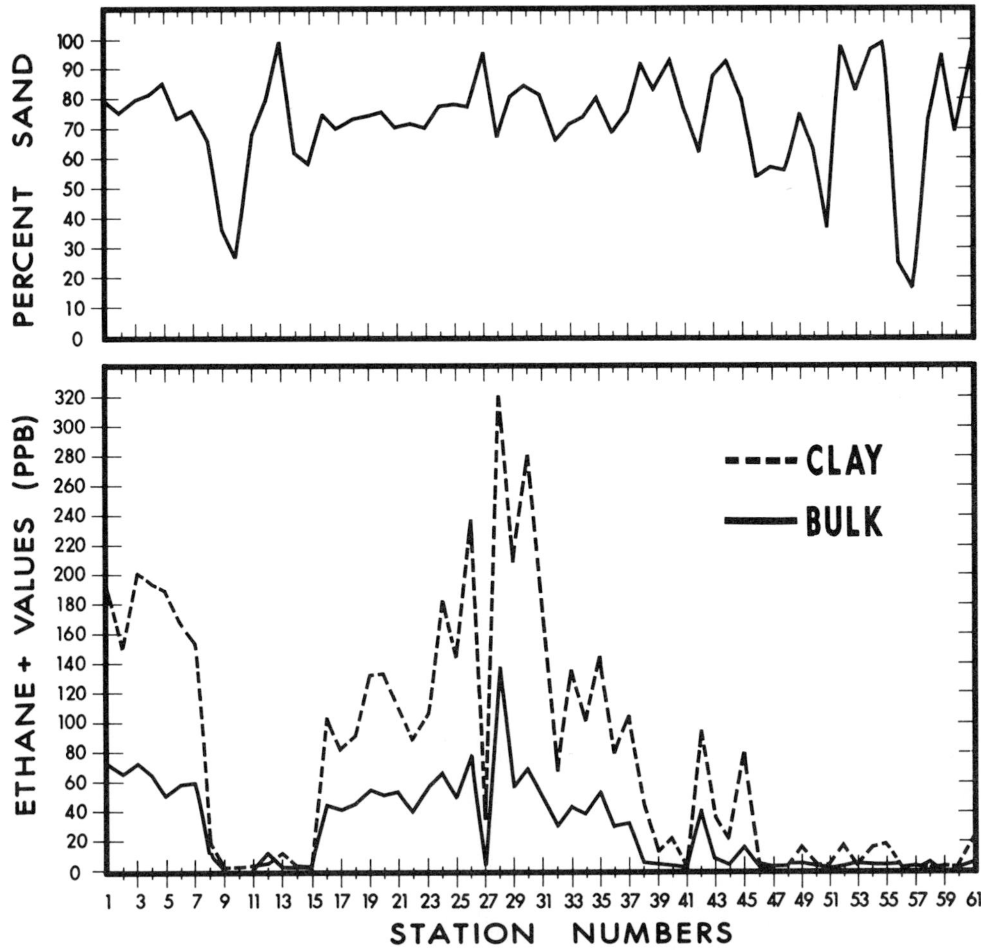

FIG. 5—Comparison of ethane and heavier hydrocarbon values from analyses of clay (<63μ) fractions (----) with those of bulk samples (———). Series of samples were taken at 12-ft (3.7-m) from Jay area. Anomalous values of clay fractions are considerably higher than those of bulk samples. Upper part of figure shows sand contents of samples.

the sediment over the gas-water or oil-water contact with low values in the near-surface soil over the intercept of the accumulation and the fault. A third pattern, in which an area of high concentration is surrounded by a background of low values, is frequently observed but it commonly reflects an accumulation that is less important than those associated with the other patterns.

Inspection of the ethane and heavier hydrocarbon map (Fig. 2) reveals a strong anomaly in the central part of the area. It centers near the townsite of Jay and is characterized by a semi-halo pattern which represents the east and south edges of an anomaly that is open to the northwest. An explanation of the failure of the northwest complement to develop will be suggested later where the subsurface geology of the Jay field is presented.

In the Flomaton area (F), groups of anomalous values, considerably lower than those associated with the anomaly at Jay, are distributed in the form of a spotty pattern. To the southeast of Jay, in the Blackjack Creek area (BC), a halo type pattern is developing. To the west of this anomaly, a strong lead (L) occurs which, on detailing, may also develop into a significant anomaly.

Figure 6 is the same as Figure 2 except that it also shows, in outline, the gas and oil fields in the area. The Flomaton gas field (F) is seen to be associated with the spotty hydrocarbon anomaly. This field, discovered in October, 1968, is located on a well defined structure and is producing from the Norphlet sand which lies directly below the Smackover section. To the west of Flomaton, is Big Escambia Creek (BEC), a Smackover gas field discovered in February, 1972, after the data in this area were acquired. The very important Jay field, located to the southeast of Flomaton, was discovered in June, 1970, about a year before the survey was started. The Jay field also produces mainly from the Smackover Formation with some production coming from Norphlet. The Blackjack Creek field, located to the southeast of the Jay field, produces from the Smackover at a depth of about 15,700

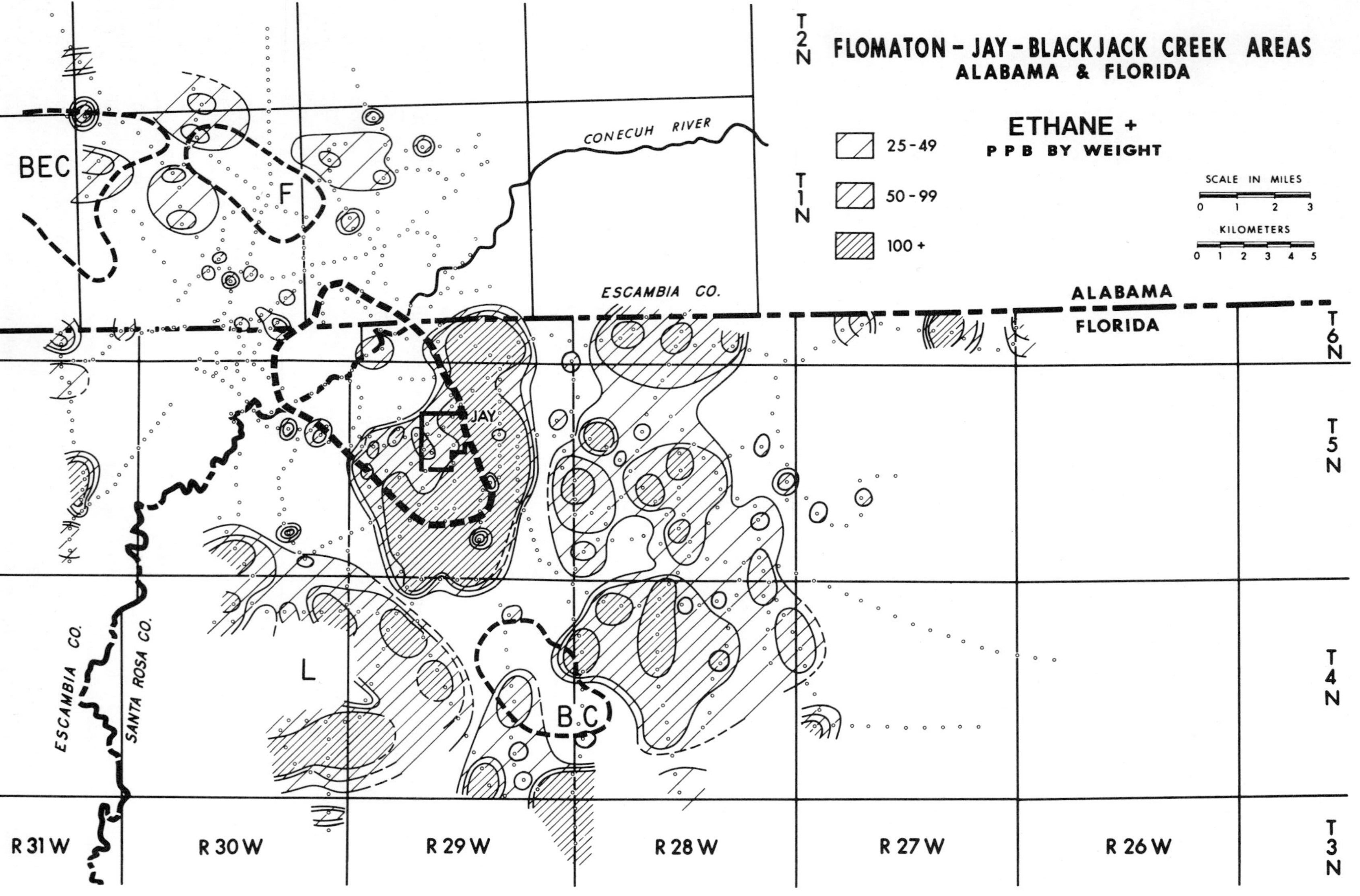

FIG. 6—Ethane and heavier hydrocarbon map containing outlines of Big Escambia Creek (BEC) and Flomaton (F) gas fields, and Jay and Blackjack Creek (BC) oil fields. Close relationships are apparent between near-surface hydrocarbons and petroleum accumulations.

ft (4,790 m). The field was discovered in January, 1972, after this part of the survey was completed.

The most important accumulation within the survey area is the Jay field which has an estimated ultimate recovery of about 350 million bbls (55 million cu m) of oil and 350 billion cu ft (10 billion cu m) of gas. The Blackjack Creek field, with an estimated ultimate recovery of more than 40 million bbls (6.3 million cu m) of oil, is probably next in importance. It is of interest to note the relationships of the anomalies to the known accumulations; the more important the accumulation, the more outstanding the ethane and heavier hydrocarbon anomaly appears to be.

Subsurface Geology of Jay and Blackjack Creek Fields

Figure 7 shows the subsurface geology of the Jay and Blackjack Creek fields. Locations of wells (from map prepared by Geological Consulting Services, Inc.) drilled to or through the Smackover Formation, as of September 1, 1977, also are shown. The subsurface study of the Jay field was made by Ottmann et al, (1973; 1976) who reported that the accumulation occurs on the south plunge of a large subsurface anticline, and the updip trap was formed by a facies change from porous dolomite to dense micritic limestone. The crest of the anticline referred to occurs to the northwest of Jay and contains the Flomaton (F) gas field which produces from the Norphlet sand.

When the subsurface geology (Fig. 7) is compared with hydrocarbon-occurrence outlines of Figure 6, an explanation of the configuration of the near-surface hydrocarbon anomaly associated with the Jay field becomes apparent. The north end of the band of high ethane and heavier hydrocarbon values, which overlie the east and south edges of the Jay producing area, terminates at the eastern end of the porosity pinchout (dashed line, Fig. 7) which represents the northwest boundary of the trap. The low hydrocarbon concentrations in the near-surface soils over the pinchout area is a phenomenon that commonly has been observed. The absence of hydrocarbons in the near-surface soil over the west side of the field is more difficult to understand but it may be related to the irregular water levels on this side of the field. The authors of the geological report cite, as an example, the oil-water contact in two adjacent wells on the west flank that differ in elevation by 100 ft (30 m) and they speculate on the possibility that this irregular oil-water contact may be due to the limited volume of water associated with the Jay reservoir. On the east side of the field, the water level appears to be regular and, presumably, it is regular on the south side. Time and time again, it has been noted that anomalous hydrocarbon concentrations occur in the near-surface soils which overlie gas-water and oil-water contacts.

An important structural feature, east of the Jay field, is the Gilbertown-Pickens-Pollard fault. This fault cannot account for the near-surface hydrocarbons associated with the Jay anomaly because it is downthrown to the west; therefore, hydrocarbons moving along the fault plane would be expected to accumulate in the near-surface soils east of the subsurface fault trace. Thus, the anomalous hydrocarbon concentrations in the near-surface soils to the east of the Jay field may be accounted for by this mechanism but the possibility also exists that these anomalous values overlie a separate petroleum deposit.

The subsurface geology of the Blackjack Creek (BC) field was reported by Burwell and Hadlow (1977). The field produces from the Smackover Formation at a depth of about 15,700 ft (4,790 m). The authors stated (p. 1235):

> "Structurally, the field is a simple northwest-southeast trending anticline with about 180 ft of closure. The flanks of the field dip gently to the west about 1½°, but steepen to about 4° on the east as the field rolls into the Foshee fault, a major geological feature in this area. There is no evidence of faulting within the productive limits of the field. The oil-water contact was determined to be at −15,746 ft subsea based on three penetrations. This oil-water contact defines a productive area of about 4,600 acres."

Because the central part of the structure is underlain by dense, impermeable limestone, the authors report that water encroachment into the reservoir will occur only around the flanks. The map of ethane and heavier hydrocarbons (Fig. 6) indicates a more complete halo pattern around the producing area than was observed over the Jay field, even though the sampling density was lower than that used at Jay. A higher sampling density might have produced an even closer relationship between the Blackjack Creek field and the data. Although the center of the hydrocarbon anomaly is displaced nearly 1.5 mi (2.4 km) to the southeast of the center of the field, the anomaly covers a substantial part of the producing area.

In view of the fact that no faults occur within the productive areas of the Jay and Blackjack Creek fields, the anomalies over these fields must be accounted for by mechanisms other than leakage along fault planes. However, as in the case of Jay, part of the relatively high concentrations found to the east of the Blackjack Creek field may be due to leakage along the Foshee fault plane which, like the Gilbertown fault, is downthrown to the west.

Methane

Although both reservoir gas and near-surface hydrocarbon anomalies contain methane as their most abundant hydrocarbon, near-surface methane (at least in part) may result from bacterial action on organic matter in the soil or sediment. For this reason, it is mapped separately from the ethane and heavier hydrocarbons. Methane is useful in interpreting near-surface hydrocarbon data and its significance is enhanced when it is determined on the finely divided

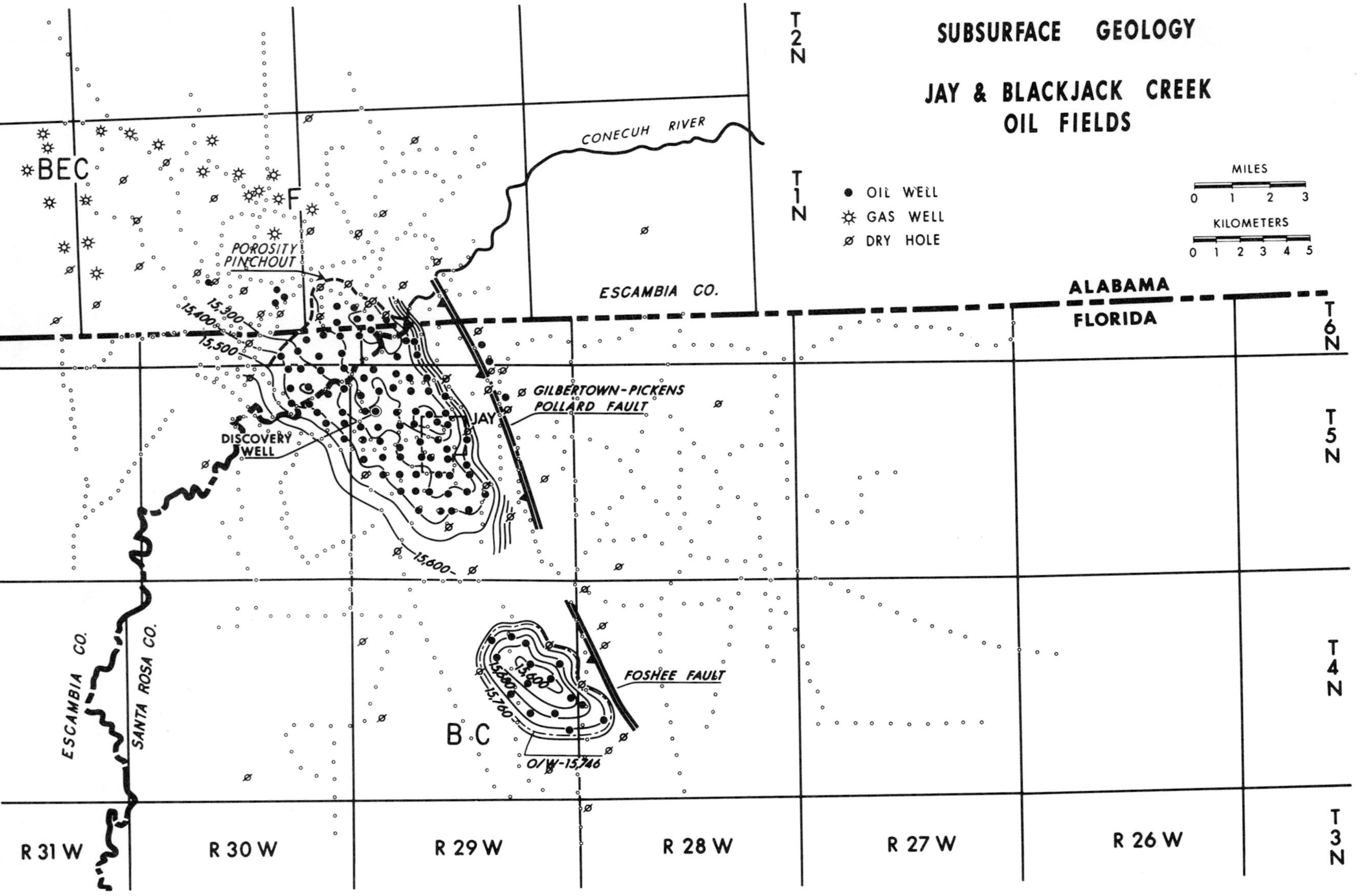

FIG. 7—Subsurface geology of Jay (after Ottmann, Keyes, and Ziegler, 1976) and Blackjack Creek (after Burwell and Hadlow, 1977) oil fields. Wells drilled to or through Smackover Formation are included on map. The Jay field occurs in a stratigraphic trap while Blackjack Creek is associated with an anticlinal structure. Geology explains near-surface hydrocarbon patterns (Fig. 6).

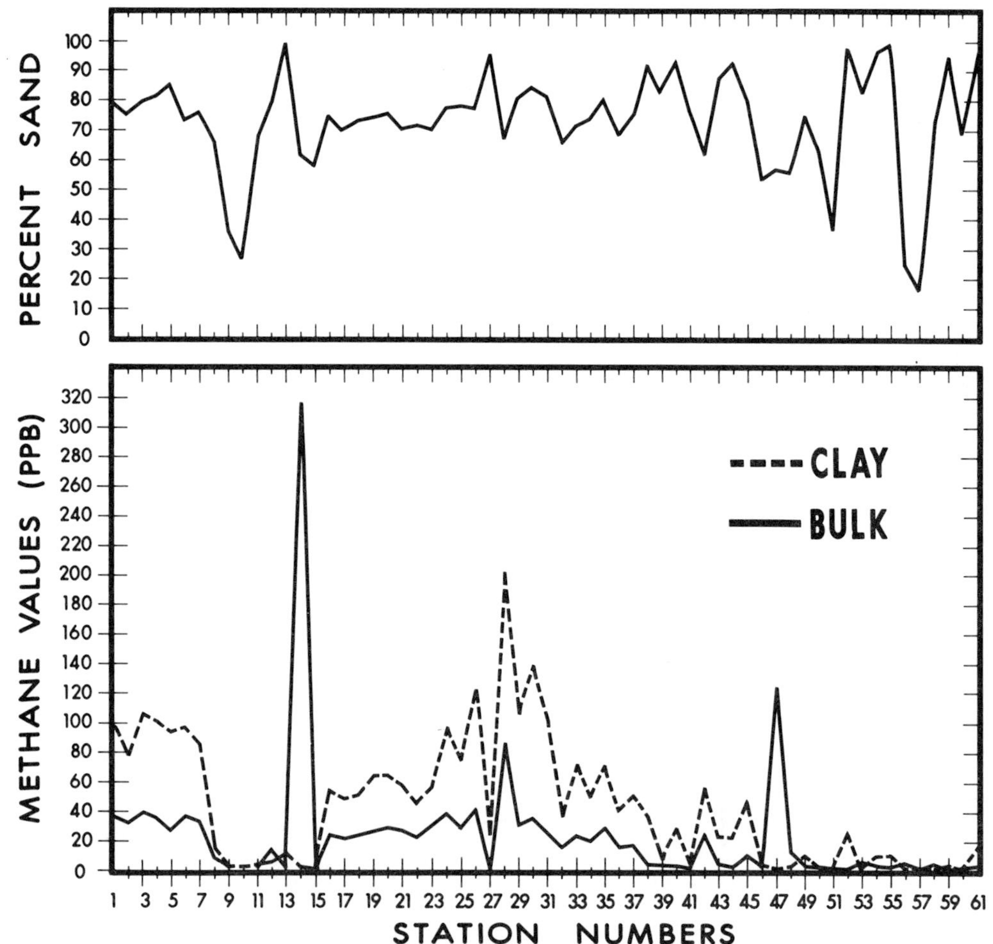

FIG. 8—Comparison of methane values in clay (<63μ) fractions (- - - -) with those of bulk (——) samples. Series of samples was taken at 12 ft (3.7 m) from Jay area. Results are similar to those obtained for ethane and heavier hydrocarbons (Fig. 5) except for two instances. At stations 14 and 47, substantially more methane was found in bulk samples than in clay fractions. The bulk methane at these two stations probably is of biogenic origin.

soil fraction which is retained after removing all coarse material larger than 63μ from the bulk sample. In Figure 8, methane data of clay fractions are compared with those obtained from analyses of bulk samples. As with ethane and heavier hydrocarbons, the methane values of the clay fractions are generally much higher than those of the bulk samples. However, in two instances, the methane in the bulk samples is higher (Fig. 8). Station 14 shows a high value, 316 ppb (parts per billion) for the bulk sample and a very low value, 1 ppb, for the clay. The bulk sample was black and apparently contained biogenic methane which was eliminated by the sand removal process. At station 47, the methane content of the bulk sample also was found to be substantially higher than that of the clay fraction. As previously noted, the greatest improvement in methane data is apparent when samples, collected from marshes and swamps, are subjected to the sand removal process prior to analysis.

Figure 9 shows a map prepared from the methane data obtained on the fine-grained parts of the samples in the Flomaton-Jay-Blackjack Creek area. The producing fields in the area are shown on the map in outline form. Values ranging from 50 to 99 ppb by weight are in the areas of light shading and values of 100 ppb and above are in the more heavily shaded areas. The methane ranges from 0 to 201 ppb by weight and the average value is 28 ppb. The ratio of the anomalous values to those in the background is 5.8.

An interesting methane pattern was produced over the Jay field and, as for the case of the ethane and heavier hydrocarbons, high concentrations were found in the near-surface soil over the east and south edges of the accumulation. Scattered anomalous methane values surround the Blackjack Creek (BC) field where a more complete pattern may have been produced if a higher sampling density had been used.

In contouring the methane data, values below 50

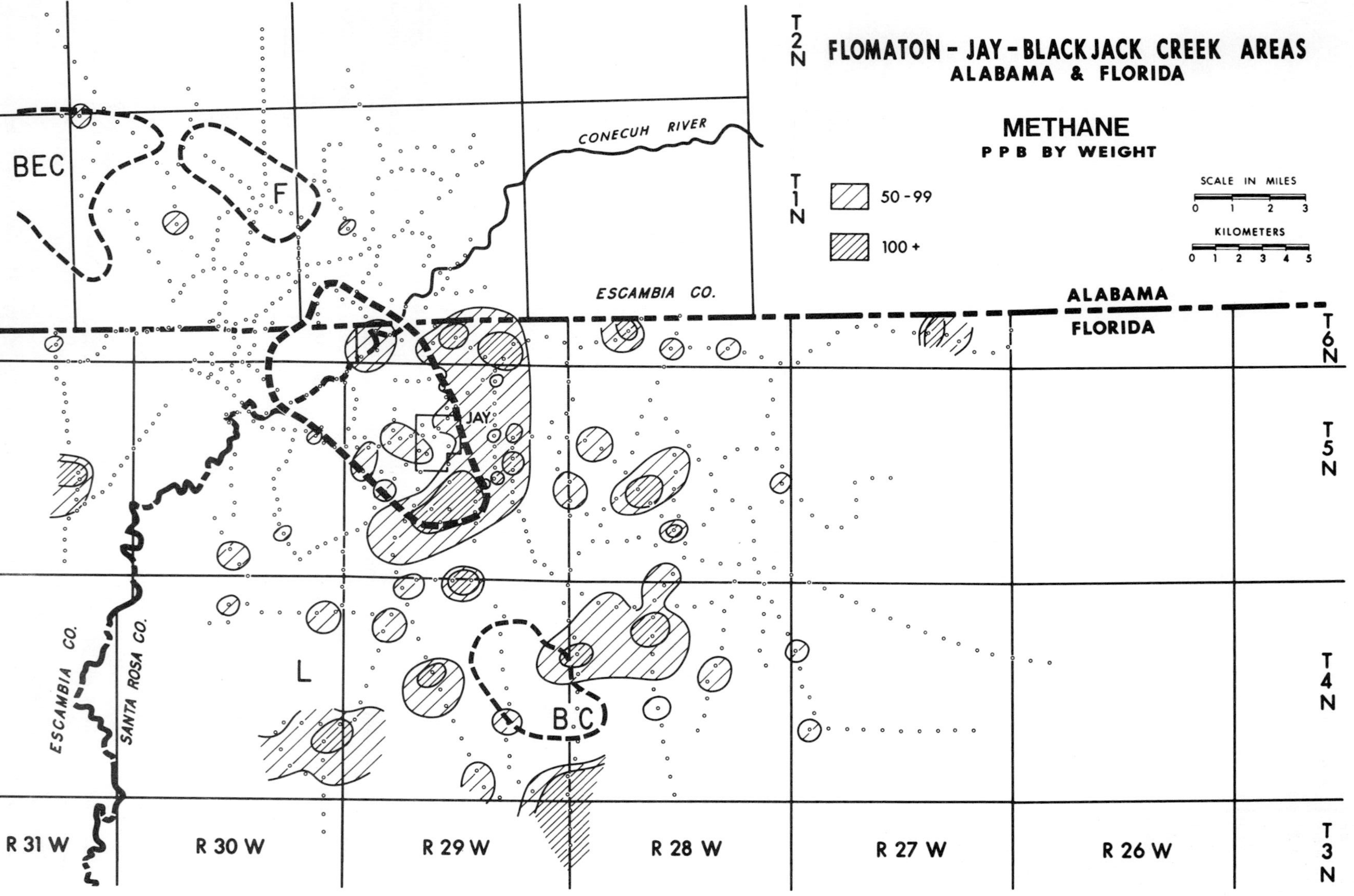

FIG. 9—Concentration patterns produced by near-surface methane in Flomaton (F), Jay and Blackjack Creek (BC) areas. Anomalies are associated with Jay and Blackjack Creek fields but no methane anomalies developed at Flomation (F) and Big Escambia Creek (BEC).

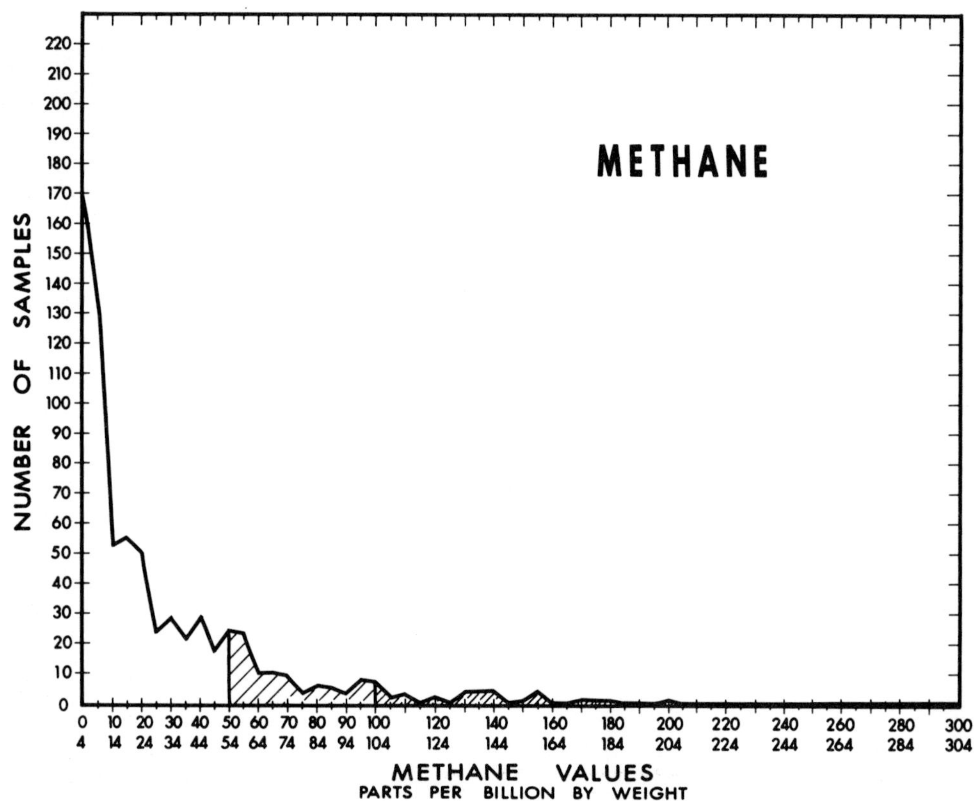

FIG. 10—Histogram prepared from methane values of Flomaton-Jay-Blackjack Creek area. Hatched areas contain values corresponding to those of Figure 9.

ppb were not considered anomalous and, on this basis, no methane anomalies developed over the Flomaton (F) or Big Escambia Creek (BEC) gas fields. According to the methane histogram (Fig. 10), a value as low as 25 ppb could be selected as anomalous. However, even when this value is used in contouring, instead of 50, a few scattered anomalous values appear but no patterns develop that resemble those produced by the ethane and heavier hydrocarbons in the vicinity of the gas fields.

The methane content of the soil of the entire survey is low relative to the ethane and heavier hydrocarbons. Normally, by weight, methane is present in amounts that are three or more times greater than those of the ethane and heavier hydrocarbon fraction and, by volume, it usually represents 85 to 95% of the C_1 to C_5 fraction. However, in the Flomaton-Jay-Blackjack Creek area the soil methane, by volume, represents only 52% of the C_1-C_5 fraction. This low ratio of near-surface methane to ethane and heavier hydrocarbons probably is related to the fact that the percentage of methane in the reservoir gas is below normal due to the presence of appreciable amounts of hydrogen sulfide, carbon dioxide, and nitrogen. For example, the gas produced by the discovery well of the Jay field contains only 69.2% of methane. Hydrogen sulfide, carbon dioxide, and nitrogen account for 13.2% and the remainder is principally ethane (10.9%) with considerably smaller amounts of propane (4.3%) and butanes (1.7%). In the near-surface soil, the principal component of the ethane through pentane fraction is also ethane.

The methane anomalies which developed over the Jay and Blackjack Creek fields, in spite of dilution of hydrocarbon gases in the reservoirs, are probably reflecting the fact that the accumulations at Jay and Blackjack Creek are much more important than the gas areas at Flomaton and Big Escambia Creek where no methane anomalies were detected.

Pentane

No pentane map was prepared for the Flomaton-Jay-Blackjack Creek area because only extremely small amounts of this hydrocarbon were found in the near-surface soil of this area. For the entire survey, the average pentane value is less than 1 ppb by weight. Therefore, it was of great interest to learn that the gas produced by the discovery well of the Jay field also contains only small amounts of pentane, 0.4% of normal, and only traces of isopentane and cyclopentane.

OFFSHORE HYDROCARBON DATA

Since January, 1967, sediment samples have been collected from many offshore areas and they, too,

ETHANE +

P. P. B. BY WEIGHT

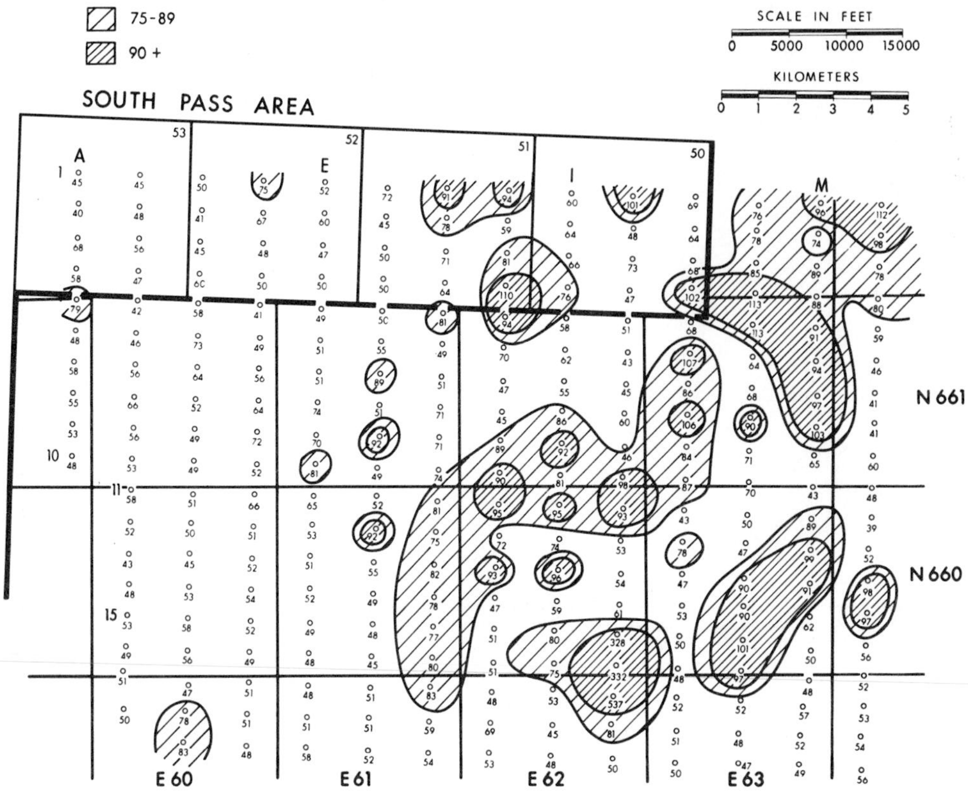

MOBILE SOUTH NO.2 AREA

OFFSHORE LOUISIANA

FIG. 11—Halo type concentration pattern produced by ethane and heavier hydrocarbons from sediment samples taken at 6 ft (1.8 m) in Mobile South No. 2 area, Gulf of Mexico. Water depth ranges from 318 ft (97 m) at northwest corner to 1,860 ft (567 m) at southeast corner of survey.

were analyzed for the light, saturated hydrocarbons, methane through pentane. Offshore samples are taken from the upper 6 ft (1.8 m) of sediment with a piston type coring device equipped with a 6 ft (1.8 m) core barrel, 3 in. (7.6 cm) in diameter, into which a plastic liner is inserted. The bottom part of the sample is removed from the liner on board the ship and packaged in plastic bags in such a manner as to allow a minimum of air to come in contact with the sample. Stored in this manner, no significant losses of adsorbed hydrocarbons have been noted even after many months.

Experiments conducted with samples taken from the same location at several intervals down to about 12 ft (3.7 m) indicated that the depth of sampling is not as critical in water-covered areas as it is onshore. Offshore samples show fairly uniform concentrations of hydrocarbons from the water-sediment interface to 12 ft (3.7 m). However, commonly the top foot (0.3 m) of sediment contains somewhat higher concentrations than does the deeper part of the hole. On land, the upper 3 to 6 ft (1 to 2 m) of soil, especially along the Gulf Coast, commonly contains very low concentrations of hydrocarbons and, therefore, samples are taken well below this depth (generally at 10 to 12 ft [3 to 3.7 m]). Apparently, the water serves as a blanket that prevents the rapid escape of hydrocarbons from offshore sediments; nevertheless, samples are taken at 6 ft (1.8 m), whenever attainable, instead of from the immediate surface because surface sediments may be moved around by water currents or storms. At 6 ft (1.8 m), the sediment is more likely to have been in place for a geologically long period of time. Sampling poses no problems in areas of mud bottom but, in areas of hard bottom, samples commonly are missed or only small quantities of rock or reef material are retrieved. In very sandy bottoms, the penetration depth is often less than 6 ft (1.8 m) and, frequently, it is only 2 to 3 ft (0.6 m to 0.9 m). However, sufficient amounts of such samples commonly are available from which adequate quantities of clay can be extracted for analysis.

ETHANE +

P. P. B. BY WEIGHT

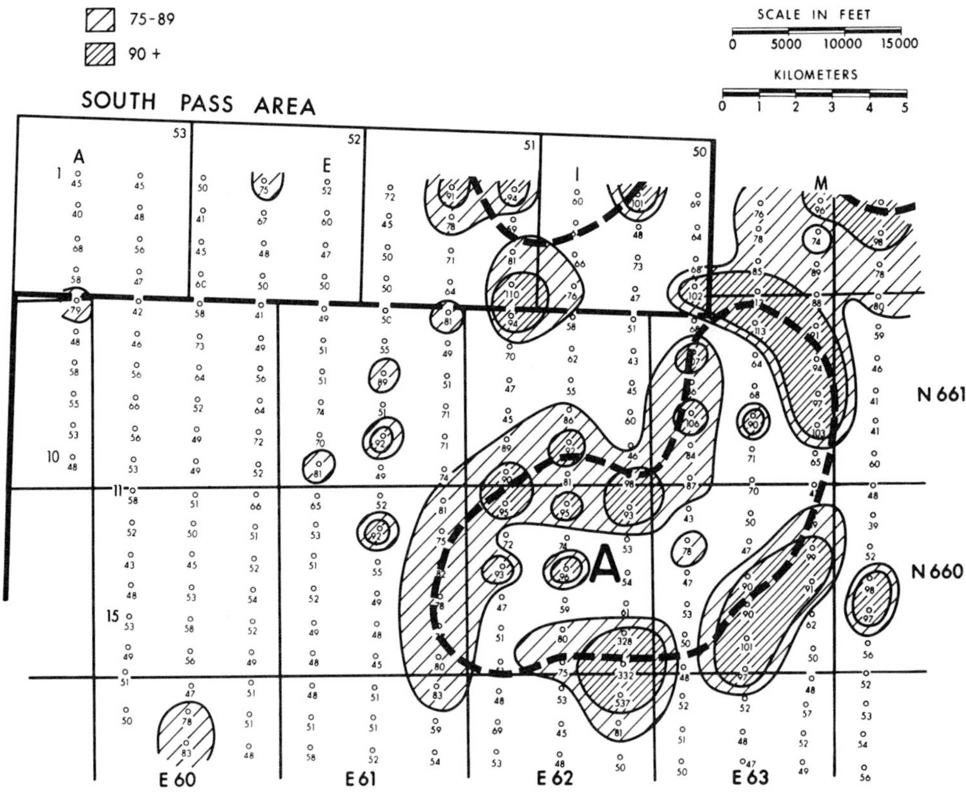

MOBILE SOUTH NO.2 AREA

OFFSHORE LOUISIANA

FIG. 12—Interpretation of ethane and heavier hydrocarbon data of Mobile South No. 2 area made prior to March, 1974, Gulf of Mexico lease sale. Area considered most prospective is outlined by heavy dashed line and graded "A".

Offshore samples are prepared for analysis by the same procedure used on onshore samples, each one being put through the sand removal process whether it is predominantly sand or clay. The same analytical techniques also are used. Probably because of the water blanket, background values in offshore petroliferous areas are much higher than those in onshore areas. In the former, background ethane and heavier hydrocarbon values are, usually, at least 35 to 50 ppb by weight while, onshore, they often are below 10 ppb. On the other hand, in nonpetroliferous areas, offshore values for the ethane through pentane fraction are also low; frequently, they are well below 10 ppb.

An additional technique has been devised for determining hydrocarbons which may be loosely held by samples composed principally of sand but are evolved during the sand removal process. These loosely bound hydrocarbons commonly are very helpful in evaluating hydrocarbon surveys. Hydrocarbons in bottom water samples collected with the sediment also yield clues that are useful in evaluating areas with sandy bottoms.

Offshore Louisiana Survey

During the Fall of 1973 and Winter of 1974, a hydrocarbon survey was conducted offshore Louisiana, in the Gulf of Mexico, to evaluate areas to be included in the March, 1974, Offshore Federal Lease Sale. One of the areas that was included is located in the Mobile South No. 2 Area, about 100 mi (160 km) southeast of New Orleans.

Figure 11 shows the locations of 256 stations from which sediment samples were collected at a depth of 6 ft (1.8 m) below the sediment surface under water ranging in depth from 318 ft (97 m) in the northwest corner of the survey to 1,860 ft (567 m) in the southeast corner. The stations sampled February 23 to March 2, 1974, were spaced at half-mile (0.8 km) intervals along north-south profiles 1 mi (1.6 km) apart resulting in a density for the area of 1 sample to 320 acres (1.3 sq km). The ethane and heavier hydrocarbon values, also shown on the map (Fig. 11) at their respective station locations, range from 39 to 537 ppb by weight and average 68 ppb. Values from 75 to 89

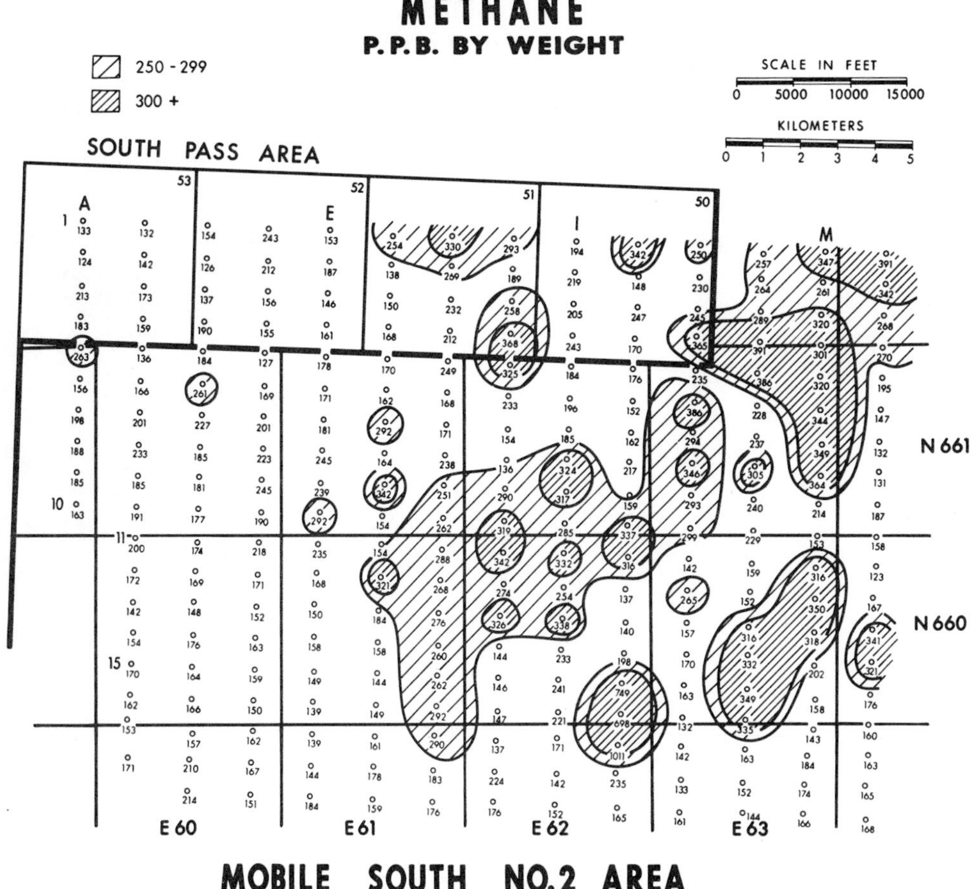

FIG. 13—Methane data in Mobile South No. 2 area produced a halo type anomaly in eastern part of survey which resembles closely the one produced by the ethane and heavier hydrocarbons (Fig. 12).

ppb are in the lightly shaded areas and values of 90 ppb and above are in the more heavily shaded areas. As in the Jay area, the contour levels were selected by visual inspection. Although the sand contents of the samples were very low, ranging from 0.1 to 18.8%, and averaging 1.1%, the sand removal process was applied to each sample prior to analysis.

The interpretation that was made on the basis of the ethane and heavier hydrocarbon data appears in Figure 12. A well defined halo type anomaly, indicated by the dashed outline, developed in the eastern part of the survey area and was graded "A". The central part of this anomaly lies under about 1,000 ft (305 m) of water.

The methane data (Fig. 13) produced a map that is very similar to that of the ethane and heavier hydrocarbons (Fig. 12). Values of 250 to 299 ppb by weight are in the lightly shaded areas and values of 300 ppb and above are in the darker areas. The methane values range from 124 to 1,011 ppb and average 220 ppb.

The map prepared from the pentane values, which range from 0 to 101 ppb and average 9.9 ppb, is shown in Figure 14. Values ranging from 12 to 14 ppb are in the lightly shaded areas and values of 15 ppb and above are in the more heavily shaded areas. The pentane pattern is also very similar to that produced by the ethane and heavier hydrocarbons (Fig. 12). The pentane data are useful in determining if the reservoir contains liquid hydrocarbons. Experience suggests that when pentane appears in the near-surface sediment, and produces an anomaly, liquid hydrocarbons may be expected in the potential reservoir. However, the absence of sedimentary pentane does not exclude the presence of liquids in the reservoir because gas associated with crude oil commonly is relatively dry and contains very little pentane. The near-surface hydrocarbons come principally from the gas phase in the reservoir and not from the heavy hydrocarbon fractions in the crude oil.

The ratios of the anomalous to the background values are 1.8 for methane and for ethane and heavier hydrocarbons, and 2.0 for pentane. Whereas these are significant ratios for offshore anomalies, they are much lower than those of most land anomalies.

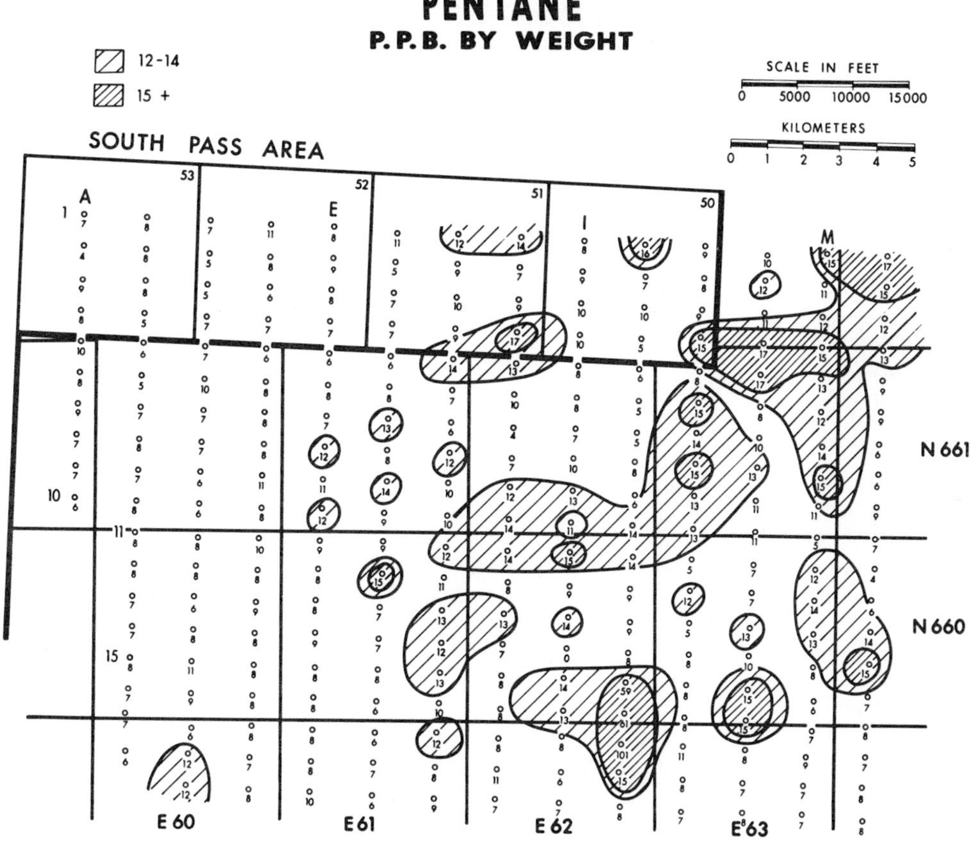

FIG. 14—Pentane data also produced a halo type anomaly in Mobile South No. 2 area which is similar to that of ethane and heavier hydrocarbons (Fig. 12).

Although the contour levels for the hydrocarbon fractions were selected by visual inspection, histograms (Fig. 15) were prepared to show the frequency distributions of the various hydrocarbons that were mapped. Parts of the curves that represent the histograms of the Mobile South No. 2 area assume shapes that recall the bell curve which in statistics represents a normal distribution of events. In the case of the methane, that part of the histogram that resembles the bell curve contains the background values, or values below 200 ppb by weight. More than 50% of the samples fall within this part of the curve. At the point where the curve deviates from the bell shape, the anomalous values are considered to begin and then continue to the right. In contouring the methane map (Fig. 13), 250 ppb was selected as the lowest anomalous value. On the basis of the histogram, a value of 200 ppb could have been used. For ethane and heavier hydrocarbons 60 ppb might have been used instead of 75 ppb and, for pentane, 10 ppb could have been used instead of 12 ppb. Minor changes in the hydrocarbon maps would have resulted had the anomalous values been selected from the histograms rather than on the basis of experience.

In July, 1975, Shell Oil Company announced the results of an exploratory well drilled at a location about 1 mi (1.6 km) west and about 300 ft (100 m) south of the northeast corner of Block N660–E62 (Fig. 16). Several gas and oil zones were penetrated which indicated that a potentially important discovery was made. Later, 11 additional exploratory wells were drilled to depths ranging from 9,000 to 13,930 ft (2,740 to 4,250 m) within the area of the dotted outline (Fig. 16). No report has been made concerning the number of wells that are potentially productive but a highly publicized, very costly production platform is being constructed to develop the area, named the Cognac field. The area containing the exploratory wells bears a very close relationship to the hydrocarbon anomaly. In fact, if the anomaly were shifted about a mile to the northwest, it would include, in entirety, the area encompassing the wells drilled to date. Such displacements of hydrocarbon anomalies from producing areas commonly are noted. Perhaps subsurface hydrody-

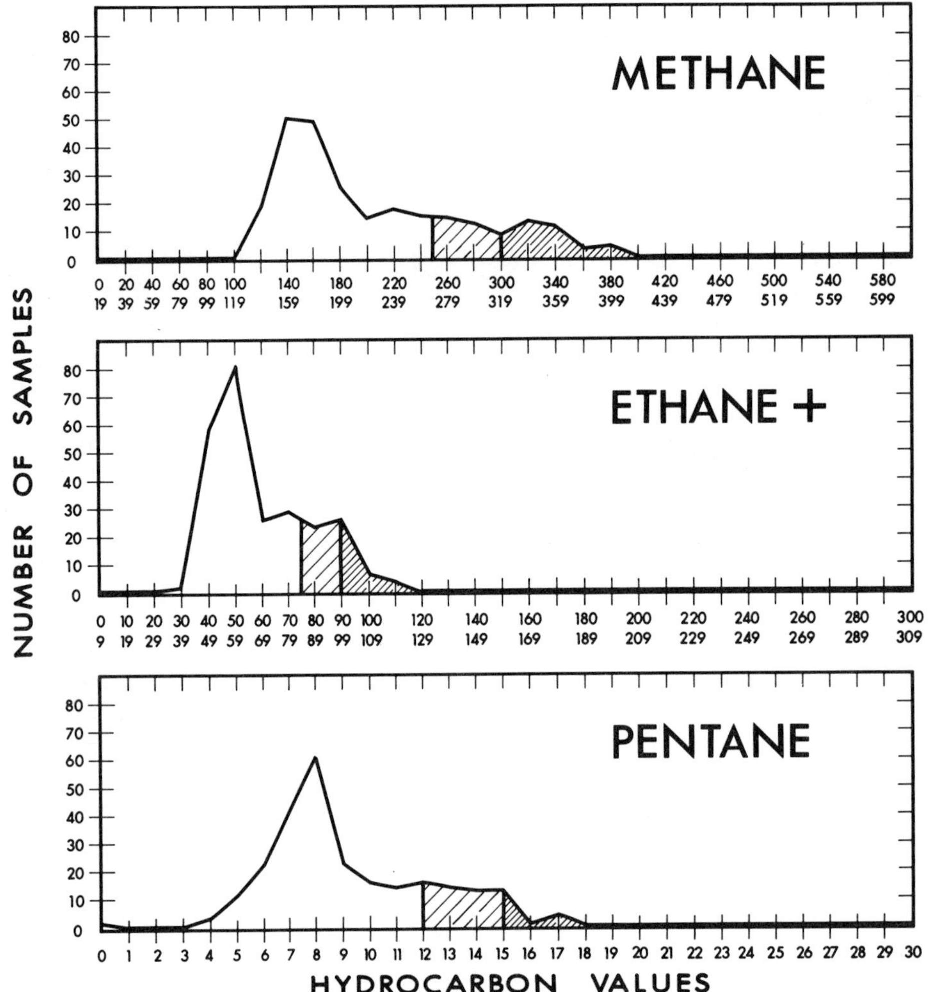

FIG. 15—Histograms prepared from methane, ethane, and heavier hydrocarbon (C_2-C_5) and pentane values of the Mobile South No. 2 area. The shaded areas contain values corresponding to those of Figures 13, 12, and 14, respectively. Based on the curves, lower values than those used in contouring the maps could have been selected as anomalous.

namic forces cause shifting in near-surface hydrocarbon anomalies. Hitchon (1974) suggested that fluid flow within the near-surface region may affect hydrocarbon movement and should be taken into consideration when interpreting geochemical anomalies. Even without taking these factors into account, it is noteworthy that the discovery well and 75% of the additional test wells drilled are located within the hydrocarbon anomaly.

Several miles west of the Cognac area, five exploratory wells were drilled within the relatively small area outlined on the map of Figure 16 but no official results have been released. Unconfirmed reports state that gas has been discovered there. However, according to the interpretation presented, no hydrocarbon anomaly developed and, therefore, any accumulation that exists there would be expected to be unimportant, especially when compared to Cognac.

The percentage of methane in the total light hydrocarbon fraction (C_1-C_5) extracted from the sediment in the Mobile South No. 2 Area is 89% and within the same range normally present in reservoir gas. Moreover, unlike the Jay area, no abnormally large amounts of carbon dioxide or hydrogen sulfide have been reported in natural gas from this part of the Gulf of Mexico.

CARBON ISOTOPE DATA LINK NEAR-SURFACE METHANE WITH DEEP SOURCES

The near-surface hydrocarbon data presented thus far bear close relationships to subsurface petroleum accumulations and are, therefore, believed to originate within them. However, the data are empirical and, although quite convincing, the statement that near-surface hydrocarbons are of ancient and not of recent origin may still be questioned. It appears that solid evidence is now available that relates near-surface hydrocarbons directly to petroleum gases at depth. Starting with Craig (1953), it has become possible to make accurate measurements of carbon isotope ratios for

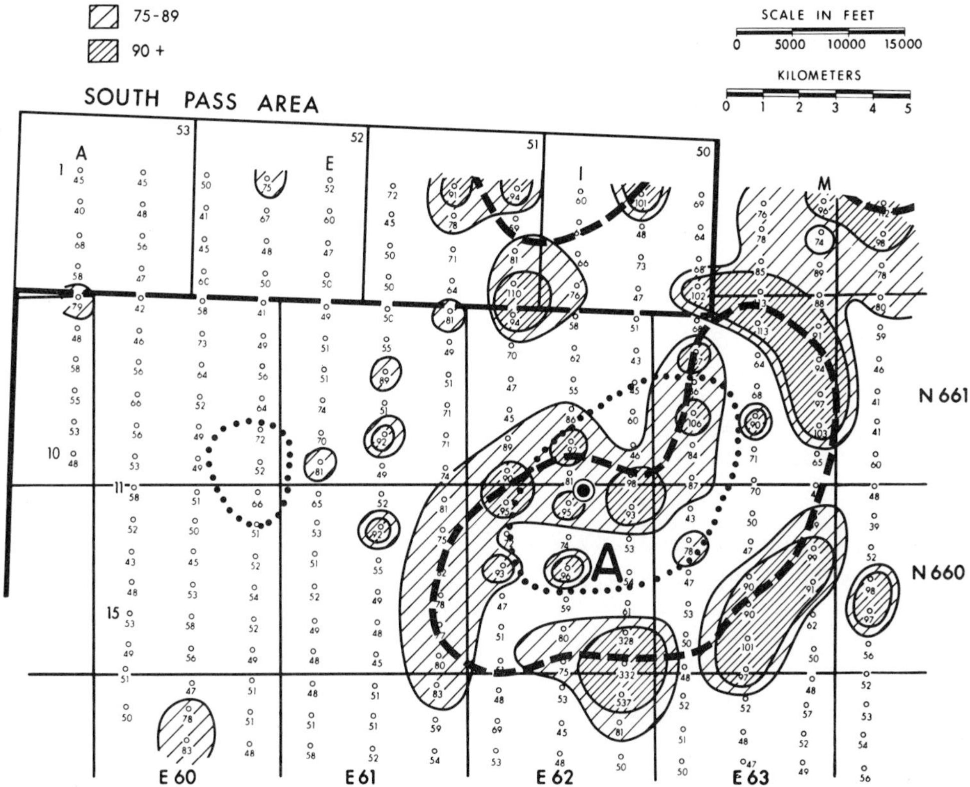

FIG. 16—Ethane and heavier hydrocarbon map (Fig. 12) to which the discovery well (●) of the new Cognac field has been added. Within the dotted outline, 11 additional exploratory wells have been drilled. The small area (dotted outline) west of the A anomaly contains 5 exploratory wells of which some may have penetrated a gas deposit. No significant hydrocarbon anomaly is apparent in this part of the survey.

the purpose of determining the source of different carbonaceous materials. For example, it is possible to determine if different petroleum samples come from the same source or from different sources if their ratios of C^{13} to C^{12} are known. Advances in mass spectrometry made these measurements possible. In recent years, through the work of many investigators, techniques have been developed for determining carbon isotope ratios of hydrocarbons in the range of concentrations that are found in the near-surface. Briefly, the technique involves removal of any carbon dioxide initially present and then converting the resulting purified hydrocarbons to carbon dioxide. Carbon isotope ratios of *this* carbon dioxide are determined in a sensitive mass spectrometer and the results are expressed as δ-values, or deviations, in parts per thousand (per mil), of the C^{13}/C^{12} ratio of a sample from that of a standard. A popular standard, abbreviated PDB, is based on the isotope ratio found in the carbonate shell of a fossil belemnite from the Peedee Formation of South Carolina. All of the δ-values reported in this paper are expressed in terms of this standard. The following equation relates the δ-value to the isotope ratios of the sample and standard respectively.

$$\delta\, C^{13} \text{ per mil} = \left(\frac{C^{13}/C^{12} \text{ sample} - C^{13}/C^{12} \text{ standard}}{C^{13}/C^{12} \text{ standard}} \right) \times 1000$$

A negative value for the δ indicates that the sample is "lighter" or contains less C^{13} than the standard gas. A positive value means that it is "heavy" or contains more C^{13} than the standard gas.

In 1973, after an experimental survey was conducted over the Francitas field in Jackson County, Texas, carbon isotope ratios were determined on adsorbed methane extracted from a soil sample collected at 12 ft (3.7 m) from a location within the hydrocarbon anomaly that was found. The C^{13}/C^{12} ratio for meth-

Table 1. Carbon Isotope Data of Samples from Various Sources

Sample	Source	δ C^{13} per mil (PDB)
Adsorbed Methane (soil)	Near-surface hydrocarbon anomaly, Francitas field,	-44.0
Interstitial Methane (soil)	Texas	-40.8
Methane from reservoir gas	Francitas field	-41.0 to -43.8
Adsorbed Methane (sediment)	Mobile South No. 2 area, Offshore Louisiana	-37.3 to -39.2
11 natural gas samples* (principally methane)	Producing wells: U.S. Canada, Trinidad, and Venezuela	-33.6 to -47.4
Gases of bacterial origin** (principally methane)	Marsh gases in U.S.S.R. and other biochemical gases	-50 to -80

* Silverman (1964)
** Lebedev, et al (1979)

ane only was determined. A second sample, from another part of the anomaly, which contained an appreciable amount of interstitial, or free, methane was also analyzed. A $\delta^{13}C$ per mil value of -44.0 was obtained for the adsorbed methane and a value of -40.8 for the interstitial methane. Samples of the gas producing from the Francitas field were also collected and their $\delta^{13}C$ per mil values for methane ranged from -41.0 to -43.8, practically the same as the near-surface methane. Silverman (1964, p. 97) analyzed 11 natural gas samples from widely separated localities in the United States, Canada, Trinidad, and Venezuela and found $\delta^{13}C$ per mil values ranging from -33.6 to -47.4. For methane of biologic origin, $\delta^{13}C$ values range from -50 to -80 (Lebedev et al, 1969). Some natural gases yield δ-values which fall in the upper range of biologic methane indicating that they, too, can be of biogenic origin.

The near-surface samples used in the Francitas carbon isotope experiments were from a survey conducted after discovery of the field and it could be argued, therefore, that the near-surface methane resulted from contamination by produced gas. Recently, however, C^{13}/C^{12} ratios were determined on adsorbed methane extracted from three sediment samples that were still available from the Mobile South No. 2 survey. Since these samples were collected prior to the discovery of the Cognac field, the possibility of contamination of the sediment by produced gas was eliminated. Two of the samples, M-3 and M-4 (Fig. 13), are from the northeastern part of the anomaly and yielded values of -38.1 and -39.2, respectively, and the third sample, M-14, yielded a $\delta^{13}C$ per mil value of -37.3 and turned out to be from a location just outside the southeastern part of the anomaly. The δ-values of all three of the near-surface adsorbed methane samples fall well within the upper range of values for methane in petroleum gases. It would be of great interest to compare these values with those of the methane from the Cognac productive zones. The carbon isotope data are summarized in Table 1.

HYDROCARBON DATA FROM WELL CUTTINGS SUPPORT VERTICAL MIGRATION

Yet more evidence which suggests upward movement of hydrocarbons from petroleum deposits is provided by analysis of cuttings collected during the drilling of wells. The cuttings, usually collected for each 30 ft (9 m), of drilling, are washed free of drilling fluid prior to analysis. The adsorbed hydrocarbons are removed from the cuttings by the same procedure used for near-surface samples but the sand removal process is not applied to these samples. The extracted gases are analyzed by the hydrogen flame chromatograph and include hydrocarbons heavier than those normally found in near-surface samples, often as heavy as nonane. Prior to 1963, the extracted gases were analyzed by low temperature fractionation techniques previously published (Horvitz, 1949; 1954).

Data obtained from cuttings collected from a large number of wells (Rosaire, 1940; Horvitz, 1949) have disclosed the existence of definite relationships between the hydrocarbons found in well cuttings and petroleum accumulations. For example, logs of producing wells show relatively low values in the upper part of the well but, at some distance above the accumulation, a definite increase in concentration is encountered. From this point on, the values gradually increase until the deposit is reached when maximum values are obtained. Logs of nonproductive wells show relatively small values throughout. The distance above the deposit, over which significantly high values appear, is dependent upon the nature of the accumulation; the lower the gravity of the deposit, the shorter is

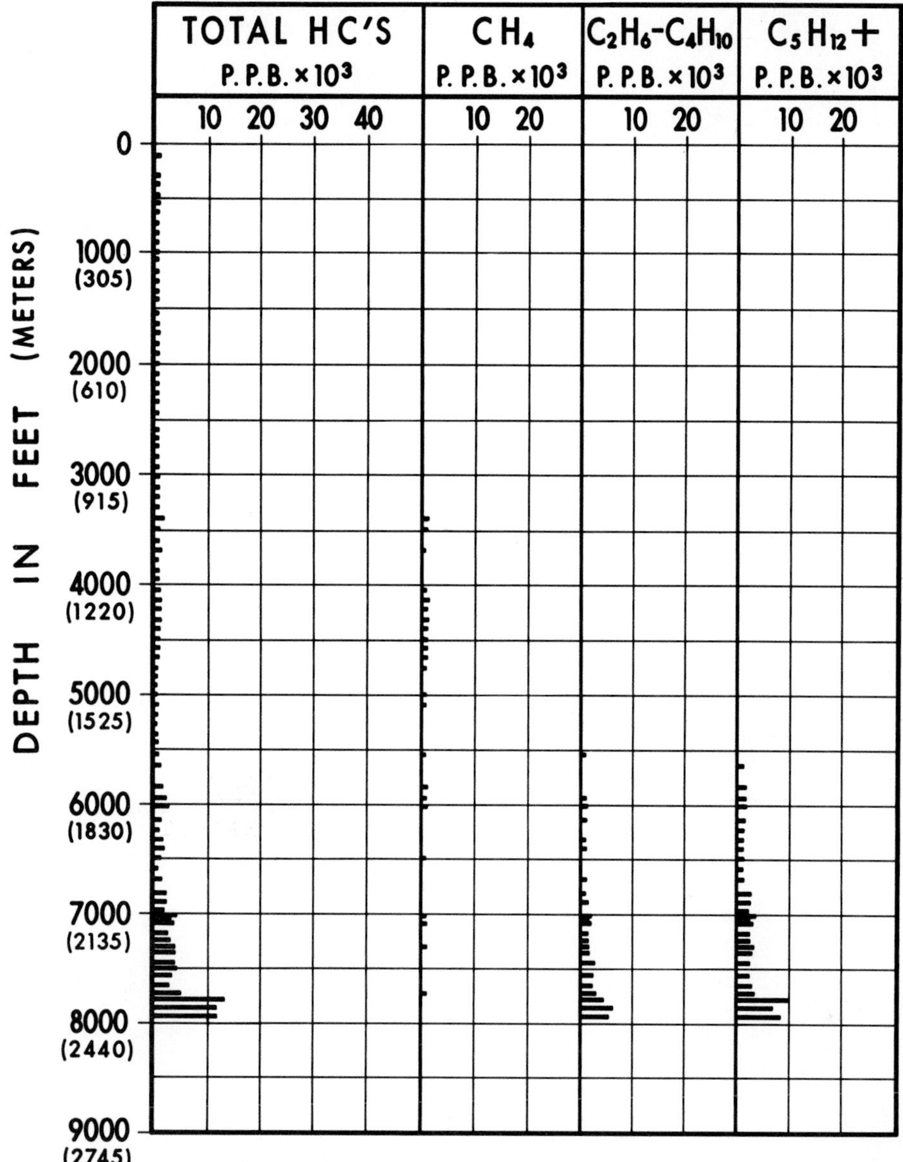

FIG. 17—Hydrocarbon log of No. 1 Leveridge Estate Well, Wharton County, Texas, at projected total depth of 8,000 ft (2,440 m). No oil or gas indications were encountered and geology was discouraging at this depth. Based on hydrocarbon buildup, the well was deepened.

this distance. The relative quantities, as well as the types of hydrocarbons found in the cuttings, bear definite relationships to the composition of the deposit. Above an oil deposit, for example, the predominant fraction is pentane and heavier hydrocarbons. For the case of gas, the predominant fraction contains methane through butane.

Figure 17 shows the hydrocarbon log prepared from cuttings collected in May, 1940, from the No. 1 Leveridge Estate, Wharton County, Texas. Data are shown to 8,000 ft (2,440 m), the projected total depth. The first column shows the total hydrocarbons, with significant values ranging from about 3,500 ppb at 7,000 ft (2,135 m) to about 13,000 ppb at 8,000 ft (2,440 m). These values are 35 to 150 times greater than are near-surface anomalous values. The ethane and heavier hydrocarbon fraction was composed of approximately equal amounts of ethane-propane-butane and of pentane and heavier hydrocarbons. This distribution of hydrocarbons typically is produced by gas-distillate accumulations. The relatively high hydrocarbon values obtained over a long distance, more than 1,000 ft (305 m), which increased rapidly over the last 200 ft (60 m), suggested that the source of the hydrocarbons may be near. At this depth, 8,000 ft (2,440 m), the geology was discouraging as the well correlated 100 ft (30 m) low to a dry hole 1 mi (1.6 km) away. However, on the basis of the hydrocarbon data

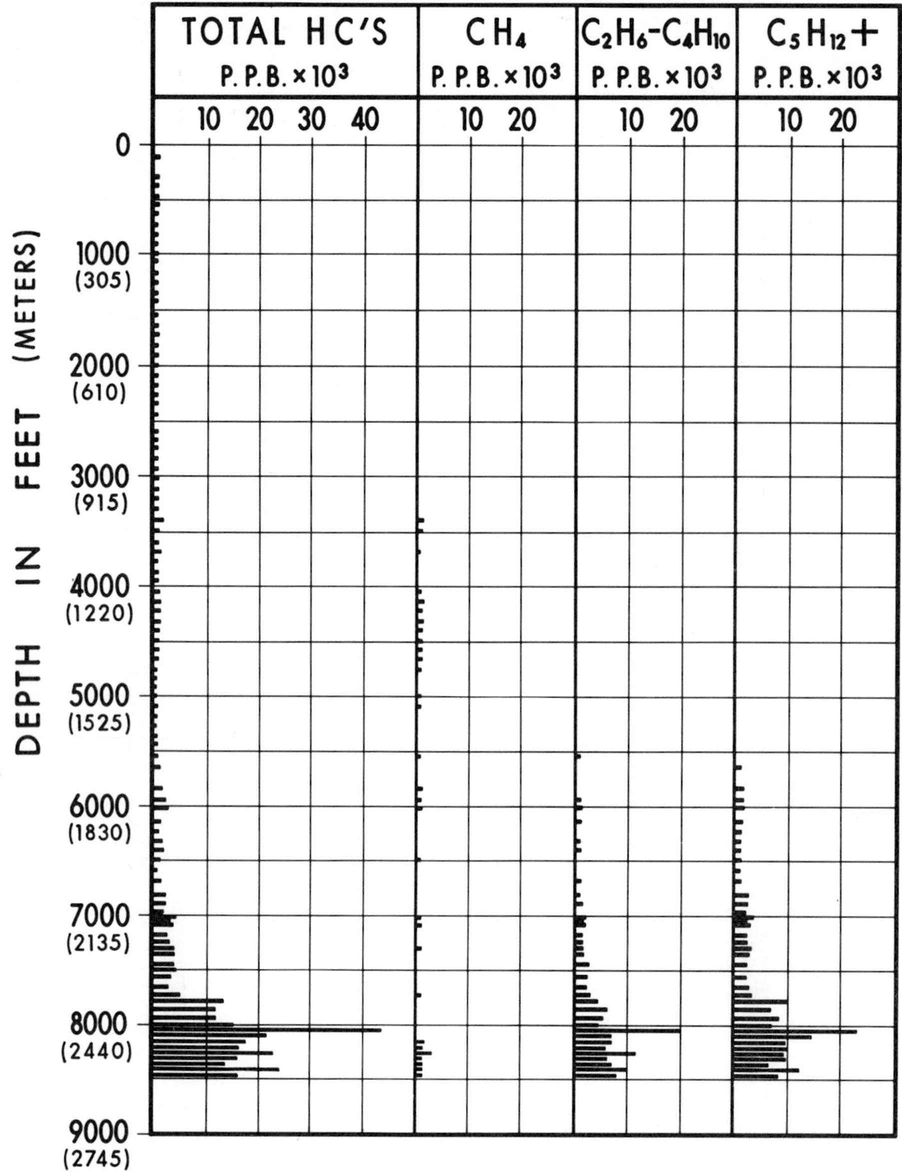

FIG. 18—Hydrocarbon log of No. 1 Leveridge Estate after deepening. Well was completed as discovery of East Bernard field after encountering a 30-ft (9-m) gas-distillate sand in the Cook Mountain Formation at 8,074 ft (2,460 m). After Horvitz, 1949.

alone, drilling of the well was resumed.

At 8,074 ft (2,460 m), a 30 ft (9 m) sand (Fig. 18) in the Cook Mountain Formation was encountered which contained gas and distillate. The well was completed as the discovery well of the East Bernard field which is still producing.

SUMMARY AND CONCLUSIONS

The mechanism involved in moving hydrocarbons from subsurface petroleum deposits to the near-surface is not yet clearly understood. However, evidence is available which substantiates the concept that this movement does occur. The following observations and data represent some of this evidence:

1. Concentration patterns produced at a depth of 12 ft (3.7 m) below ground surface over the Jay and Blackjack Creek fields of Alabama and Florida by the light saturated hydrocarbons (C_1-C_5) show close relationships to the producing areas. Where the hydrocarbon anomalies are compared to the available subsurface geology of these fields, the specific configurations of the anomalies are easily understood.

2. Hydrocarbon data from the samples collected at a depth of 6 ft (1.8 m) below the sediment surface in the Mobile South No. 2 area, offshore Louisiana, prior to the March, 1974, lease sale, produced an outstanding halo type anomaly. The anomaly now contains the discovery well of Cognac field. Of 11 additional exploratory wells drilled, locations of 75% of them fall within the anomaly.

3. Carbon isotope data of methane, extracted from near-surface samples located within a hydrocarbon anomaly over the Francitas field, Texas, yielded C^{13}/C^{12} ratios which are almost identical to those of methane in the reservoir gas of the field. Values for $\delta^{13}C$ per mil of methane extracted from 6-ft (1.8 m) sediment samples in the Mobile South No. 2 area, offshore Louisiana, are also well within the upper range of δ-values of methane in petroleum gases.

4. The composition of the light hydrocarbon fraction, in near-surface soil that is associated with an anomaly, is generally similar to that of the gas phase of the subsurface petroleum accumulation which the anomaly reflects.

5. Data obtained from well cuttings also support the concept that hydrocarbons leak from oil and gas deposits. Hydrocarbon buildups, resulting from these data, have anticipated petroleum deposits.

REFERENCES CITED

Burwell, R. B., and R. E. Hadlow, 1977, Reservoir management of the Blackjack Creek field: Jour. Petroleum Technology, v. 29, p. 1235-1241.

Craig, H., 1953, The geochemistry of the stable carbon isotopes: Geochim. et Cosmochim. Acta, v. 3, p. 53-92.

Fedynsky, V. V., E. V. Karus, and M. K. Polshkov, 1975, Geophysical and geochemical surface surveys for detection of oil and gas pools, *in* Proceedings 9th World Petroleum Congress, v. 3: Essex, England, Applied Science Publishers, Ltd., p. 279-288.

Hitchon, B., 1974, Application of geochemistry to the search for crude oil and natural gas, *in* A. A. Levinson, Introduction to exploration geochemistry: Calgary, Alberta, Canada, Applied Publishing, Ltd., p. 509-545.

Horvitz, L., 1939, On geochemical prospecting: Geophysics, v. 4, p. 210-225.

——— 1945, Recent developments in geochemical prospecting for petroleum: Geophysics, v. 10, p. 487-493.

——— 1949, Geochemical well logging, *in* A symposium on subsurface logging techniques: Norman, Oklahoma, University Book Exchange, p. 89-94.

——— 1954, Near-surface hydrocarbons and petroleum accumulation at depth: Mining Eng., v. 6, p. 1205-1209.

——— 1959, Geochemical prospecting for petroleum: Symposium on geochemical exploration: 20th Internat. Geol. Cong. (Mexico City), v. 2, p. 303-319.

——— 1969, Hydrocarbon geochemical prospecting after thirty years, *in* Unconventional methods in exploration for petroleum and natural gas: Dallas, Texas, Southern Methodist Univ., p. 205-218.

——— 1972, Vegetation and geochemical prospecting for petroleum: AAPG Bull., v. 56, p. 925-940.

Laubmeyer, G., 1933, A new geophysical prospecting method, especially for deposits of hydrocarbons: Petroleum, v. 29, no. 18, p. 1-4.

Lebedev, V. S., V. M. Ovsyannikov, and G. A. Mogilevskiy, 1969, Separation of carbon isotopes by microbiological processes in the biochemical zone: Geochemistry International, v. 6, p. 971-976.

Ottmann, R. D., P. L. Keyes, and M. A. Ziegler, 1973, Jay field—A Jurassic stratigraphic trap: Gulf Coast Assoc. Geol. Socs. Trans., v. 23, p. 146-157.

——— ——— ——— 1976, Jay field, Florida—A Jurassic stratigraphic trap, *in* North American oil and gas fields: AAPG Mem. 24, p. 276-286.

Rosaire, E. E., 1938, Shallow stratigraphic variations over Gulf Coast structures: Geophysics, v. 3, p. 96-115.

——— 1939, Discussion and communications: Geophysics, v. 4, p. 300-305.

——— 1940, Geochemical prospecting for petroleum: AAPG Bull., v. 24, p. 1418-1426.

Silverman, S. R., 1964, Investigations of petroleum origin and evolution mechanisms by carbon isotope studies, *in* Isotopic and cosmic chemistry: Amsterdam, North - Holland Publishing Co., p. 92-102.

Sokolov, V. A., 1935, Summary of the experimental work of the gas survey: Neftyanoye Khozyaystvo, v. 27, no. 5, p. 28-34.

Note: An appendix of analytical information follows on the next six pages.

Appendix: Analytical Data—Flomaton, Jay, Blackjack Creek Areas, Alabama and Florida

Sample Number	Methane ppb*	Ethane +** ppb*	Sand %***	Sample Number	Methane ppb	Ethane + ppb	Sand %
1	100	189	79	70	10	14	89
2	76	149	75	71	7	13	62
3	106	201	79	72	6	8	56
4	100	193	81	73	4	4	42
5	93	187	85	74	13	24	52
6	96	166	73	75	3	3	80
7	84	152	76	76	1	1	62
8	15	18	66	77	1	1	58
9	1	1	36	78	1	1	72
10	1	1	26	79	3	3	94
11	3	3	67	80	1	1	66
12	5	4	79	81	1	1	66
13	12	11	99	82	0	1	26
14	1	1	62	83	15	25	65
15	0	1	58	84	7	9	70
16	54	102	75	85	4	4	63
17	48	80	70	86	4	4	49
18	50	91	73	87	19	36	76
19	63	131	74	88	1	4	71
20	64	132	75	89	2	1	38
21	56	109	70	90	7	8	89
22	44	87	71	91	15	12	90
23	55	104	70	92	11	14	96
24	96	181	77	93	13	9	94
25	74	142	78	94	1	1	66
26	122	236	77	95	1	1	81
27	23	34	95	96	3	2	65
28	201	320	67	97	3	3	62
29	104	207	81	98	1	1	47
30	138	280	84	99	5	2	45
31	98	193	81	100	4	2	46
32	36	67	66	101	0	1	67
33	71	135	71	102	1	2	67
34	50	99	74	103	5	3	94
35	71	144	80	104	6	4	92
36	39	77	68	105	0	1	59
37	50	104	75	106	0	1	48
38	36	43	92	107	1	1	90
39	7	9	83	108	2	2	95
40	29	21	93	109	20	12	95
41	5	3	76	110	1	1	96
42	55	94	62	111	1	1	68
43	22	37	87	112	4	3	94
44	22	19	93	113	1	1	30
45	47	78	79	114	2	1	81
46	2	2	53	115	24	15	96
47	1	1	56	116	9	7	93
48	2	2	55	117	8	13	90
49	9	15	75	118	3	2	80
50	2	2	62	119	1	1	56
51	1	2	36	120	1	1	49
52	26	17	97	121	13	12	91
53	1	1	82	122	14	12	80
54	9	14	95	123	1	1	70
55	10	17	99	124	2	1	64
56	1	1	26	125	1	1	65
57	0	1	15	126	2	2	78
58	1	1	72	127	3	2	79
59	3	2	94	128	15	24	46
60	1	1	58	129	7	11	64
61	15	20	93	130	4	7	64
62	10	8	95	131	3	4	65
63	8	4	91	132	3	3	39
64	56	103	77	133	1	1	79
65	13	21	50	134	1	1	24
66	10	17	48	135	12	19	76
67	24	51	53	136	10	15	71
68	13	24	50	137	13	20	71
69	3	2	62	138	8	13	67

(Continued on next page)

Appendix, Continued

Sample Number	Methane ppb*	Ethane +** ppb*	Sand %***
139	10	16	51
140	27	47	64
141	10	7	59
142	2	1	53
143	3	3	80
144	5	4	81
145	1	1	63
146	2	2	82
147	2	2	85
148	9	12	76
149	13	7	94
150	5	4	76
151	17	11	93
152	7	3	81
153	7	2	63
154	50	91	72
155	13	8	94
156	47	91	83
157	46	98	76
158	33	54	76
159	20	30	65
160	95	183	77
161	68	149	76
162	133	224	83
163	133	221	87
164	13	17	76
165	7	4	78
166	2	1	48
167	3	3	91
168	2	1	49
169	18	28	72
170	20	35	63
171	18	33	58
172	19	34	55
173	14	19	90
174	50	120	79
175	34	64	54
176	16	23	83
177	16	23	97
178	12	15	80
179	7	9	83
180	6	7	81
181	12	22	58
182	18	36	55
183	19	38	64
184	7	7	53
185	3	2	82
186	2	1	62
187	5	2	70
188	1	1	70
189	34	66	60
190	72	147	61
191	58	121	73
192	33	65	99
193	59	118	94
194	19	27	93
195	17	24	90
196	23	31	94
197	19	22	94
198	6	2	67
199	77	162	68
200	17	22	82
201	19	13	91
202	33	60	65
203	17	12	96
204	3	1	19
205	3	1	20
206	4	1	30
207	3	1	17

Sample Number	Methane ppb	Ethane + ppb	Sand %
208	2	1	58
209	3	2	52
210	23	38	62
211	9	6	86
212	3	1	80
213	4	2	72
214	10	14	72
215	61	62	94
216	31	60	78
217	5	4	61
218	7	6	67
219	8	8	70
220	31	28	70
221	21	29	57
222	34	56	56
223	18	26	56
224	23	34	52
225	21	27	60
226	1	1	64
227	4	6	69
228	1	1	37
229	1	1	26
230	5	5	99
231	5	8	65
232	8	13	77
233	19	31	85
234	35	61	87
235	10	6	98
236	8	5	94
237	7	8	81
238	2	2	83
239	2	2	85
240	3	2	95
241	4	3	97
242	43	72	68
243	55	105	76
244	44	86	80
245	36	69	77
246	55	105	75
247	58	117	81
248	54	110	93
249	54	102	76
250	36	70	80
251	56	105	71
252	9	13	57
253	21	40	72
254	47	104	80
255	28	52	80
256	140	158	85
257	54	113	80
258	50	93	88
259	18	31	62
260	3	1	23
261	21	38	57
262	56	106	84
263	51	106	92
264	24	39	60
265	51	107	73
266	31	57	64
267	48	103	73
268	57	120	72
269	46	93	61
270	45	89	68
271	61	131	76
272	42	91	71
273	82	167	72
274	31	56	65
275	42	72	83
276	68	135	65

(Continued on next page)

Appendix, Continued

Sample Number	Methane ppb*	Ethane +** ppb*	Sand %***	Sample Number	Methane ppb	Ethane + ppb	Sand %
277	63	113	85	346	46	91	83
278	29	43	49	347	63	118	69
279	34	61	57	348	87	179	97
280	13	13	48	349	3	2	33
281	41	69	58	350	3	3	33
282	66	105	55	351	1	1	8
283	91	150	80	352	16	18	87
284	23	26	92	353	18	32	96
285	14	13	97	354	7	7	78
286	9	7	85	355	11	10	68
287	5	3	84	356	7	5	59
288	6	3	70	357	41	68	91
289	19	15	85	358	39	83	95
290	1	1	6	359	51	91	57
291	21	27	48	360	25	42	58
292	70	120	69	361	40	56	95
293	39	58	63	362	27	46	61
294	40	62	58	363	6	5	29
295	63	123	71	364	59	112	65
296	34	54	61	365	64	114	69
297	43	85	71	366	66	140	70
298	31	58	73	367	97	186	75
299	17	23	62	368	73	140	74
300	74	155	82	369	113	207	80
301	157	286	67	370	105	178	87
302	80	119	58	371	3	1	31
303	59	93	75	372	7	8	34
304	35	55	56	373	4	2	34
305	46	88	76	374	4	3	31
306	88	134	76	375	42	58	98
307	58	109	62	376	86	162	73
308	31	37	54	377	133	255	66
309	139	237	80	378	40	77	94
310	68	105	66	379	134	227	76
311	58	94	68	380	66	123	75
312	16	19	62	381	142	226	69
313	40	71	62	382	50	78	92
314	17	22	61	383	51	79	82
315	39	73	64	384	6	4	41
316	23	37	56	385	8	7	74
317	20	32	53	386	6	3	77
318	13	14	51	387	3	2	17
319	6	3	29	388	8	8	36
320	4	2	83	389	4	2	82
321	12	13	87	390	5	2	88
322	68	122	70	391	4	3	79
323	66	131	68	392	2	1	17
324	7	2	11	393	5	3	97
325	48	88	62	394	4	2	31
326	135	218	78	395	4	4	67
327	113	165	66	396	2	2	25
328	21	31	89	397	3	1	44
329	113	192	61	398	4	2	27
330	38	65	47	399	4	2	31
331	43	72	55	400	4	3	42
332	55	38	91	401	4	2	20
333	63	107	66	402	5	3	63
334	95	180	73	403	4	2	87
335	172	299	50	404	2	1	28
336	142	274	74	405	7	6	95
337	86	176	73	406	4	2	78
338	70	139	70	407	43	87	77
339	41	64	68	408	16	23	57
340	54	82	78	409	11	21	64
341	42	72	62	410	25	54	66
342	44	93	68	411	5	7	72
343	11	18	55	412	4	3	73
344	23	46	77	413	3	1	50
345	26	45	75	414	2	1	50

(Continued on next page)

Appendix, Continued

Sample Number	Methane ppb*	Ethane +** ppb*	Sand %***
415	23	44	64
416	7	6	54
417	32	63	67
418	5	6	81
419	5	3	91
420	4	4	60
421	4	7	91
422	3	3	96
423	6	6	92
424	7	3	57
425	9	8	57
426	23	8	78
427	3	2	59
428	14	9	88
429	5	3	36
430	9	5	87
431	6	4	69
432	29	45	78
433	7	6	82
434	10	4	74
435	21	41	83
436	9	5	95
437	40	31	68
438	3	1	64
439	5	3	47
440	12	7	77
441	15	23	81
442	9	10	63
443	47	82	78
444	3	2	33
445	3	2	48
446	6	4	91
447	4	3	90
448	12	2	27
449	16	7	94
450	7	2	36
451	16	9	74
452	8	5	77
453	4	1	12
454	3	2	58
455	1	1	72
456	3	2	67
457	3	2	49
458	6	6	54
459	4	2	77
460	14	7	67
461	7	3	24
462	10	11	38
463	6	5	65
464	3	2	30
465	3	2	64
466	3	2	70
467	9	6	84
468	3	2	75
469	5	3	72
470	5	2	63
471	5	2	71
472	3	1	48
473	3	1	72
474	3	1	27
475	4	2	12
476	4	2	87
477	3	1	15
478	2	1	10
479	3	1	76
480	4	2	92
481	2	1	21
482	10	6	96
483	4	2	70

Sample Number	Methane ppb	Ethane + ppb	Sand %
484	5	2	94
485	4	2	87
486	69	32	98
487	57	34	93
488	55	19	97
489	52	14	96
490	6	4	95
491	5	6	98
492	3	3	89
493	3	2	94
494	2	2	82
495	2	1	49
496	4	3	92
497	15	4	77
498	7	4	94
499	19	7	75
500	3	2	65
501	5	6	59
502	45	93	58
503	59	116	73
504	123	260	77
505	101	121	74
506	153	327	76
507	158	329	92
508	51	102	71
509	23	31	97
510	101	186	79
511	139	261	86
512	4	3	48
513	6	9	37
514	83	137	96
515	102	170	59
516	73	137	93
517	16	22	92
518	6	9	91
519	15	20	93
520	43	43	80
521	27	22	92
522	184	354	80
523	143	282	80
524	155	337	72
525	156	322	80
526	177	341	76
527	9	10	52
528	23	23	95
529	59	91	66
530	24	39	81
531	24	42	55
532	38	56	72
533	8	13	45
534	5	6	98
535	3	5	22
536	14	16	69
537	4	6	54
538	9	9	42
539	8	10	47
540	32	58	57
541	5	7	46
542	3	4	46
543	12	19	38
544	15	16	82
545	6	10	41
546	28	58	68
547	23	44	58
548	40	82	96
549	44	36	68
550	3	3	57
551	9	12	95
552	7	7	91

(Continued on next page)

Appendix, Continued

Sample Number	Methane ppb*	Ethane +** ppb*	Sand %***	Sample Number	Methane ppb	Ethane + ppb	Sand %
553	5	7	93	622	27	53	65
554	4	8	69	623	23	73	83
555	7	11	63	624	28	48	89
556	5	6	59	625	27	46	92
557	5	5	74	626	30	43	92
558	39	35	99	627	20	34	87
559	15	25	55	628	24	45	55
560	45	77	53	629	16	31	68
561	9	10	29	630	20	34	79
562	10	12	74	631	15	26	59
563	96	177	70	632	17	34	62
564	18	12	40	633	20	27	77
565	43	25	98	634	34	63	87
566	7	4	67	635	34	51	92
567	6	5	66	636	96	131	83
568	5	4	57	637	56	72	99
569	29	46	86	638	25	41	89
570	30	59	59	639	26	45	93
571	47	104	56	640	23	39	92
572	51	100	57	641	2	1	10
573	29	54	62	642	34	60	72
574	31	58	64	643	17	28	86
575	22	45	51	644	5	6	64
576	18	34	61	645	4	5	72
577	9	8	38	646	7	11	69
578	21	38	58	647	2	1	38
579	42	39	67	648	3	3	48
580	39	53	70	649	4	4	42
581	92	118	71	650	2	1	51
582	87	142	60	651	3	3	69
583	32	55	59	652	4	1	53
584	34	55	77	653	5	1	29
585	42	36	88	654	8	4	68
586	44	65	74	655	7	3	70
587	35	37	55	656	36	23	80
588	33	63	66	657	7	2	48
589	49	77	80	658	13	2	56
590	82	34	89	659	16	20	77
591	13	14	45	660	13	16	80
592	11	13	32	661	7	2	77
593	7	9	58	662	8	3	57
594	22	34	88	663	7	2	70
595	37	65	48	664	5	2	41
596	37	58	47	665	8	4	81
597	50	96	64	666	7	3	85
598	61	121	68	667	7	5	82
599	68	127	91	668	12	12	68
600	19	28	89	669	20	31	89
601	20	37	58	670	54	101	88
602	19	37	57	671	23	25	93
603	31	57	86	672	5	1	61
604	33	57	58	673	20	11	96
605	28	53	63	674	46	55	86
606	21	40	71	675	55	70	97
607	18	34	69	676	36	60	88
608	36	64	83	677	40	67	87
609	32	50	78	678	27	41	97
610	54	100	85	679	9	5	66
611	6	5	50	680	8	5	88
612	13	4	16	681	11	9	95
613	15	21	96	682	6	5	94
614	17	18	98	683	17	17	98
615	24	32	93	684	6	2	36
616	6	8	81	685	51	78	87
617	24	38	62	686	16	20	61
618	31	52	61	687	11	15	84
619	22	29	98	688	21	32	81
620	44	55	49	689	12	14	76
621	28	56	54	690	17	10	98

(Continued on next page)

Appendix, Continued

Sample Number	Methane ppb*	Ethane +** ppb*	Sand %***	Sample Number	Methane ppb	Ethane + ppb	Sand %
691	6	2	44				
692	21	25	78				
693	6	6	86				
694	12	11	88				
695	9	14	92				
696	5	2	81				
697	4	1	42				
698	5	10	52				
699	4	2	60				
700	22	38	86				
701	38	71	79				
702	9	9	86				
703	4	1	15				
704	4	2	52				
705	103	196	94				
706	77	133	94				
707	84	57	80				
708	15	15	90				
709	13	19	86				
710	27	37	96				
711	41	78	53				

*Parts per billion by weight (dry sample basis).
**Ethane and heavier saturated hydrocarbons, through pentane.
***Percent by weight (undried sample basis).

Explanation of Indexing

A reference is indexed according to its important, or "key" words.

Three columns are to the left of the keyword entries. The first column, a letter entry, represents the AAPG book series from which the reference originated. In this case, S stands for Studies in Geology Series. Every five years, AAPG merges all its indexes together, and the letter S will differentiate this reference from those of the AAPG Memoir Series (M) or from the AAPG Bulletin (B).

The following number is the series number. In this case, 10 represents a reference from Studies No. 10.

The last column entry is the page number in this volume where this reference will be found.

Note: This index is set up for single-line entry. Where entries exceed one line of type, the line is terminated. (This is especially evident with manuscript titles, which tend to be long and descriptive.) The reader must sometimes be able to realize keywords, although commonly taken out of context.